Animal Signals

Signalling and Signal Design in Animal Communication

Tapir Academic Press, Trondheim, Norway 2000

ISBN 82-519-1545-7

Printed by: Tapir

FRONT COVER
Top: Male bluethroat (*Luscinia svecica*). Photo courtesy by Staffan Andersson.
Bottom: Contrasting colour patterns in the reef-living angel fish *Pygoplites diacanthus*.
Photo courtesy by Justin Marshall.

Tapir Academic Press
N-7005 Trondheim
Tel.: +47 73 59 32 10
Fax: +47 73 59 32 04
E-mail: forlag@tapir.no
http://www.tapir.no/forlag

Animal Signals

Signalling and Signal Design in Animal Communication

Edited by

YNGVE ESPMARK
TROND AMUNDSEN
GUNILLA ROSENQVIST

Department of Zoology
Norwegian University of Science and Technology
Trondheim, Norway

Tapir
Academic Press

The Royal Norwegian Society of Sciences and Letters
The Foundation

This book is based on the oral presentations given at an international symposium on *Adaptive Significance of Signalling and Signal Design in Animal Communication* venued at Kongsvoll, Norway, September 15-20, 1998. The symposium was organised, financed and hosted by The Royal Norwegian Society of Sciences and Letters, The Foundation.

The following edited volumes, based on the previous Kongsvoll Symposia and published by Tapir Publishers, Trondheim, can be ordered from The Royal Norwegian Society of Sciences and Letters' Foundation, Erling Skakkesgt. 47 C, N-7491 Trondheim, Norway.

Brubakk, Kanwisher & Sundnes (Eds.) 1985. Diving in Animals and Man.

Brubakk, Hemmingsen & Sundnes (Eds.) 1988. Supersaturation and Bubble Formation in Fluids and Organisms.

Sundnes (Ed.) 1993. Human Impact on Self-recruiting Populations.

Sundnes (Ed.) 1995. Biomedical and Psychosocial Consequences of Radiation from Man-made Radionuclides in the Biosphere.

Contents

Contributors

Trond Amundsen, Department of Zoology, Norwegian University of Science and Technology, N-7491 Trondheim, Norway.
E-mail: trond.amundsen@chembio.ntnu.no

Staffan Andersson, Department of Zoology, Göteborg University, Box 463, S-405 30 Göteborg, Sweden.
E-mail: s.andersson@zool.gu.se

Alexandra L. Basolo, School of Biological Sciences, University of Nebraska-Lincoln, Lincoln, NE 68588-0118, USA.
E-mail: basolo@niko.unl.edu

Andrew T. D. Bennett, School of Biological Sciences, University of Bristol, Woodland Road, Bristol BS8 1UG, UK

Florens Bettens, Unité d'immunologie de transplantation, Centre Médical Universitaire, CH-1211 Genève 4, Switzerland

Michel Chapuisat, Department of Biochemistry and Genetics, La Trobe University, Bundoora, 3083 Victoria, Australia

Innes C. Cuthill, School of Biological Sciences, University of Bristol, Woodland Road, Bristol BS8 1UG, UK.
E-mail: i.cuthill@bristol.ac.uk

John A. Endler, Department of Ecology, Evolution and Marine Biology, University of California, Santa Barbara, CA 93106, USA.
E-mail: endler@lifesci.ucsb.edu

Susan Fitspatrick, School of Applied Biological and Chemical Sciences, University of Ulster at Jordanstown, Newtownabbey, County Antrim BT37 0QB, Northern Ireland.
E-mail: sm.fitzpatrick@ulst.ac.uk

Leo J. Fleishman, Department of Biological Sciences, Union College, Schenectady, New York 12308, USA.
E-mail: fleishman@union.edu

Sandra Füri, Unité d'immunologie de transplantation, Centre Médical Universitaire, CH-1211 Genève 4, Switzerland

Mats Grahn, Department of Biosciences, Södertörns University College, Karolinska Institute, S-141 57 Huddinge, Sweden.
E-mail: mats.grahn@cnt.ki.se

Oren Hasson, P.O. Box 11598, Ramat-Gan 52015, Israel.
E-mail: ohasson@hotmail.com

Felix R. Homberger, Section of Comparative Medicine, Yale University, School of Medicine, PO Box 208016, New Haven, CT 06520-8016, USA

Felicity A. Huntingford, Fish Biology Group, Division of Environmental and Evolutionary Biology, IBLS, Graham Kerr Building, University of Glasgow, Glasgow G12 8QQ, UK. *E-mail: f.a.huntingford@bio.gla.ac.uk*

Rufus A. Johnstone, Department of Zoology, University of Cambridge, Downing Street, Cambridge CB2 3EJ, UK.
E-mail: raj1003@hermes.cam.ac.uk

Rebecca M. Kilner, Department of Zoology, University of Cambridge, Downing Street, Cambridge CB2 3EJ, UK.
E-mail: rmk1002@hermes.cam.ac.uk

Robert F. Lachlan, School of Environmental & Evolutionary Biology, University of St Andrews, St Andrews, Fife, Scotland KY 16 9TS, UK

Naomi E. Langmore, Division of Botany and Zoology, Australian National University, Canberra, ACT 0200, Australia.
E-mail: naomi.langmore@anu.edu.au

Ervin Macas, Departement für Frauenheilkunde, Klinik für Endokrinologie, Universitätsspital Zürich, CH-8091 Zürich, Switzerland

Peter K. McGregor, Department of Animal Behaviour, Zoological Institute, University of Copenhagen, Tagensvej 16, DK-2200 Copenhagen N, Denmark.
E-mail: pkmcgregor@zi.ku.dk

Kurt A. McKean, Department of Biology, University of California, Riverside, CA 92521, USA

N. Justin Marshall, Vision Touch and Hearing Research Centre, University of Queensland, Brisbane, Queensland 4072, Australia.
E-mail: j.marshall@vthrc.uq.edu.au

Francis C. Neat, Fish Biology Group, Division of Environmental and Evolutionary Biology, IBLS, Graham Kerr Building, University of Glasgow, Glasgow G12 8QQ, UK

Stephen Nowicki, Evolution, Ecology and Organismal Biology Group, Department of Zoology, Duke University, Durham, NC 27708, USA

Ken Otter, Department of Animal Behaviour, Zoological Institute, University of Copenhagen, Tagensvej 16, DK-2200 Copenhagen N, Denmark

Julian C. Partridge, School of Biological Sciences, University of Bristol, Woodland Road, Bristol BS8 1UG, UK

Tom M. Peake, Department of Animal Behaviour, Zoological Institute, University of Copenhagen, Tagensvej 16, DK-2200 Copenhagen N, Denmark

Alexander J. Paepke, Abteilung Verhaltensökologie, Zoologisches Institut, Universität Bern, CH-3032 Hinterkappelen, Switzerland

Katharina Riebel, School of Environmental & Evolutionary Biology, University of St Andrews, St Andrews, Fife, Scotland KY 16 9TS, UK

Thomas Rülicke, Biologisches Zentrallabor, Universitätsspital Zürich, CH-8091 Zürich, Switzerland

William A. Searcy, Department of Biology, University of Miami, Coral Gables, FL 33124, USA.
E-mail: wsearcy@umiami.ir.miami.edu

Thomas Seebeck, Institut für Allgemeine Mikrobiologie, Universität Bern, Baltzerstrasse 4, CH-3012 Bern, Switzerland

Ben C. Sheldon, Edward Grey Institute, Department of Zoology, South Parks Road, Oxford OX1 3PS, UK.
E-mail: ben.sheldon@zoology.oxford.ac.uk

Peter J.B. Slater, School of Environmental & Evolutionary Biology, University of St Andrews, St Andrews, Fife, Scotland KY 16 9TS, UK. *E-mail: pjbs@st-andrews.ac.uk*

Lynne U. Sneddon, Fish Biology Group, Division of Environmental and Evolutionary Biology, IBLS, Graham Kerr Building, University of Glasgow, Glasgow G12 8QQ, UK

John P. Swaddle, Centre of Behavioural Biology, School of Biological Sciences, University of Bristol, Woodland Road, Bristol BS8 1UG, UK.
E-mail: john.swaddle@bristol.ac.uk

Glenn-Peter Sætre, Department of Biology, Division of Zoology, University of Oslo, P.O. Box 1050 Blindern, N-0316 Oslo, Norway.
E-mail: g.p.satre@bio.uio.no

Alan C. Taylor, Fish Biology Group, Division of Environmental and Evolutionary Biology, IBLS, Graham Kerr Building, University of Glasgow, Glasgow G12 8QQ, UK

George W. Uetz, Department of Biological Sciences, University of Cincinnati, Cincinnati, OH 45221-0006, USA.
E-mail: uetzgw@email.uc.edu

Sandra L. Vehrencamp, Laboratory of Ornithology,Cornell University, Ithaca, NY 14850, USA.
E-mail: slv8@cornell.edu

Claus Wedekind, Abteilung Verhaltensökologie, Zoologisches Institut, Universität Bern, CH-3032 Hinterkappelen, Switzerland.
E-mail: wedekind@esh.unibe.ch

Brian D. Wisenden, Department of Biology, Moorhead State University, Moorhead, MN 56563, USA.
E-mail: wisenden@mnstate.edu

Marlene Zuk, Department of Biology, University of California, Riverside, CA 92521, USA.
E-mail: mzuk@citrus.ucr.edu

Preface

During September 15-20, 1998, the fifth symposium in the series of Kongsvoll Symposiums, named after the venue on the Dovre mountains in central Norway, was organised and hosted by the Royal Norwegian Society of Sciences and Letters' Foundation. The Board of the Foundation decided to devote this symposium to the topic *'Adaptive Significance of Signalling and Signal Design in Animal Communication'* and invite leading international researchers in the field to attend. The chapters in this volume are based on the twenty-five oral presentations given at the meeting.

The aim of the symposium, as of the previous Kongsvoll symposia, was to create a stimulating atmosphere for open discussions on the current state of knowledge and prospects for the future of a selected topic of wide international interest. Included in this aim was also a desire to create a forum where the discussants were encouraged to put weight on general and problem-oriented approaches rather than dwelling on narrow and specific case-studies. In retrospect, we are strengthened in our belief that such aims are best achieved when a limited number of people, sharing the same scientific interest, come together in a place like Kongsvoll where nothing but a magnificent nature is distracting, and where they are allowed to use most of the time for informal talks in inspiring settings.

Even though the twenty-five chapters in this book cover a broad sprectrum, thematically as well as taxonomically, they only make up a small sample of what animal communication as a whole is about. It is about blackbirds defending their territories, peacocks informing peahens of their qualities, and wild dogs telling about their aggressive intentions. It is about salmon and blueberry advertising their ripeness, as partners or prey. And it is about human language, and information travelling across cell walls within a body. In sum, signals may communicate qualities, status, and intentions of entities at various levels. Sometimes, the communicating parties share a common interest in precise information, at other times, senders and receivers may disagree over signal design. Animal communication has also important interfaces with numerous other disciplines, such as evolutionary biology, molecular biology, neuroscience, medicine, linguistics, cognitive science, chemistry and physics. Communication may take place in different signal modalities, of which the visual, vocal and chemical are the most studied and also the most prevalent ones. Whatever modality, signals have been designed through evolution to fit the sensory abilities of receivers, and the ecological conditions. All animals use signals, but signals and communication are not confined to songbirds and ungulates displaying their qualities and intentions in elaborate displays. Signalling also abounds among plants, and even among primitive organisms like bacteria and yeast. In fact, the same principles that have been useful in explaining communication between peacock and peahen seem relevant even for understanding communication between organs within the body, and between organelles within the cell. The present volume explores some of this diversity of signalling and communication, combining theoretical and empirical approaches. We believe that the contributors have provided an exciting overview of some of the most intriguing recent findings within this fast growing field. The ambition has not been to be comprehensive and balanced in all dimensions, but rather to reflect major areas of current interest, where important advances have taken place in the last few years.

As organisers of the symposium and editors of this volume we owe a great debt of gratitude to several people. We are grateful for the stimulating and positive interactions we have had with the authors and for their endurance, and we are especially grateful to those who attended the symposium and thereby substantially contributed to make it such a stimulating meeting. We are also indebted to our graduate students Per Terje Smiseth, Stein Are Sæther, Ingebrigt Uglem, Åslaug Viken and Åsa Espmark Wibe for their devoted help with the numerous practical tasks in association with the symposium. Stein Are deserves, in addition, a special thank for indexing the book. Last, but not least, we thank Turid Fredagsvik who served as a most helpful secretary at the symposium.

Trondheim, May 2000.

The Editors

IN: ESPMARK, Y., AMUNDSEN, T. & ROSENQVIST, G. (EDS.) 2000
ANIMAL SIGNALS:
SIGNALLING AND SIGNAL DESIGN IN ANIMAL COMMUNICATION, PP. 11-46
TAPIR ACADEMIC PRESS, TRONDHEIM, NORWAY

1

Evolutionary Implications of the Interaction between Animal Signals and the Environment

JOHN A. ENDLER

Department of Zoology and Tropical Ecology, James Cook University,
Townsville QLD 4811, Australia
Present address: Department of Ecology, Evolution and Marine Biology, University of California,
Santa Barbara, CA 93106, USA

E-mail: endler@lifesci.ucsb.edu

Abstract

Animal signals are used by senders to try to increase their fitness by affecting the behaviour of receivers. Simultaneously, receivers use the same signals to try to increase their own fitness. The environment and biophysics have significant effects on the signal as it is emitted, travels through air, water or substrates, and received. There are a very large number of ways that the content of signals can be coded by varying properties or parameters of the signals. The evolution of signal parameters is also influenced by biophysics and environment because some parameters are affected more than others. Signal degradation is not necessarily a disadvantage and can be used to mitigate the tradeoff between intended and unintended receivers and can also transmit distance information. Such effects can be increased by evolution of signal processing as well as signal structure and the two probably coevolve. There are many implications of these observations.

Animal signals are used by senders to try to increase their fitness by affecting the behaviour of receivers. Simultaneously, receivers use the same signals to try to increase their own fitness (for detailed discussion, see Johnstone, this volume). In considering signal evolution, it is useful to make the distinction between the content and the structure of a signal (Endler 1993a), as well as between the content and the efficacy of a signal (Guilford & Dawkins 1991, 1993). The content is what is being communicated, or the information on which the receiver makes a decision. Information does not exist by itself, it must be carried by something (the carrier) to the receiver, and the carrier must be detected and processed to extract the information. Carriers include light, sound, substrate vibrations, electric field changes, or chemical compounds, and each has a particular structure which contains the information to be extracted by the receiver.

Efficacy is a measure of how strongly the signal affects the behaviour of the receiver, and is a measure of the success of a particular signal. Arnold (1983) made the distinction between morphology, performance and fitness. In terms of signals, structure is equivalent to morphology and efficacy is equivalent to performance. The content of a signal is that part of the signal morphology which is proportional to performance, while structure contains both content and other components. As in anatomy, some components of signals are present (and required for function) but their variation does not affect performance.

Efficacy consists of two components, efficiency and effectiveness. Efficiency is a measure of how well a transmitted signal is received, relative to the signal production cost. It could be measured by the receiver signal/noise ratio divided by sender's production costs. Efficiency is affected by sender properties, the signalling environment, and the signal receptors of the receiver. Once a signal is detected and received it will vary in effectiveness in modifying the receiver's behaviour. Effectiveness could be measured by the probability that a received signal will change the behaviour of the receiver. It could also be measured as the probability that a received signal will increase the fitness of the sender. The sender and receiver fitness aspects of effectiveness are not the same and may be in conflict (Johnstone, this volume). Effectiveness will be determined by the information in the signal as well as the internal state of the receiver, or receiver psychology (Guilford & Dawkins 1991, 1993). The combined results of efficiency and effectiveness determine how strongly the signal affects the fitness of sender and receiver (efficacy). If a signal is easily received (efficient) but ignored (low effectiveness), then its efficacy is low, but efficacy may also be low if a signal is very difficult to receive but always effects the decisions of the receiver when it is received. One has to be careful not to take these distinctions too seriously because they grade into one-another. For example, if signal structure involves added noise or other deceptive components, then structure becomes part of the signalling strategy as well. This paper will be concerned with the evolution of signal structure and how content can be carried (coded) and detected.

Signal processes and evolution

In order for signalling to work, signals must be produced, transmitted, travel through the medium (air, water, vegetation, ground), be received, processed, and decisions made based upon the received and processed signals (Endler 1992, 1993a; Endler & Basolo 1998). The processes from production to reception (efficiency) are affected by the signaller and the environment (Endler 1992, 1993a, b) while processes from reception to decision making (effectiveness) are affected by the experience and internal state of the receiver (Guilford & Dawkins 1991, 1993). Effects on the receiver will depend upon (1) the emitter's genetic quality; (2) the emitter's physiological state and ability to transmit information; (3) the environmental conditions during signal emission, transmission and reception; (4) the receiver's receptor condition during reception [physiological adaptation]; (5) the receiver's attentive status [habituation]; (6) the receiver's decision status (readiness to make a decision); (7) the receiver's decision criteria; (8) magnitude of short-term fitness consequences of decision; and (9) magnitude of lifetime fitness consequences of the decision (see also Endler & Basolo 1998). Natural or sexual selection can affect any of these factors.

Signalling involves close causal and functional relationships between the signals, signalling behaviour, the environmental conditions during signalling, sensory systems, the brain, perception of signals, and decision criteria. This means that a change in one of these factors will result in evolutionary changes in the others. This is most clearly seen by considering a population which has invaded a new habitat or whose existing habitat has changed (see Endler & Basolo 1998 for discussion). The population may experience new conditions during foraging, predator avoidance, courtship and mate choice, and the new environment will probably induce natural selection on the senses and brain. Because there are physical and biological limits to how neural systems can function in a particular environment, we expect neural systems to evolve suites of properties characteristic of the environments in which they are used (for examples see Lythgoe 1979; Dusenbery 1992). However, evolution of sensory systems may not necessarily occur (as in *Anolis* lizards, Fleishman, this volume) if genetic and developmental constraints to change are too strong, or directional selection is too weak. Changes in the neural system may alter perception and hence modify female preferences for male traits. Changes in perception can cause male traits to evolve because new or different forms of male traits are the most stimulating under the new perceptual regime (Basolo 1990, 1995, this volume; Ryan & Rand 1990, 1993). The evolution of male traits which are most stimulating to females under specific environmental conditions can lead to the evolution of habitat- and microhabitat-specific signalling, which maximizes the efficiency under those conditions (Endler 1989, 1991, 1992, 1993a, b; Endler & Thery 1996). Predation and mating rivalry will also affect the evolution of male traits and signalling microhabitats. Selection of signals that stimulate females maximally and stimulate predators and rivals minimally may favour habitat selection and evolutionary specialization in particular microenvironments and conditions. Changes in microhabitat choice will affect environmental conditions under which sensory modes are used. Such changes will affect food and predator detection. Success at prey and predator detection will further affect the evolution of the neural system. This favours neural systems which work most efficiently in these specific and specialized conditions. Evolving neural systems can then affect female preferences, continuing the cycle of evolutionary interactions. Biophysics, neurobiological principles, and the signalling environment will bias the direction of these coevolving traits in predictable directions. This cycle of evolutionary interactions is known as Sensory Drive (Endler & McLellan 1988; Endler 1989, 1992). Various components of sensory drive have been called sensory exploitation, pre-existing bias, sensory traps and others (reviews in Endler 1992; Endler & Basolo 1998). Sensory drive allows or encourages evolutionary novelty in all these suites of traits because new directions of evolution can start and the origin of new traits can occur anywhere in the cycle. For example, new directions can start as a result of new environments (range expansion or habitat change in situ), new neural processing mechanisms, mistakes in or improved decisions, new signals, new or evolving predators and mate competition strategies, new microhabitat choice mechanisms, and changes in available food.

Given that the interactions and suites of traits are affected by known principles of biophysics, neurobiology, ecology and behaviour it should be possible to predict the direction of evolutionary changes in these suites of traits (Endler 1992, 1993a, b). There is a growing literature on predicting the evolution of male traits from sensory and cognitive biases (reviewed by Endler & Basolo 1998), but the interaction with the environment, habitat choice and their effects on the evolution of sensory systems is only just starting. In this paper I will explore some of these effects. I will also discuss effects and processes which have either not been studied, or if they are studied, have not been fully integrated into studies of the evolution and design of animal signals.

Signals and signal components

It is often assumed implicitly that signals are single simple traits, but this is an ardent oversimplification. At a very basic level signals are disturbances in the environment caused by the sender, but these disturbances have much structure. It is the structure which allows the coding of the signal content. Different components of the structure can be affected by different gene loci, biophysical principles and environmental properties during emission and reception. To understand how signals can be affected by the environment and natural or sexual selection, it is useful to consider the basic parameters and components of signals (Table 1). In what follows I will concentrate on factors which most signals have in common rather than those used in a specific way by each sensory mode. Table 1 is a preliminary summary of the specific uses by each mode. For example, although wavelength is a characteristic parameter of most senses, it is used very differently, if at all, for carrying information in each sensory mode.

Carriers have four fundamental parameters, mean intensity, frequency, maximum amplitude and phase. The mean intensity, I (Fig. 1A), is the mean absolute value of the carrier, for example the mean absolute sound pressure level, the mean absolute brightness (radiance) of the light from a moving object, the mean brightness of a colour pattern, the mean current of an electrical signal, the mean tension in a spider web thread, or mean emitted pheromone concentration. Information is rarely carried by I but it is the reference for the rest of the signal.

Table 1. *Preliminary summary of the relative incidence* of signal parameters in coding information in various sensory modes*

Symbol	Description	Vision: spatial	Vision	Vision one point	Sound airborne	Vibration substrate	Electric	Chemical
I	Mean intensity						?	
λ, F, x	Wavelength, Frequency, Distance or Component				+	+	+	+
A	Amplitude	+	+	+	+	+	+	+
ϕ	Phase	+(crypsis)	+	+	+	+	+	
t_p	Pulse length	+	+	+	+	+	+	+
t_i	Inter-pulse interval	+	+	+	+	+	+	+
t_r	Rise time	+	+	+	+	+	+	?
t_f	Fall time	+	+	+	+	+	+	?
t_b	Bout length		+	+	+	+	+	+
$S(\lambda)$, $S(x)$	Spectrum or component mix	+	+	+	+	+	+	+
T	Total intensity	+	?	?	+	?		
H	λ, F, or x at maximum change in intensity	+	+	+	+	?		
C	Difference in intensity at H	+	+	+		?		
P	λ, F, x of maximum intensity	+	+		+	?		+
B	Bandwidth (usable x range)	+	+	+	+	?		+

*Table entries: + frequently used; ? equivocal, used occasionally, or not known; blank: rare or absent. Note that any one species does not necessarily use all parameters. AM (amplitude modulation) and FM (frequency modulation) are used in all modes.

Except for most chemical signals, most signals are periodic; the carrier intensity oscillates about I. This oscillation has a characteristic wavelength or frequency. The wavelength is the distance (λ) or time (t_w) over which one cycle occurs (see Fig. 1A). (Note that $\lambda = t_w v$, where v is the velocity of the carrier). The frequency is the reciprocal of the wavelength ($F = 1/t_w$). Hue and sound pitch are examples. Frequency is most often determined by biomechanics (e.g. resonance determined by size of sound generation organ; size and tension of spider web) or biochemistry (e.g. pigment hue), and is often fixed during a signal. Most animal sounds and colours are complex and consist of many different frequencies whereas substrate vibrations are relatively simple and consist of fewer or a single fundamental frequency. A spectrum is a summary of the frequencies used and their relative contribution to the total signal. In the case of chemical signals the equivalent data would be a list of the compounds and their relative proportions or concentrations.

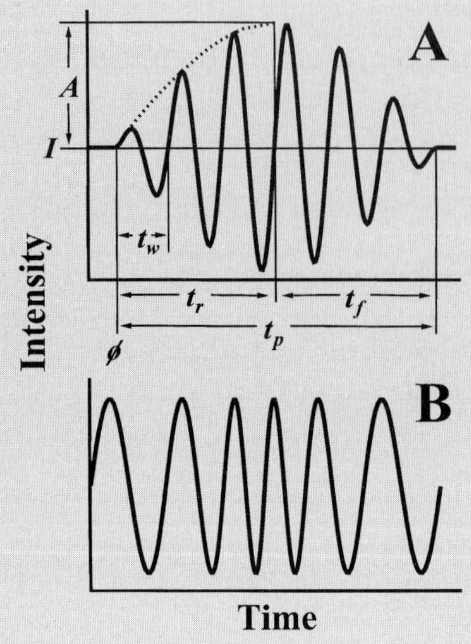

Figure 1. Basic signal parameters.
(A) a signal consisting of a single pulse of amplitude modulation.
(B) a signal consisting of a single pulse of frequency modulation.

I: mean intensity of the signal.
A: maximum amplitude.
ϕ: phase.
t_w: wavelength (time for one cycle).
t_r: rise time.
t_f: fall time.
t_p: pulse length.

The Amplitude (A) is the maximum deviation from the mean intensity (I) of the carrier (Fig. 1A), and is what we loosely think of as signal strength. Amplitude is related to the energy (or quantum flux in light) expended to produce the signal and usually varies, as in Fig. 1A.

The phase (ϕ) is the time from an arbitrary start time to the beginning of the carrier or signal (Fig. 1A). It is generally only important in communication networks where jamming is a possibility, as in frog choruses, communal firefly flashing, electric fish jamming, and vibratory communication among multiple Homoptera on the same plant. Phase is also important in duetting or countersinging or equivalents in other modes. Spatial phase is important in some cryptic animals who must align their colour pattern with that of the background.

Information (signal content) is usually coded by modulating (varying) the carrier. There are two basic kinds of modulation, Amplitude and Frequency Modulation (Fig. 1A, B, respectively). These are often abbreviated AM and FM. AM involves changes in the amplitude of the signal, and can occur in time (all modes) or space (vision). For example, in vision AM can involve intensity as a function of time, position of a part of a moving colour pattern element as a function of time, or intensity changes with position on the surface occupied by the colour pattern. FM involves changes in the wavelength or signal composition as a function of time or space. For example, in vision a single colour spot can change in colour with time (cuttlefish and some lizards) and colour varies with position in the colour pattern. Structural colours change in colour with viewing angle resulting in colour change during motion. AM is more common than FM in animal signals, and is particularly common when FM is difficult because of mechanical or biochemical limitations.

For example, pheromones of fixed compound proportions can be given out in pulses (strong AM) which vary in duration, spacing and time, as can sounds based upon fixed mechanical resonance in crickets, or flash colours in butterflies.

Modulation is often in the form of pulses (Fig. 1A, B). Information may be encoded by the shape and duration of each pulse, as well as the number, spacing and spectra of each pulse. Characteristic parameters include pulse length, interpulse interval, rise time and fall or decay time (Fig. 1). Pulse length (t_p) is the time for one signal pulse (Fig. 1A), and the inter-pulse interval (t_i) is the time between the onset of each pulse. Rise time (t_r) is the time from pulse onset to the maximum amplitude A, and fall time (t_f) is the time between time of maximum A and the end of the pulse. The bout length (t_b) is the time between the start of the first pulse and the end of the last pulse in a series of pulses. There are many other measures and modulation modes.

Information can also be carried by the particular combination of wavelengths or chemical components—the spectrum. A spectrum A(x) is defined as a specific relationship between amplitude (A) and wavelength (or component); x is wavelength and A(x) is the A at that wavelength. For example the spectrum of the signal in Fig.1A would show a single peak (at height A) at t_w, whilst the spectrum of Fig.1B would have a single peak which moved as a function of time. Real spectra are more complex (Fig. 2). For example, we perceive a particular light spectrum as a particular colour, a particular sound spectrum as a particular recognizable sound, and a particular chemical component mix as a particular smell. In Fig. 2, if the wavelength (x) axis were in nm from 300 to 700nm, then we would perceive the spectra in Fig. 2A as yellow, the spectra in Fig. 2B as varying from yellow to orange, the spectra in Fig. 2C as varying from yellowish-white to bright yellow and the spectra in Fig. 2D as varying from leaf green through yellow (this is in fact the fading sequence as a fallen live leaf dies and yellows).

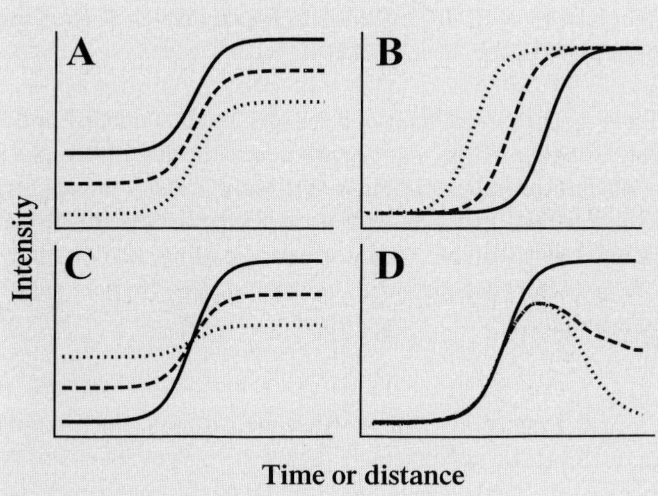

***Figure 2**. Spectra showing variation in some spectral shape parameters. (**A**) Variation in total intensity T. (**B**) Variation in cutoff wavelength H. (**C**) Variation in relative intensity around the cutoff C. (**D**) Variation in bandwidth B, but not wavelength of maximum sensitivity P (except solid line). Spectra can vary in other ways and can have many peaks, particularly in mechanical signals such as sound.*

Spectral shape (S(x)) can vary in several ways. Five useful parameters are T, H, C, P and B, and all of them can carry information. T is the total intensity in the spectrum, calculated as $\int S(x)dx$ (Fig. 2A), and is a measure of the total energy (or quanta for light) in the signal. (The integral is between the lower and upper wavelength limits in the signal, or the bandwidth of the receiver, whichever is smaller). H is the wavelength at which the largest change in intensity occurs (Fig. 2B). This is important because it determines which receptors are or are not stimulated in the receiver's sense organs (see below). Spectra which have one or more peaks (Fig. 2D) will have more than one H. In visual signals for the wavelength spectrum H is proportional to what humans perceive as hue, for the spatial pattern spectrum it would define pattern patch size, and for motion and other spectra it would define potential bandwidth boundaries. C is the difference between the total intensities above and below H, and is standardized by the total intensity T (Fig. 2C), or $C = [\int_H S(x)dx - \int^H S(x)dx]/T$. C is important because it determines the relative degree of stimulation of receptors sensitive to different wavelengths. For light and sound wavelength spectra C is proportional to chroma, 'saturation', or 'purity' of colours and sounds to humans (pink is a low chroma red); for spatial spectra C is proportional to pattern clarity and contrast, and for motion it would be the degree of regularity. P is the wavelength of maximum intensity, and is important when considering resonance with biomechanical and environmental structures (particularly in sound and substrate signals). B is the bandwidth, and if narrow, will result in several H and P. All of these parameters vary in animal signals, and each of them, or combinations of them, can carry information about the signaller.

In addition to modulation in time and spectral properties, signals can also have a spatial component which may contain information. Visual signals are particularly rich in spatial information which is found in the details of colour patterns as well as their movement. A spatial component may also be present in sound (as in hummingbirds and melanoplid grasshoppers which have complex aerial auditory displays), lateral line displays in fishes, and in firefly displays. Signals also contain a direction (location) component, and the form of the signal and carrier will affect how easy or difficult it is to locate (Wiley & Richards 1982; Dusenbery 1992; Bennet-Clark 1998).

Signalling usually occurs simultaneously in more than one sensory mode, and each mode usually involves the modification of many of the signal parameters. A good example is the complex visual and auditory displays of hummingbirds. Different sensory modes may also vary in the amount of information that they can code (see Endler 1993a for discussion). Clearly there are many dimensions available for information coding, and evolution can act on any or all of them, both within and between sensory modes. Signals should certainly not be regarded as single traits.

The different parameters of signals reveal different things about the sender, and any of these can be used by receivers to make decisions which affect their fitness. For example parameters such as dominant wavelength and spectrum are set by the biomechanics of the sender. Other parameters such as t_r, t_p, and t_i may be limited by biomechanics and energetics. For example, t_r will be larger in systems which involve successive excitation of a resonant system, such as that in frogs and gryllid crickets, whereas it will be smaller for systems in which the pulse is caused by a strong impact, such as grasshoppers. Similarly t_f will be long for resonant systems with low damping or continuous excitation and will be short where there is high damping or a single excitation (Bennet-Clark 1998). Biomechanical properties are related to size and developmental stability (Dusenbery 1992; Bennet-Clark 1998), allowing direct assessment of these aspects of the sender by the receiver. Insect, frog and bird (doves) calls are a good example; within and among species the dominant wavelength declines with body size owing to resonance (Narins 1992; Bennet-Clark 1998; Tubaro & Mahler 1998). Other parameters may be set by the environment during emission. For example, the t_w, t_r and t_f of substrate vibrations of some Homoptera, Hemiptera and spiders will depend upon the flexibility, mass, and tension of the plant stems, water surface, or silk; t_w by resonance, t_r by resistance to deformation by the animal, and t_f by dampening. In Homoptera, signalling by vibration of the plants is influenced by the plant's physiological condition, which affects stiffness and resonance. In fossorial mammals which use seismic vibrations to communicate, the signals will be affected by soil quality as well as the ability to dig (Narins et al. 1997). Consequently, resource quality on the territory may be signalled by t_w, t_r and t_f. Biomechanical and environmental effects at transmission are not necessarily hindrances to communication and can be quite useful to the receiver for signaller assessment.

The quality of the sender can also be assessed directly by I, A, t_r, t_p, and t_b, as well as total signal length and total signal energy expenditure; it takes more energy to increase I, A, t_f and t_b, and to decrease t_r, and t_i. For example, in movement it is more costly to produce a square wave (movements with rapid accelerations and decelerations) than a sine wave. Amplitude (A) is a common indicator of energetic cost and mate quality in a variety of animals (Ryan & Keddy-Hector 1992; Römer & Bailey 1998; Römer et al. 1998). The total length of the calling bout (t_b) can also indicate quality because it is proportional to the total energy used in the bout. For example, in crickets it is used in mate choice (Hedrick 1986). Quality could also be assessed by t_w if there is a cost to deviating strongly from a biomechanical resonance, but this would require an internal standard in the receiver. Spectral properties and parameters can also indicate quality; a good example is the relationship between feeding success, carotenoid input and visual spectral properties, particularly chroma (Hill & Montgomerie 1994). More than one parameter can be assessed at the same time, each indicating something different about the signaller, as in bushcrickets (Bailey et al. 1990; Römer et al. 1998). For example, in Orthoptera the AM pattern may indicate species and sex, while A indicates male fighting ability and quality (Römer et al. 1998). In principle, cost could also be indicated by the temporal complexity of the signal. Neural processing can be costly (Laughlin et al. 1998), and temporally complex signals may require complex neural processing for production. However, signal complexity is limited by the cost to the receiver in processing complex signals (Laughlin et al. 1998). There are a very large number of ways quality can be assessed but this has not been investigated systematically.

It is a pity that most studies of signal content have examined only A or t_p. Natural and sexual selection should favour the use of whichever parameters in any sensory mode which yield reliable information about the signaller. The study of parameters aside from A and t_p, multimodal signalling, and the factors determining the use of particular parameters and sensory modes, would be very interesting and valuable. Studies of signal content have also tended to concentrate on mate quality, but have ignored other functions of signalling, such as alerting the receiver that a signal will be sent (Wiley & Richards 1982; Fleishman 1986, this volume), individual recognition (Godard & Wiley 1995; Crawford et al. 1997; Balshine-Earn & Lotem 1998), age, location, sex, and species (Crawford et al. 1997; Römer et al. 1998). The study of multiple functions of signals and their evolution should yield significant new insights into animal communication and evolution.

Environmental effects on signals

The environment can have two effects on signals; it can modify signals and can mask and distort them by adding noise. Some signals and signal parameters are more sensitive to environmental effects than others, even within sensory modes. The most obvious is amplitude and all aspects of amplitude modulation. Amplitude is very sensitive to noise because many different kinds of random factors can attenuate or accentuate amplitude at any time instant. The received signal consists of the sum of signal and noise. Therefore, at any given time interval, if noise is high then the received amplitude (A) is high, and if the noise is low then the received A is low (Fig. 3). If the signal's amplitude A_s is high relative to the noise amplitude A_n, then the signal is easy to detect and little distorted (Fig. 3C), but as the signal/noise ratio (A_s/A_n) declines, distortion increases (Fig. 3D) and detection gets more difficult (Fig. 3E). One way to minimize the effects of noise is to use FM instead of AM (Fig. 1B); because noise is random, noise spectra are usually flat, and regular changes in signal spectra will be detectable as changing wavelength peaks at relatively higher signal/noise ratio than signals with AM. When wavelengths must be very small (as in small insects) or there is much noise, information may also be carried by modulation of the timing of the pulses and their widths (t_p, t_i, t_b) rather than by modulating the pulses themselves (Bennet-Clark 1998; Römer & Bailey 1998).

If the environment contains objects which reflect the signaller's signal and these reflections reach the receiver with varying time delays, the original and reflected signals can modify the received signal so much that it is unrecognizable. This is called self-interference or reverberation and is best known in sound and vibratory communication, but also can be a problem for small aquatic bioluminescent animals when the water is full of reflective particles. Fig. 3F shows that even if the direct and reflected signals are in phase the signal can still be obscured. It is greater at smaller wavelengths (higher frequencies) because objects reflect wavelengths which are less than or equal to their size and there are more small than large objects in natural environments (Bennet-Clark 1998). These effects can be reduced by using longer wavelengths (Wiley & Richards 1982; Narins et al. 1997; Bennet-Clark 1998; Römer & Bailey 1998). Using larger t_r and t_f or FM will also reduce the effects of reverberation, as will choice of signalling position relative to obstructions. However, if the body size of the animal is small (as in small insects), it will be difficult to produce long wavelengths or t_w (Bennet-Clark 1998), and the only reliable information that can be successfully transmitted is the start times of each pulse (Stephen & Hartley 1991).

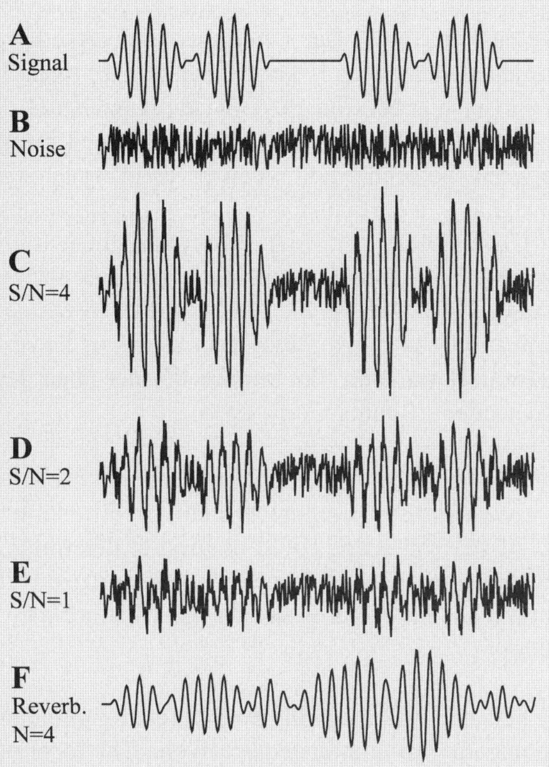

Figure 3. Effects of noise and reverberation. Horizontal axis: time. Vertical axis: signal intensity. (A) Signal without noise. (B) Noise without signal. (C-E) Signal plus noise with declining signal amplitude. S/N is the signal/noise ratio. (F) Reverberation; signal in A plus reflections of signal A from four reflectors at various distances from the receiver, resulting in randomly varying time lags to receipt. A similar pattern occurs when signals are received from several different individuals at varying distances.

In leks and other communication networks (as in frogs, orthopterans, fireflies, and fiddler crabs), the individual signals will also be obscured by phase differences between different individuals' signals (Wang & Narins 1996), unless all signal in synchrony (as in some firefly species; Moiseff & Copeland 1995). Interference from other individuals' signals may be intentional (jamming) and part of territorial interactions or anti-predator mechanisms. Examples are conspecific jamming in crickets (Jones 1966) and electric fishes (Bastian 1990), and bat signal-jamming in moths (Fullard 1982; Miller 1983).

In addition to added noise from the environment or other signallers, signals are degraded by the medium (air, water, substrate) as they travel between the signaller and receiver, and also may be degraded as they are received and converted to neural impulses by the receiver (Lythgoe 1979; Wiley & Richards 1982; Dusenbery 1992; Römer & Lewald 1992; Embleton 1996; Bennet-Clark 1998; Römer & Bailey 1998). This is most easily measured by observing how the spectrum, ϕ and other characteristics change with distance (Endler 1990). For example, the harder the substrate, the longer the minimum wavelength of vibrational signals that can be transmitted, and longer wavelengths travel further than shorter ones. Similarly, depending upon local conditions, water attenuates different light wavelengths at different rates (Lythgoe 1979; Endler 1991), and air attenuates different sound frequencies at different rates (Wiley & Richards 1982; Dusenbery 1992; Embleton 1996; Bennet-Clark 1998). Some of the effects of signal degradation can be avoided by sending signals containing wavelengths which attenuate less than other wavelengths, or wavelengths for which the background noise is minimized. Evolved wavelength selection for minimum environmental influences is known for species using light in water (Lythgoe 1979) and sound in air (Gish & Morton 1981; Wiley & Richards 1982; Römer & Lewald 1992; Römer & Bailey 1998). Effects can also be reduced by minimizing the number of obstructions between the sender and receiver (as in visual signals and in some calling frogs), aiming the signal in a particular direction to minimize spherical loss (as in mole crickets; Bennet-Clark 1998), choosing substrates which transmit most efficiently, and choosing times and microhabitats which are optimal for transmission (Wiley & Richards 1982; Endler 1991, 1993a, b). For example, insects using plants for substrate vibratory communication use twigs rather than stems or leaves because thin cylinders (nearly 1-dimensional) transmit the signals further than leaves (2-dimensional) or stems (nearly 3-dimensional; Bennet-Clark 1998).

Differential absorption of different wavelengths is not necessarily detrimental because the ratio of the intensities of two or more wavelengths of a signal will be proportional to distance and so this ratio or other aspects of spectral shape could be used to determine distance to the signaller (Morton 1982). Rapid signal attenuation may be advantageous if the average distance between signaller and intended receiver is shorter than that between signaller and eavesdroppers such as predators.

Signalling always represents a balance between efficiency in communicating to the intended receiver while inefficiently communicating to unintended receivers. The local population will evolve signal characteristics which are a function of the compromise between these conflicting selective factors (Endler 1978; Römer & Bailey 1998). Guppy *Poecilia reticulata* nuptial colours and predation are an example (Endler 1978, 1980). If predation and sexual signalling can be made to occur under different conditions or spatial scales, then environmental degradation of signals can be an advantage (Endler 1978, 1980, 1991, 1992, 1993a, b; Endler & Thery 1996).

Visual signals are subject to additional environmental influences which are not found in other sensory modes. This occurs because, aside from bioluminescent species, the signals originate in the environment (Lythgoe 1979; Endler 1990, 1993b). The light spectrum arriving at the receiver's eye from an animal's colour pattern component is a product of the ambient light spectrum striking the animal (Endler 1993b), the reflectance spectrum of the colour pattern element on the animal, the absorbence spectrum of the medium (air or water), and the spectrum of additional light (veiling light) arriving at the eye from other sources in the line-of-sight (Lythgoe 1979; Endler 1990). As in other signalling modes, the signalling animal can improve transmission conditions by choosing better microenvironments (lower air or water absorbence, less veiling light), but must additionally choose ambient light environments which maximize the radiance from the colour pattern or the difference in radiance between the colour pattern and the visual background (Endler 1991, 1993b; Endler & Thery 1996; Andersson, this volume; Fleishman, this volume; Marshall, this volume).

One of the biggest constraints on visual signalling efficiency is that changing light conditions can alter the radiance spectrum of the animal (the spectrum of the light coming from the animal; Endler 1990, 1997). Natural light environments are spectrally diverse. There are five spectrally distinct light environments associated with terrestrial vegetation (Endler 1993b), and even more in aquatic environments where water absorbence spectra are important (Lythgoe 1979). The radiance spectrum is the product of the ambient light and reflectance spectra. Therefore, in the wrong light environment, the radiance spectra can have entirely different shapes, and even major characteristics such as hue can change (Endler 1997; Fig. 4). Effects can be even greater in aquatic environments where the absorbence spectra modifies the radiance spectra even more (Lythgoe 1979; Endler 1990, 1991). Ambient light effects are greatest for reflectance spectra with smaller C (chroma), and for reflectance spectra which have a shape which is very different from the ambient light spectrum. Consequently, the best signal components should have high C and be similar in shape to the ambient light spectra used during communication (Endler 1991, 1992, 1993b, 1997). This prediction will have to be modified for maximizing contrast with the visual background (Endler 1978, 1993b, 1997; Endler & Thery 1996), for example in detecting green objects in green shade against a green background (Endler 1993b).

Given that the interaction between the ambient and reflectance spectra will happen for all the colour pattern elements of the animal, as well as the background colour patterns, there will be changes in pattern element characteristics, and in within-animal and animal-background contrast (Endler & Thery 1996). Thus, both the detectability and the apparent

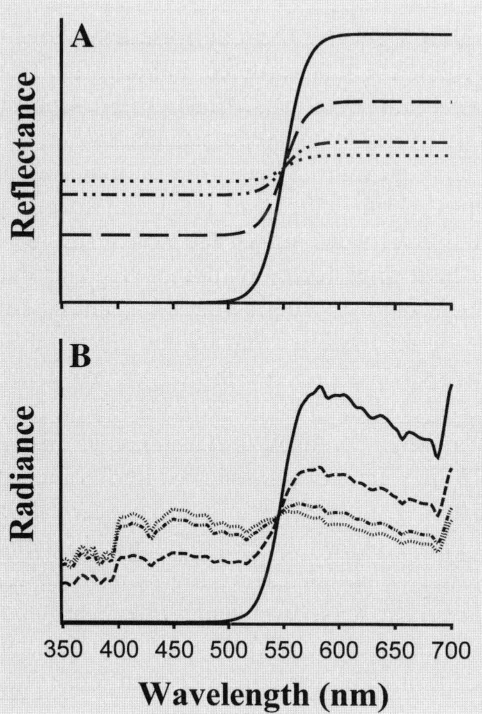

*Figure 4. Example of the effect of ambient light on radiance spectra when there is a mismatch in shape between ambient and reflectance spectra. (**A**) Reflectance spectra varying in C (chroma) but not in H (hue, H = 550nm). Spectra with declining C from solid, broken, dash-dotted, through dotted lines. (**B**) Radiance spectra resulting from illumination of each spectrum in **A** by Woodland Shade, the bluish shade near large canopy gaps in vegetation (Endler 1993b). Note how the radiance spectra are affected more when C is smaller. The spectra with the two smallest chroma (dotted and dash-dotted) will appear gray rather than yellow. Effects are less when the ambient and reflectance spectra match. For more details, see Endler (1997).*

content of signals can be affected by changes in ambient light (Lythgoe 1979; Endler 1990, 1992, 1993b, 1997; Endler & Thery 1996). For example, in guppies (Endler 1991), two manakin species and the cock-of-the-rock *Rupicola rupicola* (Endler & Thery 1996), display conditions are chosen which maximize the colour and brightness contrast for conspecifics, and minimize it at other times and places. This occurs for both single colour patches and for within animal contrast. In addition, for some of these species the animal-background contrast is very small, even during displays, and can be reduced much further as long as potential predators are far enough away for adjacent colours to mix below acuity limits (Endler 1978, 1991; Endler & Thery 1996). For greatest signal efficiency, colour pattern elements on parts of the body used for communication should have high C and spectral shapes which are similar to the ambient light used during signalling. For additional contrast, they should also have coloured patches with spectral properties very different from each other (but still with high C), and different from that of the visual background. Parts used to avoid predation by crypsis should have low C (as do most natural visual backgrounds except coral reefs) so that they change with ambient light in the same way as the visual backgrounds, leading to similar crypsis no matter what light conditions (Endler 1978, 1990, 1992, 1993a, b, 1997; Endler & Thery, 1996). Parts used to avoid predation by aposematism (visibility and distastefulness) should have high C to appear the same in all light environments. In aposematic colouration the demands on visual communication are the same for conspecifics and predators. See Endler (1978, 1980, 1991; Endler & Thery 1996) for more discussion of the balance between predation and communication, and the interaction between colour patterns, visual backgrounds, and ambient light spectra.

In aquatic or very dusty terrestrial habitats, veiling light is a form of noise analogous to noise in other sensory modes. Veiling light is light which is scattered and reflected from particles in the line-of-sight between the signaller and receiver (Lythgoe 1979). But, unlike other sensory modes, this noise will be correlated with the signal because veiling light and animal radiance both depend upon ambient light. They differ as much as the reflectance spectrum of the animal differs from the reflectance and scattering spectra of the veiling particles. As a result of the correlation between signal and noise, it is more difficult to remove veiling light by filtering than other forms or noise which are uncorrelated with the signal. When veiling light becomes too high, signalling must either temporarily stop (e.g. wait for a dust storm or smoke to clear, wait for the clearer water season), or evolve signalling in other sensory modes.

To summarize, communication efficiency is affected by many environmental factors. There are four major sources: added noise, self-interference, conspecific interference and environmental degradation. Visual communication is further complicated by correlation between the signal and noise and the fact that visual signals depend upon environmental light. These factors can be reduced by increasing amplitude, using high amplitude and AM, using FM, and choosing the times and places where environmental effects are minimised. Environmental effects may not always be a problem, and may actually be useful for predation avoidance or other forms of deception. The tradeoffs between these functions need more investigation.

Reception of signals

Signals are received by sense organs, which have the triple task of collecting the signal from the environment, decomposing the signal into various parameters, and coding the parameters into neural impulses. Sensory systems are not perfect and cannot collect, extract and code all aspects of signals, nor can they collect all possible signals within their sensory mode. Species and taxonomic groups vary not only in where their sensitivity range falls, but also in the peak and width (bandwidth) of their sensitivity spectrum. For example, each species using sounds can only hear a subset of all possible sound wavelengths; some species hear longer wavelengths (lower pitches), others intermediate and others shorter wavelengths (higher pitches). For biomechanical reasons this is usually related to body size (as in insects; Bennet-Clark 1998). Species also vary in the peaks and ranges of wavelengths received, and this is not only a function of the optics or biomechanics of the sense organ, but also the distribution and properties of the receptors within the organ.

Each sense organ consists of components or receptors which themselves sample part of the range received by the whole organ; their combined sampling range affects the organ's sensitivity spectrum. Species and higher taxonomic groups differ in their receptor distributions and hence their use of a given communication channel. For example, guppies have more cone types (photoreceptors) than humans and they cover a wider range of wavelengths (Fig. 5).

Optics, biomechanics, and biochemistry also affect the range of wavelengths or other physical stimuli which can be sensed by various sense organs. For example, the spectrum

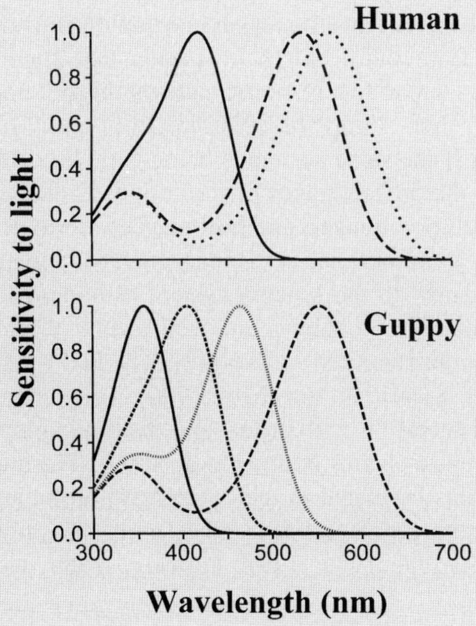

***Figure 5.** Sensitivity functions of cones of humans and guppies. (There is a polymorphism in the red-sensitive cones of guppies, but only the mean is shown here for simplicity).*

visible to humans ranges from approximately 400nm (blue) through 700nm (red). Light below about 400nm is blocked by filters in the lens and cornea, even though the blue-sensitive human cones could potentially sense UV (Fig. 5). Guppies can see well into the UV (below 400nm) because their eye optics transmit UV and there is a UV-sensitive cone. Birds and other animals have optical filters associated with different receptor classes which allow finer resolution than the photochemistry of the receptors would predict (Partridge 1989). Limits to sensory ranges occur in other sensory modes. For example, sound, vibrational, or electrical impedance mismatching between the medium (air or water) and the surface of the sense organs restricts what can be received (Bennet-Clark 1998; Römer & Bailey 1998). A knowledge of biomechanics, optics, and biochemistry can be used to predict some aspects of signals.

The restricted wavelength or component range (bandwidth) for each receptor is the basis for extracting information from signals. The first stage is simply to code the input spectrum. In most species, the output of a given receptor is proportional to the logarithm of the total intensity of the signal within its bandwidth and different receptor classes respond to different parts of the spectrum (as in Fig. 5). As a result, the array of outputs is a crude replication of the shape of the spectrum. The degree of crudeness is related to the number of receptors and their bandwidths. In vision, receptor class numbers range from as few as two cones in some primates (Osorio & Vorobyev 1996) and cloudy water-dwelling fish (Levene & MacNichol 1970) to more than ten in mantis shrimps (Cronin et al. 1994), but is typically four (particularly in fishes and birds). Sensory modes differ in the number and bandwidth of their receptors. For instance, hearing organs have many small-bandwidth receptors while other modes tend to have fewer broader-band receptors.

The evolved degree of resolution of a sensory system is dependent upon the need for resolving power in that channel as well as the kind and amount of information carried. For example, colour resolution is limited by few cones, but their spacing and small size allows high spatial resolution and their speed of response allows high temporal (movement) resolution. If light is limiting, then it must be collected from many photoreceptors, which reduces spatial resolution. If the same system is being used for different purposes, there may be a tradeoff between the different axes of resolution. Examples are the tradeoff between spatial and colour resolution in eyes, and between sensitivity to AM and spatial resolution for electric signals of gymnotid fishes (Stoddard 1998). Sometimes these tradeoffs can be mitigated by redundancy in the sensory system (Stoddard 1998). We can make some predictions about the number, positions, and bandwidths of receptor classes. For example, receptors should be concentrated where they will yield the best detection and discrimination among signals (Osorio & Vorobyev 1996). They cannot be too close together or too far apart or discrimination will suffer, and discrimination will increase with decreased overlap among receptor classes (Cronin et al. 1994). The number of classes is limited by their overlap, and this is related to the complexity of the signals processed (Chittka 1996b). Predictions about receptor form and spacing are only just beginning to be explored (Lythgoe & Partridge 1989, 1991; Partridge 1989; Chittka 1996b; Vorobyev & Osorio 1998).

The basis for extraction of spectral properties and other signal parameters (Figs. 1 and 2) is the ability to compare the outputs of different receptor classes (Fig. 6). This is widespread among animals in all sensory modes. For example, it is found in sound intensity used for signaller direction finding in insects (Römer et al. 1998) and colour, texture and movement in vertebrates (Lythgoe & Partridge 1989; Endler 1990; Vorobyev & Osorio 1998). Calculations of the differences between receptor outputs have been successful in modelling vision (Vorobyev & Osorio 1998) as well as in helping us understand sexual selection (Endler 1991; Endler & Houde 1995; Endler & Thery 1996).

Receptor outputs must be compared because the output of a single receptor does not tell the brain anything about parts of the signal outside its bandwidth. For example, consider two signals S_1 and S_2 (Fig. 6A), or a signal which varies between these two shapes as a function of time. Consider two receptors R_a and R_b which sample different parts of the signal (Fig. 6A). The output of R_a will be high for S_1 and low for S_2 in this case (Fig. 6A), but it will also be high and low if S_1 and S_2 were spectrally flat (as is S_1 in Fig. 6A) but S_2 had a much lower intensity than S_1. The output of R_b will be similar for both signals (Fig. 6A). Worse, the output of R_b cannot by itself be distinguished from any set of spectra with the same intensity within R_b's bandwidth. However, the difference between the receptor outputs $d = (R_a\text{-}R_b)$ will depend upon the shape of the spectrum, and will be zero for a flat spectrum, and larger as the bandwidths differ in intensity (Fig. 6A). Finer detail of spectral shape can be obtained by using the proportional difference $d_p = (R_a\text{-}R_b)/(R_a+R_b)$ rather than d (Fig. 6B). For example, the two signals in Fig. 6B differ in both T and C but are similar in shape. They have different d but similar d_p. The former may be desirable for assessment of signaller quality and the latter may be desirable for species, gender, or age recognition signals, which must be relatively independent of inter-individual variation within species, gender or age classes. Methods of detection need more study, especially from an evolutionary viewpoint.

The distributions of sense organ and receptor spectral sensitivities are correlated with environmental properties in vision and hearing in some taxonomic groups (Levene & MacNichol 1979; Lythgoe 1979; Wiley & Richards 1982; Dusenbery 1992; Cronin et al. 1996; Bennet-Clark 1998), but not in others (Fleishman et al. 1995, 1997; Chittka 1996a). There tend to be broader sense-organ bandwidths and more widely spaced receptors where the possible signal range is greater, as in vision in mantis shrimp (Cronin et al. 1994, 1996) and fishes (Levene & MacNichol 1979). The spacing and position of receptors is also closely related to their use in detection and discrimination (Levene & MacNichol 1979; Lythgoe 1979; Cronin et al. 1994, 1996; Chittka 1996a, b; Osorio & Vorobyev 1996; Bennet-Clark 1998; Vorobyev & Osorio 1998). For instance, the red-green opponency system in primates may have evolved for detection of fruits against leafy visual backgrounds (Osorio & Vorobyev 1996) while the cone spacing and sensitivities in pelagic fishes are nearly ideal for detecting objects against a plain background of consistent colour (Lythgoe 1979). Animal groups in which visual properties vary little are probably those with relatively invariant visual tasks such as detecting objects against leafy backgrounds in all habitats (Fleishman et al. 1997). In evolutionary terms, the sense organ mean wavelength and bandwidth are a function of parts of the spectrum which are

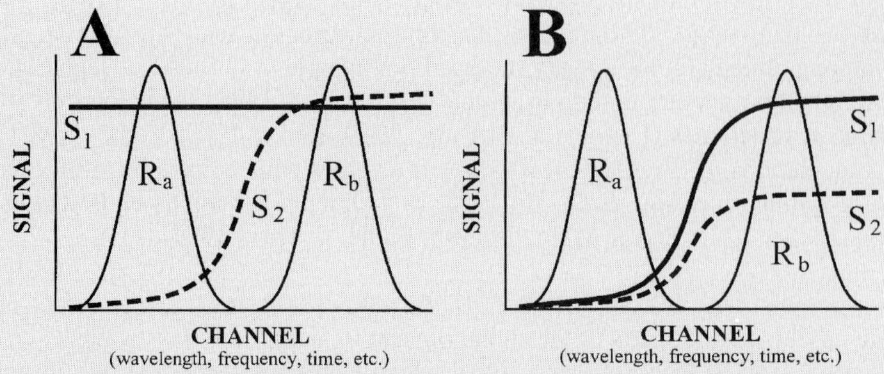

Receptor output	Signal S_1	S_2
R_a	High	Low
R_b	High	High
Difference (d)	none	large

Receptor output	Signal S_1	S_2
R_a	Low	v. Low
R_b	High	Medium
Difference (d)	large	med.large
Prop. Diff. (d_P)	large	large

Figure 6. *Generalised process of early stages of signal detection and processing. The channel (horizontal axis) is wavelength, frequency, time, distance or other units. The vertical axis is signal intensity (signal) or signal detection sensitivity (receptors).* S_1 *and* S_2 *are two signals, differing in spectral (wavelength or frequency), temporal (time), or spatial (distance) properties.* R_a *and* R_b *are two receptors differing in what part of the channel they detect, each with a maximum sensitivity and declining sensitivities away from the peak. Tables show the output of the two receptors and the output of a ganglion which responds to the difference in the output between them (lateral inhibition). (**A**) Signals differ in shape and are similar only in the upper part of the channel. (**B**) Signals similar in shape but differing in C. 'v.low' = very low, 'med.large' = medium to large.*

relatively noise- and distortion-free as well as the parts which allow signals which transmit the information most easily (Bennet-Clark 1998). Within a sense organ, receptor spacing seems to be optimized for the most important discrimination task (Lythgoe 1979; Lythgoe & Partridge 1989, 1991; Chittka 1996b; Osorio & Vorobyev 1996). Factors influencing signal and receptor bandwidths, receptor spacing and properties are only just beginning to be investigated in a systematic way, and should lead to significant new insights into the evolution of signals and signalling behaviour.

The differential receptor outputs are used only during the earliest stage of processing received signals. Yet they are capable of extracting a surprising amount of information from the signal simultaneously through different combinations of outputs. Figure 7 is an illustration of the earliest stages of signal component extraction. Outputs from the different receptor classes can be added (S, Fig. 7) to estimate I, A and other intensity-related parameters, e.g. loudness in sound or brightness (luminance) in vision (not all receptor classes are used to code brightness in many species). Outputs from different classes can also be subtracted by first level difference cells (D_1 - D_4, Fig. 7) to yield spectral information. For example, if the four receptors in Fig. 7 corresponded to the four cone classes found in fish (Fig. 5) and birds, then D_1 (a horizontal cell connecting red and green cones) might code the redness (red chroma) of an object. Outputs from the first level cells can also be combined in various ways. For example, second level difference cells (D_A in Fig. 7) taking differences between first level difference cells can extract more complex information. If Figure 7 were a retina, the D_A (Fig. 7) will be proportional to hue in part of the visual spectrum. These cells also provide information on spatial contrast and colour (or sound frequency) contrast, depending upon whether the first order difference cells work on the outputs from the same or different receptor classes. Note that all of this information is processed simultaneously, so the receiver can obtain information on multiple signal parameters at the same time ('parallel processing'). With time delays in the connections, or comparisons to biological clocks, temporal changes in the signal can also be extracted, allowing further signal parameter extraction and estimation. Higher order signal processing (mostly in the brain) extracts detailed information on the parameters shown in Figures 1 and 2, for these (and others) are the basis for behavioural and other decisions by the receiver. We are only beginning to understand how this is done in a few systems in a handful of species.

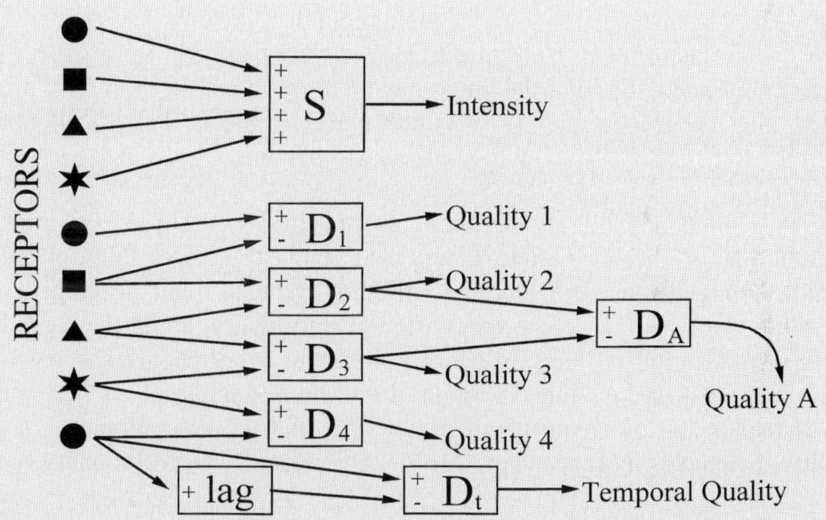

Figure 7. *Generalized process of early stage signal parameter extraction by various combinations of outputs from receptors which differ (circle, square, triangle, star) in the parts of their channel that they receive. S is a cell giving an output proportional to the sum of receptor outputs and D_1-D_4 are cells giving outputs proportional to the differences between pairs of different receptor types. D_A is a cell giving the difference between the first level difference cells. Time delays or biological clocks in the system allow comparison of outputs from any of these cells in time, allowing temporally varying parameters to be extracted.*

Lower and higher order processing in the brain can also improve the received signal quality. The analogue (continuous signal) - to - digital (nerve impulses) conversion by the receptors or ganglion cells nearby can actually amplify the signal, if the threshold is set correctly (normally by physiological adaptation and feedback systems). Other forms of amplification can be done in the brain. One way to reduce the effects of noise, reverberation and other interference is by processing and averaging repetitive signals. If the repeated components of a repetitive signal are averaged, any noise uncorrelated with the signal cancels out. For example, there is more noise at higher frequencies, and high frequency carriers tend to be repetitive (e.g. insect song, Wiley & Richards 1982; Endler 1993a; Bennet-Clark 1998). Another way to reduce noise is to have filters which are matched to the signal carrier in AM and FM; this would make the signal in Fig. 3E quite clear (Dusenbery 1992; Endler 1993a; Bennet-Clark 1998). As mentioned before, FM and FM detectors are more efficient in noise than AM and AM detectors. Amplitude sensitive filters (Römer et al. 1998), directional noise masking filters (Wang & Narins 1996), differential amplifiers, and rise-time detectors (Bennet-Clark 1998) are additional methods of increasing the signal/noise ratio and making signal parameter estimation more accurate.

Processing and signal redundancy is another mechanism of reducing noise. If the same parameters can be estimated through more than one parameter or channel, the signal/noise ratio is reduced because different parameters and channels will have different kinds of noise. These different noise sources will be uncorrelated, and can therefore be removed by comparison. For instance, A, t_p and t_b yield redundant information on energy expenditure with uncorrelated noise. Note that t_b may not measure the same aspect of energy expenditure as A and t_p because the latter may be proportional to short term muscle tension and t_b may be proportional to sustained muscle tension. This distinction may be important for some receivers, especially in sexual selection.

Selection for redundancy will favour more complex signals within sensory modes and the use of more than one sensory mode. It will also favour the use of multi-channel signals within and among modes. But, because different channels (and different signal parameters within channels) can carry very different information about the signaller, multichannel communication is likely to be very common even when noise levels are low. Whichever channels and parameters are used will be a function of environmental noise, the biomechanics and biochemistry of production, and the neuroethology of reception. The choice of channels and redundancy needs more work in an evolutionary context.

What can we say about the best signals?

On the basis of principles of signal generation, information coding, environmental effects, and reception, there are a few things we can say about the best kinds of signals, but this needs much more work (Endler 1978, 1992, 1993a, b; Andersson, this volume). Signals should be inexpensive enough to produce so that they can be given many times. Expenses can be minimised by using resonant structures (as in sound and vibration), biophysical matching between the emitting organ and the environment, shorter signals with less structure, and choosing environmental conditions with least attenuation, distortion, and interference. Some sensory modes are also cheaper than others (e.g. vision which uses environmental light). Of course some of these strategies run counter to the need to communicate energy expenditure, and more expensive signals may also contain more different kinds of information (simple pulses such as frog or insect calls can contain less information than complex pulses with both AM and FM as in bird song). Except for Andersson (this volume), this tradeoff has not been explored seriously, empirically or theoretically.

Factors facilitating detection include using signals tuned to the receiver's senses (Sensory Exploitation, Ryan & Rand 1990), using FM, larger A, t_f, t_b, and C, intermediate t_f and t_i (too long results in physiological adaptation), intermediate B (too wide gives low d, too narrow gives low T), smaller t_r, redundant signals and other noise-reducing mechanisms, complex and predictable signal structure, and multi-modal signals. Once again there is a tradeoff between efficiency and signal content although this can be at least partially mitigated by multi-channel communication and redundancy. There can also be a temporal structure allowing different kinds of information to be transmitted in sequence, as in the coqui *Eleutherodactylus coqui* (Narins 1992) and Túngara frogs *Physalaemus pustulosus* (Ryan 1985). Initial detection is often made easier with more efficient initial parts of signals, as in the rapid initial head-bobbing of *Anolis* lizards. Once the attention of the receiver is obtained, these signals continue with the less efficient but potentially more information-rich subsequent sequence (Fleishman et al. 1985; Fleishman 1986, 1992). Even if the signal itself is clear and stronger than the background, it may still be more efficient to make the signal as different as possible from the background. This is particularly true in visual communication where spatial as well as temporal information is being processed and integrated. If signals become too complex, or noise patterns require complex filtering, then the reciever's metabolic signal processing costs (Laughlin et al. 1998) may limit the amount of information that can be received. Reciever costs can be reduced by sending complex signals more slowly or reducing their complexity.

For situations in which detection by eavesdroppers (such as predators) is not desired, signals should match or be similar to that of the background under the eavesdropper's receiving conditions (Endler 1978), be used at close range, have long t_r, be non-repetitive and have other properties allowing extensive environmental influences. 'Private channels' is another method—using modes and parts of channels not used by eavesdropper. In visual signals, crypsis, low chroma colours (change with environmental light), low reflectance, mismatch with ambient light, slow movement, and indistinct boundaries between adjacent colours all reduce communication efficiency (Endler 1978, 1991, 1992, 1993b, 1997). In guppies, the use of red for courtship in areas where there are red-insensitive predators, and a combination of short courtship distance and small coloured spots below the resolution of some predators also mitigates the balance (Endler 1978). In chemical communication, reaction between the pheromones and the environment would also reduce detection by predators and parasitoids. As always there will be a tradeoff between efficient communication to conspecifics and inefficient communication with eavesdroppers, and the problem will be particularly acute when eavesdroppers are conspecifics.

There are other kinds of tradeoffs which will affect signal evolution. For example matched filters and matched tuning between signals and receivers will restrict the range of wavelengths that can be used and hence restrict the amount of information that can be transmitted. Amplification and filters also restricts the range of amplitudes that can be used, once again affecting the amount of information that can be carried. Tradeoffs between amount of information transmittable and clarity will affect many aspects of signal evolution both directly, and indirectly because evolution in signals will cause evolution in receiver characteristics (Endler 1992; Endler & Basolo 1998). The best signals may in fact be the best compromises rather than ideal in efficiency.

Evolutionary implications of natural selection of signal processing

The evolutionary effects of the environment on sensory systems, and the environment's reciprocal effect on signalling traits and behaviour means that natural selection on the sensory system can be complex. Because sensory systems contain so many different kinds of connections and subprocesses (Fig. 7), the response to selection on one part of a system may differ greatly among populations, even in the same environment. This can occur because the same selection may, by chance, sort genetic variation in disparate parts of neural pathways in different populations. Natural selection specifies only a response, not which system or part of a system is responsible for the response. This is most easily seen by means of an example. Consider a fish or bird which can perceive light over 300-700 nm (as does the guppy in Fig. 5b). Sensitivity to light of various wavelengths tends to be greatest at intermediate wavelengths (Fig. 8A). Let natural selection favour individuals with greater sensitivity to UV light (300-400 nm). Increased sensitivity to UV light can occur as a result of changes in the visual system in at least three different ways (Fig. 8), depending upon which system (Fig. 7) happens to respond to selection. The system responding to selection most strongly will be the one which has the greatest genetic variation or heritability (Falconer & Mackay 1996). (1) If the system collecting information on brightness (S, Fig. 7) includes the UV cones and has the highest heritability (h^2), then the entire sensitivity curve will rise (Fig. 8A) and the animals will be more sensitive to all wavelengths. This could also result from selection to increase clarity of the cornea and lens, or the size of the lens or eye. (2) If the UV and blue outputs have the highest h^2, then sensitivity will only increase at the UV and blue end of the spectrum (Fig. 8B). This could also happen if selection caused the removal of a yellow corneal filter found in some fish, changed some visual pigments, or (in a bird or reptile) removed UV-absorbing oil droplets within the cones. (3) If the highest h^2 were found in parts of the brain which determine the location of the entire spectral sensitivity function, or in the kind of visual pigments (rhodopsin or porphyropsin), then the location of the spectral sensitivity function will shift (Fig. 8C). If more than one system has high h^2, then selection can have one or more of these effects. The main point is that any of these responses can occur as a result of selection for increased UV sensitivity.

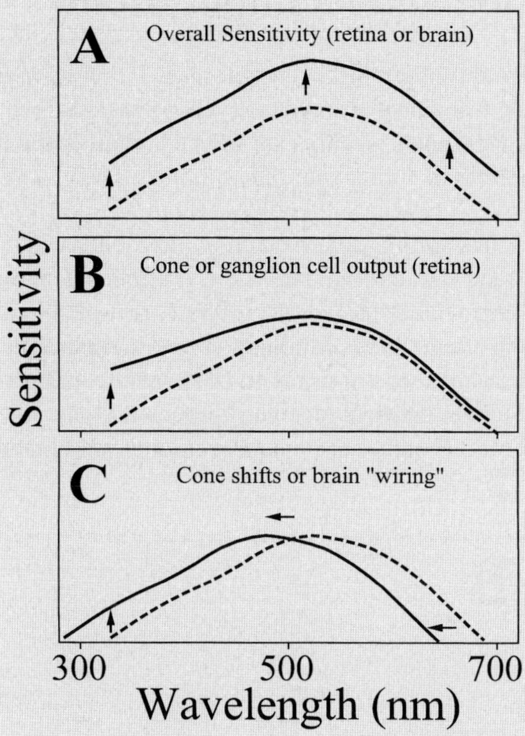

Figure 8. *Example of multiple evolutionary pathways of selection on sensory systems. (A-C): Three different possible responses to natural selection for increased sensitivity to UV light (340nm).*

These three alternative responses to the same selection have different consequences to perception, and hence can cause communication systems to evolve in very different directions. Increased sensitivity overall (Fig. 8A) would probably not change the fundamental carrier parameters, may reduce noise effects and may allow greater discrimination among lower amplitude (A) signals, but may decrease discrimination at higher A. It may also allow more information to be sent. So the main result may be the same bandwidth but greater signal complexity. Greater sensitivity to certain wavelengths (Fig. 8B) will result in carriers shifted towards those wavelengths and better discrimination at that part of the spectrum (in Fig. 8B, more use of UV and blue, and more different kinds of UV and blue colours used), and a greater range of wavelengths used (retention of some older colours). There may be a shift in the kind of information transferred as constraints change. A general shift in wavelengths used (Fig. 8C) will result in an overall shift in the carrier, but probably the same amount of information transferred and the same amount of discrimination possible. As a result there may be no or little change in signal complexity. If populations experience these three different responses to selection, then the entire process of sensory drive will result in divergence in sensory systems, signals, signalling behaviour, and decision criteria (Endler 1992). Since these processes affect mate choice, speciation is likely to be started or accelerated by these processes (Endler & McLellan 1988; Endler 1989). Given that local environments are different, even within habitats, divergences as in Figure 8 are even more likely. This may be a significant mechanism promoting biodiversity. Differences in predation intensity (e.g. Endler 1978, 1980) and other influences on the senses will further promote divergence and speciation among populations.

Evolution of perception and its implications

Once information is extracted from the signal by the sensory system, it is largely a mystery how it is used in the brain. This is the domain of Receiver Psychology (Guilford & Dawkins 1991, 1993), and of behavioural ecologists working on the behavioural and fitness consequences of signal efficacy (example in Johnstone, this volume). There is very little work on mechanisms causing or allowing shifts in preferences (Ryan & Rand 1995; Endler & Basolo 1998); to date most work has simply documented evolutionary changes in preferences and suggested that male traits can evolve after new preferences evolve (Basolo 1990, 1995; Ryan & Rand 1990, 1993; Endler & Basolo 1998). One aspect which has not been discussed in the literature is the effects of evolutionary changes in perception on decision making. Perception and decision making are likely to be disparate and probably genetically independent processes. Signals are first received, processed, perceived, compared with internal standards and the internal state (Receiver Psychology), and then a

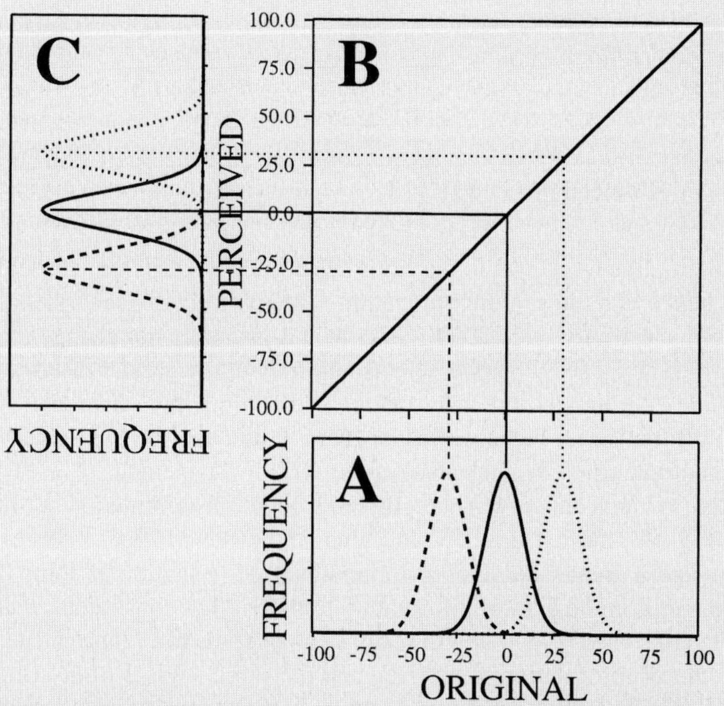

Figure 9. *Example of a linear relationship between the measured trait distribution and the perceived trait distribution. (**A**) The physical trait or signal parameter distribution (for example, tail length in mm) for three populations (shown by dashed, solid, and dotted lines). (**B**) The perceptual function—the relationship between actual value and perceived value (input-output function). (**C**) The perceived distribution of traits for the three populations. Because the perceptual function is linear, the original and perceived distributions are the same.*

decision is made whether or not to utilise the information in the signal. Just as separate enzymes in a biochemical pathway, each process in the sequence of detection, reception and decision making can be genetically independent, even if we see only the final end product. What happens when perception evolves before or faster than decision criteria?

Perception can be regarded as a transformation from the physical signal or parameter to how it is represented in the brain, and this will affect all variants of the signal. When we measure the signal and receiver response, we often implicitly assume that there is a linear relationship between the signal we measure and how it is perceived (Fig. 9). However, this is unlikely to be the case. At the very simplest level, A (amplitude) is perceived as the log(A) rather than A in all sensory modes. In addition, most sensory systems are more sensitive to some ranges of parameter values than others (tuned). For example, bush-crickets are tuned to particular frequencies (F) as well as particular amplitudes (Römer et al. 1998); sensitivities decline away from the tuned F and A. The effect of tuning and other processes of sensation and perception is to modify the distribution of perceived signals relative to their physical values (Fig. 10).

Figure 10 shows an example of the effects of tuning on perceived distributions of signals. The effect of tuning is to expand the distribution for which the system is tuned and to contract (Fig. 10D) or truncate (Fig. 10E) distributions at the edges or outside the tuned range. These effects follow from one of the main functions of sensory systems-discrimination. In order to discriminate most efficiently, the sensory system must increase the perceived variance of naturally occurring signals relative to the physical variance. This is done by transforming the signal with an input-output (perceptual) function which has a slope greater than 1.0 (Fig. 10 B and C). Increasing the perceived variance makes discrimination easier and is most efficient if the perceptual function is approximately linear within the mean±sd of the distribution of signals. (This is one stage later in processing than the logarithmic transformation at the receptor level). It becomes easier to assess relative and absolute differences among signal parameters when they are widely spaced. Note that choice on the basis of the perceived signal parameter can still favour the individuals with the highest (or lowest) values; the perceptual function is simply a way of making the decision possible or easier.

Unfortunately, increasing the slope above 1.0 means that the curve cannot be linear over the entire range of input values because it still must go through the origin and have some maximum value. The result is that it becomes sigmoid (s-shaped), and only linear in the centre of the sigmoid. The transformation (perception) is most efficient at the linear part of the curve and breaks down outside that level, and so the best systems have the longest possible linear portion. The linear portion is known as the dynamic range and is a fundamentally important parameter of signal processing. (The breakdown outside the linear portion is known as 'reciprocosity failure' in photography and video and we see it as very low contrast at the darkest and brightest parts of the picture; modern film and better video systems have large dynamic ranges). If natural selection and sexual selection have resulted in tuning of the sensory system's dynamic range to the natural variance in the signal, then discrimination will be efficient for most or all signals in that population, but will be

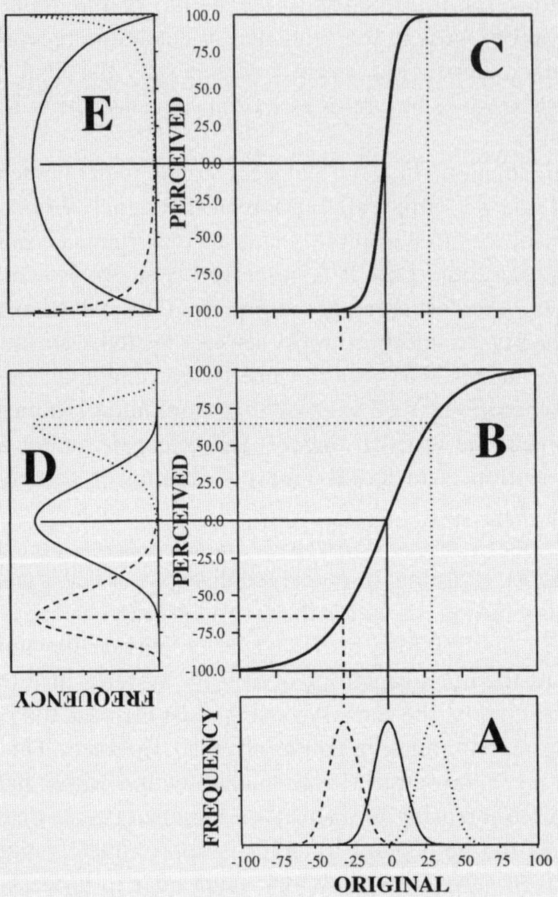

Figure 10. Examples of non-linear relationships between the measured trait distribution and the perceived trait distribution. As in Figure 9A, Figure 10A shows the distribution of the original (physical) signal parameter in three different populations (solid, dashed and dotted curves), but here the solid line represents the population for which the sensory system is tuned. (A) The physical distribution for three populations (shown by dashed, solid, and dotted lines). (B-C): Perceptual functions differing in slope (degree of amplification). (D-E): The perceived distribution for the three populations. As a result, the distribution which is centred on the linear part of the amplification curve has a larger variance and those off the linear area have smaller variance. It is therefore easier to discriminate among the individuals in the middle population (solid line) than in the other two. With very strong amplification (C), the tuned distribution becomes very easy to discriminate and the non-tuned distributions both difficult to discriminate and difficult to detect. (B and C are analogous to photography with medium and high contrast film and the distributions classes of objects differing in brightness; the dashed and broken-line objects are under and overexposed, respectively).

inefficient outside this range, as shown in Fig. 10. An incidental side-effect is that distributions of signals outside the dynamic range may not be recognized (Fig. 10E) or at least more difficult to discriminate among than those inside the range (Fig. 10D). Therefore, selection of receivers for increasing discrimination ability within a population (from Fig. 10B to 10C) will favour isolation and speciation. Even if speciation does not occur, selection for discrimination among a particular range of signal parameters will change the distribution of perceived signals in the population.

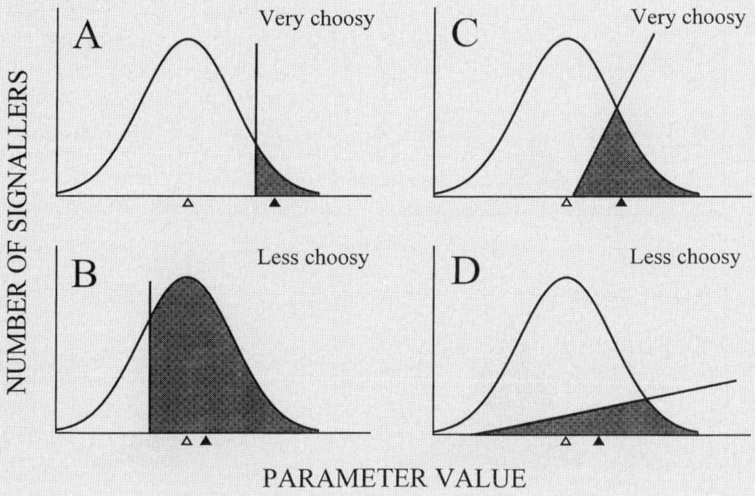

Figure 11. *The meaning of choosiness. In each panel the curve is the distribution of signal parameter values and the straight line is the mean preference function for the same population. Shaded areas represent the proportion of signallers chosen by the receiver. Triangles point to the mean signal parameter before (open) and after (closed) selection by the receiver. In (**A**) and (**B**), the preference function is a threshold, above which the higher signals are chosen. In (**C**) and (**D**), the preference function is gradual (probability of selection is an increasing function of signal value). Note how the selection differentials on the signal (distance between open and closed triangles) are less with less choosy receivers.*

Changes in the perceived signal distribution will affect receivers' decisions on the basis of the signals, and therefore will affect the evolution of the signals. Consider a population with a particular distribution of signals and a particular mean signal preference function. Preference functions relate the signal parameter to the probability that the receiver will choose the signal. Preference functions can be in the form of a threshold (Fig. 11A, B), or an increasing (Fig. 11C, D) or decreasing function of the signal parameter. Choosiness can vary in two ways, the position of the preference function relative to the signal parameter (Fig. 11A, B) or in the slope of the function (Fig. 11C, D). If the preference function is a threshold, then the slope is infinite and the position determines the choosiness (Fig. 11A, B), but if the slope is not infinite then choosiness is affected by both slope and position of the preference function (Fig. 11C, D). The result of choosiness is to displace the mean of the selected individuals relative to the mean of all individuals (Fig. 11). This difference in means before and after selection is the selection differential on the signal parameter, and will affect the distribution of signals in the next generation if the signal parameter is heritable (Falconer & Mackay 1996). The form of choosiness is of great importance in signal evolution.

Figure 12 shows an example of the evolution of perception if the preference function is fixed. This is not unrealistic because perception and choice are separate processes. The

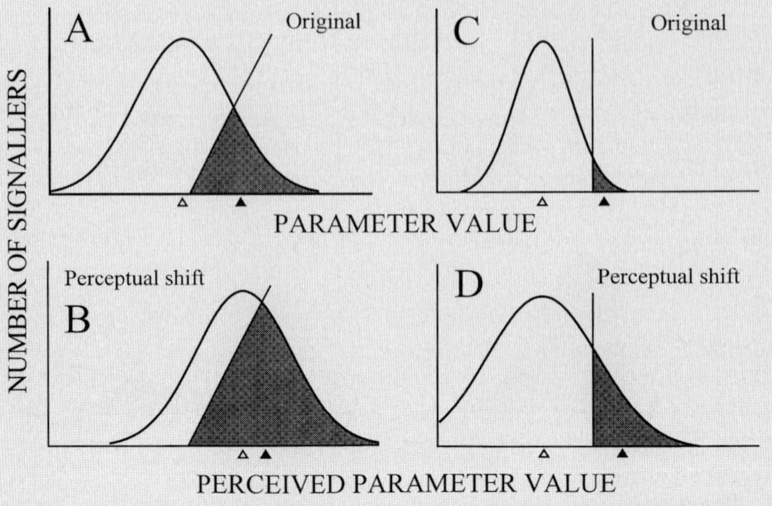

Figure 12. *Selective effect on signals of evolutionary changes in perception. (**A**) and (**C**) show the ancestral condition, before sensory system or brain has changed, while (**B**) and (**D**) show the effects of changes in the mean (B) and the variance (D) of the perceived signal. Conventions as in Figure 11. Note how selection differentials can decrease (A-B) as well as increase (C-D).*

preference function relates the perceived signal to the decision whether or not to accept the signal, and could easily be genetically and functionally independent of sensation and perception. It presumably evolves for strategic reasons (see Johnstone, this volume). If the perceptual function (Fig. 10) evolves because of natural selection for increased discrimination ability or other reasons, then the mean or the variance of the distribution of perceived signals can shift (Fig. 10). If the mean of the perceived signal distribution shifts towards the preference function, then the selection differential can decline (Fig. 12 A, B), and evolution of the signal will slow up or even stop. Of course it could also shift in the other direction, resulting in accelerated evolution of the signal. If the variance of the perceived signal distribution increases, then the selection coefficient may change only a little because the increased fraction above the threshold (which reduces the selection differential, as in Fig. 12A, B) may be counteracted by the increase in more extreme values (Fig. 12 C, D). Therefore changes in perceived variance are likely to have less effect than changes in perceived mean. Which happens will depend upon where the population is relative to the centre of the tuning curve (Fig. 10). Preference functions can also evolve, making the effects of changes more complex. However, the main point is that evolutionary changes in perception may result in counterintuitive changes (or no changes) in the results of decision processes, requiring much more careful predictions and interpretation of field and laboratory studies.

Where do we go from here?

Research by behavioural ecologists on signal evolution has so far largely concentrated on fitness consequences of signal content. We have neglected the efficiency considerations which allow detailed predictions needed for significant understanding of signalling behaviour and evolution, as well as allowing us to answer questions like 'why is a rainbow lorikeet coloured the way it is?', and 'why do we find bright colours in rainforest canopies and coral reefs'? In this paper, I have tried to identify areas and phenomena which are in need of study before we can really understand signals. There are at least twelve questions which we can probably answer:

1. Signals have many parameters which can be used for assessment. We need to find out which ones are used and how they are used. Are there certain principles which are common to all signalling systems to indicate certain properties of potential mates, territory quality, alerting signals, aposematic signals, or cryptic signals? We probably have a good idea for the last three, but not the others. We have to be careful about assumptions about coding; for example, energy expenditure can be coded in two ways by at least three different parameters: A, t_p for burst capacity and t_b for long term muscle activity.
2. What are the heritabilities of various signal parameters, and how much genetic correlation is between them? Are there particular genetic patterns characteristic of particular parameters or particular signal contents?
3. What signal properties and parameters are best for particular microenvironments? This has very great potential for making specific predictions, at least about carriers.
4. Given microenvironments, noise and sensory modes, what is the best way to send the most information? When is it better to use repetitiveness and when is it better to use redundancy? When it is better to use redundancy within or among channels or sensory modes?
5. What favours the use of different sensory modes and their proportions? What factors affect relative information content, reliability and signal/noise ratio? Which parameters give the most useful information to a receiver? Which parameters are by their nature most honest and which are easiest to cheat?
6. What are tradeoffs between signal functions, and how does this affect which parameters are used? Different parameters can be used for disparate functions within the same signal. How are tradeoffs adjusted? By relative importance? By the need for redundancy?
7. How are different signal functions coded? Examples are species, gender, age, mate quality, alerting, aposematic. How are tradeoffs between functions mitigated?
8. How are tradeoffs handled generally? How is the tradeoff between efficiency and need to signal cost mitigated? How is the tradeoff between efficiency and signal content mitigated (more noise, less content)? How is the tradeoff between eavesdroppers and efficiency handled?
9. How often are environmental effects on signals taken advantage of to avoid predation, to maximize contrast, or to work for only brief times?

10. How does natural selection on the sensory system (for food finding, predator avoidance, finding the right microenvironment) affect the evolution of the sensory system, and how does this affect perception of and discrimination between signals?
11. What are the heritabilities of various components and subsystems within sensory systems, and how much genetic correlation is there between them? How much genetic correlation is there between signal parameters and the systems which detect them? How genetically independent are signals, sensory systems, perceptual systems, and decision making systems?
12. How do evolutionary changes in perception affect decision making?

Unfortunately, most of these have not yet been addressed because they have not been explicitly asked.

Conclusions

From the discussion of signal structure, transmission environment, detection, parameter extraction and perception, we can make six general observations about the evolution of signal structure and behaviour. First, there are many ways to assess signals. Even if a signal is unmodulated, the carrier structure reveals much about the signaller. There are numerous ways to modulate or otherwise add structure to the carrier, allowing many different signal parameters to be sent to and estimated by the receiver. In spite of this, most studies have only studied one or two parameters in attempts to relate signals with decisions by the receiver. A study of the different parameters in each signalling mode, and a search for common properties within and among modes, would be valuable.

Second, environmental effects are a nuisance, because they can interfere with the detection and assessment of signals. But they can also be useful. They can be used for estimation of distance to the signaller, hiding from eavesdroppers, and mitigating the tradeoff between multiple signal functions. Although there is a good sized literature on environmental effects, the benefits of environmental influences on signals are little studied.

Third, evolutionary signal divergence is easy, even under the same environmental conditions. This is partially because there are myriad ways to modify signals, many different signal parameters which can carry information, and various sensory modes which can be used in different combinations. Divergence is also facilitated because selection on one aspect of sensory processes can result in very different kinds of responses, with very different sensory and perceptual consequences (Fig. 8), and selection on discrimination can result in changes of direction or rate of evolution of signal parameters (Fig. 10). With even minor differences in sensory environment among populations, there is an even greater potential for divergence in senses, signals, and signal structure. There is tremendous potential for divergence and speciation in signalling systems, but this has not yet been studied other than the well known empirical observation that species signal differently.

Fourth, knowledge about environmental properties, signal production biophysics, and signalling behaviour and microenvironments is getting good enough so that we can start predicting in detail the basic parameters of signal carriers. Knowledge of environmental effects allow predictions about the best signals in given environmental conditions, including some predictions about signals, conditions under which they are used, as well as which sensory modes should be used (e.g. Endler 1991, 1992, 1993b).

Fifth, it should be possible in principle to predict at least some aspects of signal content from signal efficiency principles. Currently we are mostly limited to predicting amplitude (A) because few studies have related receiver behavioural decisions to parameters other than A or t_b, except in a simple species-recognition context (exceptions are Endler 1978, 1991, 1993b; Lythgoe 1979; Andersson this volume; Fleishman, this volume; Marshall, this volume). If we knew more about how different signal parameters (Figs 1 and 2) were used in decision making, as well as the environmental degradation effects in detecting these parameters, we should be able to predict a good deal about the details of signal structure. For example, in guppies, part of the male display signal carrier is colour pattern, and environmental properties predict that orange, blue, UV and purple (blue+orange) are best for efficiency under conditions when courtship is most common (Endler 1991). Orange chroma indicates feeding success and the flicker from structural blue UV and silver may indicate physical fitness, which is required for the energetically costly oscillation during the display. Signals must evolve to be efficient in detection and information transmission, and the combination of best parameters for a given set of environmental conditions and best components for assessment parameters should provide predictions about many aspects of signals.

Sixth, it should be possible in principle to predict at least some aspects of sensory system structure and functions from knowledge of signalling environment and signal structure. The sensory system must be efficient at extracting the signal parameters from the signal, and making the variation among individual signals large enough to enable discrimination among and decisions about the different signals. We should be able to predict sensory properties from methods of detection and parameter extraction. We should also be able to predict which sensory modes and kinds of modulation are used in any given set of environmental conditions. We are still at the very early stages of this, and it will be interesting to see what emerges.

Acknowledgements

I am most grateful for comments on the manuscript from and discussions about it with Staffan Andersson, Mark Blows, Rob Brooks, Will McClintock, Leo Fleishman, and an anonymous reviewer.

References

Arnold, S. J. 1983. Morphology, performance and fitness. *American Zoologist*, **23**, 347-361.

Balshine-Earn, S. & Lotem, A. 1998. Individual recognition in a cooperatively breeding cichlid: evidence from video playback experiments. *Behaviour*, **135**, 369-386.

Basolo, A. L. 1990. Female preference predates the evolution of the sword in swordtail fish. *Science*, **250**, 808-810.

Basolo, A. L. 1995. Phylogenetic evidence for the role of a pre-existing bias in sexual selection. *Proceedings of the Royal Society of London, Series B*, **259**, 307-311.

Bailey, W. J., Cunningham, R. C. & Lebel, L. 1990. Song power, spectral distribution and female phonotaxis in the bushcricket *Requena verticalis* (Tettigoniidae, Orthoptera). *Zeitschrift für Tierpsychologie*, **70**, 320-330.

Bastian, J. 1990. Electroreception. In: *Comparative Perception. II. Complex Signals* (Ed. by W. C. Stebbins & M. Berkley), pp. 35-88. New York: Wiley.

Bennet-Clark, H. C. 1998. Size and scale effects as constraints in insect sound communication. *Philosophical Transactions of the Royal Society of London, Series B*, **353**, 407-419.

Chittka, L. 1996a. Does bee colour vision predate the evolution of flower colour? *Naturwissenschaften*, **83**, 136-138.

Chittka, L. 1996b. Optimal sets of colour receptors and opponent processes for coding of natural objects in insect vision. *Journal of Theoretical Biology*, **181**, 179-196.

Crawford, J. D., Cook, A. P. & Heberlein, A. S. 1997. Bioacoustic behaviour of African fishes (Mormyridae): potential cues for species and individual recognition in *Pollimyrus*. *Journal of the Acoustical Society of America,* **102**, 1200-1212.

Cronin, T. W., Marshall, N. J. & Caldwell, R. L. 1994. The intrarhabdomal filters in the retinas of mantis shrimps. *Vision Research*, **34**, 279-291.

Cronin, T. W., Marshall, N. J. & Caldwell, R. L. 1996. Visual pigment diversity in two genera of mantis shrimps implies rapid evolution (Crustacea Stomatopoda). *Journal of Comparative Physiology, Series A*, **179**, 371-384.

Dawkins, M. S. & Guilford, T. 1996. Sensory bias and the adaptiveness of female choice. *American Naturalist*, **148**, 937-942.

Dusenbery, D. B. 1992. *Sensory Ecology: How Organisms Acquire and Respond to Information*. New York: W. H. Freeman.

Embleton, T. F. W. 1996. Tutorial on sound propagation outdoors. *Journal of the Acoustical Society of America*, **100**, 31-48.

Endler, J. A. 1978. A predator's view of animal colour patterns. *Evolutionary Biology*, **11**, 319-364.

Endler, J. A. 1980. Natural selection on colour patterns in *Poecilia reticulata*. *Evolution*, **34**, 76-91.

Endler, J. A. 1989. Conceptual problems in species and speciation. In: *Speciation and its Consequences* (Ed. by D. Otte & J. A. Endler), pp. 625-648. Sunderland MA: Sinauer.

Endler, J. A. 1990. On the measurement and classification of colour in studies of animal colour patterns. *Biological Journal of the Linnean Society of London*, **41**, 315-352.

Endler, J. A. 1991. Variation in the appearance of guppy colour patterns to guppies and their predators under different visual conditions. *Vision Research*, **31**, 587-608.

Endler, J. A. 1992. Signals, signal conditions, and the direction of evolution. *American Naturalist*, **139** (supplement), s125-s153.

Endler, J. A. 1993a. Some general comments on the evolution and design of animal communication systems. *Philosophical Transactions of the Royal Society of London, Series B*, **340**, 215-225.

Endler, J. A. 1993b. The colour of light in forests and its implications. *Ecological Monographs*, **63**, 1-27.

Endler, J. A. 1997. Light, behaviour and conservation of forest-dwelling organisms. In: *Behavioral Approaches to Conservation in the Wild* (Ed. by J. R. Clemmons & R. Buchholz), pp. 330-356. Cambridge: Cambridge University Press.

Endler, J. A. & Basolo, A. L. 1998. Sensory ecology, receiver biases and sexual selection. *Trends in Ecology and Evolution*, **13**, 415-420.

Endler, J. A. & Houde, A. E. 1995. Geographic variation in female preferences for male traits in *Poecilia reticulata*. *Evolution*, **49**, 456-468.

Endler, J. A. & McLellan, T. 1988. The processes of evolution: towards a newer synthesis. *Annual Review of Ecology and Systematics*, **19**, 395-421.

Endler, J. A. & Thery, M. 1996. Interacting effects of lek placement, display behaviour, ambient light and colour patterns in three neotropical forest-dwelling birds. *American Naturalist*, **148**, 421-452.

Falconer, D. S. & Mackay, T. F. C. 1996. *Introduction to Quantitative Genetics*, 4th Edn. London: Longmans.

Fleishman, L. J. 1986. Motion detection in the presence and absence of background motion in an *Anolis* lizard. *Journal of Comparative Physiology, Series A*, **159**, 711-720.

Fleishman, L. J. 1992. The influence of the sensory system and the environment on motion patterns in the visual displays of Anoline lizards and other vertebrates. *American Naturalist*, **139** (supplement), s36-s61.

Fleishman, L. J., Marshall, C. J. & Hertz, P. E. 1995. Comparative study of temporal response properties of the visual system of three species of anoline lizards. *Copeia*, **2**, 422-431.

Fleishman, L. J., Bowman, M., Saunders, D., Miller, W. E., Rury, J. J. & Loew, E. R. 1997. The visual ecology of Puerto Rican Anoline lizards: habitat light and spectral sensitivity. *Journal of Comparative Physiology, Series A*, **181**, 446-460.

Fullard, J. H. 1982. Echolocation assemblages and their effects on moth auditory systems. *Canadian Journal of Zoology*, **60**, 2572-2576.

Gish, S. L. & Morton, E. S. 1981. Structural adaptations to local habitat acoustics in Carolina wren songs. *Zeitschrift für Tierpsychologie*, **56**, 74-84.

Godard, R. & Wiley, R. H. 1995. Individual recognition of song repertoires in two wood warblers. *Behavioral Ecology and Sociobiology*, **37**, 119-123.

Guilford, T. & Dawkins, M. S. 1991. Receiver psychology and the evolution of animal signals. *Animal Behaviour*, **42**, 1-14.

Guilford, T. & Dawkins, M. S. 1993. Receiver psychology and the design of animal signals. *Trends in Neurosciences*, **16**, 430-436.

Hedrick, A. V. 1986. Female preferences for calling bout duration in a field cricket. *Behavioral Ecology and Sociobiology*, **19**, 73-77.

Hill, G. E. & Montgomerie, R. 1994. Plumage colour signals nutritional condition in the house finch. *Proceedings of the Royal Society of London, Series B*, **258**, 47-52.

Jones, M. D. R. 1966. The acoustic behaviour of the bush cricket *Pholidoptera griseoaptera*. 1. Alternation, synchronism and rivalry between males. *Journal of Experimental Biology*, **45**, 31-44.

Laughlin, S. B., de Ruyter van Steveninck, R. & Anderson, J. C. 1998. The metabolic cost of neural information. *Nature Neuroscience*, **1**, 36-41.

Levene, J. S. & MacNichol Jr, E. F. 1979. Visual pigments in Teleost fishes: effects of habitat, microhabitat, and behaviour on visual system evolution. *Sensory Processes*, **3**, 95-131.

Lythgoe, J. N. 1979. *The Ecology of Vision*. Oxford: Oxford University Press.

Lythgoe, J. N. & Partridge, J. C. 1989. Visual pigments and the acquisition of visual information. *Journal of Experimental Biology*, **146**, 1-20.

Lythgoe, J. N. & Partridge, J. C. 1991. The modelling of optimal visual pigments of dichromatic teleosts in green coastal waters. *Vision Research*, **31**, 361-371.

Miller, L. A. 1983. How insects detect and avoid bats. In: *Neuroethology and Behavioural Physiology* (Ed. by F. Huber & H. Markl), pp. 271-266. Berlin: Springer-Verlag.

Moiseff, A. & Copeland, J. 1995. Mechanisms of synchrony in the North American firefly *Photinus carolinus* (Coleoptera: Lampyridae). *Journal of Insect Behaviour*, **8**, 395-407.

Morton, E. S. 1982. Grading, discreteness, redundancy, and motivation-structural rules. In: *Acoustic Communication in Birds,* Vol. 1 (Ed. by D. E. Kroodsma & E. K. Miller), pp. 183-212. New York: Academic Press.

Narins, P. M. 1992. Evolution of Anuran chorus behaviour, neural and behavioural constraints. *American Naturalist*, **139** (supplement), s90-s104.

Narins, P. M., Lewis, E. R., Jarvis, J. J. U. M. & O'Riain, J. 1997. The use of seismic signals by fossorial Southern African mammals: A neuroethological gold mine. *Brain Research Bulletin*, **44**, 641-646.

Osorio, D. & Vorobyev, M. 1996. Colour vision as an adaptation to frugivory in primates. *Proceedings of the Royal Society of London , Series B*, **263**, 593-599.

Partridge, J. C. 1989. The visual ecology of avian cone oil droplets. *Journal of Comparative Physiology, Series A*, **165**, 415-426.

Römer, H. & Bailey, W. J. 1998. Strategies for hearing in noise: peripheral control over auditory sensitivity in the bushcricket *Sciarasaga quadrata* (Austrosaginae: Tettigoniidae). *Journal of Experimental Biology*, **201**, 1023-1033.

Römer, H. & Lewald, J. 1992. High frequency sound transmission in natural habitats; implications for the evolution of insect communication. *Behavioral Ecology and Sociobiology*, **29**, 437-444.

Römer, H., Spickermann, M. & Bailey, W. 1998. Sensory basis for sound intensity discrimination in the bushcricket *Requena verticalis* (Tettigoniidae, Orthoptera). *Journal of Comparative Physiology, Series A*, **182**, 595-607.

Ryan, M. J. 1985. *The Túngara Frog: A Study in Sexual Selection and Communication*. Chicago: University of Chicago Press.

Ryan, M. J. & Keddy-Hector, A. 1992. Directional patterns of female mate choice and the role of sensory biases. *American Naturalist*, **139** (supplement), s4-s35.

Ryan, M. J. & Rand, A. S. 1990. The sensory basis of sexual selection for complex calls in the Túngara frog, *Physalaemus pustulosus* (sexual selection for sensory exploitation). *Evolution*, **44**, 305-314.

Ryan, M. J. & Rand, A. S. 1993. Sexual selection and signal evolution: the ghost of biases past. *Philosophical Transactions of the Royal Society of London , Series B*, **340**, 187-195.

Ryan, M. J. & Rand, A. S. 1995. Female responses to ancestral advertisement calls in Túngara frogs. *Science*, **269**, 390-392.

Stephen, R. O. & Hartley, J. C. 1991. The transmission of bush cricket calls in natural environments. *Journal of Experimental Biology*, **155**, 227-244.

Stoddard, P. K. 1998. Detection of multiple stimulus features forces a trade-off in the pyramidal cell network of a gymnotiform electric fish's electrosensory lateral line lobe. *Journal of Comparative Physiology, Series A*, **182**, 103-113.

Tubaro, P. L. & Mahler, B. 1998. Acoustic frequencies and body mass in new world doves. *Condor*, **100**, 54-61.

Vorobyev, M. & Osorio, D. 1998. Receptor noise as a determinant of colour thresholds. *Proceedings of the Royal Society of London, Series B*, **265**, 351-358.

Wang, J. & Narins, P. M. 1996. Directional masking of phase locking in the amphibian auditory nerve. *Journal of the Acoustical Society of America*, **99**, 1611-1620.

Wiley, R. H. & Richards, D. G. 1982. Adaptations for acoustic communication in birds: sound transmission and signal detection. In: *Acoustic Communication in Birds*, Vol. 1 (Ed. by D. E. Kroodsma & E. K. Miller), pp. 132-181. New York: Academic Press.

IN: ESPMARK, Y., AMUNDSEN, T. & ROSENQVIST, G. (EDS.) 2000
ANIMAL SIGNALS:
SIGNALLING AND SIGNAL DESIGN IN ANIMAL COMMUNICATION, PP. 47-60
TAPIR ACADEMIC PRESS, TRONDHEIM, NORWAY

2

Efficacy and Content in Avian Colour Signals

STAFFAN ANDERSSON

Department of Zoology, Göteborg University,
Box 463, S-405 30 Göteborg, Sweden

E-mail: s.andersson@zool.gu.se

Abstract

Signals must have both 'efficacy' and 'content' and the balance between these requirements is an important key to understanding signal design. Here I discuss colour signals in birds, and make predictions based on 'efficacy' (spectral contrast against the background), versus 'content' (honesty-maintaining costs). Compared to 'efficacy' designs, 'content' predicts signals that are (1) exaggerated rather than conspicuous, (2) costly to produce, (3) variable, especially in signal dimensions related to cost/quality, and (4) favoured (e.g. by mate choice) through such 'cost dimensions' rather than 'detectability dimensions'. To illustrate how these predictions might apply to colour signals, I compare three major types of extravagant bird coloration: structural UV/blue in the palearctic bluethroat Luscinia s. svecica, *and carotenoid yellow and red in the African golden bishop* Euplectes afer *and red bishop* E. orix, *respectively. (1) In agreement with 'efficacy', the UV/blue and yellow signals are the most contrasting alternatives in their respective habitats and light conditions. (2) The red signal is based on a supposedly more costly mechanism (dietary red carotenoids), and (3) shows most intraspecific variation (particularly), and thus agrees better with a 'content' design. It is also discussed whether structural UV/blue coloration might indicate 'developmental stability', through its regular nanostructure, increasing chroma with age.*

Efficacy and content in signal design

The tremendous diversity of epigamic signals is largely unexplained. Although there is now a fair understanding of sexual selection mechanisms that drive signal exaggeration and produce variations in the degree of sexual dimorphism, most variation in signal design is still a mystery (Andersson 1994). The diversity of dazzling colours in birds and reef fish (Marshall, this volume) are prominent examples, begging questions like 'why does species A have a blue cap but species B a red belly'?

One approach to signal design is to distinguish between signal 'content' (the message) and signal 'efficacy' (efficiency in conveying the message) (Guilford & Dawkins 1991; Endler 1993a, this volume; Johnstone 1997). In the case of sexual ornaments, research has usually concerned either 'efficacy' (signal conditions, receiver tuning) or 'content' (e.g. benefits of mate choice), but these two aspects have rarely been explored together, although they sometimes might be in conflict. An immediate difficulty with this approach is that the 'efficacy'/'content' distinction is not always clear. First, where in the neural pathways does signal reception end and decision-making begin? Second, are 'efficacy' and 'content' to be judged from the perspective of the signaller or the receiver, or both? However, although the most suitable definitions might differ depending on what aspects of signal evolution are in focus, it might be fruitful to explore the 'efficacy'/'content' balance in signal design.

One example of the receiver's perspective on 'efficacy' and 'content' of epigamic traits is the model by Schluter and Price (1993). They show how female preferences and signal traits can diverge dramatically in response to the combined effect of 'detectability' (related to search cost) and 'honesty' (related to direct benefits) - in essence amalgamating the 'sensory drive' (reviewed by Endler & Basolo 1998) and the 'handicap' mechanisms (Zahavi 1975; Grafen 1990; Price et al. 1993). In the light of this model, the relative match of signal designs with the demands of 'efficacy' (detectability) versus 'content' (honesty) would clearly be interesting in a study of epigamic character divergence. From the signaller's perspective, 'efficacy' also includes costs of signal production and signal exploitation (Endler 1993, this volume; Johnstone 1997, this volume), for example tradeoffs between detectability to mates and to predators (Endler 1991; Endler & Thery 1996).

The multitude of environmental and social selection pressures on signal design (see Endler, this volume) is formidable, and the 'efficacy'/'content' dichotomy is one approach to sorting them out. A main question then is 'how these two different aspects of signal evolution can be reconciled?' (Johnstone 1997).

Efficacy and content in bird colours

Extravagant and sexually dimorphic bird colours have long been a focus for sexual selection studies, converging in recent years on quality-indicating 'content' (e.g. Fitzpatrick 1994, this volume; Hill 1999), but they have only recently been investigated in terms of (ecologically determined) 'efficacy' (Marchetti 1993; Endler & Thery 1996;

Andersson et al. 1998b). Here I will compare three major types of conspicuous and sexually dichromatic plumage colours in birds:

1) UV/blue structural coloration on the chin and breast of the bluethroat *Luscinia s. svecica* (BT, Fig. 1), a small palearctic migrant, breeding in open mountain habitat. It is mostly socially monogamous but with frequent extra-pair fertilizations (Johnsen et al. 1998). Aviary and field experiments have demonstrated that the UV component of the colour is important in social and extra-pair mate choice (Andersson & Amundsen 1997; Johnsen et al. 1998).
2) Yellow carotenoid pigmentation on the crown and rump of the golden bishop *Eupletes afer* (GB, Fig. 1), a polygynous weaverbird, breeding in loose colonies in wet grasslands ('vleis') of Southern Africa (Craig 1980). The golden yellow plumes are raised and displayed in perching displays and aerial 'bumble-bee' flights over the territory (Craig 1980; J. Örnborg & S. Andersson, unpublished).
3) Red carotenoid pigmentation on the chest, crown and rump of the red bishop *Euplectes orix* (RB, Fig. 1), a species with similar ecology to the golden bishop, except that they breed in denser colonies and usually in reeds or tall grass close to water (Craig 1974, 1980).

Since these are species with striking seasonal and sexual dimorphism, I make the reasonable assumption that the brilliant colours have evolved as intraspecific signals, and not for interspecific signalling (e.g. aposematism), or non-signalling functions (e.g. abrasion-resistance or thermoregulation) (Savalli 1995). I will consider only the design aspect of colour as such (i.e. hue and chroma, but not pattern), and for the sake of this example, I make the (admittedly unreasonable) assumption that all three colours could have evolved in any of the three species, in other words that there are no phylogenetic constraints. Finally, I assume that colour perception is the same in all species, with a spectral sensitivity range from 320 to 750 nm. This need not be true, but recent results suggest that at least visual range and cone absorbance functions are similar across passerines (N. Hart, unpublished; Cuthill, this volume). Given these assumptions, I propose and discuss four predictions of 'efficacy' versus 'content' design principles (Table 1), and tentatively associate them with the three epigamic colour signal types.

Table 1. *Predictions of signal characteristics based on the relative emphasis of 'efficacy' versus 'content' design principles.*

	'EFFICACY' Signal should be:	**'CONTENT'** Signal should be:
1.	Conspicuous, but not necessarily the most exaggerated	Exaggerated, but not necessarily the most conspicuous
2.	Cheap	Costly
3.	Less variable (fixation or threshold preferences)	More variable (quality variation and open-ended preferences)
4.	Favoured through signal dimension that best relates to detectability	Favoured through signal dimension that best relates to cost

Efficacy versus content, and conspicuousness

Efficacy predicts maximal conspicuousness or 'detectability' for the signalling condition. Ideally this should include also receiver 'sensory bias' or tuning, which here is assumed not to vary (see above). (For a uniquely complete comparative analysis of colour 'efficacy', including aspects of receiver tuning, see Fleishman, this volume). A 'content' emphasizing (honest) signal, on the other hand, might be exaggerated from for example generations of mate choice, but as long as its 'efficacy' exceeds some threshold, it need not be the most conspicuous alternative. Comparing the relative detectability of signals is thus a first indication of the balance between 'efficacy' and 'content' selection (provided that this balance is not completely obscured by other selective forces on detectability such as parasitism or predation, as discussed below). Also, disregarding 'content', estimates of detectability may often be important (even sufficient) predictors of signal design (e.g. Endler 1991; Wiley 1991; Marchetti 1993; Endler & Thery 1996; Fleishman, this volume; Marshall, this volume), so this offers an obvious first approach to signal diversity.

Objective analyses of colour signals (Hailman 1977; Endler 1990, 1993; Cuthill, this volume; Fleishman, this volume; Marshall, this volume) are not yet commonplace in behavioural ecology, and I will therefore present the following comparison of the three plumage colours in some detail.

Reflectance

First, colour can be quantified by the shape of the spectral reflectance function measured at the signal surface (Endler 1990). Reflectance spectra for the colour signals in the three species are shown in Figure 1 (for spectroradiometry details see Andersson et al. 1998b). There are different ways to extract indices of spectral shape (i.e. colour) from a reflectance or radiance spectrum (Endler 1990), but what matters in the following is that the spectral position of a peak or a slope correlates strongly with 'hue', whereas peak steepness and relative height corresponds to 'chroma'.

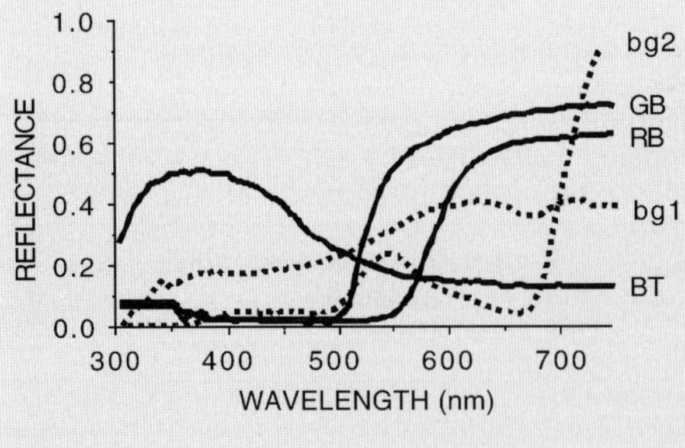

Figure 1. *Solid lines: average spectral reflectance (in relation to a Spectralon™ white standard) of the UV/blue chin of male bluethroats (BT, n = 23), the red crown of male red bishops (RB, n = 31), and the yellow crown of male golden bishops (GB, n = 44). Dashed lines: average background reflectance (radiance divided by irradiance; see text) in bluethroat alpine habitat in Øvre Heimdalen, Norway, 20 May 1998 (bg 1) and golden and red bishop open grassland habitat in Mpumalanga District, South Africa, 6 Dec 1997 (bg 2)*

Backgrounds
I measured background radiance (see Andersson et al. 1998a for details) from the two signal habitats at the time of pair formation (BT: Øvre Heimdalen, Norway, 20 May 1998. GB, RB: Volksrust, Mpumalanga Province, South Africa, 6 Dec 1997). Radiance spectra were divided by simultaneously obtained cosine-corrected ambient light spectra to produce estimates of background reflectance against which signal reflectance could be compared.

The alpine habitat during bluethroat pair formation in mid-May is a mosaic of dead grass, evergreen junipers and bare willows. An average background spectrum (Fig. 1, bg1) was computed from measures of these three radiance sources.

The background of displaying bishop birds is uniform lush grass and reeds. The reflectance from this habitat (Fig. 1, bg2) was measured on a golden bishop territory, but it is typical of any uniform green vegetation background.

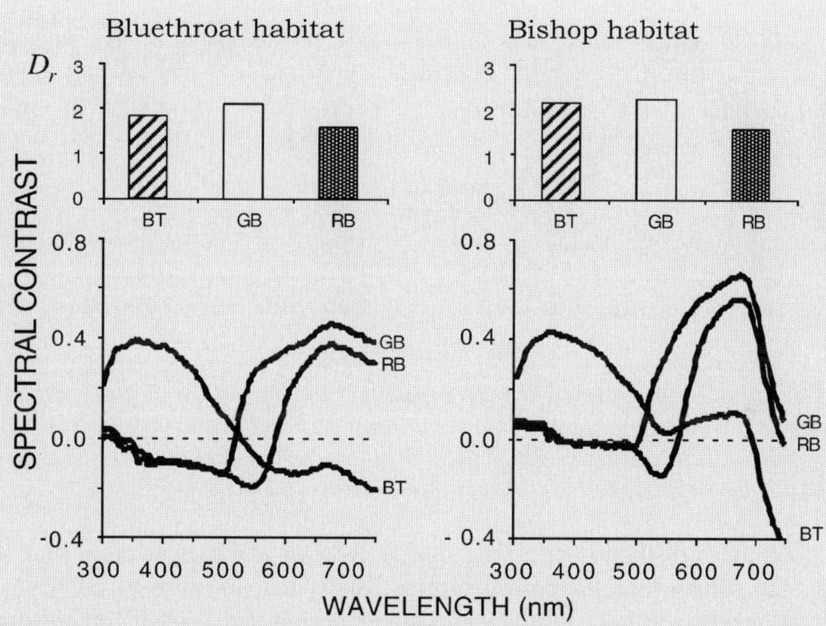

Figure 2. *Bottom panels: Spectral 'reflectance contrast' functions computed by subtracting background from plumage reflectance functions for each signal (BT,GB,RB) and both backgrounds (bg 1, bg 2). Top panels: 'Reflectance contrast' (D_r) computed by summing the (squared) reflectance contrast functions over the assumed 320-750 nm visual range.*

The mismatch between signal and background reflectances represents the 'raw' (illumination-independent) spectral contrast, shown in Figure 2 for each signal against both backgrounds. A summation of the square root of the sum of the squared differences across the spectrum (320-750 nm) yielded an overall contrast measure, D_s (capturing both colour and brightness contrast; see Endler 1990) for the three colours (Fig. 3, top panel), suggesting that, of the three alternatives, GB yellow is most contrasting in both habitats, while red is least contrasting in both. This is, however, assuming that illumination is identical for the two signal conditions, which it is not.

Illumination

I measured ambient light (details in Andersson et al. 1998a) in the two habitats, at the respective times of peak display. Bluethroats display mainly before sunrise and after sunset (Merilä & Sorjonen 1994) during the long dawn and dusk at these latitudes. In clear weather, this means a long period of relatively short-wave-rich light ('early/late' condition; Endler 1993b), as illustrated by the irradiance spectra measured 1.5 hours before sunrise (Fig. 3a; irradiance controlled for brightness by setting all spectra to a peak quantal flux of 1). I use the spectrum at thirty minutes before sunrise (-30, Fig. 3a) as the estimate of ambient light in bluethroat habitat.

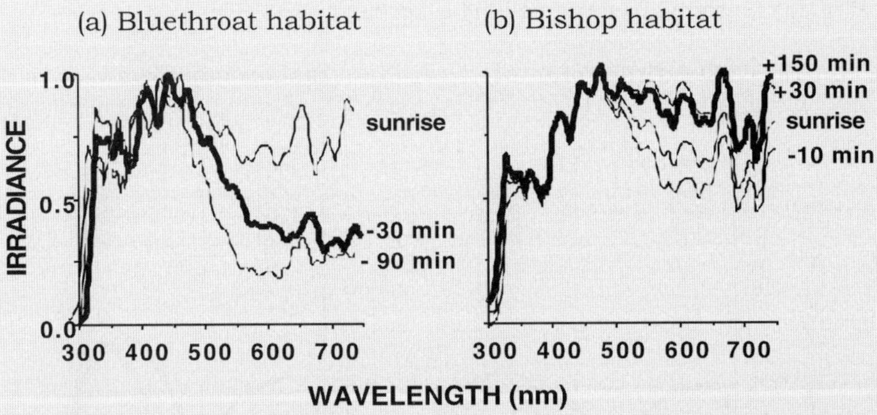

Figure 3. *Irradiant light spectra (relative quantal flux; see text and Andersson et al. 1998a) at various times (minutes) before (-), after (+) and at sunrise in (**a**) bluethroat and (**b**) bishop habitat, respectively. Thick lines denote the spectra at peak display times (relative to sunrise) of bluethroats and bishops, respectively.*

At tropical latitudes (bishop habitat) the sunrise is faster and the period of blue light very brief (Fig. 3b) (thirty minutes before sunrise it was still too dim to measure with my instrumentation). In addition, most bishop males do not arrive on their territories (from communal roosts) until well after sunrise (J. Örnborg & S. Andersson, unpublished) when the sun quickly climbs to produce the typical white daylight spectrum (Endler 1993b). The irradiance at thirty minutes after sunrise (+30, Fig. 3b) is used as the estimate of ambient light in bishop habitat.

Radiance contrast

By multiplying either of the two habitat ambient spectra with signal and background reflectances, respectively, and repeating the calculations of spectral contrast, measures of radiance contrast specific to the two signalling situations are obtained (Fig. 4). The bluethroat UV/blue now appears to be the most contrasting alternative in its brown alpine spring habitat and light, whereas golden bishop yellow is the most contrasting in bishop habitat and light.

This will suffice as a reasonable comparison of signal 'efficacy' between the three colour signals. From prediction 1 (Table 1) the conclusion is that the UV/blue and yellow agrees well with 'efficacy', and red more closely with 'content'.

Efficacy versus content, and signal cost

The 'content' signal design rule assumed in this example is the 'handicap principle' of signal selection postulating that, to be reliable, signals should be costly in relation to the information they encode, e.g. male quality as a social or genetic partner (see Johnstone 1997 for a recent review). 'Content' emphasizing signals should thus be particularly costly (prediction 2, Table 1), and might be *costly at the expense of 'efficacy'*. Conversely, signals primarily designed for 'efficacy' should be relatively less costly.

For large differences on the 'efficacy'-'content' scale of signal types (say, the peacock train versus the red chick-pecking spot on a gull's bill), the prediction of different costs seems trivial. However, for more closely related signals such as epigamic ornaments, a detailed comparison of signal costs might be informative to assess variation in 'selection for reliability' (or the detectability/honesty balance in Schluter & Price 1993). In particular, one might be able to identify signal dimensions where there is a trade-off between detectability and cost. Such tradeoffs might be rare in most metric traits, but quite possible to find in colours or sounds (see below).

Of course, there might be no such tradeoffs if it is the conspicuousness itself (e.g. through predation, as in the bat-attracting 'chuck' of the Túngara frog *Physalaemus pustulosus*; Tuttle & Ryan 1981) that is the costly signal 'content'. However, if signal conditions (environment or receiver tuning) differ between predator attraction and intraspecific

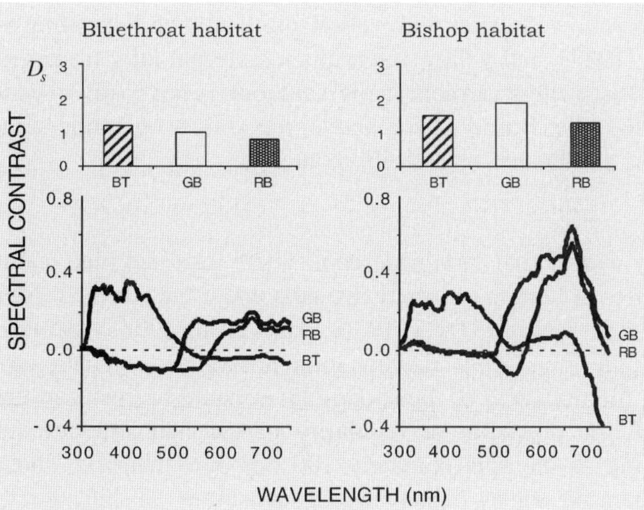

Figure 4. Bottom panels: Spectral 'radiance contrast' functions computed as in Figure 2 but after multiplying plumage and background reflectance functions with the habitat-specific irradiance spectrum (Fig. 3) Top panels: 'Radiance contrast' (D_s) computed by summing the (squared) radiance contrast functions over the assumed 320-750 nm visual range.

signalling, the signal design should strike some balance between these. That is, in addition to an 'efficacy' mediated tradeoff between crypsis to predators and conspicuousness to females (see e.g. Endler 1991, 1992, this volume; Endler & Thery 1996), 'content' should here select for detectability to predators rather than to females! A paradoxical prediction from the handicap principle that might provide an ultimate test of its generality. (For example, is the Túngara frog's 'chuck' better tuned to the bat than to frog hearing?).

Costs of avian colour production

In this limited example, I consider only production costs of coloration (i.e. not predation, social punishment, etc.). Zahavi and Zahavi (1997) suggested few costs of colour generation, but (unexpectedly) give a number of 'efficacy' explanations (e.g. black for long range signals and shape indication, white for enhancing movement, etc.). Probably, some ultimate costs of detectability or amplified morphology (Hasson 1989) were assumed, but this highlights the lack of knowledge of costs of colour production in birds. As regards the colour mechanisms included here, some implications of ontogenetic costs are therefore worth discussing.

Carotenoid pigmentation

Carotenoid pigments are excellent candidates for costly quality advertisements, since they are dietary (i.e. reflecting foraging ability: Endler 1980; Hill 1990, 1999) and critical for vitamin A and as antioxidants (i.e. reflecting health: Lozano 1994). Carotenoid plumage displays have been found to signal male quality to choosy females (recently reviewed by Hill 1999). There are, however, a wide range of carotenoid pigments that vary in availability or cost of synthesis (from ingested precursors), physiological importance, and effects on coloration (Goodwin 1984). In particular, Hill (1996) observed that whereas some yellow carotenoids (mainly lutein) are highly available in the diets of most terrestrial birds, red carotenoids are usually rare and are instead metabolically converted from the yellow precursors, supposedly at some cost (although this remains to been shown). This suggests that red displays generally might be more costly than yellow, which Hill (1996) also supported by comparative evidence that red plumage patches are smaller and more sexually dimorphic compared to yellow. Species with yellow pigmentation in nestling but not adult plumage (Partali et al. 1987), and yellow internal and external fatty tissues not used in display further suggests that yellow carotenoid pigmentation is not costly to generate, whereas red more often might be. Based on the above, the yellow epigamic signal again agrees best with 'efficacy', and the red better with 'content'.

Structural colours

The origin of the taxonomically widespread 'non-iridescent' UV, violet and blue feather colours is a fine 'spongy structure' of keratin rods and channels under the cortex of barbs (the branches off the main feather 'rachis') (Dyck 1976; Prum et al. 1998; Andersson 1999). Fitzpatrick (1998) and Andersson (1999) discuss some possible ontogenetic costs entailed by this structure. First, the thickness of the spongy layer could be costly in time or energy to produce. Second, the precision or regularity of the nanostructure (i.e. variance in diameter and spacing of the approximately 100 nm thin channels) should primarily affect chroma (Andersson 1999) which might constitute an indicator of 'developmental stability' in much the same way as proposed for fluctuating bilateral asymmetries (Møller 1990; Swaddle, this volume). The same would apply to classic thin-film iridescence (e.g. hummingbirds), in which regular spacing of the sandwiched melanin and keratin layers likewise should affect chroma.

Investigations of costs of structural colour production should be especially interesting given some recent suggestions that they play a role in 'good genes' sexual selection. First, evidence that avian structural colours are sexually selected (Johnsen et al. 1998), or at least involved in intersexual signalling (Andersson & Amundsen 1997; Bennet et al. 1997; Andersson et al. 1998b) come from species with a high incidence of extra-pair copulations (EPC). As a corollary, the champion of sexual infidelity, the superb fairy-wren *Malurus cyaneus* (e.g. Mulder et al. 1994) has one of the most intense (chromatic) structural UV/blue colours (own observations). Second, Owens and Hartley (1998) found that EPC frequency covaried with sexual dichromatism based on structural, but not melanin or carotenoid coloration. Taken together, this suggests that studies of genetic mate choice based on structural colours (including UV) are likely to be interesting.

To conclude this section, the relative cost (prediction 2, Table 1) and thus 'content' value of the UV/blue colour is an open question which might get some interesting answers in the near future.

Efficacy versus content, and signal variation

Another distinction between 'efficacy' and 'content' designs might be through differences in the amount and type of intraspecific signal variation. 'Content' signals (here, 'handicaps') are likely to show more variation than other traits because of the expected high costs and directional female preferences (see e.g. Alatalo et al. 1988). In addition, this variation should be largest in the dimension of the signal that best correlates with its honesty-maintaining cost. For ornamental tails, it could be symmetry, mass, or area, depending on whether it is 'developmental stability', production, or impaired aerodynamics that is the critical cost. In colour signals, properties like hue, chroma and brightness might have different (even opposing) relationships to signal cost, and this will vary between colour mechanisms. This has not been investigated.

Relative to 'content' signals, 'efficacy' signals should more often be subject to threshold preferences (e.g. against detectability below some optimum or maximum level), or stabilizing selection (e.g. signals used for species or sex recognition and constrained in all directions in signal space). They should therefore be less variable or fixed. Moreover, if signal production is cheap, signal variation need not be largest in the costliest dimension. The most variable dimension might instead be the one least restrained by 'efficacy' selection. (In the red bill spot in some gulls, mentioned earlier, patch size might vary most if hue is crucial for 'efficacy', even if hue is the dimension that relatively best relates to a cost).

For three colorimetric dimensions (objective indices of hue, chroma and brightness), the intraspecific coefficients of variation in the three epigamic colours are shown in Figure 5.

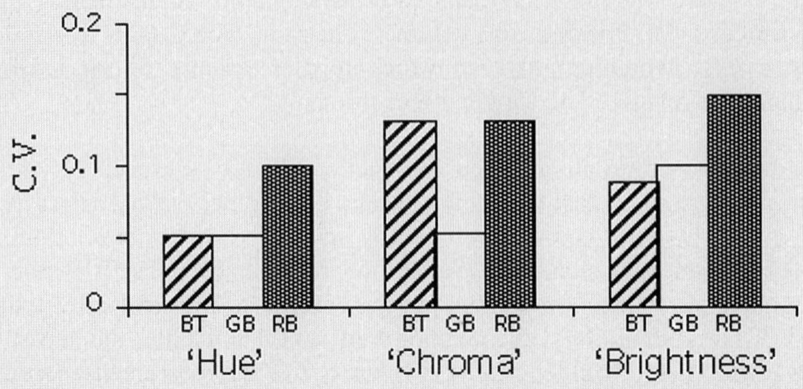

Figure 5. *Coefficients of variation (c.v.) for three objective estimates of colour signal dimensions, for each of the three signals (BT, GB, RB). 'Hue' = l(max-slope); 'Chroma' = Rmax/Rmin; 'Brightness' = R320-750).*

Once again, yellow agrees relatively better with an 'efficacy' design and red with 'content'. Adult, territory-holding golden bishop males show less variation than red bishops in hue, chroma and brightness (Fig. 5). The variation in hue, which in some saturated carotenoid displays might be the only predictor of pigment concentration (see below), was significantly variable only in the red bishop (J. Örnborg & S. Andersson, unpublished), but it was not the most variable dimension.

The variation in bluethroat structural colour variation shows another pattern, with little variation in hue and brightness, but stronger variation in chroma (Fig. 5). Interestingly, the two potentially costly aspects of a structural colour mechanism discussed above, thickness and regularity, should both primarily affect chroma rather than hue (the latter being a function of the average diameter of the keratin channels; see Andersson 1999). In addition, a difference in UV chroma between one year old and older males (S. Andersson, unpublished) also suggested that this might be a costly dimension of structural UV and blue plumage colours.

Efficacy versus content, and signal fitness

The fourth and final prediction concerns the details of the signal variation in relation to the fitness of the emitter. For 'efficacy' designs, a measure of signal fitness (e.g. mate attraction) should correlate strongest with the natural (or experimentally induced) variation in that signal dimension which most affects 'efficacy'. For definitions of 'sensory drive' (Endler & Basolo 1998) that encompass the preference itself, this might seem circular (as is 'survival of the fittest'). However, if environmental signal conditions and only 'peripheral' receiver tuning are objectively quantified (i.e. not behaviourally tested!), one should be able to pinpoint the dimension of signal variation that most affects 'detectability', and hence should predict signal success. For example, in the bluethroat a higher reflectance peak (chroma variation) would have a stronger effect on spectral contrast (Fig. 3) than would a peak shift in either direction (hue variation). Experiments with the UV component of this colour signal (Andersson & Amundsen 1997; Johnsen et al. 1998) had insufficient knowledge of receiver tuning and were not subtle enough to test chroma independently of hue, but could in principle be designed to do that, and thus test if it is primarily the main 'efficacy' dimension (chroma) that is favoured.

If 'content' (honesty) is the main target of signal selection, success should correlate best with the 'cost dimension' discussed above. Colour signals based on pigments, such as carotenoids, are interesting in this respect. The reason is that beyond saturation (100% absorbance in some spectral region), additional pigmentation will no longer increase chroma (but might cause a continued hue shift; own observations), and spectral contrast against the background is likely to decrease rather than increase in this interval. Since many carotenoid plumage colours are 'saturated' in this respect, it means that 'detectability' and 'honesty' indeed are in conflict. Consequently, if a female preference was found for males with redder hue (a variation found here to decrease detectability), this should indicate a primarily 'content' based (i.e. 'handicap') signal selection. Unfortunately, this 'efficacy'/'content' distinction does not seem to lend itself to the classic case of quality-advertising carotenoid displays, the house finch *Carpodacus mexicanus* (Hill & Montgomerie 1994; Hill 1999). In this species, male coloration varies both in the area it covers and from a pinkish (e.g. unsaturated) wash to a deep (saturated) red. Over this range of chroma variation, detectability and honesty are not likely to be in conflict, and could not be a guide to the 'efficacy'/'content' distinction (although this is hardly needed in this case where 'content' has been firmly demonstrated in other ways; Hill 1999).

To conclude, the fitness effects of the three colour signals have not been studied at all (bishops) or in enough resolution (bluethroat) as regards the details of how they are favoured by sexual selection. Analyses of the fitness consequences of different dimensions of signal variation should be a fruitful way of discriminating between 'efficacy' based and 'content' based signal selection.

Final remarks

Based on the 'efficacy' and 'content' demands on biological signals (Guilford & Dawkins 1991; Endler 1993, this volume; Johnstone 1997), I have suggested predictions as to the adaptive design of epigamic signals. In illustrating how such predictions might be tested in colour signals, I have touched upon some aspects of avian epigamic colour communication that need further study: (1) Bird vision obviously needs to be investigated in more species. Even rather crude estimates of visual range and spectral sensitivity for major ecological and taxonomic groups would greatly improve assessments of relative signal 'efficacy'. (2) Signalling microhabitats and behaviours need to be studied to determine the relevant conditions under which 'efficacy' and 'content' aspects are to be analysed (see Endler 1993a; Endler & Thery 1996). (3) More studies are needed of the costs of different colour mechanisms (see references in Hill 1999), especially structural colouration, and the details of how costs in terms of production, maintenance and signal exploitation (predation, eavesdropping) relate to different colour signal dimensions such as hue, chroma and brightness.

These kinds of information, and the distinction between 'efficacy' and 'content' in signal design, should help clarifying the evolution of the avian colour code.

Acknowledgements

First of all, thank you Yngve, Trond, and Gunilla for a beautifully organised and stimulating workshop. I am also grateful to Malte Andersson, John Endler, Mike Lawes, Peter Slater, Justin Schultz, and Jonas Örnborg for helpful comments on this manuscript. The work was funded by the Swedish Natural Sciences Research Council, and a travel grant to South Africa from the Royal Swedish Academy of Sciences.

References

Alatalo, R. V., Höglund, J. & Lundberg, A. 1988. Patterns of variation in tail ornament size in birds. *Biological Journal of the Linnean Society*, **34**, 363-374.

Andersson, M. 1994. *Sexual Selection.* Princeton, NJ: Princeton University Press.

Andersson, S. 1999. Morphology of UV and violet reflectance in a whistling-thrush: implications for the study of structural colour signalling in birds. *Journal of Avian Biology*, **30**, 193-204.

Andersson, S. & Amundsen, T. 1997. Ultraviolet colour vision and ornamentation in bluethroats. *Proceedings of the Royal Society of London, Series B*, **264**, 1587-1591.

Andersson, S., Rydell, J. & Svensson, M. G. E. 1998a. Light, predation and the lekking behaviour of the ghost swift *Hepialus humuli* (L.) (Lepidoptera: Hepialidae). *Proceedings of the Royal Society of London, Series B*, **265,** 1345-1351.

Andersson, S., Örnborg, J. & Andersson, M. 1998b. Ultraviolet sexual dimorphism and assortative mating in blue tits. *Proceedings of the Royal Society of London, Series B*, **265**, 1-6.

Bennett, A. T. D., Cuthill, I. C., Partridge J. C. & Lunau, K. 1997. Ultraviolet plumage colours predict mate preferences in starlings. *Proceedings of the National Academy of Sciences, USA*, **94**, 8618-8621.

Bradbury, J. W. & Vehrencamp, S. 1998. *Principles of Animal Communication.* Sunderland, MA: Sinauer.

Briskie, J. V., Naugler, C. T. & Leech, S. M. 1994. Begging intensity of nestling birds varies with sibling relatedness. *Proceedings of the Royal Society of London, Series B*, **258**, 73-78.

Craig, A. J. F. K. 1974. Reproductive behaviour of the male red bishop bird. *Ostrich,* **45**, 149-160.

Craig, A. J. F. K. 1980. Behaviour and evolution in the genus *Euplectes. Journal of Field Ornithology*, **121**, 144-161.

Dyck, J. 1976. Structural colours. *Proceedings of the International Ornithological Congress*, **16**, 426-437.

Endler J. A. 1980. Natural and sexual selection on color patterns in *Poecilia reticulata*. *Evolution,* **34**, 76-91.

Endler, J. A. 1990. On the measurement and classification of colour in studies of animal colour patterns. *Biological Journal of the Linnean Society*, **41**, 315-352.

Endler, J. A. 1991. Variation in the appearance of guppy color patterns to guppies and their predators under different visual conditions. *Vision Research,* **3**, 587-608.

Endler, J. A. 1993a. Some general comments on the evolution and design of animal signalling systems. *Philosophical Transactions of the Royal Society of London, Series B*, **340**, 215-225.

Endler, J. A. 1993b. The color of light in forests and its implications. *Ecological Monographs,* **63**, 1-27.

Endler, J. A. & Basolo, A. L. 1998. Sensory ecology, receiver biases and sexual selection. *Trends in Ecology and Evolution*, **13**, 415-420.

Endler, J. A. & Thery, M. 1996. Interacting effects of lek placement, display behaviour, ambient light, and color patterns in three neotropical forest-dwelling birds. *American Naturalist,* **148**, 421-452.

Fitzpatrick, S. 1994. Colorful migratory birds: evidence for a mechanism other than parasite resistance for the maintenance of 'good genes' sexual selection. *Proceedings of the Royal Society of London, Series B*, **257**, 155-160.

Goodwin, T. W. 1984. *The Biochemistry of Carotenoids*. New York: Chapman & Hall.

Guilford, T. & Dawkins, M. S. 1991. Receiver psychology and the evolution of animal signals. *Animal Behaviour*, **42**, 1-14.

Guilford, T. & Dawkins, M. S. 1995. What are conventional signals? *Animal Behaviour*, **49**, 1689-1695.

Hailman, J. P. 1977. *Optical Signals: Animal Communication and Light*. Bloomington: Indiana University Press.

Hasson, O. 1989. Amplifiers and the handicap principle in sexual selection: a different emphasis. *Proceedings of the Royal Society of London, Series B*, **235**, 383-406.

Hill, G. E. 1996. Redness as a measure of the production cost of ornamental coloration. *Ethology, Ecology and Evolution*, **8**, 157-175.

Hill, G. E. 1999. Mate choice, male quality, and carotenoid-based plumage coloration. In: *Proceedings of the 22nd International Ornithology Congress* (Ed. by N. Adams & R. Slotow), pp 1654-1668. Johannesburg: Birdlife South Africa.

Hill, G. E. & Montgomerie, R. 1994. Plumage colour signals nutritional condition in the house finch. *Proceedings of the Royal Society of London, Series B*, **258**, 47-52.

Johnsen, A., Andersson, S., Örnborg, J., & Lifjeld, J.T. 1998. Ultraviolet plumage ornamentation affects social mate choice and sperm competition in bluethroats (Aves: *Luscinia s. svecica*): a field experiment. *Proceedings of the Royal Society of London, Series B*, **265**, 1313-1318.

Johnstone, R. A. 1997. The evolution of animal signals. In: *Behavioural Ecology; an Evolutionary Approach* (Ed. by J. R. Krebs & N. B. Davies), pp. 155-178. Oxford: Blackwell Scientific Publications.

Lozano, G. A. 1994. Carotenoids, parasites, and sexual selection. *Oikos,* **70**, 309-311.

Marchetti, K. 1993. Dark habitats and bright birds illustrate the role of the environment in species divergence. *Nature*, **362**, 149-152.

Merilä, J. & Sorjonen, J. 1994. Seasonal and diurnal patterns of singing and song-flight activity in bluethroats (*Luscinia svecica*). *Auk*, **111**, 556-562.

Mulder R. A., Dunn, P. O., Cockburn, A., Lazenby-Cohen, K. A. & Howell, M. J. 1994. Helpers liberate female fairy-wrens from constraints on extra-pair mate choice. *Proceedings of the Royal Society of London, Series B*, **255**, 223-229.

Møller, A. P. 1990. Fluctuating asymmetry in male sexual ornaments may reliably reveal male quality. *Animal Behaviour*, **40**, 1185-1187.

Owens, I. P. F. & Hartley, I. R. 1998. Sexual dimorphism in birds: why are there so many different forms of dimorphism? *Proceedings of the Royal Society of London, Series B*, **265**, 397-407.

Partali, V. S., Liaaen-Jensen, S., Slagsvold, T. & Lifjeld, J. T. 1987. Carotenoids in food chain studies – II. The food chain of *Parus* spp. monitored by carotenoid analysis. *Comparative Biochemical Physiology, Series B*, **87**, 885-888.

Savalli, U. M. 1995. The evolution of bird coloration and plumage elaboration. In: *Current Ornithology* (Ed. by D. M. Power), pp. 145-190. New York: Plenum Press.

Schluter, D. & Price, T. 1993. Honesty, perception and population divergence in sexually selected traits. *Proceedings of the Royal Society of London, Series B*, **253**, 117-122.

Tuttle, M. D. & Ryan, M. J. 1981. Bat predation and the evolution of frog vocalizations in the neotropics. *Science*, **214**, 677-678.

Wiley, R. H. 1991. Associations of song properties with habitats for territorial oscine birds of eastern North America. *American Naturalist*, **138**, 973-993.

Williams, G. C. 1966. *Adaptation and Natural Selection: A Critique of Some Current Evolutionary Thought.* New Jersey: Princeton University Press.

Zahavi, A. 1975. Mate selection - a selection for a handicap. *Journal of Theoretical Biology*, **53**, 205-214.

Zahavi, A. & Zahavi, A. 1997. *The Handicap Principle*. Oxford: Oxford University Press.

IN: ESPMARK,Y., AMUNDSEN,T. & ROSENQVIST,G. (EDS.) 2000
ANIMAL SIGNALS:
SIGNALLING AND SIGNAL DESIGN IN ANIMAL COMMUNICATION, PP. 61-82
TAPIR ACADEMIC PRESS, TRONDHEIM, NORWAY

3

Avian UV Vision and Sexual Selection

**INNES C. CUTHILL*,
JULIAN C. PARTRIDGE & ANDREW T. D. BENNETT**

School of Biological Sciences, University of Bristol, Woodland Road, Bristol BS8 1UG, UK

* *E-mail: i.cuthill@bristol.ac.uk*

Abstract

The bright colours and sometimes elaborate ornamentation of male birds have long been a focus for students of sexual selection. Yet it is only recently that scientists have started to investigate avian signalling from the avian perspective. Avian colour vision differs from that of humans in having a broader spectral range (including the near ultraviolet waveband), more photoreceptor types, and coloured oil droplets which filter the light incident on these photoreceptors. This implies that birds will perceive objects as coloured in different ways from humans, and see colours we cannot experience directly. Recent behavioural research has concentrated on the implications of ultraviolet vision for avian signalling, as these are the wavelengths to which humans are blind. We review the increasing number of studies, using different methods on different species, which all demonstrate that ultraviolet is important in avian signalling. Whilst examples of hidden UV sexual dimorphism exist, UV is also important in the colour signals of species which are colourful to humans. The current evidence suggests that UV is a normal part of the avian hue perception mechanism, so should not be ignored in future studies of avian coloration.

The widespread sexual dimorphism in plumage colours of birds, and their often extravagant ornamentation, made them a natural example for Darwin to use when illustrating the effects of sexual selection (Darwin 1871). Since then, behavioural ecologists have often turned to birds when undertaking empirical investigations of Darwin's ideas and the many other proposed mechanisms of sexual selection and signalling (see e.g. Andersson 1994). However, what Darwin did not realise, and most of his successors have failed to consider, is that birds have a very different colour vision system from humans (reviews by Bennett & Cuthill 1994; Burkhardt 1996; Bowmaker et al. 1997; Cuthill et al. 1999b). As the perceived colour of an object is fundamentally dependent on the visual system of the animal viewing the object, the use of human colour judgements, or 'objective' standards based on human colour vision (e.g. video, photographs, or commercial colour standards such as Munsell chips), is inappropriate (Endler 1990; Cuthill & Bennett 1993; D'Eath 1998; Fleishman et al. 1998). The implications of this 'mismeasure of colour', and its widespread occurrence, has been discussed in detail before (Endler 1990; Bennett et al. 1994) and we consider the point well made. The purpose of this paper is to review the increasing number of studies which explicitly consider avian colour perception in the context of signalling and sexual selection.

The differences between avian and human vision

An object's colour depends on the relative amounts of different wavelengths of light it reflects, emits or transmits. However, colour is not the same as wavelength, nor can it be directly inferred from the spectrum of light coming from the object; colour is a psychophysical property (see e.g. Jacobs 1981; Wyszecki & Stiles 1982; Thompson et al. 1992). To see colour, an animal must have, as a basic minimum, (i) photoreceptors sensitive to different wavelengths of light and (ii) the ability to compare the output of those photoreceptors (see e.g. Jacobs 1981). Thus human vision at low light levels (scotopic vision) is achromatic, because it is mediated by a single class of photoreceptor (the retinal rod cells) the output of which is not compared with any other photoreceptor type. At higher light levels (photopic vision), human colour vision is effected by the neural comparison of the output of the three types of cone cell. The three types of cones are maximally sensitive to different wavelengths and are referred to as our 'blue' (shortwave), 'green' (mediumwave) and 'red' (longwave) cones. The equating of each cone type with a colour is somewhat misleading, as the sensation of colour does not result from the action of these cells in isolation, but from the neural comparison of their outputs. High stimulation of the longwave cone relative to the mediumwave cone results in the sensation of red, and the opposite pattern of stimulation is perceived as green. High stimulation of both long- and mediumwave cones relative to the shortwave cone is seen as yellow, the reverse pattern as blue. Summation of the three cone types' output, rather than comparison, is equated with the brightness of the object. This is a gross simplification, as the perceived colour of an object is also influenced by the colour of other objects in the visual field. This relates to the mechanism(s) which maintain colour constancy, whereby objects are perceived as the same colour despite changes in the illuminant. Thus the perceived colour at any one point on the retina is not a simple function of the signals from the photoreceptors at that location alone. Nevertheless, the primary stage in colour vision is the transformation, by the nervous system, of the input channels of the three cone types into three output channels: an achromatic brightness channel, and two hue channels, red-green and yellow-blue (Table 1). A fuller description of the relationship between wavelength and colour can be found in texts on visual perception and colour science (e.g. Wyszecki & Stiles 1982). The purpose of this brief caricature is to set the scene for the implications of the way we, and other animals, see colour (full and excellent accounts of the logic can be found in Endler 1990 and Thompson et al. 1992).

Table 1. *The input channels of the human colour vision system are the three cone types, sensitive to short- (SW), medium- (MW) and long-wave (LW) regions of the human-visible spectrum. These are transformed, via neural summation (+) or comparison (-) into an achromatic brightness channel and two opponent colour (hue) channels: blue-yellow and red-green.*

	Cone type SW	MW	LW	Sensation	Colour channel (Output)
Transformation	+	+	+	White*	Brightness
	0	-	+	Red*	Red-Green
	-	+	+	Yellow*	Yellow-Blue

*With the opposite sign of transformation/comparison, the colours experienced would be black (- - -), green (0 + -) and blue (+ - -). '0' indicates the lack of contribution to a colour channel.

The first corollary of the mechanism of colour vision described above is that objects with different reflectance spectra will be perceived as the same colour as long as the three cone types are stimulated in the same proportions. A colour television takes advantage of this mechanism to mimic the colours of objects, without actually reproducing their reflectance spectra. The coloured phosphors of a TV screen emit relatively narrow band light. Natural objects rarely reflect narrow wavebands of light, but by appropriate mixing of narrow band light of the three wavebands to which human cones are sensitive (blue, green and red), the objects on television appear to be the same colour as in real life (see Fleishman et al. 1998). Second, species with different receptor sensitivities, or comparing receptors in different ways, will see the colour of objects differently from humans (Bennett et al. 1994; Endler 1990). We cannot, of course, know what the animal's subjective experience of the colour is, but we can expect their abilities to detect and discriminate coloured objects to differ from that of humans. Third, the dimensionality of a species' colour vision depends on the number of receptor types being compared. Humans have three-dimensional, or trichromatic, colour vision. Most mammals, other than Old World and some New World primates, have only two cone types, blue-violet (or UV) and green, so are dichromatic (Jacobs 1993). Goldfish *Carassius auratus*, with four cone types being compared, have four-dimensional, or tetrachromatic, colour vision (Neumeyer 1992). An interesting consequence of higher-than-two-dimensional colour vision is the existence of, so-called, non-spectral colours. These are colours which are not part of the rainbow and cannot be simulated by any monochromatic light source; they arise from stimulation of two or more photoreceptors which are sensitive to non-adjacent wavebands. Purple is the one human non-spectral colour, perceived when red and blue cones are stimulated (i.e. purple objects are those which reflect long- and short-, but not medium-, wave light). A tetrachromat, with four receptor types being compared, would have several non-spectral colours, corresponding to the different possible combinations of non-adjacent wavebands of light to which it was sensitive.

Bird colour vision differs from that of humans in three key ways that are liable to produce differences in their colour perception (Bennett & Cuthill 1994). First, they can see near-ultraviolet light, wavelengths below 400 nm which the human lens and other optical media block from the retina (Goldsmith 1990; Jacobs 1992). As they can also see the entire human-visible waveband (ca. 400-700 nm), they have a wider spectral range than humans. Ultraviolet vision is widespread amongst both vertebrates and invertebrates (Bowmaker 1991a; Jacobs 1992; Goldsmith 1994; Tovee 1995), and is even found in some mammals (Jacobs et al. 1991; Jacobs & Deegan 1994), so this capacity is far from unusual. The first evidence for UV vision in birds came from behavioural discrimination experiments on hummingbirds (Huth & Burkhardt 1972) and pigeons (Wright 1972). Since then, electrophysiological and behavioural experiments have shown UV vision to be the norm in diurnal birds (reviewed by Bennett & Cuthill 1994) although the first avian UV visual pigment was only detected by microspectrophotometry in 1993 (Maier & Bowmaker 1993). UV vision has now, by one means or another, been demonstrated in over 35 bird species (see Bennett & Cuthill 1994; Bowmaker et al. 1997; Hart et al. 1998, 1999; Cuthill et al. 1999b), with negative evidence for two nocturnal species, both owls (Bowmaker & Martin 1978; Koivula et al. 1997).

The second key difference between birds and humans is that birds have more than three retinal cone types, and so have the potential for at least four-dimensional colour vision. Some caution is warranted here, as the colour-mixing experiments necessary to establish the dimensionality of avian colour space have yet to be published, but it would seem a reasonable working hypothesis, given goldfish tetrachromacy. Birds actually have six cone classes: four single cones, and the dissimilar members of the 'double cones', which are also found in fish and turtles (Liebman & Granda 1971; Ohtsuka 1985), but are lacking in humans. The function of double cones is as yet unconfirmed, intriguing given that they can constitute the majority cone type in a bird's retina (Hart et al. 1998; Wilkie et al. 1998). The broad spectral sensitivity of the double cones suggest that they would not be useful as part of a hue discrimination mechanism (Vorobyev & Osorio 1998; Vorobyev et al. 1998), and spectral sensitivity experiments (e.g. Maier & Bowmaker 1993) indicate that only the single cones are involved in colour perception (Vorobyev & Osorio 1998; Vorobyev et al. 1998). The avian single cones span the avian-visible spectrum fairly evenly, but there is some variation between species in the wavelengths of maximum sensitivity (λ_{max}) of their visual pigments. Most variable of these is the pigment which confers UV sensitivity, and there appear to be broadly two types, based on microspectrophotometry (MSP). Those passerines which have been investigated (Pekin robin *Leiothrix lutea*, Maier & Bowmaker 1993; zebra finch *Taeniopygia guttata,* Bowmaker *et al.* 1997; starling *Sturnus vulgaris*, Hart et al. 1998; blue tit *Parus caeruleus* and blackbird *Turdus merula*, both N. S. Hart, unpublished); and the budgerigar *Melopsittacus undulatus,* Bowmaker et al. 1997) have a

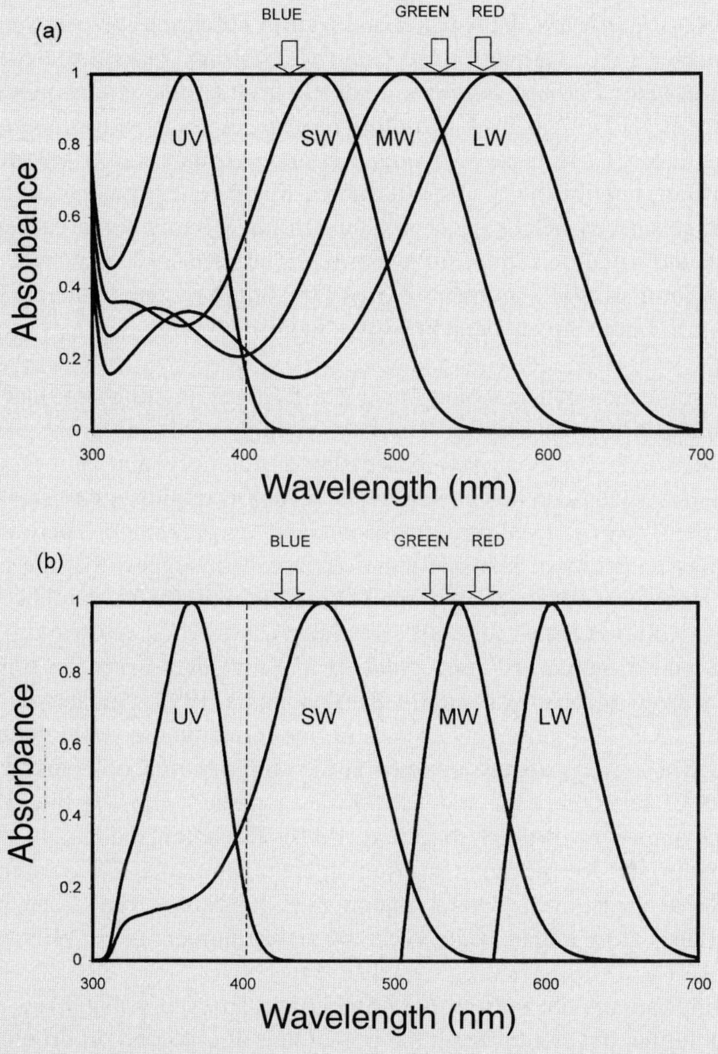

Figure 1. ***(a)*** *Normalised absorbance of the visual pigments from the four types of single retinal cone cells in the European starling (UV=ultraviolet, SW=shortwave, MW=mediumwave, LW=longwave). Curves are the best-fitting visual pigment templates for average difference spectra based on microspectrophotometry (data and details in Hart et al. 1998). For comparison, the wavelengths of maximum spectral sensitivity of the blue, green and red cones of the human are indicated by arrows at the top of the figure.* ***(b)*** *Normalised spectral sensitivity of the starling's four cone types, after correcting for wavelength-specific absorption by the different oil droplet types in each cone, and the optical media in the light path (data and details in Hart et al. 1998). Note that the wavelengths of maximum sensitivity of even the SW, MW and LW cones differ notably from their human counterparts (the arrows labelled 'blue', 'green' and 'red').*

'true UV' visual pigment, with λ_{max} around 355-370 nm. Other species can see UV by virtue of violet visual pigments with λ_{max} values in the range 400-420 nm (Humboldt penguin *Spheniscus humboldti*, Bowmaker and Martin 1984; mallard *Anas platyrhyncos*, Jane and Bowmaker 1988; Japanese quail *Coturnix coturnix japonica*, Bowmaker et al. 1993; Manx shearwater *Puffinus puffinus*, domestic fowl *Gallus gallus*, pigeon *Columba livia*, Bowmaker et al. 1997; ostrich *Struthio camelus*, emu *Dromaius novaehollandiae*, Wright and Bowmaker 1998; turkey *Meleagris gallopavo*, Hart et al. 1999; peacock *Pavo cristatus*, N. S. Hart, unpublished). The other three single cone pigments are more similar in their spectral sensitivity across species, although not without some variation. Typical for passerines is the situation found in the starling (Hart et al. 1998) which, as shown in Figure 1a, in addition to its UV pigment (λ_{max} = 362 nm) has cone pigments with λ_{max} at 449 nm ('blue'), 504 nm ('green') and 563 nm ('red').

The adaptive significance of the violet/true-UV dichotomy is unknown, and it would be unwise to speculate while research on the visual ecology of the different groups is in its infancy. The wider range of avian taxa possessing a violet pigment, and the fact that the two palaeognathus species also possess the violet receptor, might suggest this is the ancestral state for birds. However, even this conclusion is as yet premature. There is variation in the λ_{max} of violet receptors of non-passerines (recent data suggest that the pigeon really has a near-UV receptor, with λ_{max} 393 nm; Yokoyama et al. 1998) and the λ_{max} of other vertebrates' UV/violet receptors can also be either true-UV or violet (e.g. rat 358 nm, chameleon 358 nm, Xenopus 425 nm, goldfish 359 nm; data based on regeneration of visual pigments from cDNA; Kawamura & Yokoyama 1998; Starace & Knox 1998; Yokoyama et al. 1998). Of those few species of vertebrate whose opsin genes have been sequenced, and whose λ_{max} values are known, the current evidence is that the violet/UV visual pigments of birds, reptiles and fish cluster together with the blue/UV cones of mammals (Yokoyama et al. 1998; Wilkie et al. 1998). The avian and human blue pigments are not homologous, nor are the avian and human green pigments. The primate green cone pigment is derived from the ancestral mammalian longwave (red) cone pigment, the sequence of which is close to that of the avian red visual pigment gene (Wilkie et al. 1998).

Birds differ from humans not only in the number and spectral sensitivities of the visual pigments in the retina, but also through the possession of coloured oil droplets which lie, within the cone cells, between the incident light and the visual pigment. Photoreceptor oil droplets occur in many vertebrates, including a few fish, amphibians, reptiles and even marsupial mammals (Walls 1963), but in marsupials and nocturnal birds they are not densely pigmented. In diurnal birds, and other groups with densely pigmented oil droplets, their function appears to be to filter the light entering the cones (Partridge 1989; Bowmaker 1991b). Each visual pigment is twinned with a particular oil droplet type which cuts out wavelengths somewhat less than the λ_{max} of the visual pigment (Goldsmith & Collins 1982; Goldsmith et al. 1984; Partridge 1989; Bowmaker 1991b; Bowmaker et al. 1997; Kawamuro et al. 1997; Hart et al. 1998). Thus, in single cones, the red pigment is paired with a red droplet, the green with a yellow, the blue with a UV-blocking clear droplet, and the UV or violet pigment with a fully transparent droplet. The net effects are to increase the effective λ_{max} of the longer wavelength cones, to narrow the waveband to which each cone is sensitive and to reduce the overlap in their spectral sensitivities (Fig. 1b), a striking difference compared to the considerable overlap of human (particularly red and green) cone spectral sensitivities. The suggested advantages of photoreceptors with narrowed spectral sensitivity and reduced overlap in waveband, as in birds and turtles, is enhanced discrimination of certain classes of spectra (Goldsmith 1990; Vorobyev et al. 1998) and improved colour constancy (Vorobyev et al. 1998; see also Osorio et al. 1997). This comes at the cost of reduced quantal catch (particularly in the longwave cones, where the densely pigmented oil droplets block much light), so the system is most efficient at high light intensities. Reduced overlap in spectral sensitivities of photoreceptors may reduce discriminability of monochromatic lights (as used in psychophysical tests), but narrowed spectral tuning is best suited to discrimination of objects with sharply varying spectra, such as those of many natural objects (see Vorobyev et al. 1998). Modelling the separation, in avian colour space, of natural reflectance spectra from a diversity of avian plumage colours, Vorobyev et al. (1998) concluded that tetrachromacy and possession of oil droplets each improved the discriminability of these colours, as compared to trichromatic systems. With this in mind, it is important to remember that birds are not only liable to differ from humans in their detection of differences in ultraviolet reflectance. The ability of birds to detect and discriminate even those spectral differences which humans can detect, will differ from the human experience. In summary, humans do not see the colours of objects in the same way as birds, and vice versa.

UV vision and signalling

Given the profound differences between bird and human colour vision, it would seem obvious that studies of colour based on human judgements, when birds are the natural receivers, should be considered unreliable (Bennett et al. 1994). This would include not only sexually selected plumage colours, but also studies of insect crypsis, aposematism and mimicry when birds are the principal predator, or flower or fruit coloration when birds are the major pollinators or dispersers. However, there has been an understandable resistance to the idea that human colour standards are invalid, perhaps because birds are so colourful to us and because (our sense of) colour dominates the way we perceive the world. In addition, in those instances where people have looked for 'hidden' UV-only colour patches in birds, they have rarely found them (Finger 1989; Radwan 1993; Andersson 1999). Also, one might expect some degree of convergence between the ways in which humans/primates and birds perceive colour. This is because a plausible adaptive explanation of primate trichromacy is that it aids in the detection and discrimination of fruits (Mollon 1989; Osorio & Vorobyev 1996). The shift from ancestral dichromacy to blue-green-red trichromacy has happened more than once, independently, in different primate lineages (Jacobs 1996; Jacobs et al. 1996; Bowmaker 1998; Hunt et al. 1998a; Bowmaker & Hunt 1999; Bowmaker et al. 1999). In this sense, the primate red-green colour channel may have converged on a colour discrimination system already employed by birds. This would put the onus on visual ecologists to demonstrate that differences between avian and human colour vision actually result in differences in the way birds respond behaviourally to colour patterns, and that human colour judgements distort our interpretation of avian visual signals (see Vorobyev et al. 1998).

Recent research on avian colour vision and signalling has concentrated on the possible role of UV information, simply because this is the most obvious difference between bird and human colour perception. The initial focus has been on colour and signalling because sexual selection has been such a dominant theme in behavioural ecology over the last 15 years, but UV vision also has a plausible role in orientation (Coemans et al. 1994; Vos Hzn et al. 1994) and has recently been shown to be involved in foraging (Viitala et al. 1995; Church et al. 1998b). Certainly, there is UV information in many species' plumage reflectance (e.g. Burkhardt 1989, 1996; Finger 1989, 1995; Finger et al. 1992; Bennett & Cuthill 1994; Finger & Burkhardt 1994). Furthermore, many common backgrounds (e.g. leaves, earth, bark) reflect very little in the UV (see e.g. Chittka et al. 1994) so UV-reflecting plumage would often be conspicuous. However, just because a species' plumage reflects in the UV does not mean that this is a visual signal, as it may be an unselected consequence of its chemical or physical structure (Andersson 1996). High UV reflection from plumage that is otherwise dark is perhaps more convincing evidence of a role in signalling (Finger et al. 1992; Andersson 1996), as is sexual dichromatism in the UV but not human-visible wavelengths (e.g. in the blue tit, Andersson et al. 1998; Hunt et al. 1998b). There were differences between the two latter studies, in that Hunt et al. (1988) found peak reflectance from the 'blue' crest actually in the UV (ca. 350 nm, Hunt et al. 1998b) whilst Andersson et al. (1988) found peak reflectance at 430 nm, in the human-visible blue. This may have been due to a difference in measurement protocols or genuine geographical variation (Andersson et al. 1998; Hunt et al. 1998b), but in both studies male crests were brighter than those of females. Mate choice experiments indicated that this region of plumage may play a role in female mate choice since all females tested chose the male with the brightest crest (Hunt et al. 1998b). Furthermore, Andersson et al. (1998) found that there is assortative mating in the field on the basis of crest reflectance.

The conclusion that the UV reflection from the blue tit's crest is used in mate choice may seem reasonable, but the evidence is only correlational. The only true test of whether UV cues actually matter to the birds is experimental manipulation of the wavelengths available to them. The first such experiments used filters to block ultraviolet light (Maier 1993; Bennett et al. 1996, 1997; Hunt et al. 1997), or application of UV-blocking chemicals directly to the plumage (Andersson & Amundsen 1997; Johnsen et al. 1998). In each case, mate choice has been assayed using association preference in choice chambers or aviaries which, in at least some species, has been shown to correlate with real mating preferences (e.g. zebra finches, Burley et al. 1982, 1994, 1996; Swaddle & Cuthill 1994b; Swaddle 1996). There are pros and cons to each method of manipulating the UV information available to a bird viewing a conspecific. Filters do not interfere directly with the bird, and are of precise, homogeneous and relatively unvarying optical quality. However, they change the appearance of both the stimulus bird and the background. Chemical application is much more selective, allowing manipulation of specific plumage areas, but may alter the plumage's physical properties and the behaviour of the target bird (e.g. preening). However, whichever method has been used, very similar results have been obtained: removal of UV cues reduces the attractiveness of stimulus birds treated in this way. Maier (1993) was the first to use such behavioural choice tests, showing that the roosting preference of Pekin robins for companions of either sex, was reduced when a UV-blocking filter was

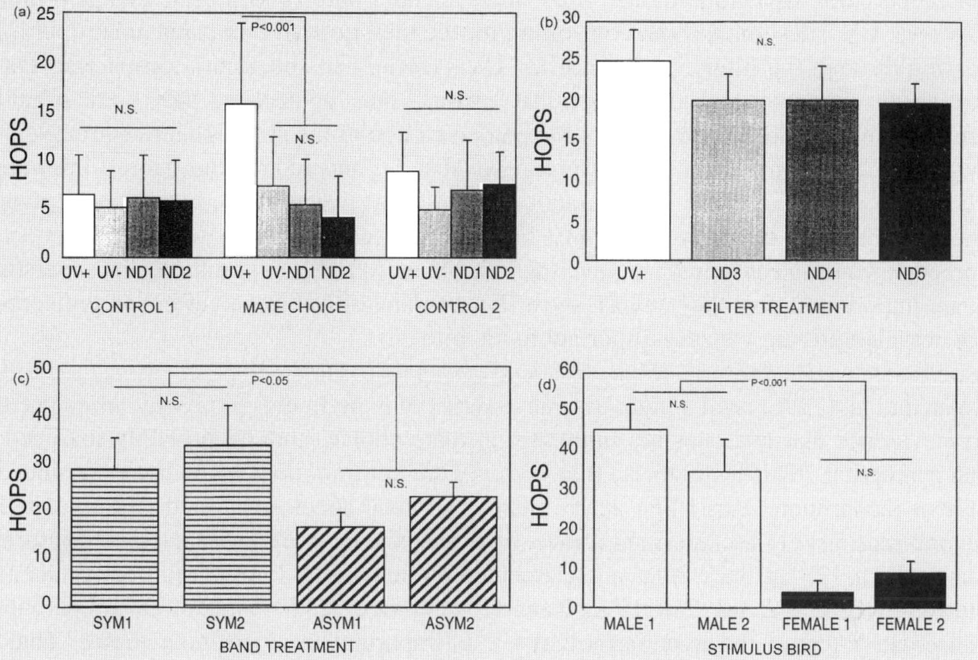

***Figure 2. (a)** The mean number of hops (+SE) by female zebra finches facing stimulus cages fronted by filters of four types: UV-transmitting (UV+) or UV-blocking with varying degrees of total light transmission (UV- > ND1 >ND2; ND = Neutral Density). There were three phases to the experiment: 'mate choice', during which randomly assigned males were in the stimulus cages, and two control phases, during which no males were present. These controls tested for a preference for a particular 'coloured view' irrespective of mate choice. (Details in Bennett et al. 1996). Females preferred males behind UV+ filters and showed no preferences when the males were absent. **(b)** Mean hops (+SE) by females choosing between males behind filters differing only in the total quantity of light transmitted. Filters were UV+ (as in a), and three UV-transmitting neutral density filters which reduced light transmission (UV+ brighter than ND3 > ND4 > ND5; details in Bennett et al. 1996). Females showed no significant preferences, indicating that the preference in (a) is due to the effect of UV-removal on hue rather than brightness. **(c)** Mean hops (+SE) by females facing stimulus cages containing males with UV-symmetric or UV-asymmetric leg bands (Bennett et al. 1996). Two of the four leg band arrangements were asymmetrical (ASYM1, ASYM2) and two were symmetrical (SYM1, SYM2). Females preferred symmetrically banded males, even though the ornaments' symmetry was only visible in the UV. **(d)** Mean hops (+SE) by females given a choice between two males and two females, each placed individually in one of the four arms of the choice chamber (see Bennett et al. 1996). All filters were identical and UV-transmitting: females preferred males, indicating that the experiments were measuring sexual preferences, rather than some non-sexual flocking response. Data replotted with permission from Nature (Bennett et al. 1996), copyright (1999) Macmillan Magazines Ltd.*

placed between test and stimulus bird. However, as Maier (1993) used a binary choice (UV+ vs. UV-) design, without controlling for the total quantal flux (total amount of light passing through the filter), preference for UV+ (full avian spectrum) conspecifics could have been either because they were the 'correct' hue, or because they were brighter. Removal of UV should both change the perceived hue of the stimulus bird (through altering the relative proportions of different wavebands) and reduce the overall brightness (through a reduction in the total amount of light of all wavebands). Furthermore, avoidance of the UV- birds could simply have been because they looked 'odd' or even unrecognisable as conspecifics. Nevertheless, Maier's (1993) experiment, and others like it, are important in that they at least show that removal of UV has an effect on preference; UV wavelengths are visually important to the bird.

Bennett et al. (1996) used a more extensive series of experiments to provide more persuasive evidence that UV cues are important in mate choice, and are probably involved in hue perception. (This is not to say that UV cues are not important in other conspecific signals or recognition tasks.) The zebra finch is an ideal species for such experiments for several reasons: (1) as described above, choice-chamber preference correlates with real mate choice, (2) 'colour', at least as perceived by humans, is known to affect choice (reviewed by Collins & ten Cate 1996; Zann 1996; Hunt et al. 1997), (3) they are perhaps a conservative test of the hypothesis that UV is important in avian mate choice. That is, they are colourful and sexually dichromatic to humans, with greatest plumage reflectance at long wavelengths (Bennett et al. 1996). For this reason, we would expect a demonstration of an effect of UV on choice in such a species to be particularly convincing to a sceptic. Bennett et al. (1996) used a four-arm choice chamber, with different filters between the test female and the stimulus males, to manipulate both presence/absence of UV and the overall amount of light passing through. Females could chose between four males, each one viewed through a different filter: UV+ and UV-, each with identical transmission in the human-visible waveband but the UV- cutting out all light below 400 nm, and two 'neutral density' filters, which transmitted similar proportions of each wavelength to the UV- filter, but different total amounts (i.e. they were darker). Bennett et al. (1996) found no detectable effect of total light transmission, but a highly significant reduction in preference for birds behind UV-blocking filters (Fig. 2a). Furthermore, during two control phases before and after the stimulus males were present, females exhibited no preferences for any stimulus chamber. This indicates their preference was not due to a non-specific bias toward a particular 'coloured view' irrespective of an effect on mate choice. In a second experiment, where only total light flux was manipulated, Bennett et al. (1996) again found no effect on female preference (Fig. 2b). Obviously at some level (e.g. a totally opaque filter), reducing available light must affect choice, but it did not over a wider range of quantal fluxes than those where blocking UV affected choice. The obvious explanation is that blocking UV has its effect by changing the perceived hue of the stimulus bird behind such a filter. The simplest explanation of the result, as with Maier (1993), is that blocking UV makes a stimulus bird look strange or even unrecognisable. To see if UV affects choice even where the stimulus bird has natural (full avian spectrum) plumage, Bennett et al. (1996) conducted a third experiment in which they manipulated only a specific sexual ornament, and all filters were of the UV-transmitting (UV+) type. The ornament was an artificial one, known to affect choice-chamber preference and actual breeding success (Swaddle & Cuthill 1994b; Swaddle 1996): the symmetry of colour bands placed on the legs. The advantage of using an artificial ornament is that all birds so-treated look equally 'odd' and their natural plumage is unaltered (it is worth noting that zebra finches also prefer symmetrical natural plumage traits, Swaddle & Cuthill 1994a). Whilst Swaddle & Cuthill (1994b) had used bilaterally symmetrical or asymmetrical arrangements of orange and green bands, Bennett et al. (1996) used similar arrangements of UV+ and UV- bands. Thus band patterns were invisible to humans and, to birds, the symmetry of the arrangement was all that differed between stimulus males; the total amount of UV-reflecting and UV-absorbing material on the legs was identical. In this experiment, females showed a significant preference for symmetrical males (Fig. 2c), confirming the symmetry preference of Swaddle & Cuthill (1994b) and, importantly, showing that UV cues can affect choice in a more subtle way than obliterating species recognition. A final experiment, giving females a simple choice between males and females, confirmed the assumption, based on considerable independent research (reviewed by Zann 1996), that the choice chamber measured sexual preference and not some flocking response (Fig. 2d).

In the first experiment to manipulate a natural plumage ornament in isolation, Andersson & Amundsen (1997) compared pairwise preferences of female bluethroats (*Luscinia s. svecica*) for males whose human-blue (but highly UV reflective) throat patches had been treated with either preen oil+sunblock (no UV) or preen oil+black pigment. The latter treatment, as with Bennett et al.'s (1996) use of neutral density filters, was an attempt to control for the effects of UV-removal on brightness, as the preen oil+black pigment reduced overall light reflection to similar levels to the UV+sunblock treatment. Although one cannot be sure that the brightness of the two treatments matched to avian eyes, the mixtures were chosen so that total quantal flux (amount of reflected light) was as similar as possible, so that it seems that here too, the effect of removing UV reflectance was to alter the perceived plumage hue. Johnsen et al. (1998), using the same manipulation in the field, found that UV-reduced males had a lower success in attracting mates and lower success in achieving extra-pair fertilisations.

The Andersson & Amundsen (1997), Maier (1993) and Bennett et al. (1996, experiment 1, but not experiment 3) experiments are all fairly gross alterations of the birds' appearance. The birds' colours may look unaltered to human eyes, but the effect to an avian observer should be like cutting out a whole waveband: for example, changing a turquoise/cyan object (green + blue) to green (blue removed). However, an experiment by Bennett et al. (1997) on starlings showed that female preferences persisted in the absence of UV cues, but that these preferences were different from those the females exhibited when UV was present. The reflectance spectra of iridescent starling feathers are complex, with peaks of reflectance in both the UV and human-visible wavebands (Fig. 3a; see Cuthill et al. 1999a), so Bennett et al. (1997) effectively also investigated whether the UV cues were redundant in female choice. They did this by comparing the preference rankings of pairs of females for quartets of males when the latter were viewed through UV-transmitting filters, with the preference rankings of the same females viewing the same males through

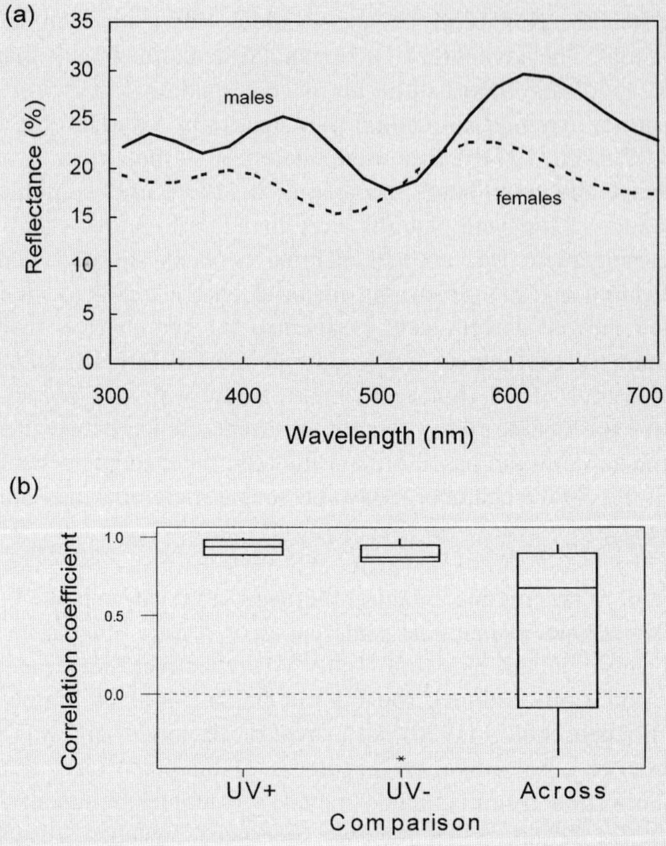

***Figure 3.** (**a**) Mean reflectance spectra of the iridescent throat feathers of male (n=6) and female (n=6) starlings in breeding plumage. Spectra are the mean of the bird means across 20 randomly located measurements, relative to that of a 99% Spectralon white standard (Data replotted with permission from the American Naturalist (Cuthill et al. 1999, copyright (1999) The University of Chicago Press). (**b**) Correlations in preferences of pairs of female starlings independently viewing quartets of males through filters which either transmitted (UV+) or blocked (UV-) ultraviolet wavelengths (Data replotted with permission from Bennett et al. (1997) PNAS 94: 8618-8621. Copyright (1997) National Academy of Sciences USA). Boxplots represent the median and inter-quartile range of correlations for eight sets of females/ males, correlations being calculated from the numbers of hops facing each male. Under UV+ conditions, preferences of pairs of females were highly significantly correlated. They were also highly correlated under UV- conditions, although less so than under UV+ conditions. Preferences across UV+ and UV- treatments were not correlated (see Bennett et al. 1997).*

UV-blocking filters (Fig. 3b). Within any one lighting condition (i.e. UV+ or UV-), female preferences for the stimulus males were highly correlated, indicating that females agreed on which males were attractive, and that species-recognition and motivation for choice were intact. However, the preference rankings under the two filter conditions were

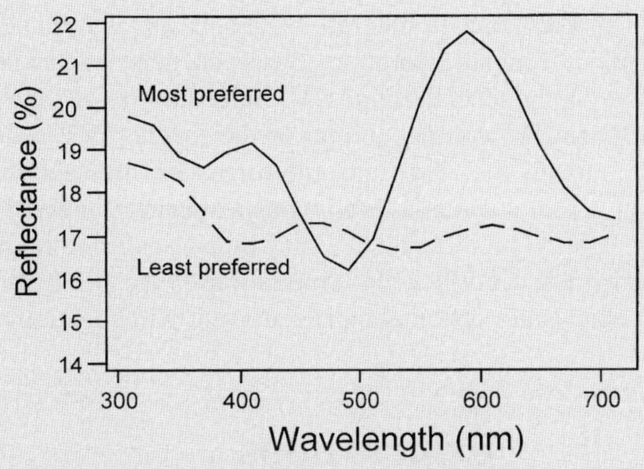

Figure 4. *Reflectance spectra of the iridescent throat feathers of the most preferred and least preferred starling males in the UV+ treatment of Bennett et al. (1997; see also Figure 3). Each spectrum is the mean of 10 measurements from eight males, relative to a Spectralon 99% white standard. Preferred males have spectra with more sharply defined peaks (and thus probably more saturated colours, Endler 1990), and the location of the peaks suggests they are purplish (red+blue) whilst non-preferred males are 'UV+green'. (Data replotted with permission from Bennett et al. (1997) PNAS 94: 8618-8621. Copyright (1997) National Academy of Sciences USA).*

significantly different from each other; that is, removal of UV changed the criteria for choice. Bennett et al. (1997) showed that under UV+ (full avian spectrum) conditions the female preferences correlated with the shape of the reflectance spectra from the males' throat and covert feathers (Fig. 4), regions displayed in mate attraction (Feare 1984; Eens et al. 1990). However, no measured variable, spectrophotometric or morphometric, predicted choice under UV- conditions. This experiment therefore shows that natural variation in UV reflectance is important in starling mate assessment and that the prevailing light environment can have profound effects on observed mating preferences. That is, experiments conducted under UV deficient conditions (e.g. under many types of artificial light, or with UV-blocking glass or plastic) can produce repeatable results which are, none-the-less, unrelated to the patterns of choice in daylight. A clear example of this effect is shown by Hunt et al. (1997). Burley's (1982) demonstration of band colour preferences in zebra finches has been replicated under some conditions and not others (Collins & ten Cate 1996). Hunt et al. (1997) found they could replicate Burley et al.'s (1982) results under wide spectrum lighting (UV + human visible), but that the preference disappeared in the absence of UV cues. A subsequent experiment showed that this was not due to a change in the colour of the leg bands, but an effect of UV-blocking filters on the overall appearance of stimulus birds (Hunt et al. 1997). This is an interesting result because it potentially explains the inconsistent results in earlier investigations, where different types of artificial light or one-way glass have been used (Collins & ten Cate 1996; Hunt et al. 1997).

So far, research effort has been directed at determining whether, and in what species, UV information is used in choice. This is natural, as this is the waveband that has been ignored in previous studies of avian coloration. Based on these experiments, one could not claim that there is anything special about UV. Indeed, none of the experiments described above concern 'pure UV' colours and all of the species' UV-reflecting plumage patches are coloured to humans too. This only reinforces the message that one must be wary of applying human colour standards, even when a species is colourful to us. Wave-dependent properties of light do predict that UV may be particularly important for certain types of signalling (Bennett & Cuthill 1994; Andersson 1996, 1999), but fresh experiments, where other wavebands are systematically removed, will be required to investigate whether this is so.

Conclusions

Vision physiologists would not be surprised at experiments indicating that UV can be important in avian signalling and mate choice. The different design of the avian retina, with its four types of single cone, one of them maximally sensitive in the violet or ultraviolet, would suggest that many behaviours mediated by colour vision could be influenced by the ultraviolet reflectance of the object and/or its background. When the effect of removing UV on colour-based behavioural decisions has been investigated, such as in foraging or signalling tasks, positive influences have been found. The obvious exceptions (e.g. Koivula et al. 1997) have concerned nocturnal species, which may not possess UV-sensitive cones (Bowmaker & Martin 1978) and, *a priori*, are unlikely to rely on colour vision when hunting at night.

It is perhaps not that surprising that UV has been implicated in mate choice in species, such as the blue tit, which are highly UV reflective and hard for humans to sex visually (Andersson et al. 1997; Hunt et al. 1997). But it is notable that in several studies, blocking the UV has affected sexual preferences even in species which are clearly colourful and dimorphic in the human-visible spectrum, such as the zebra finch (Bennett et al. 1996) and bluethroat (Andersson & Amundsen 1997). This is a reminder that we should not think of UV as a colour, because it is the relative amounts of different wavelengths (UV and human-visible) that a visual system interprets as a colour. Burkhardt (1996) has catalogued examples of red, yellow, green, blue, black and white feathers which vary in their UV reflectance, and thus, whilst classed as similar colours by humans, will vary in hue to a bird. Even within the human-visible waveband, because the λ_{max} values and spectral overlap of the avian cones differ from those of humans, the discriminability of different coloured objects will differ between humans and birds (Vorobyev et al. 1998). This is not to say that many dull cryptic species will prove to be spectacularly coloured in the UV. In an attempt to address this question, Vorobyev et al. (1998) modelled a range of species' plumage reflectances in human and avian colour space. For some species, human and avian 'colourfulness' correlated well, in others it did not. Analogously, spectral analysis of cryptic caterpillars on oak (*Quercus robor*) leaves showed most to be cryptic in the UV as well as the green (i.e. the caterpillars matched their leaf background, Church et al. 1998a). However, some 'green' caterpillars are conspicuous in the UV (Church et al.

1998a), and so too may be some bird species (Burkhardt 1989; Finger et al. 1992; Bennett & Cuthill 1994; Andersson 1996, 1999). It is also clear that within some species there are colour differences which have gone unnoticed by humans (Andersson et al. 1998; Hunt et al. 1998b; Cuthill et al. 1999a). This implies that any comparative analysis which uses human judgements to score sexual dichromatism will underestimate, and perhaps distort, the distribution of the trait. It also suggests that racial and sub-species differences may have gone unnoticed, a significant consideration for taxonomic and conservation purposes.

The extent to which sensory systems, and the neural and cognitive machinery behind them, direct, bias, or constrain signal evolution is one of the most active areas in evolutionary and behavioural ecology (Endler & McLellan 1988; Guilford & Dawkins 1991; Endler 1992a, b, 1993; Ryan & Keddy-Hector 1992; Ryan & Rand 1993; Basolo & Endler 1995; Endler & Basolo 1998). Because we humans are primarily visual creatures, it is visual signals, such as the spectacular colours and ornaments of birds, which attract the attention of humans. Yet to study these traits from the narrow perspective of human colour vision is to miss the true diversity of colours seen through avian eyes and is, in fact, erroneous. By greater interaction between visual physiologists and behavioural ecologists, not only will a better understanding of animal coloration be gleaned, but also, perhaps, an insight into why different colour vision systems have evolved in the first place.

Acknowledgments

I.C.C. would particularly like to the thank the Royal Norwegian Society of Sciences and Letters, the Foundation, for the opportunity to attend the 5th International Kongsvoll symposium. And Yngve Espmark, Trond Amundsen, Gunilla Rosenqvist, and Turid Fredagsvik, for organising the excellent and stimulating meeting which spawned this paper. Our research has been supported by grants from the BBSRC, NERC, the Royal Society and the Nuffield Foundation. We are grateful to other members of the Bristol vision group for their contributions, notably Stuart Church, Sarah Hunt and Nathan Hart, and to Daniel Osorio, Misha Vorobyev and Staffan Andersson for discussion and access to unpublished material.

References

Andersson, M. 1994. *Sexual Selection.* Princeton, NJ: Princeton University Press.

Andersson, S. 1996. Bright ultraviolet coloration in the Asian whistling-thrushes (*Myiophonus* spp.). *Proceedings of the Royal Society of London, Series B,* **263**, 843-848.

Andersson, S. 1999. Morphology of UV reflectance in a whistling-thrush: implications for the study of structural colour signalling in birds. *Journal of Avian Biology,* **30**, 193-204.

Andersson, S. & Amundsen, T. 1997. Ultraviolet colour vision and ornamentation in bluethroats. *Proceedings of the Royal Society of London, Series B,* **264**, 1587-1591.

Andersson, S., Örnborg, J. & Andersson, M. 1998. Ultraviolet sexual dimorphism and assortative mating in blue tits. *Proceedings of the Royal Society of London, Series B,* **265**, 445-450.

Basolo, A. L. & Endler, J. A. 1995. Sensory biases and the evolution of sensory systems. *Trends in Ecology and Evolution,* **10**, 489.

Bennett, A. T. D. & Cuthill, I. C. 1994. Ultraviolet vision in birds: what is its function? *Vision Research,* **34**, 1471-1478.

Bennett, A. T. D., Cuthill, I. C. & Norris, K. J. 1994. Sexual selection and the mismeasure of color. *American Naturalist,* **144**, 848-860.

Bennett, A. T. D., Cuthill, I. C., Partridge, J. C. & Lunau, K. 1997. Ultraviolet plumage colors predict mate preferences in starlings. *Proceedings of the National Academy of Sciences, USA,* **94**, 8618-8621.

Bennett, A. T. D., Cuthill, I. C., Partridge, J. C. & Maier, E. J. 1996. Ultraviolet vision and mate choice in zebra finches. *Nature,* **380**, 433-435.

Bowmaker, J. K. 1991a. The evolution of vertebrate visual pigments and photoreceptors. In: *Vision and Visual Dysfunction*, Vol. 2 (Ed. by J. R. Cronly-Dillon & R. L. Gregory), pp. 63-81. Boston: CRC Press Inc.

Bowmaker, J. K. 1991b. Visual pigments, oil droplets and visual photoreceptors, In: *Vision and Visual Dysfunction,* Vol. 6 (Ed. by P. Gouras), pp. 108-127. Boston: CRC Press Inc.

Bowmaker, J. K. 1998. Evolution of colour vision in vertebrates. *Eye,* **12**, 541-547.

Bowmaker, J. K., Heath, L. A., Wilkie, S. E. & Hunt, D. M. 1997. Visual pigments and oil droplets from six classes of photoreceptor in the retinas of birds. *Vision Research,* **37**, 2183-2194.

Bowmaker, J. K. & Hunt, D. M. 1999. Molecular biology of photoreceptor spectral sensitivity. In: *Adaptive Mechanisms in the Ecology of Vision* (Ed. by S. N. Archer, M. B. A. Djamgoz, E. R. Loew, J. C. Partridge & S. Valerga), pp. 439-464. London: Chapman & Hall.

Bowmaker, J. K., Kovach, J. K., Whitmore, A. V. & Loew, E. R. 1993. Visual pigments and oil droplets in genetically manipulated and carotenoid deprived quail: a microspectrophotometric study. *Vision Research,* **33**, 571-578.

Bowmaker, J. K. & Martin, G. R. 1978. Visual pigments and colour vision in a nocturnal bird, *Strix aluco* (Tawny Owl). *Vision Research,* **18**, 1125-1130.

Bowmaker, J. K. & Martin, G. R. 1984. Colour vision in the penguin, *Spheniscus humboldti*: a microspectrophotometric study. *Vision Research,* **24**, 1702-1702.

Bowmaker, J. K., Wilkie, S. E. & Hunt, D. M. 1999. Photoreceptors and molecular genetics of visual pigments. In: *Proceedings of the 22nd Ornithological Congress*, (Ed. by N. Adams & R. Slotow). Pp. 2729-2742. Johannesburg: Birdlife South Africa.

Burkhardt, D. 1989. UV vision: a bird's eye view of feathers. *Journal of Comparative Physiology, Series A,* **164**, 787-796.

Burkhardt, D. 1996. Ultraviolet perception by bird eyes and some implications. *Naturwissenschaften,* **83**, 492-497.

Burley, N., Krantzberg, G. & Radman, P. 1982. Influence of colour-banding on the conspecific preferences of zebra finches. *Animal Behaviour,* **30**, 444-455.

Burley, N. T., Enstrom, D. A. & Chitwood, L. 1994. Extra-pair relations in zebra finches: differential male success results from female tactics. *Animal Behaviour,* **48**, 1031-1041.

Burley, N. T., Parker, P. G. & Lundy, K. 1996. Sexual selection and extrapair fertilization in a socially monogamous passerine, the zebra finch (*Taeniopygia guttata*). *Behavioral Ecology,* **7**, 218-226.

Chittka, L., Shmida, A., Troje, N., & Menzel, R. 1994. Ultraviolet as a component of flower reflections, and the colour perception of Hymenoptera. *Vision Research,* **34**, 1489-1508.

Church, S. C., Bennett, A. T. D., Cuthill, I. C., Hunt, S., Hart, N. S. & Partridge, J. C. 1998a. Does lepidopteran larval crypsis extend into the ultraviolet? *Naturwissenchaften,* **85**, 189-192.

Church, S. C., Bennett, A. T. D., Cuthill, I. C. & Partridge, J. C. 1998b. Ultraviolet cues affect the foraging behaviour of blue tits. *Proceedings of the Royal Society of London, Series B,* **265**, 1509-1514.

Coemans, M. A. J. M., Vos Hzn, J. J. & Nuboer, J. F. W. 1994. The relation between celestial colour gradients and the position of the sun, with regard to the sun compass. *Vision Research,* **34**, 1461-1470.

Collins, S. A. & ten Cate, C. 1996. Does beak colour affect female preference in zebra finches? *Animal Behaviour,* **52**, 105-112.

Cuthill, I. C. & Bennett, A. T. D. 1993. Mimicry and the eye of the beholder. *Proceedings of the Royal Society of London, Series B,* **253**, 203-204.

Cuthill, I. C., Bennett, A. T. D., Partridge, J. C. & Maier, E. H. 1999a. Plumage reflectance and the objective assessment of avian sexual dichromatism. *American Naturalist*, **153**, 183-200.

Cuthill, I. C., Partridge, J. C. & Bennett, A. T. D. 1999b. UV vision and its functions in birds. In *Proceedings of the 22nd International Ornithological Congress* (Ed. by N. Adams & R. Slotow). Pp. 2743-2758. Johannesburg: Birdlife South Africa.

Darwin, C. 1871. *The Descent of Man and Selection in Relation to Sex*. London: John Murray.

D'Eath, R. B. 1998. Can video images simulate real stimuli in animal behaviour experiments? *Biological Reviews,* **73**, 267-292.

Eens, M., Pinxten, R. & Verheyen, R. F. 1990. On the function of singing and wing-waving in the European starling *Sturnus vulgaris*. *Bird Study,* **37**, 48-52.

Endler, J. A. 1990. On the measurement and classification of colour in studies of animal colour patterns. *Biological Journal of the Linnean Society,* **41**, 315-352.

Endler, J. A. 1992a. Sensory drive: does sensory biology bias or constrain the direction of evolution? Introduction to the symposium. *American Naturalist.* **139**, S 1-S 3.

Endler, J. A. 1992b. Signals, signal conditions, and the direction of evolution. *American Naturalist,* **139**, S 125-S 153.

Endler, J. A. 1993. Some general comments on the evolution and design of animal communication systems. *Philosophical Transactions of the Royal Society of London,* **340**, 215-225.

Endler, J. A. & Basolo, A. L. 1998. Sensory ecology, receiver biases and sexual selection. *Trends in Ecology and Evolution,* **13**, 415-420.

Endler, J. A. & McLellan, T. 1988. The processes of evolution - toward a newer synthesis. *Annual Review of Ecology and Systematics,* **19**, 395-421.

Feare, C. 1984. *The Starling*. Oxford: Oxford University Press.

Finger, E. 1989. *Spektrale Reflexion von Vogelgefieder*. Regensburg: Unpublished Diploma thesis, University of Regensburg.

Finger, E. 1995. Visible and UV coloration in birds: Mie scattering as the basis of color in many bird feathers. *Naturwissenschaften,* **82**, 570-573.

Finger, E. & Burkhardt, D. 1994. Biological aspects of bird coloration and avian colour vision including ultraviolet range. *Vision Research,* **34**, 1509-1514.

Finger, E., Burkhardt, D. & Dyck, J. 1992. Avian plumage colors. Origin of UV reflection in a black parrot. *Naturwissenschaften*, **79**, 187-188.

Fleishman, L. J., McClintock, W. J., D'Eath, R. B., Brainard, D. M. & Endler, J. A. 1998. Colour perception and the use of video playback experiments in animal behaviour. *Animal Behaviour,* **56**, 1035-1040.

Goldsmith, T. H. 1990. Optimization, constraint, and history in the evolution of eyes. *Quarterly Review of Biology,* **65**, 281-322.

Goldsmith, T. H. 1994. Ultraviolet receptors and color vision: evolutionary implications and a dissonance of paradigms. *Vision Research,* **34**, 1479-1487.

Goldsmith, T. H. & Collins, J. S. 1982. Cone oil droplets of bird retinas. *Investigative Ophthalmology and Visual Science,* **22**, 189.

Goldsmith, T. H., Collins, J. S. & Licht, S. 1984. The cone oil droplets of avian retinas. *Vision Research,* **24**, 1661-1671.

Guilford, T. & Dawkins, M. S. 1991. Receiver psychology and the evolution of animal signals. *Animal Behaviour,* **42**, 1-14.

Hart, N. S., Partridge, J. C. & Cuthill, I. C. 1998. Visual pigments, oil droplets and cone photoreceptor distribution in the European starling (*Sturnus vulgaris*). *Journal of Experimental Biology,* **201**, 1433-1446.

Hart, N. S., Partridge, J. C. & Cuthill, I. C. 1999. Visual pigments, come oil droplets, ocular media and predicted spectral sensivity in the domestic turkey (*Meleagris gallopavo*). *Vision Research,* **39**, 3321-3328.

Hunt, D. M., Dulai, K. S., Cowing, J. A., Julliot, C., Mollon, J. D., Bowmaker, J. K., Li, W. H. & Hewett-Emmett, D. 1998a. Molecular evolution of trichromacy in primates. *Vision Research,* **38**, 3299-3306.

Hunt, S., Bennett, A. T. D., Cuthill, I. C. & Griffiths, R. 1998b. Blue tits are ultraviolet tits. *Proceedings of the Royal Society of London, Series B,* **265**, 451-455.

Hunt, S., Cuthill, I. C., Swaddle, J. P. & Bennett, A. T. D. 1997. Ultraviolet vision and band colour preferences in female zebra finches, *Taeniopygia guttata. Animal Behaviour,* **54**, 1383-1392.

Huth, H. H. & Burkhardt, D. 1972. Der Spektrale sehbereich eines Violetta Kolibris. *Naturwissenschaften,* **59**, 650.

Jacobs, G. H. 1981. *Comparative Color Vision*. New York: Academic Press.

Jacobs, G. H. 1992. Ultraviolet vision in vertebrates. *American Zoologist,* **32**, 544-554.

Jacobs, G. H. 1993. The distribution and nature of colour vision among the mammals. *Biological Reviews,* **68**, 413-471.

Jacobs, G. H. 1996. Primate photopigments and primate color vision. *Proceedings of the National Academy of Sciences, USA,* **93**, 577-581.

Jacobs, G. H. & Deegan, J. F. 1994. Sensitivity to ultraviolet light in the gerbil (*Meriones unguiculatus*): characteristics and mechanisms. *Vision Research,* **34**, 1433-1441.

Jacobs, G. H., Neitz, J. & Deegan, J. F. 1991. Retinal receptors in rodents maximally sensitive to ultraviolet light. *Nature,* **353**, 655-656.

Jacobs, G. H., Neitz, M., Deegan, J. F. & Neitz, J. 1996. Trichromatic color vision in New World monkeys. *Nature,* **382**, 156-158.

Jane, S. D. & Bowmaker, J. K. 1988. Tetrachromatic colour vision in the duck (*Anas platyrhynchos* L.): microspectrophotometry of visual pigments and oil droplets. *Journal of Comparative Physiology, Series A,* **162**, 225-235.

Johnsen, A., Andersson, S., Örnborg, J., & Lifjeld, J. T. 1998. Ultraviolet plumage ornamentation affects social mate choice and sperm competition in bluethroats (Aves: *Luscinia s. svecica*): a field experiment. *Proceedings of the Royal Society of London, Series B,* **265**, 1313-1318.

Kawamura, S. & Yokoyama, S. 1996. Phylogenetic relationships among short wavelength sensitive opsins of American chameleon (*Anolis carolinensis*) and other vertebrates. *Vision Research,* **36**, 2797-2804.

Kawamuro, K., Irie, T. & Nakamura, T. 1997. Filtering effect of cone oil droplets detected in the P-III response spectra of Japanese quail. *Vision Research,* **37**, 2829-2834.

Koivula, M., Korpimäki, E. & Viitala, J. 1997. Do Tengmalm's owls see vole scent marks visible in ultraviolet light? *Animal Behaviour,* **54**, 873-877.

Liebman, P. A. & Granda, A. M. 1971. Microspectrophotometric measurements of visual pigments in two species of turtle *Pseudemys scripta* and *Chelonia mydas*. *Vision Research,* **11**, 105-114.

Maier, E. J. 1993. To deal with the invisible: on the biological significance of ultraviolet sensitivity in birds. *Naturwissenschaften,* **80**, 476-478.

Maier, E. J. & Bowmaker, J. K. 1993. Colour vision in the passeriform bird, *Leiothrix lutea*: correlation of visual pigment absorbency and oil droplet transmission with spectral sensitivity. *Journal of Comparative Physiology, Series A,* **172**, 295-301.

Mollon, J. D. 1989. "tho' she kneel'd in the place where they grew…". The uses and origins of primate colour vision. *Journal of Experimental Biology,* **146**, 21-38.

Neumeyer, C. 1992. Tetrachromatic color vision in goldfish: evidence from color mixture experiments. *Journal of Comparative Physiology, Series A,* **171**, 639-649.

Ohtsuka, T. 1985. Relation of spectral types to oil droplets in cones of turtle retina. *Science,* **229**, 874-877.

Osorio, D., Marshall, N. J. & Cronin, T. W. 1997. Stomatopod photoreceptor tuning as an adaptation to colour constancy underwater. *Vision Research,* **37**, 3299-3309.

Osorio, D. & Vorobyev, M. 1996. Colour vision as an adaptation to frugivory in primates. *Proceedings of the Royal Society of London, Series B,* **263**, 593-599.

Partridge, J. C. 1989. The visual ecology of avian cone oil droplets. *Journal of Comparative Physiology, Series A,* **165**, 415-426.

Radwan, J. 1993. Are dull birds still dull in UV? *Acta Ornithologica,* **27**, 125-130.

Ryan, M. J. & Keddy-Hector, A. 1992. Directional patterns of female mate choice and the role of sensory biases. *American Naturalist,* **139**, 4-35.

Ryan, M. J. & Rand, A. S. 1993. Sexual selection and signal evolution: the ghost of biases past. *Philosophical Transactions of the Royal Society of London, Series B,* **340**, 187-195.

Starace, D. M. & Knox, B. E. 1998. Cloning and expression of a *Xenopus* short wavelength cone pigment. *Experimental Eye Research,* **67**, 209-220.

Swaddle, J. P. 1996. Reproductive success and symmetry in zebra finches. *Animal Behaviour,* **51**, 203-210.

Swaddle, J. P. & Cuthill, I. C. 1994a. Female zebra finches prefer males with symmetrical chest plumage. *Proceedings of the Royal Society of London, Series B,* **258**, 267-271.

Swaddle, J. P. & Cuthill, I. C. 1994b. Preference for symmetrical males by female zebra finches. *Nature,* **367**, 165-166.

Thompson, E., Palacios, A. & Varela, F. J. 1992. Ways of coloring: comparative color vision as a case study for cognitive science. *Behavioral and Brain Sciences,* **15**, 1-74.

Tovee, M. J. 1995. Ultraviolet photoreceptors in the animal kingdom: their distribution and function. *Trends in Ecology and Evolution,* **10**, 455-460.

Viitala, J., Korpimäki, E., Palokangas, P. & Koivula, M. 1995. Attraction of kestrels to vole scent marks visible in ultraviolet light. *Nature,* **373**, 425-427.

Vorobyev, M. & Osorio, D. 1998. Receptor noise as a determinant of colour thresholds. *Proceedings of the Royal Society of London, Series B,* **265**, 351-358.

Vorobyev, M., Osorio, D., Bennett, A. T. D., Marshall, N. J. & Cuthill, I. C. 1998. Tetrachromacy, oil droplets and bird plumage colours. *Journal of Comparative Physiology, Series A,* **183**, 621-633.

Vos Hzn, J. J., Coemans, M. A. J. M. & Nuboer, J. F. W. 1994. The photopic sensitivity of the yellow field of the pigeon's retina to ultraviolet light. *Vision Research,* **34**, 1419-1425.

Walls, G. L. 1963. *The Vertebrate Eye and its Adaptive Radiation*. New York: Hafner.

Wilkie, S. E., Vissers, P. M. A. M., Das, D., DeGrip, W. J., Bowmaker, J. K. & Hunt, D. M. 1998. The molecular basis for UV vision in birds: spectral characteristics, cDNA sequence and retinal localization of the UV-sensitive visual pigment of the budgerigar (*Melopsittacus undulatus*). *Biochemical Journal,* **330**, 541-547.

Wright, A. A. 1972. Psychometric and psychophysical hue discrimination functions for the pigeon. *Vision Research,* **12**, 1447-1464.

Wright, M. W. & Bowmaker, J. K. 1998. Visual pigments and oil droplets in the retinae of palaeognathus birds. *Investigative Ophthalmology and Visual Science,* **39**, S1059.

Wyszecki, G. & Stiles, W. S. 1982. *Color Science, Concepts and Methods, Quantitative Data and Formulae*. New York: John Wiley.

Yokoyama, S., Radlwimmer, F. B. & Kawamura, S. 1998. Regeneration of ultraviolet pigments of vertebrates. *FEBS Letters,* **423**, 155-158.

Zann, R. 1996. *The Zebra Finch: A Synthesis of Field and Laboratory Studies*. Oxford: Oxford University Press.

IN: ESPMARK,Y., AMUNDSEN,T. & ROSENQVIST,G. (EDS.) 2000
ANIMAL SIGNALS:
SIGNALLING AND SIGNAL DESIGN IN ANIMAL COMMUNICATION, PP. 83-120
TAPIR ACADEMIC PRESS, TRONDHEIM, NORWAY

4

The Visual Ecology of Reef Fish Colours

N. JUSTIN MARSHALL

Vision Touch and Hearing Research Centre, The University of Queensland,
Brisbane, Queensland 4072, Australia

E-mail: j.marshall@vthrc.uq.edu.au

Abstract

In an attempt to understand the uses of colour on coral reefs, the colours of over 200 species of reef fish from the Pacific and the Caribbean have been measured using 'Sub-Spec', a submersible spectroradiometer. Light from a variety of fish microhabitats has also been characterised. Such non-subjective quantification of reef fish colours and their environment is an essential first step towards understanding colour vision, colour communication and the functions of colour on the reef. Possibilities emerging from this data include the following: (a) Colours and colour patterns both within a fish and between a fish and its surroundings are often highly contrasting. Reasons for this may be sexual display, camouflage and apposematism. (b) Many fish colours may include a UV or a far red component. UV body regions are often selectively shown in certain circumstances. Most reef fish however are UV blind. (c) Colours can be broadly categorised into around 21 major classes but their position in the spectrum is variable.Using available data, colours and colour patterns will be discussed in the context of visual ecology.

A coral reef is perhaps the most colourful of all the world's ecosystems. During the day it throngs with multicoloured fish and invertebrates, most of which are communicating with each other almost incessantly. The only human equivalent is a football match or a stock exchange; however, where we have to shout and gesticulate to be taken notice of in a crowd, reef fish have evolved a far more elegantly tuned channel to communicate with, that of colour. This language of colour on the reef has engendered much supposition but, certainly compared to freshwater fish, has been the subject of surprisingly few studies. As a result, why reef fish are colourful remains a largely unanswered question.

Problems exist on two levels with those asking this question. Firstly, as frequently stated but often ignored, there is likely to be more than one answer (for a good mini-review of this see Helfman et al. 1997, Box 21. 1). Unfortunately a single grand answer is often sought for reef fish or large sections of them, perhaps because they all appear at one locality. On the Great Barrier Reef alone there are well over 2000 species living in hundreds of habitats and with many different life strategies. It is almost inconceivable for only one evolutionary force to be behind the colours of such a diverse assemblage, the final answer most probably being a compromise of many (Endler 1991).

Secondly, as humans are primarily visual animals we tend to assume that what we see is the way it is. However it is certain that no reef fish sees itself, or its neighbours, the way we do. This has led to a variety of misconceptions, as it has in the avian literature (Bennett et al. 1994), in spite of over 20 years of caveats from a few (e.g. Hailman 1977; Lythgoe 1979; Endler 1990, 1993). In order to prevent such mistakes, a non-subjective measure of colour is essential and one of the purposes of this work is to present such data from reef fish for the first time.

Reasons for colours in any group of animals fall into three distinct but not necessarily mutually exclusive categories. They may be non-adaptive, a by-product of some other aspect of the organism's life. Although this is often dismissed as un-Darwinian, it is a very real possibility for many marine animals which live in light restricted habitats where colours or components of colours do not appear as they do at the surface. Several surface dwelling fish are also coloured to prevent sun damage (e.g. Lowe & Goodman-Lowe 1996). Secondly, colours, including the 'bright' patterns of reef fish, may be for camouflage (Cott 1940). Lastly, they may be adapted to convey information, a signal to con-specific, predator or prey (e.g. Fine et al. 1977; Andersson 1994).

To examine these possibilities this chapter begins with a brief review of past literature on reef fish and then presents some new data which fall into three main categories: colour reflection measurements from 200 species of reef fish, initial estimates of visual performance in 182 species and some measurement of light at the microhabitat level (U. Siebeck & N. J. Marshall, unpublished; see Marshall 1996 for some methods). These three blocks of data are part of a four-way approach to understanding vision from another animals perspective. The four necessary components to achieve this are: spectral sensitivities of the colour vision system, non-subjective colour measurement, microhabitat light environment measurements and, critically, the behavioural use of colour (Hailman 1977; Lythgoe 1979).

A brief look at some past literature

Almost everyone's first visit to a coral reef results in an expression of astonishment at all the colours and the number of different fish. As Konrad Lorenz put it: 'There is in all the world, no other biotope which has produced, in so short a time or, which means the same thing, in so closely allied groups of animals, an equal number of extremely specialized forms.' (Lorenz 1962). In other words, there is nowhere on earth where one can find so many colourful animals packed into such a small space. Tropical birds such as the parrots are perhaps the nearest rival group of colourful animals on land, but the diversity of their colours and the fact that the species are often spread over a wide area makes them no match for the stunning splendour of a reef.

Lorenz was one of the first to try and explain the wonderful diversity of 'poster' colours observable in reef fish and suggested that their colours 'serve the ecological need for 'spacing out' the individuals of each species while preventing unnecessary aggression'. This is two related ideas in one sentence. Firstly, colours are used in 'flagging' territories and setting up of resource boundaries between fish. Secondly, the colours 'release' or prevent aggressive interactions through signaling intent (Lorenz 1962). Though a good first attempt, this is now recognised as an oversimplification (Brockman 1973; Ehrlich et al. 1977). Most butterflyfish (Chaetodontidae), for instance, are non-territorial during the day but are very boldly coloured (Neudecker 1989).

Highlighting weapons with colours is frequently seen in the animal kingdom as a part of aggressive displays (Lorenz 1962). A good example in reef fish are the caudal 'knives' of the acanthurids (surgeonfishes), a generally aggressive family (Lorenz 1962; Barlow 1974). Other examples on the reef include the distinctively marked dorsal fin rays of some Scorpaenidae (the lionfish) and coloured teeth and other mouth tissues in labroids (sensu Greenwood et al. 1966 - the wrasse and parrotfish) and some haemulids (sweetlips or grunts).

A central tenet of the flag hypothesis presumes reef fish do not change colour. In fact most reef fish are capable of colour change on one or several time scales; within seconds, diurnally or ontogenetically (Townsend 1929; Crook 1997). Many reef fish, as with their freshwater counterparts (Barlow 1963; Neil 1964; Leong 1969), use colour changes as part of their signaling repertoire (e.g. Townsend 1929; Ehrlich et al. 1977; Booth 1990; Dawkins & Guildford 1994; Crook 1997). However, the idea that part of the job of reef fish colours is for maintaining territory remains sound and may even rely on colour change (Burkhardt 1967; Brockmann 1973; Myreberg & Thresher 1974).

A related hypothesis suggests that the diversity of colours is for species recognition (Breder 1959; Zumpe 1965; Fricke 1966; Reese 1975). The large number of reef fish species is almost certainly due to the abundant resources, both in terms of food and living spaces. Each niche, the reef top, the reef slope, under coral ledges, or midwater, for example, can support many animals (Lorenz 1962; Randall 1967). Within this diverse assemblage, how do fish recognise the right species to mate with, which are worth competing with for food and which to escape from? Colour coding, or the species recognition hypothesis (Zumpe 1965; Fricke 1966; Reese 1975; Harrington 1993), assigns each fish its own 'professional dress' (Burkhardt et al. 1967) and is a plausible function of colour, particularly among sexually monomorphic species (Baylis 1979; Rowland 1979; Thresher & Moyer 1983).

Sex recognition is an important function of signals in many animals (Verrell 1991; Pfennig 1998) and following this, choosing a partner or sexual selection has proved a fruitful area of study in recent years (reviews in Ryan 1990; Andersson 1994; Hauser 1997; Bradbury & Vehrencamp 1998). It is interesting that, with some spectacular exceptions, like the ontogenetic sex changers (Fricke 1973; Fricke & Fricke 1977; Thresher 1984), many reef fish are monomorphic (for colour pattern at least). Whether this rules out sexual selection by vision in these species is debatable and certainly merits further investigation.

Parrotfish provide a particularly striking example of sexual dimorphism in reef fish with the terminal phase male (once a female) possessing a beautiful livery of colours and leading a harem of drab females. Robertson and Hoffman (1977) suggested that female choice is responsible for the often more colourful male. In accordance with this, where species of parrotfish possess permanent harems, the sexes have more similar colours (to human eyes) while conspicuous males are found in species which attract roving females. This study also highlights rapid colour change, some male parrotfish rapidly changing from colourful to drab during contact with females and males respectively (see Crook 1997 for a careful study of colour change in juvenile parrotfish and Townsend 1929 for a general account in other reef fish).

The beaugregory damselfish *Eupomacentrus leucosticus* provides an interesting example of sex reversal in conspicuousness, the females, non-breeding males and juveniles being bright blue and yellow while the males are a drab brown. The reason for this is likely to be that breeding males guard the nest and offspring and therefore do not want to attract attention. The other morphs all guard food territories and may 'flag' these with their colours (Brockman 1973; Barlow 1974). Many juvenile and adult reef fish possess strikingly different colour patterns. This is also probably connected with territory and food resources; their different colours enabling the juvenile to coexist in the same territory as the adult as its juvenile colours do not label it as a threat (Neal 1993; Mahon 1994).

Individual recognition has also been suggested as a function for some of the colour markings in fish (Reese 1981). As several coral reef species are territorial and may be relatively long lived (Choat & Bellwood 1991; Shapiro 1991), it is quite likely that retaining a memory of battles won and lost with near neighbours would be of value.

Camouflage is perhaps the last function of colour which comes to mind when one sees the brilliantly coloured 'coffee-table-book' close-up photographs of some reef fish. Almost all possess some degree of countershading, an adaptation to render animals evenly lit in environments illuminated predominantly from above (Cott 1940). This is good evidence that, no matter how conspicuous, all reef fish are still concerned about being camouflaged from predators (Hamilton & Peterman 1971; Fine et al. 1977; see Cott 1940 for a good description of countershading in a variety of animals). This style of camouflage is likely to be most effective at a distance, where the bright colour patterns have been attenuated by the intervening water, but the general outline of the fish remains.

Bold body patterns in reef fish have for some time been sited as a good example of another form of crypsis, disruptive camouflage (Longley 1917; Cott 1940; Barlow 1967, 1972, 1991; Ehrlich et al. 1977). Here, the contrasting pattern, often combined with unusual body shape, serve to break up the outline of the fish against its also highly contrasting background (Hamilton & Peterman 1971; Neudecker 1989).

Colours of reef fish may be used in a variety of 'cooperational' interactions with other fish, the quotes to indicate that the dynamics of the interactions are not always clear (e. g. Losey 1979, 1987). High contrast markings, especially on the head and tail, are suggested to help coordinate schooling (Hemmings 1966) and colour changes have been implicated in maintaining school integrity (Topp 1970). They may also help reef fish which form pairs to find each other in the complex reef background (Barlow 1991).

By far the most attention in cooperative relationships has gone to interspecific cleaning. Examples of this remarkable relationship are present in all the oceans and also in many freshwater systems (Potts 1973; Fine et al. 1977; Losey 1979, 1987, 1993; Grutter 1999). Cleaner fish come from polyphyletic origins and, although the details of the cleaner-host relationship may vary, the cleaner invariably removes parasites and dead skin from the host, may advertise this service, and may be solicited to clean by the hosts (Losey 1987). Although there is debate as to the details of this relationship (Losey 1979, 1993) two things are clear. Firstly, initiation of cleaning involves a highly developed form of signaling between cleaner and host. Secondly, the conspicuous markings of unrelated cleaners from different oceans are often very similar. *Gobiosoma oceanops* from the Caribbean and the Pacific species *Labroides dimidiatus*, for instance, are both longitudinally striped with black and a brilliant blue. As a result, it has been suggested that this colour pattern is a 'guild marking' (Eibl-Eibesfeldt 1955) advertising the same service.

New data from the coral reef

The colours of reef fish

The birds are the most frequently studied animal group with regard to colour (e. g. Burkhardt & Finger 1991; Finger et al. 1992; Bennett et al. 1997; Andersson & Amundsen 1997; Cuthill et al., this volume) and flowers have the best understood colour signals in terms of receiver physiology (Kevan 1972; Chittka & Menzel 1992; Chittka 1996; Dyer 1998). Colour design of flowers and birds are therefore cited for comparison with reef fish.

To describe colours, three parameters are generally used: hue, chroma (saturation) and brightness (Wyszecki & Styles 1982; Endler 1990, 1991). Each colour can then be ascribed a position in a three-dimensional space with each of the parameters as one of the axes (Hailman 1979). Although seeming desirable in principle, in practice there are problems associated with such analysis (Hailman 1979; Endler 1990). Therefore, I restrict the description of colour to a largely qualitative one for the time being. A simple model of quantitative differences in colour signals between human and reef fish visual systems is presented below.

Over 1000 different areas of fish skin in 200 individual species from 36 families have been characterised (Fig. 1 and Appendix). As with similar studies of flower colours (Chittka & Menzel 1992) and birds (Burkhardt and Finger 1991), some trends are notable; these are detailed below.

a) Reef fish colours can be subjectively divided into 21 types. The frequency and distribution of occurrence of these colours is detailed in Appendix and illustrated in Figure 1. The colours of any fish listed in the Appendix can be approximately reconstructed by cross reference to this figure (see, however, caveat in (c) below). Some colour categories which occur infrequently are not included here. Actual colours of all fish are available upon request to the author.

   It is vital to note that this is an entirely human visual system based subjective categorisation, based on what the colours look like to us, and the regions in the spectrum where they reflect. Their appearance to the fish is unknown. Previously, attempts have been made to give flower colours names based on how bees might see them (Chittka et al. 1994) and bird colours names relating to bird visual systems (Finger & Burkhardt 1994). Not enough is known about reef fish visual systems to attempt this yet and what evidence there is suggests it is likely to be different for each species (Bowmaker 1990; Partridge 1990). These categories are therefore useful only in identifying and discussing the different 'colour' types.

b) The colours can be divided into two broad categories: simple and complex (Appendix). Simple colours reflect strongly in one region of the spectrum only. Examples of these are short wavelength ultraviolet (UV), blue and green which are peaked or Gaussian-like in shape and yellow, and orange and red which are cut-on step-like functions (Fig. 1). Some complex colours are additive mixtures of these and are therefore bi-phasic, for example UV/yellow, UV/orange and UV/red. The Labroid Purple Complex colours contain two-four peaks and/or a cut-on step at 700 nm. Black, brown and light 'violet' are not assigned to either of these groups.

c) Highlighting one of the problems of categorising colours, within any of the above colour groups of step or peak functions actual spectra may lie up to 20 nm either side of the step or peak. In complex colours this may occur in either direction for any peak or step.

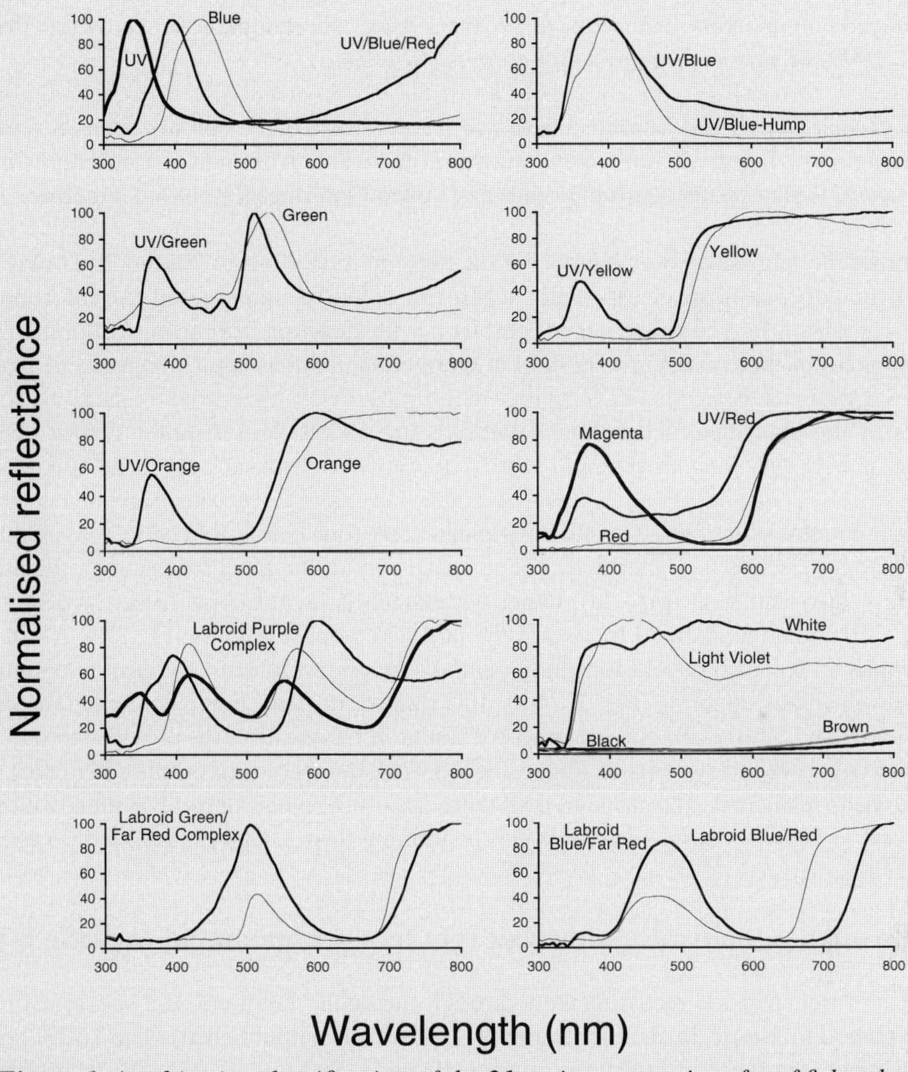

Figure 1. *A subjective classification of the 21 major categories of reef fish colours. Each panel contains two-four examples of fish reflectance spectra. The three panels containing labroid examples are counted as one category, each panel contains colours which form a complex all of which look similar to humans. With the exception of black and brown, the spectra have been normalised to emphasize the differences in hue. Colours are assigned a name based on the appearance of the colour to humans and the regions in the spectrum where the colour reflects. By cross reference with the Appendix, the colours of any reef fish listed can be approximated using these curves.*

d) As mentioned previously, colours can be quantified by three attributes, their hue, chroma and brightness (e.g. Wyszecki & Styles 1982; Endler 1990). As might be suspected from their appearance to our eyes, reef fish colours are often highly saturated. This may either be the result of very regular layering where colours are produced by interference phenomena (Denton & Land 1971), regular particle size in scatter-produced colours (Fox 1953; Fox & Vevers 1960), coherent particle scattering (Prum et al. 1999) or very pure pigmentary colours (Fox 1953).

e) The complexity of the labroid colours is greater than other colours seen in nature (although see Figure 2 for a new example in birds). Although the mechanism is unknown, it may be the result of combining several interference colours together.

f) Many of the labroid colours have a cut-on step function close to 700nm. This may be of particular significance, especially when the colour appears green, as chlorophyll also contains a far-red reflectance (combined with fluorescence) around 700nm. This component of these colours may help in camouflage (as has been suggested in frogs; see Schwalm & Starrett 1977) against chlorophyll-rich backgrounds such as coral reefs or the algal beds where many labroids are found. This requires colour vision systems with very long wavelength sensitivity, however, and such are as yet unknown.

g) Colour patterns of reef fish are often complementary (one colour reflecting in a region of the spectrum where the other does not and vice versa), the most frequent combination being yellow and blue (Fig. 2a). Other combinations include blue, green and orange (Fig. 2c), and a pair formed by two of the labroid complex colours (Fig. 2e). All these examples in reef fish have astonishing parallels in colourful birds, suggesting common themes in colour signaling (Fig. 2b, d, f). It is easy to see how simple combinations such as yellow and blue or red and green (often found in freshwater fish) provide strong signals to colour vision systems (Lythgoe 1968, 1976), but the more complex examples are yet to be investigated. The success of all these as strong signals depends critically on the spectral sensitivities of the visual systems examining them, and these are hardly known.

Ocular media transmission: most reef fish ocular media exclude UV

Examples of transmission measurements through the ocular media of 182 species of reef fish are plotted in Figure 3a (U. Siebeck & N. J. Marshall, unpublished). The T50% point for each curve, the point of 50% transmission (Douglas & McGuigan 1989), is useful to describe them and varies from 340 nm to 425 nm in ocular media which look clear to humans and extends to 480 nm in the notably yellow tinted eyes of some labrids (wrasse), balistids (triggerfish) and monocanthids (filefish). Only 30% of reef fish examined so far possess ocular media capable of transmitting UV wavelengths below 380nm. This means that many of the UV components of colours (Fig. 1; Appendix) are not visible to most reef fish (for previous data and a comparison with freshwater: see Douglas & McGuigan 1989; Thorpe et al. 1993). Species on the reef that do transmit UV fall into three ecological categories within which there may be exceptions: planktivorous fish (Loew & McFarland 1990), crepuscular or nocturnal fish, and the pomacentrids (damselfish and allies). Yellow tinted eyes in labrids, balistids and monocanthids are caused by yellow corneas, the lenses on their own transmitting well to around 400 nm (Kondrashev et al. 1986; Douglas & Marshall 1999; U. Siebeck & N. J. Marshall, unpublished).

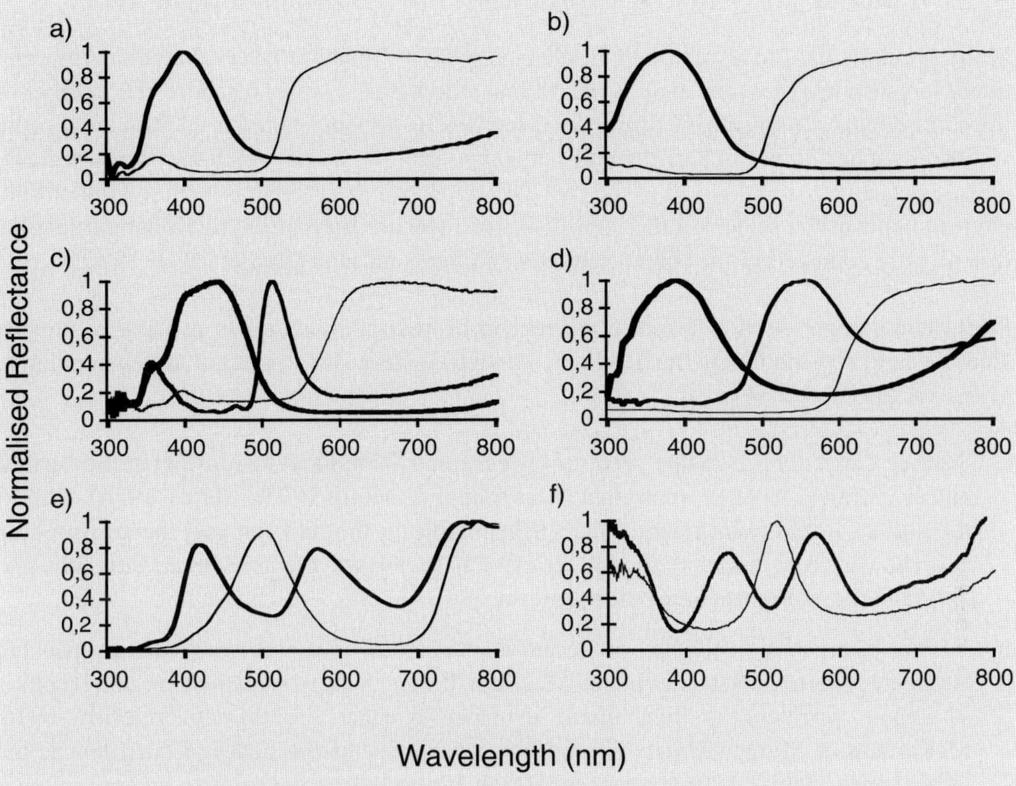

Figure 2. *Complementary colours in reef fish and birds.*

(*a*) *UV/blue hump (thick) and UV/yellow (thin) of the angelfish* Pygoplites diacanthus *appear blue and yellow to human observers.*

(*b*) *UV/blue (thick) and yellow (thin) in the budgerigar* Melopsittacus undulatus.

(*c*) *UV/blue hump (thick), UV/green (medium) and orange (thin) of the wrasse* Coris gaimard.

(*d*) *UV/blue (thick), green(medium) and orange (thin) of the red-capped parrot* Purpureicephalus spurius.

(*e*) *Labroid purple (thick) and green/far red (thin) of the parrotfish* Scarus spinus.

(*f*) *Complex combination interference colours of the king bird of paradise* Cicinnurus regius.

Four or more microhabitat light environments on the reef

Many previous light measurements in the ocean have concentrated on characterising relatively large bodies of water such as reef versus midwater or water offshore at a number of different depths (Jerlov 1976; Smith & Tyler 1976). Though valuable in their own right, such approaches do not quantify small scale spatial fluctuations which may be significant in the three-dimensional world of the reef. My approach therefore has been to look at light environments at the scale of individual fish and measure light at the microhabitat level, a task already achieved in the forest light environments on land (Endler 1993, 1996).

Four main categories of different coral reef light microhabitats taken at various sites on Heron Island (on the Great Barrier Reef, Australia) have been selected as representative of the variations seen (Fig. 3b).

a) Surface - At only 0.5 m this spectrum is essentially identical to sunlight at the surface and is identical to other such data (McFarland & Munz 1975b; Jerlov 1976). This is the sort of light environment which fish inhabiting the reef flat and the surrounding surface waters would experience. It is rich in all wavelengths including the UV, although these drop rapidly below 350 nm.

b) 10 m - Characteristically for moderate depths, both long and short wavelengths are attenuated (Jerlov 1976; Levine & McNicol 1982). A peak transmission at 510 nm is at longer wavelengths than found in more oceanic reef systems (Jerlov 1976; McFarland & Munz 1975b). The relative proximity of the coast of Australia is the likely cause of this, the water around Heron Island falling in the Type 1 Coastal water as categorised by Jerlov (1976), a relatively clear coastal water. It is worth noting here that while plotting light levels on a linear rather than log scale is good for highlighting differences, it tends to over accentuate minima. Even at this depth in this type of water enough UV and red photons exist for visual purposes (Blaxter 1988); however, vision at blue/green wavelengths will be optimal.

c) Long wavelength dominated microhabitats exist and have not been described for coral reefs before. They are found within deep holes or crevices in the reef, the sort of habitat where nocturnal fish often lurk during the day, along with certain retiring diurnal fish. The rapid increase in light level at 700 nm is the result of reflectance from the inner surface of the holes and light that is filtered through coral. It is largely the light reflected or transmitted through the chlorophyll of algae, such as coralline algae or red algea, red sponges and from the zooxanthellar constituents of coral. Some of this peak may also be the result of fluorescent emission by corals and other reef organisms (Mazel & Fux 1997). Orange sponges and bryozoans which frequently encrust cave / crevice habitats may also reflect long wavelengths.

d) The fourth category of light microhabitat has been quantified using the more restricted view of a radiance probe looking straight up. This is judged a robust category for fish living in surface waters feeding against downwelling light. This microhabitat is dominated by UV and blue wavelengths of light.

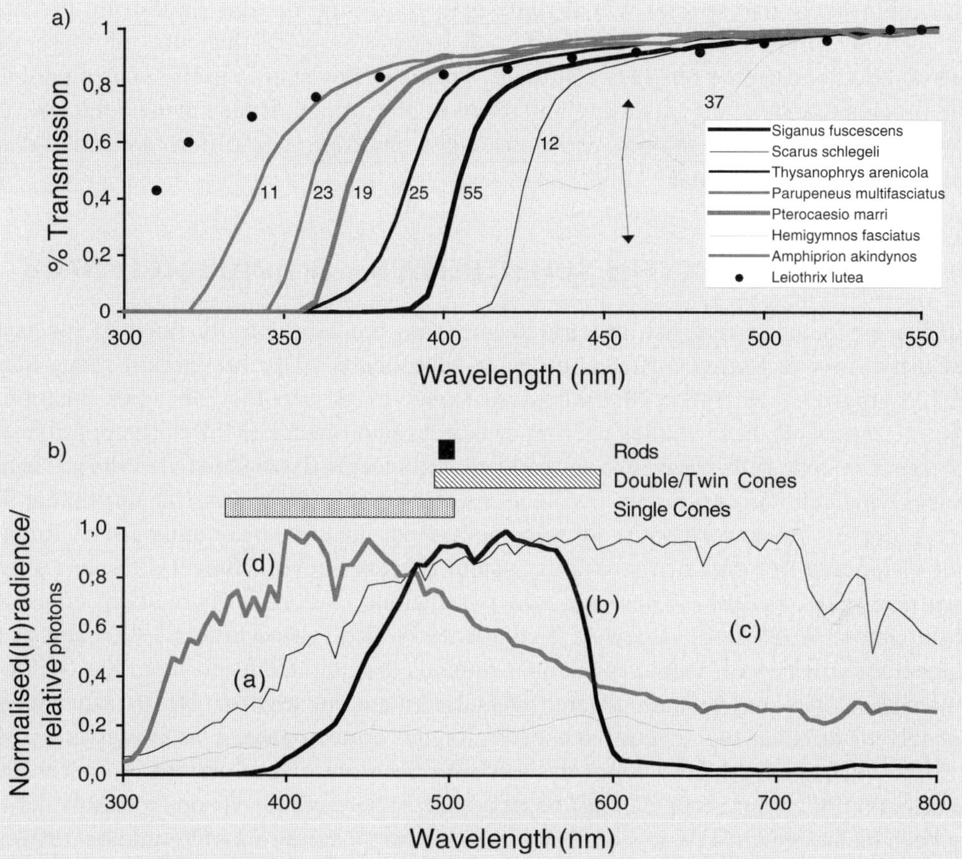

***Figure 3. (a)** Examples of ocular media transmission. All curves are normalised to 550 nm and summarise data from 185 species (U. Siebeck & N. J. Marshall, unpublished). The numbers by each curve are the numbers of species with transmission characteristics close to the curve shown. The arrows signify that the position of the step in this transmission may change throughout the day or in different corneal regions (see text for details). Data for the Pekin robin* Leiothrix lutea *are from Maier & Bowmaker 1993). **(b)** Microhabitat light environments from Heron Island and reef fish spectral sensitivities. Curves are normalised radiance and irradiance measurements in five microhabitats at Heron Island on the Great Barrier Reef. These are typical of reef environments measured elsewhere on the Barrier Reef and in the Caribbean. All measurements were taken close to mid-day under clear blue sky. Type (a) Irradiance, 0.5 m on reef crest. Type (b) Irradiance, 10 m on a sandy bottom. Type (c) Irradiance, 6 m deep under an overhang. Type (d) Radiance (25° acceptance angle looking directly up), 3.0 m depth on a small coral head. The bars at the top of the figure cover the range of the 20 spectral sensitivities so far known for reef fish rods, single cones and double cones (data from: Loew & Lythgoe 1978; Levine & MacNichol 1979; McFarland 1991; Shand 1993; Lythgoe et al. 1994; McFarland & Loew 1994).*

It is highly likely that species which swim actively during the day, hide from predators and feed at different sites on the reef, will encounter more than one of these light environments and gradations between them. They are, however, a useful starting point to look for adaptations of signal design and receiver physiology. It may also be the case that more sedentary species seldom encounter light environments different from the one typical of their microhabitat.

Reef fish spectral sensitivities - a neglected area

Reliable estimates of reef fish spectral sensitivities are available for only 20 species of reef fish (Loew & Lythgoe 1978; Levine & MacNichol 1979; McFarland 1991; Shand 1993; Lythgoe et al. 1994; McFarland & Loew 1994; see Fig. 3b). For anatomical reasons, and as all these studies used microspectrophotometry (MSP), it is possible that UV sensitive cone cells have been missed in some species (Lythgoe et al. 1994). It is also unclear which photoreceptors are involved in colour vision. Reef fish, like most other fish, have a 'duplex' retina with both rods and cones. Rods are almost certainly purely for dim-light vision and not part of the colour system (McFarland & Munz 1975a, c; Lythgoe 1979, 1984; Nicol 1989). There are two populations of cones, double/twin cones and single cones (Walls 1942; Loew & Lythgoe 1978, 1985; Bowmaker 1990) and at the moment the function of double/twin cones remains obscure. Both cell members of double or twin cone pairs may contain different visual pigment but are often electrically coupled, probably to broaden the spectral sensitivity of the doublet (Maier & Bowmaker 1993; Vorobyev et al. 1998). This is not desirable for a colour vision system; however, their participation in colour sense can not be ruled out. Where colour vision in freshwater fish has been examined closely, it seems mediated by single cones only (Neumeyer 1991a, b) and this may also be the case for reef fish.

The spectral positions of the double cones and rods in known reef fish are very different to the single cones, also suggesting a different function (Fig. 3b). The conservative placement of rods maximum sensitivity at a mean of 493nm (Fig. 3b; McFarland 1991 - however, if older visual pigment extract data from McFarland & Munz 1975a, are included the full range is 480-502 nm but these data are less reliable than those gathered by MSP) has engendered much debate, and is almost certainly in order to maximise sensitivity during low light periods or at depth (McFarland 1991). In reef fish this 'sensitivity hypothesis' (Lythgoe 1979) has a special corollary: the 'twilight hypothesis' (McFarland 1991), which specifically suggests that most reef fish rods are optimal for vision at dawn and dusk. These are the periods of high predation on the reef (McFarland & Munz 1975b; McFarland 1986) and sensitive vision at these times is clearly advantageous.

Object contrast enhancement against background, rather than colour vision, is one suggestion for two cone sensitivities in fish. This 'visual pigment offset hypothesis' (McFarland & Munz 1975b; Lythgoe 1979; Levine & McNicol 1982) proposes a visual pigment matched to background spacelight is good for detecting dark objects against the background, while a visual pigment offset from the spacelight is better for detecting bright objects. The two spectral sensitivities therefore combine to give the fish both capabilities. Many of the non-reef marine fish possess blue and green spectral sensitivities well suited for this mode of vision (Lythgoe et al. 1994). Of the reef fish examined, however, only five have been confirmed with two single cone spectral sensitivities, and these are rather variable in their position. Also, background spacelight may vary more widely on the reef, making broad generalisations difficult.

For relatively deep-living reef fish, the possession of very short or long wavelength sensitivity photoreceptors is precluded by the narrowing of environmental light around the blue wavelengths (Fig. 3b; McFarland & Munz 1975b; see below). Those inhabiting surface waters, however, may experience a broad wavelength range and there is no reason why photoreceptors at 325 nm and 600 nm might not be useful. No vertebrate is yet known with a peak sensitivity or 'λ-max' below 350 nm, although these very short wavelength sensitivities do exist in arthropods (Marshall & Oberwinkler 1999). Of those reef fish measured, the goatfish *Upeneus tragula* has the longest λ-max at 580 nm, but this is found in a double cone and may not be involved in colour discrimination (Shand 1993).

It has been suggested for fish from a variety of habitats that the position of cone spectral sensitivities is also matched to environmental light, in particular the double cones (Levine & MacNichol 1979, 1982; Lythgoe 1979). If double cones mediate luminance vision rather than colour vision one would expect their sensitivities to be correlated with the predominant light available. It is therefore interesting that the range of reef fish double cones, 487-580 nm, tends towards rather long wavelength sensitivity with an average around 550 nm. This is in fact not optimal for high sensitivity in the waters of most reef systems which typically possess maximum transmission between 480 and 520 nm (Fig. 3b; McFarland & Munz 1975b; Jerlov 1976). It remains possible that double cones may be good for 'offset' λ-max contrast enhancement as just described.

Combining old and new data - new insights on reef fish colours

Colour contrast: the key for communication and crypsis

A glance at any reef fish identification book quickly shows that there are certain recurring colour themes in reef fish patterns. Many reef fish contain adjacent yellow and blue markings for instance. In fact in reef fish these may be any colour containing yellow and blue according to the nomenclature in the Appendix, UV/yellow and blue, UV/yellow and UV/blue, or a variety of other similar combinations. Of the 200 species examined here, 47% of all families contain members with blue and yellow colouration. Other relatively frequent combinations are patterns containing black and yellow, white and yellow or blue and green (Appendix).

Yellow and blue is a frequent colour pattern combination throughout the animal world, examples to be found in insects (several butterfly species), crustaceans (the stomatopods), molluscs (the blue-ringed octopus *Hapalochlaena* spp.), fish (reef fish), amphibians (poison arrow frogs and the red-eyed treefrog *Agalychnis callidryas*), reptiles (chamelions) and birds (parrots, the courting gifts of Australian fairy wrens *Malurus* spp. and satin bowerbird *Ptilonorhyncus violaceus* bower decorations). This combination may be used for advertisement signals, and is successful as such because the colours are complementary or opponent colours (Fig. 2; Backhaus et al. 1998). That is, where blue reflects in the spectrum, yellow does not and where yellow reflects in the spectrum blue, does not. This provides a very strong contrast signal to any colour vision system with receptors with λ-maxima in the blue and yellow portions of the spectrum, and many animals, including humans, possess such receptors (Fig. 4). Yet again we are left with the fact that not enough reef fish spectral sensitivities are known to draw any firm conclusions; however, one might predict that such sensitivities exist, especially as these would lie within the envelope of blue water lighting typical of the marine environment (Lythgoe 1968).

It has been pointed out a number of times, but most elegantly by John Lythgoe (1968, 1979) that yellow and blue are suitable signal colours in oceanic water as they are transmitted well. Similarly, in freshwater, red and green are often used, again complementary colours, but also as they transmit best in organic compound laden, red-shifted freshwater environments (Lythgoe 1968; Endler 1991; Endler & Houde 1995).

As an aside from complementary colours at this point, colours which are transmitted poorly in aquatic environments potentially provide useful close range, private signals, made invisible to possible eavesdroppers by the attenuation of intervening water. It is therefore possible that UV and red wavelengths are used for this purpose by reef fish, and indeed, many sexual display areas, i.e., fins, mouth tissue or eyes have such colours (Appendix; N. J. Marshall, unpublished).

Yellow and blue are not the only complementary colours used by reef fish. Complementarity in the UV, blue and green spectral regions are less obvious to our eyes than blue and yellow; however, to a short wavelength weighted visual system, such combinations may be highly contrasting also. The pomacentrids often use these colour combinations (Fig. 4; Appendix).

The labroid colours of wrasse and notably the parrotfish are elegant exponents of this complementary system and have the potential for high contrast (Fig. 2e). The colours to human eyes of this combination are the intense green of, for instance, parrotfish, with reflectance peak at 500 nm, often adjacent to a purple colour with a trough at 500 nm and peaks at 410 and 580 nm. There is an astonishingly similar colour combination to be found in the complex structural colours of the birds of paradise (Fig. 2f; Frith & Bechler 1998; N. J. Marshall, unpublished) and indeed similar examples of all these types of complementary colours are found in birds (Fig. 2b, d, e; N. J. Marshall, unpublished).

Whether these colours which are complementary at the level of reflectance are actually perceived as highly contrasting is entirely dependant on the visual system viewing them.

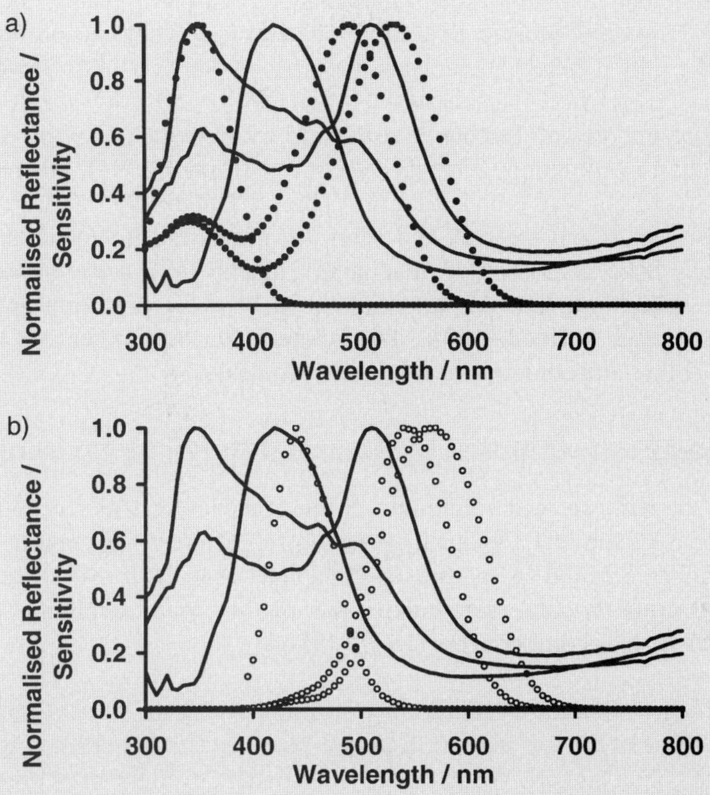

Figure 4. *Visual pigment sensitivities and colours of damselfish compared with sensitivities of humans' three cone mechanisms. (**a**) Normalised reflectance of three colours in* Chromis atripectoralis *(solid lines). On top of these the three spectral sensitivities of a close relative* Pomacentrus coelstis *are plotted (large dots) taken from McFarland and Loew (1994) and calculated using the visual pigment tempelate of Stavenga et al. (1993). (**b**) As (a) but with the three cone sensitivities of humans overlaid on the colours of* C.atripectoralis. *Note how much better the cone sensitivities of* P.coelestis *'fit' the colours.*

It is quite possible that such fine tuning of colour could be extremely apparent to one visual system while remaining cryptic to another. Highly contrasting patterns may also inherently be of use in crypsis if they form part of disruptive patterning (Cott 1940). Therefore, surprising as it may seem, the stunning blue and yellow stripes and bars of the angelfishes (pomacanthids - Appendix; Allen et al. 1998), for instance, may help hide these fish in their natural background as much as they are used in communication.

Contrast within the patterns on a fish, and between fish and their background, are both worth considering in camouflage and display. Unfortunately, yet again few observations exist for reef fish. In freshwater, one of the most interesting examples is the stickleback *Gasterosteus aculeatus* in which both the background displayed against (McDonald & Hawryshyn 1995) and the colour of the male are important in female choice. The spectral sensitivity of the females even changes in the mating season to 'match' the reddening of a male's belly (Cronley-Dillon & Sharma 1968) and is variable according to the photic regions inhabited by different stickleback populations (McDonald & Hawryshyn 1995).

Fish patterns and choice of lighting environment within which displays occur have been carefully studied in the guppy *Poecilia reticulata* (Endler 1991), where females often prefer the most conspicuous colour pattern (Endler & Houde 1995). Through studying genetic variability in colour patterns and carefully measuring light microhabitats, Endler and colleagues have demonstrated that guppy display is a compromise of conspicuousness and camouflage in order to simultaneously avoid predation and attract mates (Endler 1991; Endler & Houde 1995). The light environment on the reef and its possible influence on colour and colour pattern is discussed next.

Constraints imposed by microhabitat light environment

Structurally, the reef can be dividing into different niches such as reef top, under ledges, in holes and caves, and reef slope, which each offer different light regimes (Fig. 3b). Far greater variation is found in marine microhabitats than in forests (Endler 1993), the contributory factors on coral reefs being water quality, water depth and reflection from nearby objects such as coral or algae (Jerlov 1976).

Endler has suggested that different light environments on the forest floor (Endler 1993; Endler & Thery 1996), or indeed in forest streams (Endler 1991), will change the appearance of colour signals. While true to some extent, even the most green biased forest habitat ('forest shade' - Endler 1993) or the bluest light due to time of day ('early/late' - Endler 1993) contain plenty of photons for colour vision at all wavelengths. Furthermore, many animal visual systems, including that of humans, have colour constancy mechanisms which, within certain limits, correct colour signals for illumination changes (Osorio et al. 1997; Vorobyev et al. 1998). For these reasons, great caution is needed to prevent over-interpretation of results which do not include the behavioural response of the animal in question. On the other hand, processes such as mate choice and camouflage breaking may rely on very small incremental differences (Andersson 1994; Bennett et al. 1996) so subtle lighting differences may be enough to provide the basis for such decisions.

Reef microhabitat light environments are very different both by comparison to terrestrial habitats and between themselves (Fig. 3b). They vary from UV/blue dominated open water to predominantly red cave and crevice habitats. The lighting regimes that most reef fish experience, however, fall somewhere between the two extremes of broadband surface light (essentially light at the ocean surface) and the long and short wavelength attenuated environment of deeper waters (Fig. 3b; McFarland & Munz 1975b; Jerlov 1976; Lythgoe 1979).

Many examples exist of photoreceptor sensitivities correlated with increasing depth (Lythgoe 1984; Partridge 1990; Douglas & Partridge 1997) and this trend is likely to be found also in reef fish. The snappers (lutjanids), a widespread fish taxon from reef, shore and open water habitats, show a trend in visual pigment sensitivity with water type (Lythgoe et al. 1994). It is a feature of fish visual systems that they are more variable than those of, for instance, birds (Cuthill et al., this volume) or mammals (Jacobs 1993) this is a direct result of their varied habitats and lighting regimes therein (Bowmaker 1990; Denton 1990; Partridge 1990).

For reef fish living below a moderate depth and depending to an extent on water quality (Jerlov 1976), the light available for effective colour vision will be restricted to a range from 400-600 nm. Two things are likely in such conditions. Firstly, many of the colours reflected from fish in these depths may be visually irrelevant or 'chosen' to appear dark by not reflecting in the 400-600 nm range. Many deep-sea animals are red for precisely this reason and are therefore well camouflaged against their dark backgrounds. Secondly, the number of spectral sensitivities which will be effective for a colour vision system is restricted. As initially suggested by Barlow (1983) to explain why many land animals operational in the 400-700 nm range are trichromats, the reason for this is the relatively broad band sensitivity of most photoreceptors. The possession of more than three photoreceptors in a 300 nm spectral range results in redundancy of information. Within a 200 nm range, only two spectral sensitivities are worthwhile, so fish living in such conditions are likely to be dichromats. Most reef fish examined, including many which spend at least part of their life at such depths, as far as is known at present, possess a maximum of two colour photoreceptors.

Birds (Bowmaker et al. 1997; Vorobyev et al. 1998) and turtles (Ohtsuka 1985) typically possess four spectral sensitivities but make this worthwhile by extending the range of their spectral sensitivity an extra 100 nm or so into the UV, and also by narrowing the bandwidth of their photoreceptors with colour filters (Bowmaker 1980; Govardovskii 1983; Vorobyev et al. 1998); therefore avoiding the problems identified by Barlow (1983). On the reef, the masters of this are the stomatopod crustaceans (mantis shrimps) which manage to pack in 12 spectral sensitivities in the 300-700 nm range by extensive filtering (Marshall 1988; Cronin & Marshall 1989; Marshall & Oberwinkler, 1999). It is unknown whether reef fish are capable of such spectral tuning but, as detailed below, some have remarkable lens and corneal filters.

Compared to bird colours (Finger et al. 1992; Finger & Burkhardt 1994; Bennett et al. 1996, 1997; Andersson & Amundsen 1997; Vorobyev et al. 1998; N. J. Marshall, unpublished), it is interesting that very few reef fish colours reflect in the 300-330 nm range (Fig. 1). This may be due to the chemical nature of what produces colours in reef fish; however, there are at least two other possibilities. These short UVa wavelengths are the most heavily attenuated in seawater, due primarily to scatter caused by particulate matter (Jerlov 1976). As a result, these wavelengths transmit poorly in any water and this may explain the paucity of colour information in this bandwidth. Alternatively, all reef fish ocular media so far examined do not transmit below 330 nm, making vision here impossible. This is discussed below. For comparison, the ocular media transmission of a bird (*Leiothrix lutea*) is included in Fig. 3a and transmits well down to 300 nm. *L. lutea* is also known to possess a spectral sensitivity peaking at 355 nm (Maier & Bowmaker

1993) and is therefore capable of utilising the full spectrum from 300 nm upwards. Bee UV sensitivities average around 345 nm and, like bird colours, some flower colours also reflect to 300 nm (Chittka et al. 1994), again suggesting that it is a constraint of the aquatic environment which prevents use of the 300-330 nm region for fish colours.

Correlations can be identified between reef fish colours and habitat or light environment. For instance, many cave/crevice/under overhang dwellers are predominantly red, while open water reef fish are often a uniform blue or silvery colour. Fish which spend the day-light hours in close association with the reef are often striped or boldly patterned. There are many exceptions to these generalisations and confounding factors such as activity cycle and phylogenetic trends cloud the relationship (N. J. Marshall, unpublished). For example, nocturnal reef fish spend the day in holes and are often red. It is unclear if their colours are adaptations to day or night or both, but it has previously been pointed out that they are rarely very patterned (Cott 1940).

Relationships between colours and ambient light are likely to optimise camouflage, display or both (Fine et al. 1977; Endler 1991; Endler & Thery 1996). Silvery blue fish are well camouflaged in open water because they match the background spacelight by colour or, if very silvery, may reflect the light of their niche to match the background (Denton & Nicol 1966). Holes and crevices on the reef are often lined with red algae and a variety of encrusting sponges which give these environments their predominantly long wavelength character (Fig. 3b). It is possible that the red fish often found in such habitats are attempting to blend in with the background. Alternatively they may match their skin colour to the predominant wavelengths of light available for communication purposes. Inevitably, until more work is done in this area, such suggestions remain speculative. Two specific examples of reef fish colours and the light type of their predominant environments are discussed below (Fig. 5). First, however, the ocular mediar, lens and cornea are described, as these greatly influence the light available for vision in reef fish.

The use and non-use of UV: what ocular media tell us

Measuring reef fish colours and the light environments in which they exist only tell us what information reaches the eye. An important next step is to determine what portion of this is passed through the cornea, lens and other ocular components to the photoreceptors. Previous to this study, some freshwater fish (Douglas & McGuigan 1989) and reef fish (Thorpe et al. 1993) were known to possess lens and/or corneal filters which attenuate short wavelengths especially in the UV. It has been suggested that this may help prevent photodamage from the high energy UV photons which could make vision in this wavelength range too costly (Werner & Spillmann 1989). Birds again provide an interesting antithesis to this as many of them possess UV photoreceptors and UV transmitting ocular media (Fig. 3b; Maier & Bowmaker 1993; Hart et al. 1998). UV transmitting ocular media of course do not necessarily indicate that the animal is capable of UV vision. The retina must contain photoreceptors capable of absorbing in this light range. It is also possible that short wavelength light is actually filtered out by the retina, as is the case for macular pigment in humans (Smith & Pokorny 1975).

Faced with the wealth of reef fish colours which contain significant UV reflectance (9 out of the 21 categories identified here - Fig. 1 and Appendix) it is tempting to speculate that these wavelengths are somehow used. This is especially the case as UV is invisible to us and therefore mysterious (for de-mystification of UV: see Cuthill et al., this volume). The majority of reef fish ocular media are like those of humans, however, and possess a sharply defined cut off at 400 nm (Fig. 3a). Only around 30% of the 182 species of reef fish examined so far allow wavelenghts of light below 385 nm to reach their retinae. This means that many of the UV patterns and colours of reef fish species are not visible to their conspecifics and may be visually irrelevant. Clearly, caution is warranted here as animal colours are not only designed through evolution for the eyes of the species displaying them. It is quite possible that UV colouration in one species may be part of an antipredatory camouflage system or for territorial advertisement, both messages directed at other species with UV vision.

Three patterns exist on examination of the reef fish which may be capable of UV vision. Firstly, all damselfish (pomacentrids) from a variety of habitat and life styles transmit UV to the retina. Secondly, many planktivorous fish appear capable of UV vision, which may be advantageous for imaging small particles against the scattered UV portion of space light in the ocean (Browman et al. 1994; Douglas et al. 1995; Loew et al. 1996). It is interesting with regard to both these observations that several species of pomacentrids are planktivorous (others may be herbivorous or detritivorous) and that most reflect UV (Appendix). Lastly, many nocturnal and crepuscular fish may see UV. The relative amount of short wavelenght light increases at dawn and dusk (McFarland & Munz 1975b), so perhaps a 'desire' to match this light regime may explain this trend. Alternatively, as these fish are operating under light limited condition, it may just be an attempt to maximise light input to the eye. There is no correlation observable as yet between fish with UV patterns and the potential for UV vision.

Two specific examples and a model of reef fish vision

With knowledge of reef fish colours, environmental lighting and ocular media, we can begin to attempt to view the world as a fish does and gain insight into the constraints governing signal design. Two things are worth re-iterating from the outset. Firstly, our lack of knowledge of spectral sensitivities is a gap that urgently needs filling. Without this, whether qualitatively or quantitatively (Endler 1991), all our suggestions remain speculative. Secondly, although the descriptions that follow pick on a typical light habitat for each species, these and many other reef fish are highly motile and therefore may experience many different lighting regimes in a short space of time It is unknown if reef fish display in certain light microhabitat types; another question requiring urgent investigation.

The angel fish *Pygoplites diacanthus* is one of the reef's most beautiful and frequently photographed inhabitants (Fig. 5a; Randall et al. 1997). Its colours are violet-blue, yellow, black and white and therefore, to our eyes, it possesses a highly contrasting pattern. Of the six colours identifiable with reflectance measurements (Fig. 5a), four have significant UV reflection as is clear in UV images of the fish. It is certain, however, that this species is unaware of its UV colours as its ocular media pass only wavelengths above

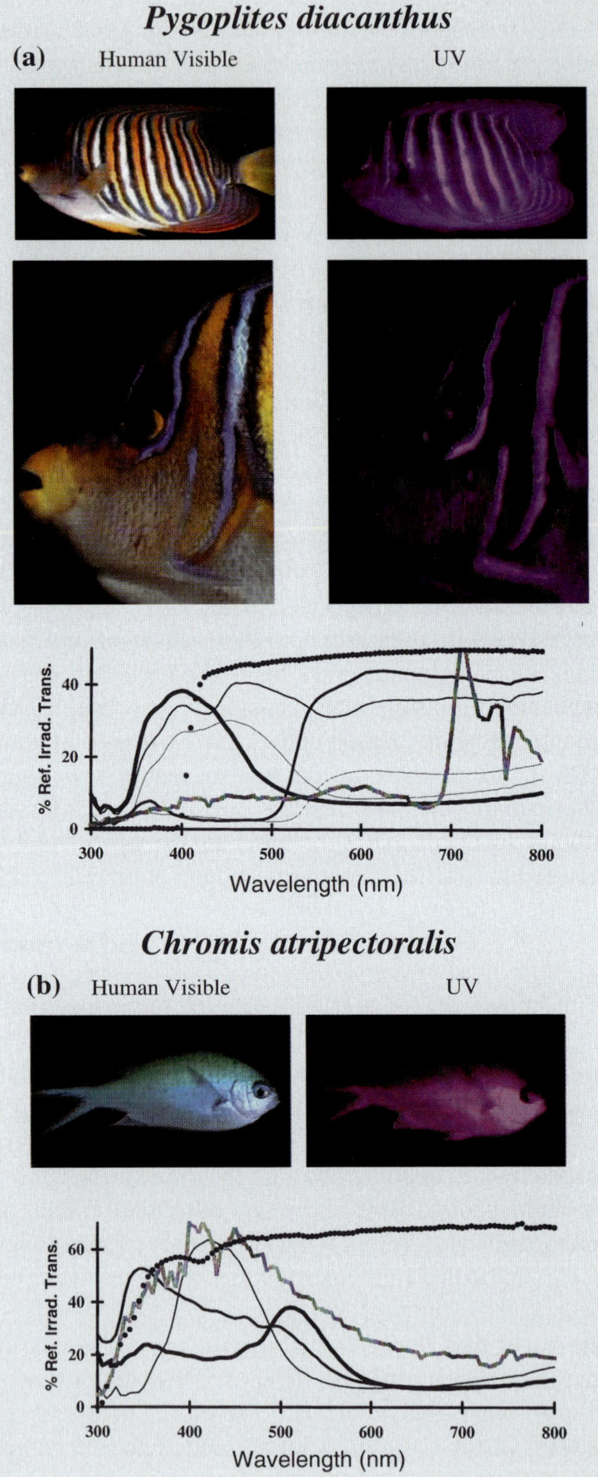

***Figure 5.** Two species of reef fish, as seen by humans, and in UV light. Digitised video images of the fish were taken with a broadband colour video camera and one limited to the UV from 350-400 nm UV. The graphs plot % reflectance of fish colours (solid lines of various thickness), % transmission of the ocular media normalised to the peak reflectance in each graph (dots), and typical microhabitat(s) irradiance similarly normalised (light grey lines). (**a**) The angelfish Pygoplites diacanthus. (**b**) The damselfish (**b**) Chromis atripectoralis.*

400 nm (Fig. 5a). Furthermore, at least around Heron Island where light measurements of *P. diacanthus* habitat were taken, the two typical lighting environments in which this fish is frequently found are low in UV. Remember, however, that although Figure 5a appears to show low radiation in the 300-400 range, these curves are plotted on a linear scale for ease of spectral comparison, and this underemphasises low intensity spectral regions. Clear blue oceanic water transmits visually useable UV down to 400 m (Frank & Widder 1996).

If the light environment as plotted in Figure 5a is a good representation of useable light, the crevice/cave light habitat where *P. diacanthus* is frequently found, provides an interesting possibility. The main long wavelength peak in irradiance is beyond 700 nm (Fig. 5a) and although it is unknown as yet if this spectral region is visible to P. diacanthus, the use of colour in this spectral region cannot be ruled out. It would require long wavelength sensitivity not yet reported in reef fish but known in freshwater fish (Neumeyer et al. 1991a, b). Three of the colours of *P. diacanthus* reflect substantial amounts at long wavelengths but, as we have seen, in the UV this is not necessarily a good indication of visual activity in a particular spectral region.

Movements from one light habitat into another is one way in which reef fish can exploit the wide variability in the light environments available to them to 'show off' different colours for different purposes, or for appearance and disappearance. Examples of using light microhabitats as a component of display and camouflage do exist both in the forest and in freshwater (Endler 1991, 1993). It is time now to look for similar examples on the reef.

Where does this leave us with regard to the biology of this beautiful fish? What *P. diacanthus* uses its colours and its colour vision system for are still completely unknown. Based on the evidence presented here, it is likely that long wavelength components of *P. diacanthus* colours, perhaps even those beyond 700 nm, are more important than UV wavelengths to other *P. diacanthus*. The UV component of its colours may, however, be visually relevant to other reef fish species.

The second specific example chosen is the damselfish *Chromis atripectoralis*, the colours of which are an irridescent blue/green to our eyes and reflect very little beyond 600 nm (Fig. 5b). *C. atripectoralis* lives in loosely associated schools often gathering above coral heads, darting upwards repeatedly to feed on suspended particles. Its light environment and line of sight is therefore dominated by the downwelling light from the surface (Fig. 5b). Its colours are largely the result of structural mechanisms (Fox 1953; Fox & Vevers 1960; Katsukawa et al. 1987) and change peak reflectance with angle of view and illumination. These changes are, however, all within the UV to green region. Sex specific colour changes are also known in this species (Thresher 1984), the males adding a black dorsal fin to their signalling repertoire during breeding season.

That *C. atripectoralis* is probably 'concerned' with short wavelengths including the UV is also clear from the transmittance of the ocular media which extends down to 300 nm.

Spectral sensitivity of this species is not known but, as detailed below, it is likely to have a UV sensitive channel. All pomacentrids so far examined possess a UV transmitting lens and cornea (U. Siebeck & N. J. Marshall, unpublished) and many have UV reflectance (Appendix), making it likely that members of this family use the short wavelength portion of the spectrum for communication. Many, like *C. atripectoralis,* are also particulate feeders, a mode of feeding enhanced by UV vision (Browman et al. 1994; Loew et al. 1996).

The spectral sensitivities of two species of reef dwelling pomacentrids are known (McFarland & Loew 1994; Fig. 4) and include visual pigments with λ-max values at 360, 490 and 516 or 532 nm. This confirms the short wavelength bias in these species at least. Making the assumption that *C. atripectoralis* has the same spectral sensitivities, a simplistic model can be constructed of the relative efficiency of the visual system and colours of the fish in its 'chosen' light environment. This is now described, including for comparative purposes, the relative efficiency for the human visual system. Just a glance at Figures 4 or 5, however, is enough to predict the outcome as the colours of *C. atripectoralis* are clearly better suited to 'analysis' by the pomacentrid colour vision system.

Assumptions for this model are as follows:

a) The UV/blue radiometrically measured light environment is an accurate description of the light in which *C. atripectoralis* is frequently viewed by conspecifics.
b) Spectral sensitivities at 360, 490 and 532 nm (data from *Pomacentrus coelestis* - McFarland & Loew 1994) are present in *C. atripectoralis*. Note that the sensitivity at 532 nm is from a double cone and is therefore possibly not involved in colour vision, but kept here in order to compare two apparently trichromatic systems.
c) Spectral sensitivities are scaled to give equal area and therefore equal relative output. This presumes von Kriesian type photoreceptor physiology (Vorobyev et al. 1998).
d) A set of these probable chromatic signals resulting from cone opponency mechanisms are calculated for both humans and *C. atripectoralis*. If we label the three sensitivities as S - short, M - medium and L - long these are (S-M), (M-L), (S-(M+L)); combinations which are known to exist in humans (Mollon 1982; Wyszecki & Styles 1982; Gouras 1991). These are in fact unknown for *C. atripectoralis* but provide a more rigorous comparison by presuming they are the same as humans.
e) *C. atripectoralis* sensitivities are calculated to include the filtering effect of the ocular media. Human sensitivities are from Smith & Pokorny (1975).
f) Chromatic sensitivities calculated from receptor sensitivities alone and ignoring neural processing beyond immediate opponent mechanisms are an accurate description of colour vision mechanisms. Good evidence for this exists (Vorobyev & Osorio 1998; Vorobyev et al. 1998).
g) Calculations follow the basic principles of Wyszecki and Styles (1982), Endler (1990) or Vorobyev et al. (1998) and include the following steps. (1) The signal reaching the eye is the product of each colour multiplied by the illumination at each wavelength between 300 and 800nm. (2) Cone signals viewing each colour are the result of (1) multiplied by each cone sensitivity and summed for each from 300-800 nm. (3)

Chromatic signals (the results of the opponent processes required for colour vision) are then calculated for each of the three opponent mechanisms in assumption (d). These give a direct measure of the appearance of colours to the visual system. (4) Chromatic signals from all three chromatic opponent mechanisms are summed for each colour and are plotted in Fig. 6.

C. atripectoralis has clearly evolved to be better at viewing its own colours in its own environment than humans (Fig. 6). For a variety of reasons this is no surprise; perhaps the most obvious reason is that we lack a UV sensitive mechanism. Nevertheless, this is a useful exercise, as it shows the possibilities for investigating signals and visual systems when all four components of the four way approach to visual ecology are known.

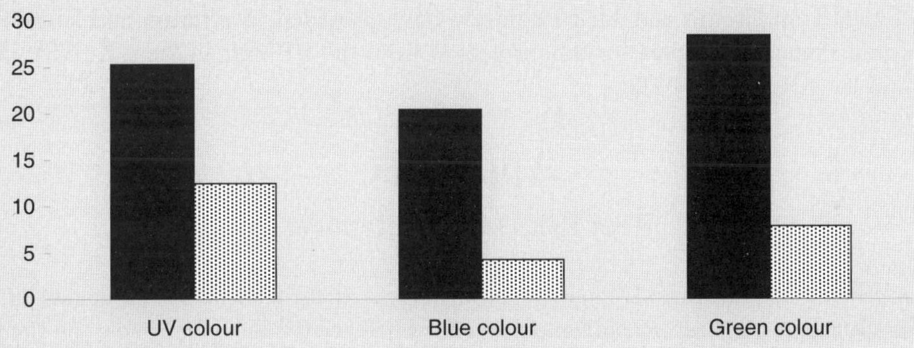

Figure 6. *A comparison of summed chromatic signals for human (shaded bars) and* P. coelestis *(black bars) 'looking at'* C. atripectoralis *colours. See text for details of calculation. For all colours* C. atripectoralis *outperforms humans.*

Conclusions and the future

Once more data exist for the spectral sensitivities of other reef fish, the kind of approach attempted here should enable similar calculations from a variety of habitats and lifestyles on the reef. Critically this will allow non-subjective experiments to be designed to examine sexual selection, territoriality, aggression, feeding, camouflage and the other factors mutually responsible for the stunning colours of reef fish.

One of the greatest challenges is to examine not just individual colours, but those colours in the context of the pattern within the fish and the interaction of the fish colours and their background. Also, the language of colour change is a fascinating area hardly touched on. On land we are impressed by the rapidity and subtlety of the changes a chameleon performs. Most reef fish are capable of colour change, some far more rapid and with just as extensive a repertoire as the chameleon. The combination of this and the potential of 'behavioural colour change' that is exploiting different light microhabitat environments for different purposes (Endler & Thery 1996) again sets quite a challenge. At least now we are beginning to identify and quantify the components of the system.

Acknowledgements

Many thanks to Yngve Espmark, Gunilla Rosenqvist, Trond Amundsen and Turid Fredagsvik for organising the fascinating conference which has resulted in this volume. Thanks to Heron Island and Lizard Island research station in Australia, The NOAA/NURC underwater laboratory Aquarius on Key Largo in the USA, and to George Losey at the Hawaiian Institute of Marine Biology, who have made the work described here a pleasure. Andor Technology built Sub-Spec, a wonderful spectrometer which has lasted almost a decade without fault. Thanks to Uli Siebeck for providing some preliminary ocular media results and to Kylie Jennings for tireless help with most things. Mike Land, Tom Cronin and Jack Pettigrew have provided inspiration and stimulating discussion. Financial support for this project is from the BBSRC in the UK, NSF in the USA and the ARC in Australia.

Appendix

Reef fish colour categories

Colours can be categorised into 21 human subjective types. In the table below, the colours of each reef fish species are classified according to this scheme. (S) and (C) stand for Simple and Complex colour type and are explained further in the text:

UV(S): a rare colour in which narrow band reflectance occurs in the UV only. UV is defined here as 300-400 nm. **UV/Blue** (S): appears 'deep blue/violet' to humans. Contains wavelengths in both the UV and blue region of the spectrum. **UV-hump/Blue** (S): similar to UV/Blue but invariably the UV component is depressed, forming a hump on the short wavelength limb. **UV/Blue/Red** (C): similar to UV/Blue but a significant although non-step-like rise in reflectance occurs at longer wavelengths. **Blue** (S): principle reflectance occurs only in the blue region of the spectrum (400-500 nm). **UV/Green** (C): reflection occurs in both the UV and green region of the spectrum as separate reflectance peaks. **Green** (S): a peak of reflection occurs in the green region of the spectrum (500-600 nm). **UV/Yellow** (C): a cut-on step reflectance occurs at around 500 nm giving this colour its yellow appearance to human eyes, however a further peak of reflectance occurs in the UV. **Yellow** (S): a cut-on step of reflectance occurs at around 500 nm. **UV/Orange** (C): as UV/Yellow but the cut-on is shifted to around 550 nm and may be less steep. **Orange** (S): as Yellow but the cut-on is shifted to around 550 nm and may be less steep. **UV/Red** (C): similar to UV/Yellow or UV/Orange but with the cut-on shifted to around 600 nm. This colour category tends to be both less saturated and brighter overall than UV/Yellow, Magenta or Red making it appear pink or light-red to human eyes. **Red** (S): a cut-on step function at around 600-650 nm. **Magenta** (C): a mixture of UV/Blue and Red. The exact position of the UV/Blue peak and the Red cut-on in this colour determines whether it appears red or purple to humans. **White** (S): a sharp cut-on occurs between 330 and 390 nm. It therefore always contains a UV component. **Black:** low reflectance over the whole spectrum.

Brown: low reflectance over the whole spectrum but with a varying degree of gradual rise past 500 nm. **Light 'Violet':** colours in this category appear a washed out violet to the human eye. They are like White but with depressed reflectance in the long wavelengths beyond 500 nm. They could also be viewed as bright UV/Blues. **Labroid Blue/Red, Blue/Far-Red complex**: almost invariably found in labroids (wrasse and parrotfish) only (for some exceptions: see Appendix). These colours may appear anything from dark blue to turquoise to the human eye. They contain a peak in reflectance at 450-500 nm and a cut-on step reflectance at 650-730 nm. This Red or Far-Red relectance is invariably not apparent to the human eye. **Labroid Green, Far-Red Complex**: like the Labroid Blue/Red, Blue/Far-Red Complex but with the peak reflectance component sharper and beyond 500 nm. They appear very green and distinct from turquoise to the human eye. **Labroid Purple Complex**: these colours are a beautiful nightmare! They contain 2-4 peaks in the spectrum but invariably have a dip at 500 nm (see relationship with Labroid Green, Far-Red Complex below), and peaks at around 410 and 575 nm. There may be a further peak in the UV and/or a cut-on step function at around 700 nm.

Family	Species	Common Name	State	UV	UV/Blue	UV-Hump/Blue	Blue	Green	Yellow	Orange	Brown	Red	Black	White	Light 'Violet'	UV/Blue/Red	Labroid Blue/Red, Blue/Far-Red	Magenta	UV/Green	UV/Yellow	UV/Orange	UV/Red	Labroid Green, Far-Red	Labroid Purple
Dasyatididae	*Taeniura lymma*	Blue Spotted Fantail Ray	Mature			■					■													
Holocentridae	*Holocentrus rufus*	Longspine Squirrelfish	Mature									■		■										
	Sargocentron melanospilos	Blackspot Squirrelfish	Mature							■		■		■										
	Sargocentron spiniferum	Sabre Squirrelfish	Mature						■	■		■	■	■										
Aulostmidae	*Aulostomus chinensis*	Trumpetfish	Mature						■															
Fistulaeriidae	*Fistularia commersonii*	Smooth Flutmouth	Mature			■					■													
Syngnathidae	*Corythoichthys intestinalis*	Banded Pipefish	Mature				■																	
Scorpaenidae	*Pterois antennata*	Ragged-finned Firefish	Mature								■	■		■										
	Pterois volitans	Red Firefish	Mature								■			■							■			
	Pterois volitans	Red Firefish	Juvenile								■			■										
	Scorpaenodes guamensis	Guam Scorpionfish	Mature							■	■	■		■										
Serranidae	Cephalopholis cyanostigma	Blue-spotted Rockcod	Mature			■				■	■													
	Diplorprion bifasciatum	Barred Soapfish	Mature						■															
	Epinephelus adscensionis	Rock Hind	Mature																			■		
	Hypoplectrus gemma	Blue Hamlet	Mature			■																		
	Hypoplectrus guttavarius	Shy Hamlet	Mature						■											■	■			
	Hypoplectrus indigo	Indigo Hamlet	Mature		■									■										
	Hypoplectrus unicolour	Butter Hamlet	Mature		■															■				
	Plectropomus leopardus	Coral Trout	Mature			■					■	■									■			
	Pseudanthias squamipinnis	Scalefin Anthias	Mature-Male									■												■
	Pseudanthias squamipinnis	Scalefin Anthias	Mature-Female						■	■														
Grammatidae	*Gramma loreto*	Fairy Basslet, Royal Gramma	Mature						■									■						
Pseudochromidae	*Pseudochromis diadema*	Diadem Basslet	Mature															■		■				
	Pseudchromis paccagnellae	Royal Dottyback	Mature						■									■						
	Pseudochromis porphyreus	Strawberry / Magenta Dottyback	Mature															■						
Priacanthidae	*Priacanthus hamrur*	Crescent-tail Bigeye	Mature									■												
Apogonidae	*Apogon aureus*	Ring-tailed Cardinalfish	Mature									■								■				
	Apogon cookii	Cook's Cardinalfish	Mature											■	■					■				
	Apogon cyanosoma	Yellow-striped Cardinalfish	Mature						■		■			■	■									
	Apogon doederleini	Doederlein's Cardinalfish	Mature								■			■								■		
	Apogon fragilis	Fragile Cardinalfish	Mature	■																■				
	Apogon leptacanthus	Lonespine Cardinfish	Mature		■	■															■			
Lutjanidae	*Lutjanus apodus*	Schoolmaster	Mature						■					■	■									
	Lutjanus bohar	Red Bass	Mature						■	■		■		■				■						
	Lutjanus bohar	Red Bass	Juvenile								■	■		■										
	Lutjanus fulviflamma	Black Spot Snapper	Mature											■				■		■				
	Lutjanus gibbus	Paddletail	Mature						■					■				■						
	Lutjanus quinquelineatus	Five-lined Seaperch	Mature		■				■			■												
Caesionidae	*Caesio cuning*	Red-bellied Fusilier	Mature			■						■							■	■				■
	Pterocaesio marri	Marr's Fusilier	Mature											■	■			■		■	■			■

Lethrinidae	*Lethrinus miniatus*	Sweetlip Emperor	Mature							■														
Nemipteridae	*Scolopsis bilineatus*	Bridled Monocle Bream	Mature						■		■			■						■				■
	Scolopsis monogramma	Monocle Bream	Mature								■			■	■					■				
Mullidae	*Parupeneus barberinoides*	Bicolour Goatfish	Mature							■		■		■										
	Parupeneus barberinus	Dash-dot Goatfish	Mature			■			■					■	■			■						
	Parupeneus cyclostomus	Goldsaddle Goatfish	Yellow Phase						■	■														
	Parupeneus heptacanthus	Cinnabar Goatfish	Mature								■				■					■				
	Pseudupeneus maculatus	Spotted Goatfish	Resting Phase									■								■				
	Parupeneus multifasciatus	Manybar Goatfish	Mature											■	■							■		
Pempheridae	*Pempheris analis*	Bronze Sweeper	Mature								■													■
Chaetodontidae	*Chaetodon aureofasciatus*	Golden-striped Butterflyfish	Mature							■	■			■						■				
	Chaetodon auriga	Threadfin Butterflyfish	Mature						■					■						■				
	Chaetodon baronessa	Triangular Butterflyfish	Mature		■					■										■				
	Chaetodon capistratus	Foureye Butterflyfish	Mature											■						■				
	Chaetodon ephippium	Saddle Butterflyfish	Mature		■				■			■		■						■				
	Chaetodon flavirostris	Dusky Butterflyfish	Mature						■	■										■				
	Chaetodon melannotus	Blackback Butterflyfish	Mature						■					■										
	Chaetodon ornatissimus	Ornate Butterflyfish	Mature						■	■				■										
	Chaetodon oxycephalus	Spotnape Butterflyfish	Mature						■	■				■	■									
	Chaetodon pelewensis	Dot & Dash Butterflyfish	Mature							■	■			■						■				
	Chaetodon plebeius	Bluespot Butterflyfish	Mature		■				■					■										
	Chaetodon rainfordi	Rainford's Butterflyfish	Mature						■					■										
	Chaetodon sedentarius	Reef Butterflyfish	Mature																					
	Chaetodon trifasciatus	Redfin Butterflyfish	Mature		■					■	■			■						■	■			
	Chaetodon ulietensis	Pacific Doublesaddle Butterflyfish	Mature						■					■										
	Chaetodon unimaculatus	Teardrop Butterflyfish	Mature						■					■										
	Chaetodon vagabundis	Criss-cross Butterflyfish	Mature		■					■				■						■				
	Chelmon rostratus	Beaked Coralfish	Mature						■	■				■										
	Coradion altivelis	Highfin Coralfish	Mature							■				■	■					■				
	Forcipiger flavissimus	Forcepsfish	Mature						■					■	■									
	Heniochus acuminatus	Longfin Bannerfish	Mature						■					■										
	Heniochus diphreutes	Schooling Bannerfish	Mature											■						■				
	Heniochus monoceros	Masked Bannerfish	Mature						■					■										
	Heniochus singularis	Singular Bannerfish	Mature						■					■	■									
	Heniochus varius	Humphead Bannerfish	Juvenile								■			■						■				
Pomacanthidae	*Centropyge argi*	Cherubfish	Mature			■			■															
	Centropyge bicolor	Bicolor Angelfish	Mature				■		■	■														
	Centropyge bispinosus	Two-spinned Angelfish	Mature						■							■								
	Chaetodontoplus meredithi	Queensland Yellowtail Angelfish	Mature		■				■	■	■													
	Chaetodontoplus mesolucucs	Vermiculated Angelfish	Mature						■					■	■									
	Holocanthus bermudensis	Blue Angelfish	Mature		■				■															
	Holocanthus ciliari	Queen Angelfish	Mature			■		■	■															
	Holocanthus (bermudensis) isabelita	Blue Angelfish	Juvenile			■			■											■				
	Holocanthus tricolour	Rock Beauty	Mature			■			■			■												
	Holocanthus tricolour	Rock Beauty	Juvenile				■													■				
	Pomacanthus (Euxiphipops) navarchus	Blue-girdled Angelfish	Juvenile			■								■						■				
	Pomacanthus paru	French Angelfish	Mature						■					■						■				
	Pomacanthus semicirculatus	Semi-circle Angelfish	Mature				■	■												■				
	Pomacanthus sexstriatus	Six-banded Angelfish	Mature				■							■	■					■				
	Pygoplites diacanthus	Regal Angelfish	Mature			■			■	■				■										
			Mature																					
Pomacentridae	*Abudefduf bengalensis*	Bengal Sergeantfish	Mature											■	■									
	Amblyglyphidodon curacao	Staghorn Damselfish	Mature												■				■					
	Amblyglyphidodon leucogaster	White-belly Damsel	Mature						■					■										
	Amphiprion melanopus	Clown Anemonefish	Mature							■					■									
	Amphiprion percula	Pink Anemonefish	Mature-Female																					
	Amphiprion perideraion	Red & Black Anemonefish	Mature			■				■				■										
	Chromis amboinensis	Ambon Chromis	Mature												■									
	Chromis atripectoralis	Black-axil Chromis	Mature			■	■							■					■					
	Chromis cyanea	Blue Chromis	Mature				■																	
	Chromis nitida	Barrier Reef Chromis	Mature																					
	Chromis viridis	Blue-green Chromis	Mature		■	■																		
	Chrysiptera cyanea	Blue Devil	Mature-Male		■	■				■														
	Chrysiptera parasema	Azure Damselfish	Mature			■		■	■															
	Dascyllus aruanus	Humbug	Mature											■										
	Dischistodus prosopotaenia	Honey-head Damselfish	Mature		■						■													
	Microspathodon chrysurus	Yellowtail Damselfish	Mature						■		■													
	Neoglyphidodon melas	Black Damselfish	Juvenile		■	■			■					■										

Family	Species	Common Name	State	UV	UV/Blue	UV-Hump/Blue	Blue	Green	Yellow	Orange	Brown	Red	Black	White	Light 'Violet'	UV/Blue/Red	Labroid Blue/Red, Blue/Far-Red	Magenta	UV/Green	UV/Yellow	UV/Orange	UV/Red	Labroid Green, Far-Red	Labroid Purple
	Neoglyphidodon melas	Black Damselfish	Juvenile		■	■			■					■										
	Neopomacentrus azsyron	Yellowtail Demoiselle	Mature				■													■				
	Pomacentrus bankanensis	Speckled Damselfish	Mature			■				■	■				■									
	Pomacentrus chrysurus	Whitetail Damselfish	Mature											■										
	Pomacentrus moluccensis	Lemon Damselfish	Mature							■										■				
	Pomacentrus nagasakiensis	Sandy Damsel	Mature											■										
	Pomacentrus wardi	Ward's Damselfish	Juvenile		■	■			■															
	Premnas biaculeatus	Spine-cheek Anemonefish	Mature							■				■						■				
	Stegastes apicalis	Australian Gregory	Mature		■	■			■															
	Stegastes partitus	Bicolour Damselfish	Mature						■															
	Stegastes variabilis	Cocoa Damselfish	Intermediate						■															
	Unidentified	Blue Diamondhead	Juvenile			■								■										
Labridae	*Anampses geographicus*	Geographic Wrasse	Mature							■	■	■					■		■					■
	Anampses neoguinacus	New Guinea Wrasse	Mature								■	■		■			■						■	■
	Bodianus axillaris	Axilspot Hogfish	Mature						■	■		■												■
	Bodianus diana	Diana's Hogfish	Mature						■			■												
	Bodeanus rufus	Spanish Hogfish	Juvenile						■								■							
	Cheilinus diagrammus	Cheeklined Maori Wrasse	Mature									■			■		■					■	■	■
	Cheilinus fasciatus	Red-breasted Maori Wrasse	Mature								■	■			■									■
	Cheilinus trilobatus	Tripletail Maori Wrasse	Initial								■						■				■		■	
	Choerodon cyanodus	Blue Tuskfish	Mature			■				■		■		■	■									
	Choerodon fasciatus	Harlequin Tuskfish	Mature			■			■			■			■		■			■	■			■
	Cirrhilabrus punctatus	Dotted Wrasse	Terminal			■						■								■				
	Coris aygula	Clown Coris	Terminal		■																			■
	Coris gaimard	Yellowtail Coris	Juvenile			■		■		■				■						■		■		
	Coris gaimard	Yellowtail Coris	Mature			■			■										■		■			
	Gomphosus varius	Bird Wrasse	Mature			■											■						■	
	Halichoeres garnoti	Checkerboard Wrasse	Mature											■					■	■	■			
	Halichoeres garnoti	Tailspot Wrasse	Mature														■						■	
	Halichoeres hortulanus	Ornate Wrasse	Mature								■			■	■									
	Halichoeres melanurus	Twotone Wrasse	Mature																					
	Halichoeres ornatissimus	Threespot Wrasse	Terminal						■					■					■					■
	Halichoeres prosopeion	Yellowhead Wrasse	Terminal-Female						■			■			■									
	Halichoeres trimaculatus	Yellowhead Wrasse	Juvenile-Male																					
	Hemigymnus fasciatus	Barred Thicklip Wrasse	Mature											■							■			■
	Hemigymnus melapterus	Blackeye Thicklip	Mature												■						■		■	
	Hologymnosus doliatus	Pastel Ringwrasse	Initial			■				■													■	
	Labrichthys unilineatus	Tubelip Wrasse	Mature			■																		
	Labroides bicolor	Bicolor Cleaner Wrasse	Mature-Female											■			■							
	Labroides dimidiatus	Cleaner Wrasse	Juvenile					■									■							
	Labroides dimidiatus	Cleaner Wrasse	Mature														■							
	Labropsis australis	Southern Tubelip	Mature							■							■							
	Labropsis xanthorata	Yellow-backed Tubelip	Mature								■					■								
	Macropharyngodon choati	Choat's Wrasse	Mature						■	■				■	■									
	Stethojulis strigiventer	Stripebelly Wrasse	Mature								■			■	■			■		■				
	Stethojulis strigiventer	Stripebelly Wrasse	Juvenile																					
	Thalassoma bifasciatum	Bluehead	Terminal			■													■					
	Thalassoma lunare	Moon Wrasse	Terminal			■											■	■					■	
	Thalassoma lutescens	Sunset Wrasse	Initial						■										■	■	■			
	Thalassoma lutescens	Sunset Wrasse	Terminal						■						■						■		■	
Scaridae	*Cetoscarus bicolor*	Bicolor Parrotfish	Juvenile						■	■				■						■				

	Scarus microrhinos	Steephead Parrotfish	Initial-Female																		■		■	■
	Scarus microrhinos	Steephead Parrotfish	Terminal-Male				■										■	■					■	■
	Scarus niger	Swarthy Parrotfish	Terminal														■						■	■
	Scarus niger	Swarthy Parrotfish	Initial												■		■				■		■	
	Scarus psittacus	Pale-nosed Parrotfish	Mature				■										■						■	■
	Scarus rivulatus	Surf Parrotfish	Mature						■					■			■						■	■
	Scarus spinus	Greensnout Butterflyfish	Mature														■						■	■
	Scarus taeniopterus	Princess Parrotfish	Terminal						■			■			■		■						■	
	Scarus taeniopterus	Princess Parrotfish	Initial																					
	Sparisoma aurofrenatum	Redband Parrotfish	Initial		■							■			■									
	Sparisoma viride	Stoplight Parrotfish	Initial									■		■										
Pinguipedidae	*Parapercis cylindrica*	Sharpnose Sandperch	Mature								■			■						■				
	Parapercis hexophtalma	Speckled Sandperch	Mature																					
Blenniidae	*Aspidontus taeniatus*	Mimic Blenny	Mature		■	■								■										
	Petroscirtes fallax	Deceiver Fangblenny	Mature						■		■			■										
	Plagoitremus rhinorhynchos	Bluestriped Fangblenny	Mature		■		■	■																
Gobiidae	*Amblygobius phalaena*	Banded Goby	Mature								■									■				
	Gobiosoma oceanops	Neon Goby	Mature														■							
	Valenciennea longipinnis	Long-finned Goby	Mature					■						■	■						■	■		
	Valencienna strigata	Blueband Goby	Mature		■									■						■				
			Adult																					
Microdesmidae	*Nemateleotris decora*	Elegant Firefish	Mature							■							■	■						
			Adult																					
Acanthuridae	*Acanthurus chirurgus*	Doctorfish	Mature		■									■										
	Acanthurus dussumieri	Eyestripe Surgeonfish	Mature			■			■						■									
	Acanthurus japonicus (aliala)	Orange Tang	Mature							■	■			■						■				
	Acanthurus leucosternon	Powder Blue Tang	Mature						■															
	Acanthurus lineatus	Striped Surgeonfish	Mature				■			■										■				
	Acanthurus nigrofuscus	Brown Surgenfish	Mature						■															
	Acanthurus olivaceus	Orangebond Surgeonfish	Mature		■		■			■					■									
	Ctenochaetus strigosus	Goldring Bristletooth	Mature						■						■									
	Naso lituratus	Orangespine Unicornfish	Frightened-Mature																					
	Naso unicornis	Bluespine Unicornfish	Mature			■																		
	Naso vlamingi	Vlaming's Unicornfish	Mature				■				■			■										
	Paracanthurus hepatus	Palette Surgeonfish	Mature		■				■															
	Zebrasoma flavescens	Yellow Sailfin Tang	Mature						■															
	Zebrasoma scopas	Brushtail Tang	Mature											■										
Zanclidae	*Zanclus cornutus*	Moorish Idol	Mature											■										
Siganidae	*Siganus corallinus*	Coral Rabbitfish	Mature				■		■					■										
	Siganus doliatus	Barred Rabbitfish	Mature						■	■	■			■	■									
Balistidae	*Balistapus undulatus*	Orange-lined Triggerfish	Mature		■				■	■	■													
	Cantherhines macrocerus	Whitespotted Filefish	Orange Phase							■														
	Rhinecanthus aculeatus	Whitebanded Triggerfish	Mature		■	■					■			■										
	Rhinecanthus assasi		Mature						■					■	■						■			
	Sufflamen bursa	Scimitar Triggerfish	Mature						■		■			■	■									
	Sufflamen chrysopterus	Flagtail Triggerfish	Mature				■		■		■			■	■									
Monacanthidae	*Cantheschenia grandisquarmis*	Large-scale Leatherjacket	Mature		■				■	■														
	Oxymonacanthus longirostris	Beaked Leatherjacket	Mature							■				■									■	
	Paraluterus prionurus	Mimic leatherjacket	Mature		■				■		■			■						■				
Ostraciidae	*Lactophrys quadricornis*	Scrawled Cowfish	Mature		■																			
	Ostracion cubicus	Yellow Boxfish	Juvenile						■															
	Ostracion cubicus	Yellow Boxfish	Mature			■			■											■				
	Ostracion meleagris	Spotted Boxfish	Mature-Male		■					■	■			■										
Tetraodontidae	*Arthron hispidus*	Stars and Stripes Puffer	Mature						■		■			■						■				
	Canthigaster rostrata	Sharpnose Puffer	Mature			■			■															
	Canthigaster valentini	Black-saddled Toby	Mature					■		■	■			■	■									
Embiotocidae	*Brachyistius frenatus*	Kelp Surf Perch	Mature-Female					■	■		■	■												
	Embiotoca jacksoni	Black Surf Perch	Mature-Female																		■			
	Embiotoca lateralis	Striped Surf Perch	Mature-Male			■			■															■
	Embiotoca lateralis	Striped Surf Perch	Juvenile							■	■	■												
	Hypsurus caryi	Rainbow Surf Perch	Mature-Male			■	■			■														
	Hypsurus caryi	Rainbow Surf Perch	Juvenile							■					■	■								■
	Hypsurus caryi	Rainbow Surf Perch	Mature-Female		■	■	■			■	■													
	Rhacochilus vacca	Pile Surf Perch	Mature											■										

References

Allen, G. R., Steene, R. & Allen, M. 1998. *A Guide to Angelfishes and Butterfly-fishes*. Perth: Odyssey Publishing/Tropical Reef Research.

Andersson, M. 1994. *Sexual Selection*. Princeton, NJ: Princeton University Press.

Andersson, S. & Amundsen, T. 1997. Ultraviolet colour vision and ornamentation in bluethroats. *Proceedings of the Royal Society of London, Series B,* **264**, 1587-1591.

Backhaus, W. G. K., Kliegl, R. & Werner, J. S. 1998. *Colour Vision*. Berlin, NY: Walter de Gruyter.

Barlow, G. W. 1963. Ethology of the asian teleost *Badis badis* II. Motivation and signal value of the colour patterns. *Animal Behaviour,* **11**, 97-105.

Barlow, G. W. 1967. Social behavior of a South American leaf fish, *Polycentrus schomburgkii*, with an account of recurring pseudofemale behavior. *American Midland Naturalist,* **78**, 215-234.

Barlow, G. W. 1972. The attitude of fish eye-lines in relation to body shape and to stripes and bars. *Copeia,* **1**, 4-12.

Barlow, G. W. 1974. Contrasts in social behavior between central American cichlid fishes and coral-reef surgeon fishes. *American Zoologist,* **14**, 9-34.

Barlow, G. W. 1991. Mating systems amoung cichlid fishes. In: *Cichlid Fishes: Behaviour, Ecology and Evolution* (Ed. by M. H. A. Keenleyside), pp. 173-190. London: Chapman and Hall.

Barlow, H. B. 1983. What causes trichromacy? A theoretical analysis using comb-filtered spectra. *Vision Research,* **22**, 635-643.

Baylis, J. R. 1979. Optical signals and interspecific communication. In: *The Behavioral Significance of Color* (Ed. by E. H. J. Burtt), pp. 359-378. New York & London: Garland STMP Press.

Bennett, A. T. D., Cuthill, I. C. & Norris, K. J. 1994. Sexual selection and the mismeasure of color. *American Naturalist,* **144**, 848-860.

Bennett, A. T. D., Cuthill, I. C., Partridge, J. C. & Maier, E. J. 1996. Ultraviolet vision and mate choice in zebra finches. *Nature,* **380**, 433-435.

Bennett, A. T. D., Cuthill, I. C., Partridge, J. C. & Lunau, K. 1997. Ultraviolet plumage colors predict mate preferences in starlings. *Proceedings of the National Academy of Sciences, USA,* **94**, 8618-8621.

Blaxter, J. H. S. 1975. Do fish have an absolute sense of light intensity? In: *Vision in Fishes* (Ed. by M. A. Ali), pp. 719-729. New York: Plenum.

Booth, C. L. 1990. Evolutionary significance of ontogenetic colour change in animals. *Biological Journal of the Linnean Society,* **40**, 125-163.

Bowmaker, J. K. 1980. Colour vision in birds and the role of oil droplets. *Trends in NeuroSciences,* **199**, 196-199.

Bowmaker, J. K. 1990. Visual pigments of fishes. In: *The Visual System of Fishes* (Ed. by R. H. Douglas & M. B. A. Djamgoz), pp. 81-107. London: Chapman and Hall.

Bowmaker, J. K., Dartnall, H. J. A. & Herring, P. J. 1988. Longwave-sensitive visual pigments in some deep-sea fishes: segregation of 'paired' rhodopsins and porphyropsins. *Journal of Comparative Physiology, Series A,* **163**, 685-698.

Bowmaker, J. K., Heath, L. A., Wilkie, S. E. & Hunt, D. M. 1997. Visual pigments and oil droplets from six classes of photoreceptor in the retinas of birds. *Vision Research,* **37**, 2183-2194.

Bradbury, J. W. & Vehrencamp, S. L. 1998. *Principles of Animal Communication.* Sunderland, MA: Sinauer.

Breder, C. M. J. 1959. Studies on social groupings in fishes. *Bullentin of the American Museum of Natural History,* **117**, 397-481.

Brockmann, H. J. 1973. The function of poster-coloration in the Beaugregory, *Eupomacentrus leucostictus* (Pomacentridae, Pisces). *Zeitschrift für Tierpsychologie,* **33**, 13-34.

Browman, H. I., Novales-Flamarique, I. & Hawryshyn, C. W. 1994. Ultraviolet photoreception contributes to prey search behaviour in two species of zooplanktivorous fishes. *Journal of Experimental Biology,* **186**, 187-198.

Burkhardt, D., Schleidt, W. & Altner, H. 1967. *Signals in the Animal World.* London: Allen & Unwin.

Burkhardt, D. & Finger, E. 1991. Black, white and UV: how birds see birds. *Naturwissenschaften,* **78**, 279-280.

Chittka, L. 1996. Does bee colour vision predate the evolution of flower colour? *Naturwissenschaften,* **83**, 136-138.

Chittka, L. & Menzel, R. 1992. The evolutionary adaptation of flower colours and the insect pollinators' colour vision. *Journal of Comparative Physiology, Series A,* **171**, 171-181.

Chittka, L., Shmida, A., Troje, N. & Menzel, R. 1994. Ultraviolet as a component of flower reflections, and the colour perception of *Hymenoptera. Vision Research,* **34**, 1489-1508.

Choat, J. H. & Bellwood, D. R. 1991. Reef fishes: their history and evolution. In: *The Ecology of Fishes on Coral Reefs* (Ed. by P. F. Sale), pp. 39-66. San Diego: Academic Press.

Cott, H. B. 1940. *Adaptive Coloration in Animals.* London: Methuen.

Cronin, T. W. & Marshall, N. J. 1989. A retina with at least ten spectral types of photoreceptors in a mantis shrimp. *Nature,* **339**, 137-140.

Cronly-Dillon, J. & Sharma, S. C. 1968. Effect of season and sex on the photopic spectral sensitivity of the threespined stickleback. *Journal of Experimental Biology,* **49**, 679-687.

Crook, A. C. 1997. Colour patterns in a coral reef fish. Is background complexity important? *Journal of Experimental Marine Biology and Ecology,* **217**, 237-252.

Dawkins, M. S. & Guilford, T. 1994. Design of an intention signal in the bluehead wrasse (*Thalassoma bifasciatum*). *Proceedings of the Royal Society of London, Series B,* **257**, 123-128.

Denton, E. J. & Nicol, J. A. C. 1966. A survey of reflectivity in silvery teleosts. *Journal of the Marine Biological Association of the United Kingdom,* **46**, 685-722.

Denton, E. J. & Land, M. F. 1971. Mechanism of reflexion in silvery layers of fish and cephalopods. *Proceedings of the Royal Society of London, Series A,* **178**, 43-61.

Denton, E. J. 1990. Light and vision at depths greater than 200 metres. In: *Light and Life in the Sea* (Ed. by P. J. Herring, A. Campell, K. M. Whitfield & L. Maddock), pp. 127-148. Cambridge: Cambridge University Press.

Douglas, R. H. & Marshall, N. J. 1999. A review of vertebrate and invertebrate ocular filters. In: *Adaptive Mechanisms in the Ecology of Vision.* (Ed. by S.Archer, M. Djamgoz, E. Lowe, J. Partridge & S. Vallerga). London & New York: Chapman and Hall.

Douglas, R. H. & McGuigan, C. M. 1989. The spectral transmission of freshwater teleost ocular media - an interspecific comparison and a guide to potential ultraviolet sensitivity. *Vision Research,* **29**, 871-879.

Douglas, R. H., Bowmaker, J. K. & Kunz, Y. W. 1995. Ultraviolet vision in fish. In: *Seeing Contour and Colour* (Ed. by F. F. Kulikowski, C. M. Dickinson & I. F. Murray), pp. 601-616. Oxford: Pergamon.

Douglas, R. H. & Partridge, J. C. 1997. On the visual pigments of deep-sea fish. *Journal of Fish Biology,* **50**, 68-85.

Dyer, A. G. 1998. The colour of flowers in spectrally variable illumination and insect pollinator vision. *Journal of Comparative Physiology, Series A,* **183**, 203-212.

Ehrlich, P. R., Talbot, F. H., Russell, B. C. & Anderson, G. R. V. 1977. The behaviour of chaetodontid fishes with special reference to Lorenz's 'poster colouration' hypothesis. *Journal of Zoology, London,* **183**, 213-228.

Eibl-Eibesfeldt, I. 1955. Über Symbiosen, Parasitismus, und andere besondere zwischenartliche Beziehungen tropischer Meeresfische. *Zeitschrift für Tierpsychologie,* **12**, 203-219.

Endler, J. A. 1990. On the measurement and classification of colour in studies of animal colour patterns. *Biological Journal of the Linnean Society,* **41**, 315-352.

Endler, J. A. 1991. Variation in the appearance of guppy color patterns to guppies and their predators under different visual conditions. *Vision Research,* **31**, 587-608.

Endler, J. A. 1993. The color of light in forests and its implications. *Ecological Monographs,* **63**, 1-27.

Endler, J. A. & Houde, A. E. 1995. Geographic variation in female preferences for male traits in *Poecilia reticulata. Evolution,* **49**, 456-468.

Endler, J. A. & Thery, M. 1996. Interacting effects of lek placement, display behaviour, ambient light, and color patterns in three neotropical forest-dwelling birds. *American Naturalist,* **148**, 421-452.

Fine, M. L., Winn, H. E. & Olla, B. L. 1977. Communication in fishes. In: *How Animals Communicate* (Ed. by T. A. Sebeok), pp. 472-518. Bloomington: Indiana University Press.

Finger, E. & Burkhardt, D. 1994. Biological aspects of bird colouration and avian colour vision including ultraviolet range. *Vision Research,* **34,** 1509-1514.

Finger, E., Burkhardt, D. & Dyck, J. 1992. Avian plumage colors. *Naturwissenschaften,* **79**, 187-188.

Fox, D. L. 1953. *Animal Biochromes and Structural Colours. Cambridge: Cambridge University Press.*

Fox, H. M. & Vevers, G. 1960. *The Nature of Animal Colours.* London: Sidgwick & Jackson.

Frank, T. M. & Widder, E. A. 1996. UV light in the deep-sea: *in situ* measurements of downwelling irradiance in relation to the visual threshold sensitivity of UV-sensitive crustaceans. *Marine and Freshwater Behavioral Physiology,* **27**, 189-197.

Fricke, H. 1966. Zum Verhalten des Putzerfisches, *Labroides dimidiatus. Zeitschrift für Tierpsychologie,* **23**, 1-3.

Fricke, H. 1973. Behaviour as part of ecological adaptation. In situ studies in the coral reef. *Helgolander Wissenschafligen Meeresuntersuchungen,* **24**, 120-144.

Fricke, H. W. & Fricke, S. 1977. Monogamy and sex change by aggressive dominance in coral reef fish. *Nature,* **266**, 830-832.

Frith, C. M. & Bechler, B. M. 1998. *The Birds of Paradise*. Oxford: Oxford University Press.

Gouras, P. 1991. Precortical physiology of colour vision. In: *Evolution of the Eye and Visual System*, Vol. 2 (Ed. by J. R. Cronly-Dillon & R. L. Gregory), pp. 163-178. London: Macmillan.

Govardovskii, V. I. 1983. On the role of oil drops in colour vision. *Vision Research,* **23**, 1739-1740.

Greenwood, P. H., Rosen, D. E., Weitzman, S. H. & Myers, G. S. 1966. Phyletic studies of teleostean fishes, with a provisional classification of living forms. *Bullentin of the American Museum of Natural History,* **131**, 341-455.

Grutter, A. S. 1999. Cleaner fish really do clean. *Nature* **398**, *672-673*.

Hailman, J. P. 1977. *Optic Signals: Animal Communication and Light.* Bloomington: Indiana University Press.

Hailman, J. P. 1979. Environmental light and conspicuous colors. In: *The Behavioral Significance of Color* (Ed. by E. H. J. Burtt), pp. 289-354. New York & London: Garland STMP Press.

Hamilton, W. J. I. & Peterman, R. M. 1971. Countershading in the colourful reef fish *Chaetodon lunula*: concealment, communication or both? *Animal Behaviour,* **19,** 357-364.

Harrington, M. E. 1993. Aggression in damselfish: adult -juvenile interactions. *Copeia* **1993**, 67-74.

Hart, N. S., Partridge, J. C. & Cuthill, I. C. 1998. Visual pigments, oil droplets and cone photoreceptor distribution in the European starling (*Sturnus vulgaris*). *Journal of Experimental Biology,* **201**, 1433-1446.

Hauser, M. D. 1997. *The Evolution of Communication*. Cambridge, MA: MIT Press.

Helfman, G. S., Collette, B. B. & Facey, D. E. 1997. *The Diversity of Fishes*. Oxford: Blackwell.

Hemmings, C. C. 1966. Olfaction and vision in fish schooling. *Journal of Experimental Biology,* **45**, 449-464.

Jacobs, G. H. 1993. The distribution and nature of colour vision among the mammals. *Biological Reviews,* **68**, 413-471.

Jerlov, N. G. 1976. *Marine Optics*. Amsterdam: Elsevier.

Kasukawa, H., Oshima, N. & Fujii, R. 1987. Mechanism of light reflection in blue damselfish motile iridophore. *Zoological Science,* **4**, 243-257.

Kevan, P. G. 1972. Floral colors in the high arctic with reference to insect-flower relations and pollination. *Canadian Journal of Botany,* **50**, 2289-2316.

Kondrashev, S. L., Gamburtseva, A. G., Gnjubkina, V. P., Orlov, O. J. & My, P. T. 1986. Colouration of corneas in fish: a list of species. *Vision Research,* **26**, 287-290.

Leong, C. Y. 1969. Quantitative effects of releasers on the attack readiness of the fish *Haplochromis burtoni* (Cichlidae, Pisces). *Zeitschrift für Vegleichende Physiologie,* **65**, 29-50.

Levine, J. S. & MacNichol, E. F. Jr. 1979. Visual pigments in teleost fishes: effects of habitat, microhabitat, and behaviour on visual system evolution. *Sensory Processes,* **3**, 95-131.

Levine, J. S. & MacNichol, E. F. Jr. 1982. Colour vision in fishes. *Scientific American,* **246**, 140-149.

Loew, E. R. & Lythgoe, J. N. 1978. The ecology of cone pigments in teleost fishes. *Vision Research,* **18**, 715-722.

Loew, E. R. & Lythgoe, J. N. 1985. The ecology of colour vision. *Endeavour,* **14**, 170-174.

Loew, E. R. & McFarland, W. N. 1990. The underwater visual environment. In: *The Visual System of Fishes* (Ed. by R. H. Douglas & M. B. A. Djamgoz), pp. 1-43. London: Chapman & Hall.

Loew, E. R., McAlery, F. A. & McFarland, W. N. 1996. Ultraviolet sensitivity in the larvae of two species of marine atherinid fishes. In: *Zooplankton Sensory Ecology and Physiology* (Ed. by P. H. Lenz, D. K. Hartline, J. E. Purcell, & D. L. Macmillan), pp. 195-209. Sydney: Gordon and Breach.

Longley, W. H. 1917. Studies upon the biological significance of animal colouration. *American Naturalist,* **51**, 257-285.

Lorenz, K. 1962. The function of colour in coral reef fishes. *Proceedings of the Royal Institute of Great Britain,* **39**, 282-296.

Losey, G. S. J. 1979. Fish cleaning symbiosis: proximate causes of host behaviour. *Animal Behaviour,* **27**, 669-685.

Losey, G. S. J. 1987. Cleaning symbiosis. *Symbiosis,* **4**, 229-258.

Losey, G. S. J. 1993. Knowledge of proximate causes aids our understanding of function and evolutionary history. *Marine Behavioral Physiology,* **23**, 175-186.

Lowe, C. & Goodman-Lowe, G. 1996. Suntanning in hammerhead sharks. *Nature,* **383**, 677.

Lythgoe, J. N. 1968. Red and yellow as conspicuous colours underwater. *Underwater Association Report*, 51-53.

Lythgoe, J. N. 1976. The ecology, function and phylogeny of irridescent multilayers in fish corneas. In: *Light as an Ecological Factor*, Vol. 2 (Ed. by G. C. Evans, R. Bainbridge & O. Rackham), pp. 211-247. Oxford: Blackwell .

Lythgoe, J. N. 1979. *The Ecology of Vision.* Oxford: Clarendon Press.

Lythgoe, J. N. 1984. Visual pigments and environmental light. *Vision Research,* **24**, 1539-1550.

Lythgoe, J. N., Muntz, W. R. A., Partridge, J. C., Shand, J. & Williams, D. M. 1994. The ecology of the visual pigments of snappers (Lutjanidae) on the Great Barrier Reef. *Journal of Comparative Physiology, Series A,* **174**, 461-467.

Mahon, J. L. 1994. Advantage of flexible juvenile coloration in two species of *Labroides* (Pisces: Labridae). *Copeia,* **2**, 520-524.

Maier, E. J. & Bowmaker, J. K. 1993. Colour vision in the passeriform bird, *Leiothrix lutea*: correlation of visual pigment absorbance and oil droplet transmission with spectral sensitivity. *Journal of Comparative Physiology, Series A,* **172**, 295-301.

Marshall, N. J. 1988. A unique colour and polarization vision system in mantis shrimps. *Nature,* **333**, 557-560.

Marshall, N. J. 1996. Measuring colours around a coral reef. *Biophotonics International,* July/August, 52-56.

Marshall, N. J. & Oberwinkler, J. 1999. The colourful world of the mantis shrimp. *Nature,* **401**, 873-874.

Mazel, C. H. 1999. Coral fluorescence characteristics: excitation-emission spectra, fluorscence efficiencies, and contribution to apparent reflectance. *Society for Photo-optical Instrumentation Engineers*, **2953**, 240-245.

McDonald, C. G. & Hawryshyn, C. W. 1995. Intraspecific variation of spectral sensitivity in threespine stickleback (*Gasterosteus aculeatus*) from different photic regimes. *Journal of Comparative Physiology, Series A,* **176**, 255-260.

McFarland, W. N. 1986. Light in the sea - correlations with behaviours of fishes and invertebrates. *American Zoologist,* **26**, 389-401.

McFarland, W. N. 1991. The visual world of coral reef fishes. In: *The Ecology of Fishes on Coral Reefs.* (Ed. by P. F. Sale), pp. 16-36. New York: Academic Press.

McFarland, W. N. & Munz, F. W. 1975a. Part I: Presumptive cone pigments extracted from tropical marine fishes. *Vision Research,* **15**, 1045-1062.

McFarland, W. N. & Munz, F. W. 1975b. Part II: The photic environment of clear tropical seas during the day. *Vision Research,* **15**, 1063-1070.

McFarland, W. N. & Munz, F. W. 1975c. Part III: The evolution of photopic visual pigments in fishes. *Vision Research,* **15**, 1071-1080.

McFarland, W. N. & Loew, E. R. 1994. Ultraviolet visual pigments in marine fishes of the family Pomacentridae. *Vision Research,* **34**, 1395-1396.

Mollon, J. D. 1982. Colour vision and colour blindness. In: *The Senses* (Ed. by H. B. Barlow & J. D. Mollon), pp. 165-191. Cambridge: Cambridge University Press.

Myreberg, A. A. & Thresher, R. E. 1974. Interspecific aggregation and its relevance to the concept of territoriality in reef fishes. *American Zoologist,* **14**, 81-96.

Neal, T. J. 1993. A test of the function of juvenile color patterns in the pomacentrid fish *Hypsypops rubicundus* (Teleostei: Pomacentridae). *Pacific Science,* **47**, 240-247.

Neil, E. H. 1964. An analysis of color changes and social behavior of *Tilapia mossambica. University of California Publications in Zoology,* **75**, 1-58.

Neudecker, S. 1989. Eye camouflage and false eyespots: chaetodontid responses to predators. *Environmental Biology of Fishes,* **25**, 143-157.

Neumeyer, C., Wietsma, J. J. & Spekreijse, H. 1991a. Separate processing of 'colour' and 'brightness' in goldfish. *Vision Research,* **31**, 537-549.

Neumeyer, C. 1991b. Evolution of colour vision. In: *Vision and Visual Dysfunction: Evolution of the Eye and Visual System*, Vol. 2 (Ed. by J. R. Cronly-Dillon & R. L. Gregory), pp. 284-305. London: Macmillan.

Nicol, J. A. C. 1989. *The Eyes of Fishes.* Oxford: Clarendon.

Ohtsuka, T. 1985. Spectral sensitivities of seven morphological types of photoreceptors in the retina of the turtle, *Geoclemys reevesii. Journal of Comparative Neurology,* **237,** 145-154.

Osorio, D., Marshall, N. J. & Cronin, T. W. 1997. Stomatopod photoreceptor spectral tuning as an adaptation for colour constancy in water. *Vision Research,* **37**, 3299-3309.

Partridge, J. C. 1990. The colour sensitivity and vision of fishes. In: *Light and Life in the Sea* (Ed. by P. J. Herring, A. K. Campbell, M. Whitfield & L. Maddock), pp. 167-184. Cambridge: Cambridge University Press.

Pfennig, K. S. 1998. The evolution of mate choice and the potential for conflict between species and mate-quality recognition. *Proceedings of the Royal Society of London, Series B,* **265**, 1743-1748.

Potts, G. W. 1973. The ethology of *Labroides dimidiatus* (Cuv. & Val.) (Labridae, Pisces) on Aldabra. *Animal Behaviour,* **21**, 250-291.

Prum, R. O., Torres, R., Williamson, S. & Dyke, J. 1999. Two-dimensional fourier analysis of the spongy medullary keratin of structurally coloured feather barbs. *Proceedings of the Royal Society of London, Series B,* **226**, 13-22.

Randall, J. E. 1967. Food habits of reef fishes of the West Indies. *Studies of Tropical Oceanography,* **5**, 665-847.

Randall, J. E., Allen, G. R. & Steene, R. C. 1997. *The complete Diver's and Fishermen's Guide to Fishes of the Great Barrier Reef and Coral Sea*, 2nd edn. Bathurst: Crawford House Publishing.

Reese, E. S. 1975. A comparative field study of the social behaviour and related ecology of reef fishes of the family Chaetodontidae. *Zeitschrift für Tierpsychologie,* **37**, 37-61.

Reese, E. S. 1981. Predation on corals by fishes of the family Caetodontidae: implications for conservation and management of coral reef ecosystems. *Bulletin of Marine Science,* **31,** 594-604.

Robertson, D. R. & Hoffman, S. G. 1977. The roles of female mate choice and predation in the mating systems of some tropical labroid fishes. *Zeitschrift für Tierpsychologie,* **45,** 289-320.

Rowland, W. J. 1979. The use of color in intraspecific communication. In: *The Behavioral Significance of Colour* (Ed. by E. H. J. Burtt), pp. 379-421. New York & London: Garland STMP Press.

Ryan, M. J. 1990. Sexual selection, sensory systems and sensory exploitation. *Oxford Surveys in Evolutionary Biology,* **7,** 157-195.

Schwalm, P. A. & Starrett, P. H. 1977. Infrared reflectance in leaf-sitting neotropical frogs. *Science,* **196,** 1225-1227.

Shand, J. 1993. Changes in the spectral absorption of cone visual pigments during the settlement of the goatfish *Upeneus tragula*: the loss of red sensitivity as a benthic existence begins. *Journal of Comparative Physiology, Series A*, **173**, 115-121.

Shapiro, D. Y. 1991. Intraspecific variability in social systems of coral reef fishes. In: *The Ecology of Fishes on Coral Reefs* (Ed. by P. F. Sale), pp. 754. San Diego: Academic Press.

Smith, V. C. & Pokorny, J. 1975. Spectral sensitivities of human foveal cone photopigments between 400 and 500 nm. *Vision Research,* **15**, 161-171.

Smith, R. C. & Tyler, J. E. 1976. Transmission of solar radiation into natural waters. In: *Photochemical and Photobiological Reviews*, Vol. 1 (Ed. by K. C. Smith), pp. 117-155. New York & London: Plenum Press.

Stavenga, D. G., Smits, R. P. & Hoenders, B. J. 1993. Simple expontential functions describing the absorbance bands of visual pigment spectra. *Vision Research,* **33**, 1011-1017.

Thorpe, A., Douglas, R. H. & Truscott, R. J. W. 1993. Spectral transmission and short-wave absorbing pigments in the fish lens. I. Phylogenetic distribution and identity. *Vision Research,* **33**, 289-300.

Thresher, R. E. 1984. *Reproduction in Reef Fishes.* Hong Kong: T. F. H. Inc. Ltd.

Thresher, R. E. & Moyer, J. T. 1983. Male success, courtship complexity and patterns of sexual selection in three congeneric species of sexually monochromatic and dichromatic damselfishes (Pisces: Pomacentridae). *Animal Behaviour,* **31**, 113-127

Topp, R. W. 1970. Behavior and color change of the rudderfish, *Kyphosus elegans,* in the Gulf of Panama. *Copeia,* **1970**, 763-765.

Townsend, C. H. 1929. Records of changes in color among fishes. *Zoologica,* **9**, 321-378.

Verrell, P. A. 1991. Illegitimate exploitation of sexual signalling systems and the origin of species. *Ethology, Ecology and Evolution,* **3**, 273-283.

Vorobyev, M. & Osorio, D. 1998. Receptor noise as a determinant of colour threshold. *Proceedings of the Royal Society of London, Series B,* **265,** 351-358.

Vorobyev, M., Osorio, D., Bennett, A. T. D. & Cuthill, I. C. 1989. Tetrachromacy, oil droplets and bird plumage colours. *Journal of Comparative Physiology, Series A,* **183**, 621-633.

Walls, G. L. 1942. *The Vertebrate Eye and its Adaptive Radiation*. Michigan: Cranbrook Institute of Science.

Webster, M. A. & Mollon, J. D. 1997. Adaptation and the colour statistics of natural images. *Vision Research,* **37,** 3283-3298.

Werner, J. S. & Spillman, L. 1989. UV-absorbing intraocular lenses: safety, efficacy and consequences for the cataract patient. *Graefes Archiv für Klinische und Experimentelle Opthalmologie,* **227,** 248-256.

Wyszecki, G. & Stiles, W. S. 1982. *Colour Science: Concepts and Methods, Quantitative Data and Formulae.* 2nd edition. New York: Wiley.

Zumpe, D. 1965. Laboratory observations on the aggressive behaviour of some butterfly fishes (Chaetodontidae). *Zeitschrift für Tierpsychologie,* **22,** 226-236.

IN: ESPMARK,Y., AMUNDSEN,T. & ROSENQVIST,G. (EDS.) 2000
ANIMAL SIGNALS:
SIGNALLING AND SIGNAL DESIGN IN ANIMAL COMMUNICATION, PP. 121-132
TAPIR ACADEMIC PRESS, TRONDHEIM, NORWAY

5

A Signalling Tail

SUSAN FITZPATRICK

School of Applied Biological and Chemical Sciences,
University of Ulster at Jordanstown, Newtownabbey, Co. Antrim BT 37 0QB, Northern Ireland

E-mail: sm.fitzpatrick@ulst.ac.uk

Abstract

Types of tail signals and their costs are outlined, and data on western Palaearctic birds are used to describe the correlations between tail signals. Shape, length and markings vary independently, apparently conveying multiple messages. Selection for elaboration of signalling traits and natural selection against tail elongation are explored within this multiple signalling system. Natural selection costs of elongated tails to migratory birds have greatly reduced the exaggeration of tail length, but markings (without aerodynamic costs) do not provide a simple substitute for length in the signalling system; complex interactions occur for some marks. The various tail markings also seem to convey multiple messages. Differential mating success has selected for greatly elongated tails only in resident polygynous birds, while migratory polygynous species are relatively short-tailed. In contrast, 'good parent' selection from differential reproductive success seems to have produced a degree of tail elongation even in migrants, but has not produced highly exaggerated tails in residents. Little sexual dimorphism in tail length occurs in species with high levels of paternal care, suggesting that selection for quality-signalling arising from differential reproductive success acts similarly on males and females. Tail shape and tail markings are also mainly monomorphic traits. Sexual selection on both sexes for quality-signalling and the associated near-monomorphic signalling traits deserve greater research effort.

Tail signals and their costs

Most birds probably signal with their tails during agonistic and/or sexual contexts; in the western Palaearctic avifauna nearly 80% of species with well-described behaviour possessed tail displays (Fitzpatrick 1998a). While these displays were identified by movement and posture, tail signalling traits are also morphological, involving the length, shape, markings and coloration of the tail.

These morphological signalling traits can affect the aerodynamic functions of tails during flight, generally reducing efficiency and influencing manoeuvrability and agility. Length and shape have serious effects on flight economy: the area of tail from the base to the point of maximum width increases both lift and drag, but any further extension increases drag without increasing lift (Thomas 1993). In ornamental shapes some rectrices (tail feathers) are elongated beyond the point of maximum width, increasing drag and hence the power required for flight (Balmford et al. 1993; Thomas 1993). Increased total length also increases drag and raises total body weight by adding mass to the tail. These factors increase the energetic cost of flight and, through effects on the turning moment and moment of inertia, alter the manoeuvrability and agility possible during flight (Evans & Thomas 1992; Thomas 1993, 1996; Thomas & Balmford 1995). Additional length thus has more potential for aerodynamic costs than ornamental shape.

Signalling through tail elongation and ornamental shape thus imposes costs to general flight economy, efficiency and ability. These tail traits have been variously interpreted as arbitrary, exaggerated by the Fisher process until these costs prevent further elaboration, or as quality-indicating signals kept 'honest' by the handicap of the cost (Zahavi 1975). There is at least some evidence that tail traits are quality-dependent (review in Andersson 1994), and thus constitute handicaps.

Handicap theory requires the handicap to be specific, acting against the very quality being signalled (Grafen 1990; Hasson 1997; Zahavi & Zahavi 1997). The general aerodynamic costs of tail elongation thus necessitate tail elongation be interpreted as a general viability indicator. However, only a small proportion of Palaearctic birds have extremely elongated tails (Fitzpatrick 1999); the majority, although using tail displays, do not appear to be greatly handicapped by their signal. Neither elongation above average nor an ornamental shape (the viability handicaps) seem to be essential components of the tail signal (Fitzpatrick 1998a). An alternative to the viability handicap therefore seems to be required for maintenance of signal 'honesty' at apparently low aerodynamic costs. So, could there be a specific quality advertised by tails which can be effectively signalled without extreme ornamentation, more in accordance with handicap theory?

Absence of melanin increases feather abrasion and weakens feather strength (Burtt 1986; Bonser 1995), so tail markings could affect the ability of the tail to withstand abrasion and breakage. Feather quality, defined as resistance to damage and abrasion, encapsulates the effects of diet, prior stresses, and environmental conditions on growth at the time of the moult (discussed further in Fitzpatrick 1998b). Growth marks on rectrices are used in ptilochronology for assessment of individual condition (Grubb 1995). Feathers can also potentially demonstrate developmental stability – Hasson (1991) has suggested that melanin barring markings amplify these growth marks as a signal of growth regularity. Bands could be amplifiers of growth regularity across feathers (O. Hasson, pers. comm.). Feathers have such major importance to birds, for energy balance as well as flight, that signals of feather quality would seem highly adaptive.

Tail markings, length and shape could indicate feather quality by improving the perceptibility of any damage (an amplifier signal) while at the same time making damage and abrasion more likely to occur (a specific handicap effect). This combination is a type of signal referred to as an amplifying handicap (Fitzpatrick 1998a). The 'honesty' of amplifiers is maintained by their effects: exaggeration of the amplifier improves discrimination by receivers and reduces the success of poor-quality individuals as their deficiencies are made more apparent (Hasson 1989, 1997). An amplifying handicap combines this cost of an amplifier with an additional specific handicap cost - the signal reduces the ability of the individual to achieve excellence in the quality being signalled by the amplifier. A handicap acting on feather quality is provided for pale markings by absence of melanin, while increased length and the extended feathers of ornamental tail shapes may increase the frequency of tail contact with objects in the environment, increasing abrasion and risk of damage.

If length, shape and markings are amplifying handicap signals of feather quality, they impose costs specific to this quality which should ensure 'honest' expression. However, by producing an aerodynamically inefficient tail, length and shape entail additional general handicapping costs which are absent in the case of markings. Tail markings may therefore provide 'backup' low-cost signals, or be part of a 'multiple messages' signalling system. A 'backup' signal should be associated consistently with the signal being duplicated, whereas multiple messages need not necessarily correlate with each other (Møller & Pomiankowski 1993; Johnstone 1996).

The results from a broad comparative study of these tail traits in the western Palaearctic avifauna are reviewed and slightly extended here to address these questions concerning the nature of tail signals and the signalling system they comprise. The index of tail elongation used was termed longtailedness to distinguish it from other measures of relative tail length. Longtailedness was measured by the residuals of a log/log regression of tail on wing length, or of tail on tarsus length, for each sex separately (Fitzpatrick 1997, 1998a, 1999).

Tail shapes were categorized as graduated, lyre, pin, rounded, square, deep forks, shallow forks and notches. In deep forks, the fork occupied a quarter or more, in shallow forks less than a quarter, and in notches less than a tenth, of the total tail length. For square tails, all rectrices were of equal length. In tails where rectrix length increased from outer to inner feathers, small differences of less than an eighth of tail length were categorized as rounded, and greater incremental differences as graduated; in pin tails central rectrices only were elongated. Lyre tails had curved rectrices (Fitzpatrick 1997, 1998a, 1999).

Tail markings were classified as having one or more of the following markings: plain – no tail markings; tip – pale terminal portion on all rectrices; bars – alternating pale and narrow dark marks perpendicular to the rectrix rachis, bars aligned on all rectrices; band – single broad dark mark perpendicular to the rectrix rachis, aligned on all rectrices; spots – one or more separated pale areas on individual rectrices; edge – pale outer (± inner) web of the outermost rectrix; stripe – inner rectrices darker than middle rectrices (Fitzpatrick 1998a).

Other factors categorized for each species were (i) Migration: resident - non-migratory in all Palaearctic populations; partial migrant - migratory in some Palaearctic populations, resident in others, or some populations including both migratory and resident individuals; migrant - all individuals of Palaearctic populations migratory. (ii) Mating system: monogamy - >90% males monogamous, any polygynous males with two mates; polygyny - >10% males polygynous, maximum number of mates >2. (iii) Male parental care. For each of the following breeding activities, male parental contribution was assessed as absent (0) or present (1): nestbuilding, incubation/incubation-feeding of female, care of young chicks, and care of older chicks. These scores were summed to give an index of male parental care, range 0-4. For further details of the study see Fitzpatrick (1997, 1998a, 1999).

Correlations between tail signalling traits

In western Palaearctic birds, displayed tails were on average longer (relative to body size) than non-displayed tails, and ornamental shaped tails were displayed in a greater proportion of species than aerodynamically less costly shapes (Fitzpatrick 1998a). Tail markings, however, were not consistently associated with any particular ornamental tail shape; in fact, less ornamental shapes such as square and notched tails more frequently carried markings than the more ornamental graduated tails (Fitzpatrick 1998a). For some tail shapes marked tails showed significantly greater tail elongation than unmarked tails (e.g. graduated + tip, graduated + bars, square + band – see Fitzpatrick 1998a), but only one marking had a consistent relationship with longtailedness across all tail shapes (stripe, being associated with short tails in all shapes). Length, shape and markings thus varied independently and appear to convey 'multiple messages' to a greater extent than 'backup' signals. The various markings also showed complex patterns of co-occurrence, varying across tail shapes (Fitzpatrick 1998a), suggesting that markings alone comprise a multiple signalling system.

Balance between selection for and against costly tail traits

Multiple signals, some with and some without general additional costs, are therefore available on a tail. It is possible that the impact of these general viability costs might shape the evolution of the signalling system. Species in which the general costs have particularly high fitness consequences might be selected to utilize the 'cheaper' signals such as markings to a greater extent and to reduce the elaboration of the more expensive traits. Such species would still be able to produce an efficacious and informative signal of quality.

The strength of sexual selection for trait elaboration depends on two factors. The first is differential mating success, when signals can increase the number of mates obtained by males through success in mate choice and/or intrasexual competition. From this source of selection male trait elaboration is expected to be greater in regularly polygynous species than in monogamous species. Secondly, differential reproductive success per breeding attempt, related to parental quality, also produces sexual selection (Hoelzer 1989). This selection will be strongest in species in which reproductive success is limited by parental care. In such species males can greatly enhance their fitness by providing parental care, and the quality of both the male and the female is likely to influence reproductive success. When both sexes can improve fitness by choosing and/or competing for a high quality mate, both will be selected to signal quality.

Through the costs of the tail signals, natural selection will oppose the tail elaboration produced by these sexual selection pressures. Inefficient flight is likely to have a particularly high viability cost in migratory birds, many of which make long non-stop flights over hostile habitats such as sea and desert, covering distances very close to their range limits (Bairlein 1992; Biebach 1992; Pennycuick et al. 1996). A second cost of an aerodynamically inefficient tail could affect reproductive success: the additional flight costs during the nesting period might reduce the amount of food parents can deliver to the nestlings, and thus reduce the number and/or quality of fledglings produced. This cost, discussed previously by Fitzpatrick et al. (1995), should be greatest for birds in which the amount or quality of parental care determines reproductive success – the situation in which 'good parent' sexual selection from differential reproductive success is greatest.

Table 1. *Relationship between longtailedness, tail markings and migration in males. Probabilities of distribution of markings with migration (χ^2), and probabilities from 2-way ANOVA relating longtailedness to tail markings and migration.*

	MARKINGS				
	tip	band	bars	edge	stripe
occurrence χ^2	0.846	0.1856	0.5804	0.1333	0.0114
ANOVA factor					
markings	0.0931	0.0039	0.0758	0.0023	0.0021
migration	0.0001	0.0006	0.0001	0.0002	0.0147
interaction	0.0531	0.206	0.0001	0.012	0.8841

Whereas polygynous mating systems produced considerably more tail elongation than monogamous mating systems in residents, this tail elongation did not occur in migrants (Fig. 1A). For polygynous migrants, the aerodynamic costs of tail elongation during migration therefore completely counteracted the sexual selection for tail elongation

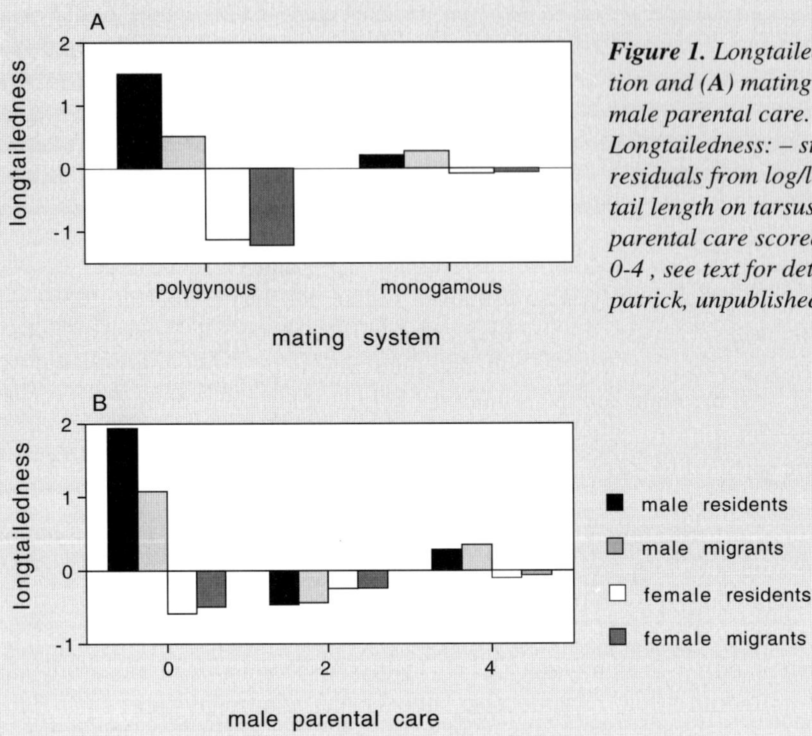

Figure 1. *Longtailedness, migration and (**A**) mating system, (**B**) male parental care. Longtailedness: – standardized residuals from log/log regression of tail length on tarsus length; male parental care scored on scale of 0-4 , see text for details. S. Fitzpatrick, unpublished.*

arising from differential mating success. Monogamous migrants, however, had considerably longer tails than polygynous migrants (Fig. 1A), suggesting additional sexual selection for tail elongation in monogamous species – presumably that due to differential reproductive success. For both males and females, migrant species with high levels of male parental care were more longtailed than those with little parental care (Fig. 1B). In con-trast, residents providing moderate amounts of male care had shorter tails than those providing either high or low levels of male care. These patterns have been interpreted as indicating that in migrants sexual selection arising from differential reproductive success, acting on species with high levels of male parental care, is sufficiently strong to overcome the natural selection costs of tail elongation (S. Fitzpatrick, unpublished). Tail elongation in residents with high male care may be constrained from extreme elaboration by the effects of aerodynamic costs of the tail on parental care.

Tail markings and tail costs in migrants

As migration has such a strong effect on signalling via tail length, it is possible that markings could provide a substitute, lower-cost signal. However, the occurrence of most markings was not significantly related to migration (Table 1), with the exception of stripe which occurred more frequently among migrants.

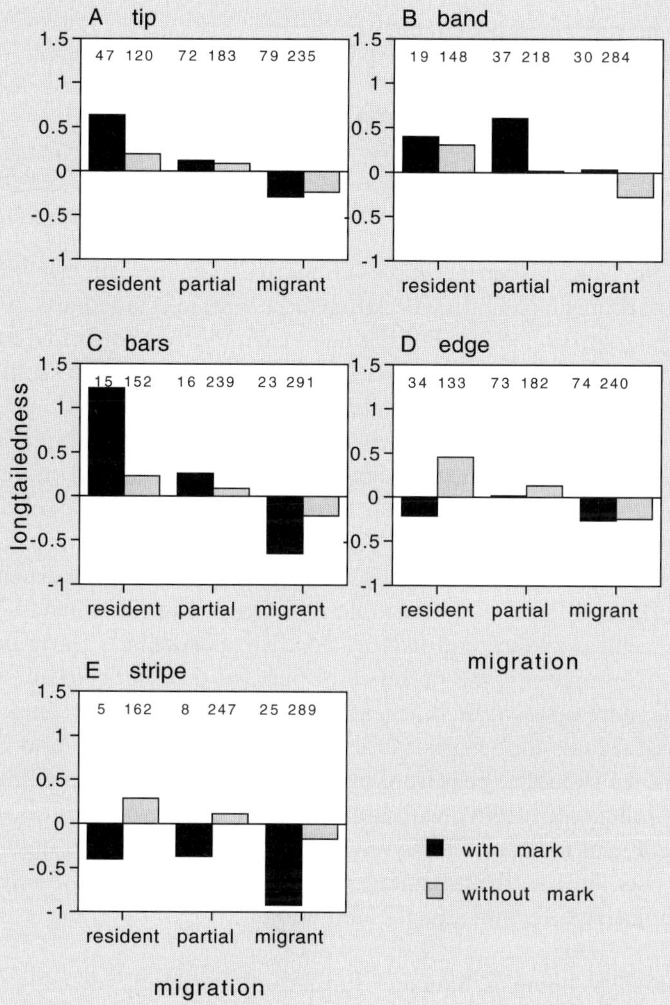

Figure 2. *Mean longtailedness of birds with and without tail marks and with different levels of migration. (**A**) tip; (**B**) band; (**C**) bars; (**D**) edge; (**E**) stripe. Number of species shown above columns. For descriptions of migration and markings see text. Longtailedness: see Figure 1.*

Longtailedness, migration, and markings had a complex relationship. Longtailedness varied significantly with presence or absence of the mark for band, edge and stripe, and tended to do so for the remaining marks (tip, bars) (Table 1). MarkingXmigration interactions were significant in bars, edge and nearly significant in tip. In migrants, tails with the markings tip, bars and stripe were shorter than those without these marks (Fig. 2A, C, E). For tip and bars in resident birds, marked tails were, however, longer than those without these markings (Fig. 2A, C), producing the interaction effect. For band and edge (Fig. 2B, D), unmarked tails showed a general reduction in tail length with migration, whereas the marked tails did not – banded tails were longer than those without bands whatever the migratory behaviour, and edged tails were short even in residents.

The hypothesis that markings, as signals without a general handicap, could be substituted for aerodynamically costly tail elongation in migratory birds is therefore not upheld.

These results indicate that markings, rather than being substitutes for tail elongation, can show complex interactions with this signal element.

Signalling implications

Markings

It is clear that the tail signalling system is very complex. Length and shape can vary independently, and the different markings have various relationships with longtailedness, with shape, and with each other (Fitzpatrick 1998a). For the most part, there is no simple substitution of one signal element for another within this system. Tail signals thus seem to provide 'multiple messages' rather than 'backup signals' (Møller & Pomiankowski 1993; Johnstone 1996). The great importance of signalling costs in the evolution of multiple signalling systems (Johnstone 1996) is clearly illustrated by the effects of migration on longtailedness, but the results are not exactly as expected from Johnstone's (1996) models.

The differing relationships between longtailedness, migration and the presence or absence of markings suggest that markings also convey multiple messages. The relationship between bars (and tips), longtailedness and migration (Fig. 2A, C) is particularly intriguing. At high cost of longtailedness in migrants, the presence of bars (or tip) seems to allow a reduction in expression of the more costly trait, whereas at lower cost of longtailedness in residents, the presence of both signals together is associated with more extreme trait elaboration. Perhaps this is what we should expect from multiple-message amplifier signals, which were not modelled by Johnstone (1996); with high costs the presence of the second amplifier allows some reduction in the more costly signal without detriment to quality-discriminability, but when costs are low the advantages of easier perception of multiple qualities perhaps have a cumulative effect on signal elaboration.

Band was interpreted as possibly being a 'damage limitation' marking, reducing the potential extent of rectrix damage (Fitzpatrick 1998a). As such, it would be an anti-handicap, ameliorating the adverse effects of pale feather tips or tail elongation. Band width should be negatively correlated with quality, as was found by Wiehn (1997). The association of bands with longer tails across all levels of migration (Fig. 2B) is consistent with this possibility, but is also consistent with bands being amplifiers of rectrix growth regularity (O. Hasson, pers. comm.). In either case, this association of bands with longer tails indicates bands to differ functionally from bars or tips. Whether the band itself is a quality-signal, or whether quality is actually signalled by the size of the paler area of rectrix proximal to the band, requires experimental study.

The signal function of edge is enigmatic. This mark has the potential to be an amplifying handicap for abrasion/breakage of the outer rectrices, and is associated with longtailedness in some shapes of tail (Fitzpatrick 1998a). However, the relationship between presence of edge, migration and longtailedness (Fig. 2D) is very different from that of tip (Fig. 2A), a marking interpreted as an amplifying handicap for end-of-tail abrasion

(Fitzpatrick 1998a). A difference in signal function might therefore be expected; perhaps a flight-related signal is involved, as residents with edge are so short-tailed. Brooke (1998), however, found no correlation between his scoring of tail 'flash' marks and flocking behaviour in waders, but the scoring system used rated barred tails, edges and completely white tails along a single scale, thus confounding edge with markings having potentially different signal functions.

Stripe was associated with relatively short tails across migration categories (Fig. 2E) and also across all shapes of tail, the only mark to have such a consistent relationship (Fitzpatrick 1998a). Together with the disproportionately frequent occurrence of stripe on migrants, this suggests a link with long-distance flight for this mark. It could be adaptive coloration rather than a signal; Burtt (1986) suggested that melanin in central rectrices would reduce abrasion from airborne particles during flight. The outer rectrices, being held furled under the central feathers and thus protected, have less need for melanin.

Sexual dimorphism of tail signals

Sexual dimorphism of these tail signals is limited. The sexes had similar tail length except in polygynous residents (Fig. 1A) and in residents providing no paternal care (Fig. 1B). In both cases, this dimorphic tail elongation in males is attributable to sexual selection from differential male mating success. Tail elongation above the natural selection optimum occurs in both sexes of residents and migrants with high paternal care and has been attributed to 'good parent' sexual selection arising from differential reproductive success (S. Fitzpatrick, unpublished). The sexes differed in tail shape in very few species, in which males possessed one of the more ornamental shapes (e.g. pin, lyre), while the females had less costly shaped tails (e.g. square) (S. Fitzpatrick, unpublished). In other ornamental shapes (e.g. graduated, deep forks) little sexual dimorphism of tail shape occurs. Dimorphism of tail markings, too, is uncommon (S. Fitzpatrick, unpublished).

The frequent occurrence of monomorphic tail traits suggests that tail signalling is advantageous for both sexes, more especially in species which provide biparental care. An alternative possibility is that female traits are a non-functional correlated response to selection on male traits, only slowly modified to produce sexual dimorphism (Lande 1980; Lande & Arnold 1985). This seems unlikely as a general explanation, especially for tail markings, as sexual dimorphism in plumage coloration is evolutionarily labile and changes in female colouring occur frequently (Irwin 1994; Peterson 1996; Price & Birch 1996; Omland 1997; Burns 1998; see also Amundsen, this volume). Where both sexes benefit from a high-quality mate, i.e. quality of biparental care influences reproductive success, the direct benefits of choice apply to both females and males, producing sexual selection on both sexes, mutual mate choice, and resultant patterns of assortative mating (Johnstone et al. 1996; Johnstone 1997). Assortative mating in near-monomorphic birds, implying mutual sexual selection, has been studied in very few species (Jones & Hunter 1993; Fitzpatrick & Price 1997; Andersson et al. 1998). Both inter- and intra-sexual selection can act on female birds (Trail 1990; Slagsvold & Lifjeld 1994; Monaghan et al. 1996; Amundsen et al. 1997) but sexual selection on females is under-researched (see

Amundsen, this volume), and little is known about quality-signalling by female birds. The widespread occurrence of a variety of tail signalling traits in females as well as males suggests they are an important aspect of quality-signalling – in both sexes. The prevalent usage of sexual dimorphism as an index of the strength of sexual selection has led to an unfortunate bias in the study of sexually selected signalling traits, that is, an emphasis on selection due to differential mating success. This appears to be weaker than 'good parent' sexual selection for quality-signalling and thus represents only a partial evolutionary explanation of signalling traits. Increasing our understanding of selection on near-monomorphic signalling traits, such as those present in many birds' tails, is an essential next step for research into the evolution of signals, and poses a considerable challenge.

References

Amundsen, T., Forsgren, E. & Hansen, L. T. T. 1997. On the function of female ornaments: male bluethroats prefer colourful females. *Proceedings of the Royal Society of London, Series B,* **264**, 1579-1586.

Andersson, M. 1994. *Sexual Selection.* Princeton, NJ: Princeton University Press.

Andersson, S., Örnborg, J. & Andersson, M. 1998. Ultraviolet sexual dimorphism and assortative mating in blue tits. *Proceedings of the Royal Society of London, Series B,* **265**, 445-450.

Bairlein, F. 1992. Recent prospects on trans-Sahara migration of songbirds. *Ibis* **134** (supplement), 41-46.

Balmford, A., Thomas, A. L. R. & Jones, I. L. 1993. Aerodynamics and the evolution of long tails in birds. *Nature*, **361**, 628-631.

Biebach, H. 1992. Flight-range estimates for small trans-Sahara migrants. *Ibis* **134** (supplement), 47-54.

Bonser, R. H. C. 1995. Melanin and the abrasion resistance of feathers. *Condor*, **97**, 590-591.

Brooke, M. de. L. 1998. Ecological factors influencing the occurrence of 'flash marks' in wading birds. *Functional Ecology*, **12**, 339-346.

Burns, K. J. 1998. A phylogenetic perspective on the evolution of sexual dichromatism in tanagers (Thraupidae): the role of female versus male plumage. *Evolution*, **52**, 1219-1224.

Burtt, E. H. J. 1986. An analysis of physical, physiological, and optical aspects of avian coloration with emphasis on wood-warblers. *Ornithological Monographs*, **38**, 1-126.

Evans, M. R. & Thomas, A. L. R. 1992. The aerodynamic and mechanical effects of elongated tails in the scarlet-tufted malachite sunbird: measuring the cost of a handicap. *Animal Behaviour*, **43**, 337-347.

Fitzpatrick, S. 1997. Patterns of morphometric variation in birds' tails: length, shape and variability. *Biological Journal of the Linnean Society*, **62**, 145-162.

Fitzpatrick, S. 1998a. Birds' tails as signaling devices: markings, shape, length, and feather quality. *American Naturalist*, **151**, 157-173.

Fitzpatrick, S. 1998b. Colour schemes for birds: structural coloration and signals of quality in feathers. *Annales Zoologici Fennici*, **35**, 67-77.

Fitzpatrick, S. 1999. Tail length in birds in relation to tail shape, general flight ecology, and sexual selection. *Journal of Evolutionary Biology*, **12**, 49-60.

Fitzpatrick, S. & Price, P. 1997. Magpies' tails: damage as an indicator of quality. *Behavioral Ecology and Sociobiology*, **40**, 209-212.

Fitzpatrick, S., Berglund, A. & Rosenqvist, G. 1995. Ornaments or offspring: costs to reproductive success restrict sexual selection processes. *Biological Journal of the Linnean Society*, **55**, 251-260.

Grafen, A. 1990. Biological signals as handicaps. *Journal of Theoretical Biology*, **144**, 517-546.

Grubb, T. C. 1995. Ptilochronology: A review and prospectus. *Current Ornithology,* **12**, 89-114.

Hasson, O. 1989. Amplifiers and the handicap principle in sexual selection: a different emphasis. *Proceedings of the Royal Society of London, Series B*, **235**, 383-406.

Hasson, O. 1991. Sexual displays as amplifiers: practical examples with an emphasis on feather decorations. *Behavioral Ecology*, **2**, 189-197.

Hasson, O. 1997. Towards a general theory of biological signaling. *Journal of Theoretical Biology*, **185**, 139-156.

Hoelzer, G. A. 1989. The good parent process of sexual selection. *Animal Behaviour*, **38**, 1067-1078.

Irwin, R. E. 1994. The evolution of plumage dichromatism in the new world blackbirds: social selection on female brightness? *American Naturalist*, **144**, 890-907.

Johnstone, R. A. 1996. Multiple displays in animal communication: 'backup signals' and 'multiple messages'. *Philosophical Transactions of the Royal Society of London, Series B*, **351**, 329-338.

Johnstone, R. A. 1997. The tactics of mutual mate choice and competitive search. *Behavioral Ecology and Sociobiology*, **40**, 51-59.

Johnstone, R. A., Reynolds, J. D. & Deutsch, J. C. 1996. Mutual mate choice and sex differences in choosiness. *Evolution*, **50**, 1382-1391.

Jones, I. L. & Hunter, F. M. 1993. Mutual sexual selection in a monogamous seabird. *Nature*, **362**, 238-239.

Lande, R. 1980. Sexual dimorphism, sexual selection, and adaptation in polygenic characters. *Evolution*, **34**, 292-305.

Lande, R. & Arnold, S. J. 1985. Evolution of mating preferences and sexual dimorphism. *Journal of Theoretical Biology*, **117**, 651-664.

Møller, A. P. & Pomiankowski, A. 1993. Why have birds got multiple sexual ornaments? *Behavioral Ecology and Sociobiology*, **32**, 167-176.

Monaghan, P., Metcalfe, N. B. & Houston, D. C. 1996. Male finches selectively pair with fecund females. *Proceedings of the Royal Society of London, Series B*, **263**, 1183-1186.

Omland, K. E. 1997. Examining two standard assumptions of ancestral reconstructions: repeated loss of dichromatism in dabbling ducks (Anatini). *Evolution*, **51**, 1636-1646.

Pennycuick, C. J., Einarsson, O., Bradbury, T. A. M. & Owen, M. 1996. Migrating whooper swans *Cygnus cygnus*: satellite tracks and flight performance calculations. *Journal of Avian Biology*, **27**, 118-134.

Peterson, A. T. 1996. Geographic variation in sexual dichromatism in birds. *Bulletin of the British Ornithologists' Club*, **116**, 156-172.

Price, T. & Birch, G. L. 1996. Repeated evolution of sexual color dimorphism in passerine birds. *Auk*, **113**, 842-848.

Slagsvold, T. & Lifjeld, J. T. 1994. Polygyny in birds: the role of competition between females for male parental care. *American Naturalist*, **143**, 59-94.

Thomas, A. L. R. 1993. On the aerodynamics of birds' tails. *Philosophical Transactions of the Royal Society of London, Series B*, **340**, 361-380.

Thomas, A. L. R. 1996. Why do birds have tails? The tail as a drag reducing flap, and trim control. *Journal of Theoretical Biology*, **183**, 247-253.

Thomas, A. L. R. & Balmford, A. 1995. How natural selection shapes birds' tails. *American Naturalist*, **146**, 848-868.

Trail, P. W. 1990. Why should lek-breeders be monomorphic? *Evolution*, **44**, 1837-1852.

Wiehn, J. 1997. Plumage characteristics as an indicator of male parental quality in the American kestrel. *Journal of Avian Biology*, **28**, 47-55.

Zahavi, A. 1975. Mate selection - a selection for a handicap. *Journal of Theoretical Biology*, **53**, 205-214.

Zahavi, A. & Zahavi, A. 1997. *The Handicap Principle.* Oxford: Oxford University Press.

IN: ESPMARK,Y., AMUNDSEN,T. & ROSENQVIST,G. (EDS.) 2000
ANIMAL SIGNALS:
SIGNALLING AND SIGNAL DESIGN IN ANIMAL COMMUNICATION, PP. 133-154
TAPIR ACADEMIC PRESS, TRONDHEIM, NORWAY

6

Female Ornaments: Genetically Correlated or Sexually Selected?

TROND AMUNDSEN

Department of Zoology, Norwegian University of Science and Technology,
N-7491 Trondheim, Norway

E-mail: trond.amundsen@chembio.ntnu.no

Abstract

Studies of sexual selection have so far focussed almost solely on males, and have been highly successful in explaining their often extravagant adornments. However, females of a variety of animal species are also beautifully decorated, but this fact has been largely ignored by evolutionary biologists until very recently. The traditional explanation for female ornamentation dates back to Darwin, and posits that it has evolved as a correlated by-product of selection on the same traits in males (the correlated response hypothesis). Alternatively, female ornamentation may be a consequence of selection acting directly on the females, through female-female competition, male choice or other processes. The two hypotheses are, however, not mutually exclusive. In fact, a genetic correlation is inevitable since the two sexes share most of the genome, and may act as a 'pre-adaptation' for functional female decorations. Recent comparative evidence suggests that, in birds, female plumage often evolves independent of changes in the male, indicating that female traits are relatively unconstrained by the genetic correlation. Regarding the hypothesis that selection may act directly on female ornamentation, recent theoretical research predicts that sexual selection on females should be expected under certain social and ecological circumstances. At the same time, empiricists have provided evidence to support both intrasexual and intersexual selection acting on female traits. For instance, in bluethroats Luscinia svecica, males seem to prefer more colourful and symmetric females as mates. However, current evidence on the functions of female ornaments is limited and to some extent contradictory. Many more studies, on a variety of taxa, are needed in order to clarify the importance of male choice and female-female competition in the evolution of female showiness.

Even before Darwin, biologists had recognised that conspicuous ornaments occur much more frequently among male than female animals. Darwin's work (1859, 1871) made it clear that the splendour of, for instance, male peafowl *Pavo cristatus* and birds-of-paradise (Paradisaeidae) requires special explanation because of the obvious uselessness, if not costliness, of exaggerated tails or plumes in 'the struggle for life'. Accordingly, the search for explanations of male extravaganza was the main focus of *The Descent of Man, and Selection in Relation to Sex* (Darwin 1871) and has remained the central theme of the renewed interest in sexual selection during recent decades (e.g. Andersson 1994).

In current studies of sexual selection on males, the term 'ornament' is often restricted to traits that have been shown to be influenced by intersexual selection (female choice). In contrast, traits evolved by intrasexual selection are termed 'armaments' by some authors (e.g. Berglund et al. 1996), while extravagant characters that are either functionless or of unknown function have no specific term. This combination of description and function in terminology complicates the discussion of traits for which the function is unknown (i.e., the majority of 'ornament-like' traits), either due to a lack of investigation or conflicting or ambiguous results. The practice of some authors (e.g. Møller & Zamora-Munoz 1997) to use sexual dimorphism of a trait as a criterion for it being a sexually selected ornament (thereby neglecting monomorphic showiness due to mutual sexual selection) adds to the terminological confusion. In this paper, I use the term 'ornament' in a broad sense, to describe any trait (in males or females) that is conspicuous and apparently useless or costly in utilitarian (natural selection) terms. Thus, I make no distinction between ornaments and armaments.

The forgotten beauty of females

Being generally the least ornamented of the two sexes, females have largely been forgotten or neglected in studies of secondary sex traits. Focusing their attention on the more frequently ornamented males, behavioural ecologists have overlooked the fact that female animals are sometimes beautifully ornamented. In fact, extravagant female characters are not uncommon. Many species of seabirds (e.g. auks Alcidae and cormorants Phalacrocoracidae) have conspicuous crests or colourful beaks in both sexes. Conspicuous but monomorphic appearance also occurs frequently in passerines (e.g. tanagers Thraupidae; Burns 1998) and several other avian orders. Among fish, identical colourful appearances in the two sexes are very common, for instance, among butterflyfishes Chaetodontidae and angelfishes Pomacanthidae on coral reefs, and also in many other taxa (e.g. Lampriformes and wrasses Labridae). Similarly ornamented males and females are likewise found in a number of invertebrate taxa (e.g. dragonflies Odonata and several species of scarabiid beetles). Among mammals, many ungulates (Cervidae and Bovidae) have horns or antlers in both sexes, either equally well developed or with a less extreme expression in females. The occurrence of ornaments in females similar to those in males, but somewhat less striking, is common both in birds (e.g. bluethroats *Luscinia svecica*; below) and fish (e.g. Arctic charr *Salvelinus alpinus*). In some species of fish, females carry ornaments that are different from those of the males (e.g. two-spotted goby *Gobiusculus flavescens* and several labrids including peacock wrasse *Labrus ossifagus* and Gaimard wrasse *Coris gaimard*). The above examples are provided simply to illustrate the widespread occurrence of female ornaments and represent no systematic survey; many more cases could have been provided from most major taxa.

Seemingly redundant and useless beauty requires biological explanation, whether it occurs in males or females. However, it was not until very recently that theoreticians and empiricists started to approach female ornaments from a functional perspective. The traditional focus on male showiness has been founded in a belief that competition for mates is usually stronger among males than females. Although probably true at a general level, we now need to ask under which circumstances significant competition may also take place among females, for access to the best male partners. For example, females might compete for mates if there are benefits to pairing with a high quality mate, if there is a shortage of territory-holding males, or if they have to compete with other females for male parental care (e.g. Langmore et al. 1996). Like males (Darwin 1871; Andersson 1994), females may compete for mates through advertising their attractiveness (male choice; intersexual selection) or in direct female-female contests (female-female competition; intrasexual selection).

The study of female ornamentation is still in its infancy; only the first few steps have been taken towards a general understanding. At the heart of the current scientific debate over female ornaments is the question of whether they are due to selection acting directly on the females, or only genetically correlated by-products of sexual selection on males. The aim of the present paper is (i) to sketch the history of the study of female ornaments, (ii) to assess the current state of knowledge, and (iii) to provide suggestions for future research. I do not claim to provide a comprehensive review of the issue. Instead, I illustrate the various suggested functions for female ornaments primarily using birds as examples, and with a particular emphasis on my own studies of bluethroats. Birds are the animal taxon I know the best; somewhat regrettably, it is also a fact that the vast majority of detailed studies regarding female ornamentation have been conducted on birds. Hence, there is more to report than for other taxa. Moreover, I concentrate on species with 'conventional sex roles' and pay only little attention to animals with reversed sex roles. This is simply because female ornaments in such species can be explained by analogy to male ornaments in species with conventional sex roles (see Rosenqvist 1990); hence, their presence represents no particular challenge to traditional theory. Finally, I have focussed solely on visual ornaments, primarily because this is the class of secondary sex traits that has been subject to most studies on females. However, evidence is now accumulating on the functions of female vocalisations in birds; I refer the reader to recent reviews by Langmore (1998, this volume) on the issue.

The legacy of Darwin and Wallace

Female animals can be visually ornamented by having gaudy colours, extravagant structures (for example, long tails and crests in birds), or patterns and markings (see Fitzpatrick, this volume). While female ornaments have received very little attention during the recent 'explosion' of sexual selection studies in the 1970s and 80s, the phenomenon did not pass untreated by Darwin and Wallace. Darwin (1871) was very well aware of the fact that females were sometimes ornamented, and discussed the phenomenon of female traits in detail. In the *Descent*, he treats cases where females are more ornamented than males, where the two sexes are equally or similarly ornamented, and where males are ornamented and females quite drab. Darwin's discussion of female showiness is embedded in his discussion of sexual dimorphism, with a main focus on explaining male ornaments by sexual selection. Like the present paper, Darwin's (1871) treatment of the issue was bird biased. In fact, the whole theory of sexual selection in the *Descent* is to a very large extent based on the extensive knowledge that existed already at Darwin's time of the natural history of birds. Darwin (1871) was, however, extremely careful in considering all evidence available, including invertebrates, amphibia, reptiles, fish and mammals. He generally uses the same kind of arguments in explaining female ornaments in the various taxa; hence, his statements regarding birds can be regarded as of general relevance.

It is one of the *Descent*'s fundamental insights that, normally, males are much more elaborately ornamented than females. Nevertheless, Darwin (1871) also showed interest in exceptions from this rule. For instance, he reports that female painted snipes *Rostratula benghalensis* and phalaropes *Phalaropus* spp. are brighter than the males. However, Darwin (1871) points out that 'the difference in colour between the sexes is far from conspicuous' (II, p. 203) and much less pronounced than when males are the more ornamented sex. Darwin's understanding of such cases was remarkably modern. He emphasised that males of these and some other species undertook 'the duty of incubation', and also that a female-biased sex ratio might render it 'not improbable that the females would have been led to court the males, instead of being courted by them' (II, p. 207). Thus, Darwin (1871) attributes female ornamentation in such cases to sexual selection.

Darwin's view of female coloration being sexually selected in what we now call 'sex-role reversed species' contrasts strongly with his view of female ornaments when equal to or less developed than those of the males. In such cases, Darwin refrained from adaptive explanations, and argued that 'the differences in colour between the sexes, whether great or small, may to a large extent be explained on the principle of the successive variations, acquired by the males through natural selection, having been from the first more or less limited in the their transmission to the females' (1871, II, p. 177). Thus, he considered variation in 'vestigial' or 'rudimentary' female expression of 'male ornaments' to be a non-adaptive consequence of different 'modes of transmission'. It is unclear why Darwin (1871) considered the 'mode of transmission' an inflexible constraint, instead of adopting his general adaptationist view that any trait could be modified by evolution if relevant selection pressures are present. Having based his whole theory of sexual selection on the utilitarian cost of exaggerated ornaments, the cost to the female from inheriting 'male ornaments', if these were of no use to her, must have been very clear to Darwin. The implication, not addressed by Darwin, that selection might act to de-couple male and female traits under such circumstances seems close at hand. Perhaps, Darwin's (1871) reluctance to explain the various degrees of sexual dimorphism (and, hence, of female ornaments) by selection was, in some way, related to his mistaken metaphysical hypotheses on the mechanisms of heredity; strongly contrasting his materialistic understanding of selection and evolution (Lewontin 1974; Mayr 1982).

It is also unclear why Darwin *a priori* attributes ornaments equally expressed in the two sexes as solely the result of selection in one sex (males). Apparently, he finds simultaneous sexual selection on both sexes theoretically unlikely, as demonstrated in his discussion of certain hummingbirds with both male and female ornamentation (Darwin 1871, II, pp. 220-222). Darwin's suggested explanation is that both sexes have been subject to sexual selection, but during different periods, due to changes in sex ratio. Darwin's difficulties with mutual sexual selection in animals are puzzling in the light of his treatment of sexual selection in man. Here, Darwin (1871, Ch. XIX-XX) very clearly acknowledges that both men and women are choosy, although apparently on different cues.

Not only did Darwin reject selection to explain female ornaments similar to those of males; he also rejected selection as an explanation of the female drabness typical of many species. Again, his explanation was the 'laws of inheritance'. Cronin (1991) has pointed out that Darwin's denial of selection for protection in females is hard to explain; it is seemingly inconsistent with the general selectionist approach that penetrates most of his work. Wallace (1889), on the other hand, argued strongly that female drabness was a result of selection for protection, but denied sexual selection explanations for male ornamentation. In Cronin's (1991) words, it was 'males for Darwin, females for Wallace'; each of them explained one sex but not the other in selective terms. However, neither of them considered conspicuous female ornaments, when equal or weaker than similar ornaments in the males, to require a selective explanation.

Darwin's theory of sexual selection was long rejected or ignored by evolutionary biologists (Cronin 1991). In particular, the idea of sexual selection through female choice received little if any support in the scientific debate following his book. During a time when Darwin's major arguments regarding sexual selection were questioned, it is not surprising that less central issues like that of ornaments occurring in both sexes, or mostly in females, were largely forgotten. When Fisher (1915, 1930) took up Darwin's idea, his focus was on evaluating the validity of Darwin's (1871) main and most disputed claim: that female preferences could shape the appearance of males. To the extent that Fisher (1930) considered female traits, he emphasised that the genome was mostly shared between the sexes and thus could lead to 'correlated characters' in females.

At first glance, it may seem odd that a phenomenon treated so explicitly and in such detail by Darwin (1871) has largely remained outside the arena of sexual selection studies until the present decade. I suggest that there are two main reasons for the long-lasting lack of interest in female ornaments, related to the sociological and historical dynamics of the scientific process.

First, the predominant theoretical position has not stimulated empirical work. From Darwin (1871) through Fisher (1930) and Lande (1980, 1987), leading theoreticians have interpreted female ornaments as a correlated result of selection acting on males. In principle, genetic correlations may explain not only female ornaments (vocal, visual or of any kind); even male choosiness could be a by-product of selection for female choosiness (Lande 1980; Halliday & Arnold 1987). This dominating theoretical view has probably inhibited empirical research on functional explanations of female ornaments. However, the view is based on the assumption that sexual selection acts only on the males. Although a reasonable assumption in many cases, it need not always be true. Recent theoretical studies suggest that, under certain circumstances, male choosiness and mutual sexual selection should be expected (e.g. Dewsbury 1982; Parker 1983; Owens & Thompson 1994, Deutsch & Reynolds 1995; Johnstone et al. 1996). I believe that this more pluralistic theoretical approach to sexual selection has been, and will increasingly be, important in stimulating research on the potential functions of female showiness.

Second, until recently, empirical research on sexual selection had simply not reached a stage that encouraged influential scientists to focus at anything but the major questions. After Darwin (1871) followed 'one hundred years without sexual selection' (Cronin 1991), despite the efforts of Fisher (1915, 1930) and a few others. When sexual selection again became fashionable in the early 60s, and primarily during the 70s and 80s, the focus was on the two big questions: (1) Can the process of sexual selection be supported theoretically, through formal evolutionary models? (2) Can sexual selection in the wild be demonstrated by experiment? In fact, it was not until the beginning of the 90s that science had provided a reasonable set of convincing examples of sexual selection (on male traits) in the wild, and a similar set of models describing sexual selection through Fisherian, good genes or good parent processes. With the pressing questions relating to the evolution of male extravaganza (admittedly much more common than female showiness) unsolved, it is no wonder that only few scientists showed an interest in female ornaments. This situation has changed dramatically during the 90s. More and more scientists now realise that, in order to understand the dynamics of mate choice and reproduction, knowledge on the female party is essential (e.g. Rosenqvist & Berglund 1992; Cunningham & Birkhead 1997, 1998; Gowaty 1997; Parker & Burley 1998). Still, there has been some reluctance to address female showiness from a functional perspective, and the present knowledge is scant. I would not be surprised if female ornamentation and mutual sexual selection are among those areas in sexual selection that will provide the most important new insights in the forthcoming years.

Genetic correlations and phylogenetic analyses

In most cases, ornaments occurring in females are similar or identical to those occurring in males of the same species. As argued already by Darwin (1871), this fact suggests that traits in the two sexes are not inherited independently. Lande and co-workers (Lande 1980, 1987; Lande & Arnold 1985) have elaborated this idea in formal genetic models. Basically, they argue that the expression of 'male traits' in females is the inevitable consequence of pleiotropy and linkage of genes affecting the characters of both sexes, and that the evolution of sexual dimorphism is a very slow process. When females show 'subequal' development of a male sexual character, the evolution of sexual dimorphism is still in progress, according to Lande (1980). Implicit in this scenario is the assumption that the character benefits only one sex, an assumption that is not necessarily true, for example in the case of mutual mate choice (Jones & Hunter 1993; Johnstone et al. 1996). The 'genetic correlation' idea has been highly influential, and its widespread acceptance is probably one of the main reasons why researchers have been very reluctant to address female ornaments from a perspective of selection. However, just as with Darwin, the scientific community adopted Lande's main message ('female ornament expression is correlated to that of the male'), but largely ignored the less strongly emphasised point that selection *can* and apparently *does* affect female traits directly. In a way, the genetic insights of Lande and colleagues acted as an excuse for empiricists not to address adaptive explanations of extravagant female characters.

One way to assess whether female extravaganza is due to selection on females is to consider evolutionary changes in character states of the two sexes by means of comparative methods. If the ancestral state of a taxon is conspicuous monomorphism, the existence of female drabness in derived states suggests selection on females for crypsis, which apparently has been strong and consistent enough to break up the male-female correlation. Likewise, a change in female character state from drab to conspicuous without a parallel change for the male indicates sexual or natural selection for female conspicuousness. Two recent, phylogenetically based, comparative analyses have confirmed such scenarios. In New World blackbirds (Passeriformes: Icterinae) (Irwin 1994) and tanagers (Burns 1998), changes in character state between drab and colourful have occurred much more frequently in females than in males. These changes have been in both directions; not infrequently has the female state changed from drab to colourful without any change in the male (Irwin 1994; Burns 1998). This strongly suggests that selection has acted directly on the females in favour of more extreme appearances ('ornamentation'). Price & Birch (1996) arrived at similar conclusions from a broader analysis including all passerine species. In their data, changes from dichromatism to monochromatism and from monochromatism to dichromatism were both frequent, suggesting that transitions between character states are fairly unconstrained by genetic correlations (Price & Birch 1996). In fact, dichromatism was lost much more often than gained (Price & Birch 1996). This pattern was also found in a survey of geographical variation in avian dichromatism (Peterson 1996), and in a phylogenetic analysis of dichromatism in dabbling ducks (Anatini) (Omland 1997). Taken together, these studies show that, although ornaments in males and females normally share a common genetic basis and hence are genetically correlated, selection has the potential to break this correlation and can act efficiently on one sex independent of the other.

Phylogenetic analyses have, however, limited potential in establishing what kind of selection is shaping the sometimes colourful or otherwise extravagant appearance of females. This requires focussed empirical work, including experiments to test the importance of male choice and female-female competition. Moreover, comparative studies analysing the ecological correlates of female ornamentation would be highly valuable; such knowledge is almost completely lacking today. Using a verbal comparative approach, Bleiweiss (1992) described variation in female plumage iridescence among and within species of sunangel hummingbirds (genus *Heliangelus*), and suggested that its occurrence in females is related to defence of feeding territories (see also Wolf 1969, 1975; Bleiweiss 1985). Focussing on nest predation as a selective agent for crypsis, Martin & Badyaev (1996) recently provided comparative evidence that nest predation negatively affects female but not male plumage brightness in warblers (Parulinae) and finches (Carduelinae) (but see Stutchbury & Howlett 1995 for experimental evidence that predation is unrelated to female colour in hooded warblers *Wilsonia citrina*). Martin & Badyaev's (1996) comparative data basically support Wallace's (1889) idea that female drabness is due to 'selection for protection', but does not address whether showiness may sometimes confer benefits on females.

One aspect of female showiness that has received very little attention is variation *within* species, geographically or individually. However, Peterson's (1996) recent survey suggests that geographically separated populations of the same species often vary in dichromatism, due to plumage differences in one or both sexes. For instance, in red-winged blackbirds *Agelaius phoeniceus*, Cuban females are black and rather male-like, in contrast to the typically brown females in North America (Whittingham 1992). At the individual level, females of certain species display highly variable plumage (varying from drab to conspicuous 'male-like'; e.g. bluethroats; below, European blackbirds *Turdus merula*; F. Götmark, pers. comm.) whereas, in other species, females are invariably drab (e.g. chaffinches *Fringilla coelebs*; F. Götmark, pers. comm.). The importance of various selection pressures in explaining such differences between populations and species remains to be explored.

Bluethroats - a case study of ornamentation

Bluethroats are amongst the most brilliantly ornamented European bird species. The male carries a multiply coloured throat patch; a glossy patch of structural blue surrounds a chestnut (northern subspecies *svecica*) or white (southern subspecies *cyanecula*) spot, and is bordered below by black, white (sometimes) and chestnut bands (e.g. Fig. 1 in Amundsen et al. 1997). In contrast, most field guides have reported the female to be quite drab in coloration, with only traces of male-like colour. However, when starting to work on bluethroats (northern sub-species) in the early 90s, I soon realised that females were extremely variable in throat coloration. From 1992 on, we have systematically described and photographically documented all females caught in our population. This has revealed a striking variability: the drabbest females lack any trace of blue or chestnut colours, while the most colourful females resemble males in appearance (Amundsen et al. 1997, Fig. 1). The female ornament is clearly the same as that of the male, but with a more variable and normally lesser degree of expression. Most likely, the ornament in both sexes shares a common genetic basis. In females, the blue and chestnut colours vary partly independently of each other: some individuals have blue but not chestnut, others chestnut without blue. However, individuals with a strong expression of either element would almost always express the other to some degree. The degree of ornamentation is unrelated (Smiseth & Amundsen, in press) or only weakly related (Rohde et al. 1999, T. Amundsen, unpublished) to female age. Female ornamentation appears to be relatively stable for each individual across seasons (Rohde et al. 1999, T. Amundsen, unpublished), despite the annual moulting of the ornament.

Why, then, are bluethroat females sometimes so colourful? Although there is no exact knowledge of the costs related to throat ornamentation in bluethroats, the mere fact that both sexes moult to a drabber appearance during winter and thus incur the costs of a second, pre-breeding moult, demonstrates that the ornament is not cost-free. Hence, some kind of benefits from extravagant female ornamentation must be invoked in order to explain the maintenance of female colour in the population. Since the female (and the male) ornament is expressed exclusively during the breeding season, lasting only about two months in our study population, the function of the ornament must be sought during this period. Basically, the ornament could be functional in three different contexts: (i) attraction of male mates, (ii) female-female competition over mates or breeding territories, or (iii) female-female competition over non-sexual resources such as food.

In bluethroats, like in most other small passerines, encounters between same-sex or opposite-sex individuals are infrequently observed in the wild. When seen, they are often very swift and the behavioural details hard to record. Moreover, interpreting the role of each party in such natural encounters is not straightforward (Amundsen et al. 1997). For instance, when an advertising male is visited by a female that does not settle, it cannot *a priori* be concluded that it is the female that has rejected the male, and not *vice versa* (although this is very often the inference drawn). Because of these difficulties with field studies, we designed aviary experiments to test the three alternative (though not mutually exclusive) functions.

Male preferences for female ornamentation were tested in mate choice aviaries erected on the breeding grounds, in prime habitat. This approach had previously proven successful in demonstrating female preferences for male characteristics (symmetry; Fiske & Amundsen 1997, ultraviolet ornamentation; Andersson & Amundsen 1997). Basically, we recorded the preference of males when given a choice between one colourful and one drab female that were matched with respect to age. Preferences were assessed from the time spent in proximity to each of the two females, and from sexual displays. The results revealed a significant male preference for colourful females, both in terms of time (association) and sexual behaviours (Amundsen et al. 1997). Thus, our experiment indicated that female coloration might be subject to sexual selection acting directly on the females, through male choice. However, it remains to be demonstrated that the preference expressed in the standardised mate choice set-up really affects mate choice in the wild (Amundsen et al. 1997). We have argued that high-quality males might benefit from being choosy early in the season (when there are many potential female partners), but that males should be indiscriminate if of poor quality or late in the season (Amundsen et al. 1997).

The male preference for colourful females raises the question of what qualities might be signalled by female coloration. Potentially, female quality variation could either result in direct (e.g. parental care) or indirect (genetic) benefits of male mate choice. Since direct benefits can be substantial in species with extensive parental care (including passerine birds), and may incur stronger selection than genetic (indirect) effects in such systems (Kirkpatrick & Barton 1997), the bluethroat studies have so far concentrated on testing whether female coloration reflects parental skills. Using a non-experimental approach, Rohde and co-workers (1999) found no relationship between female colour and provisioning or nestling growth in bluethroats. However, assessing female parental quality in species with bi-parental care is complicated by the fact that the optimal level of care by each sex is dependent upon the amount of care provided by the other sex (e.g. Houston & Davies 1985), rendering non-manipulative studies inconclusive. We therefore performed an experiment in which male bluethroats were temporarily removed (for about ten hours), so that the relationship between female colour and provisioning of nestlings could be evaluated independent of male effects. Our results basically confirm Rohde et al.'s (1999) findings and lend no support to the suggestion that colourful females are better at providing nestlings with food (Smiseth & Amundsen, in press). Coloration in female bluethroats also seems to be unrelated to defence of nestlings against (artificial) predators (T. Amundsen & M. Myklebust, unpublished). Thus, it appears that any male benefit of mating with colourful females must be indirect, either through producing attractive offspring (the Fisher process) or offspring of superior genetic quality (good genes). Unfortunately, studies of intergenerational effects are almost impossible in bluethroats, because only a very small proportion of the offspring return to settle in their natal area (T. Amundsen et al., unpublished). However, estimating the immune response of females (see Saino et al. 1997) in relation to coloration would be a valuable first step in assessing the potential for genetic benefits related to immunocompetence (Folstad & Karter 1992).

As stated above, male attraction and female competition functions of female ornaments are not mutually exclusive. In fact, both of these functions are fundamentally aspects of competition among females to mate with the best males available. We are currently testing if female coloration is related to dominance in competition over sexual (mates) or non-sexual (food) resources. Bluethroats are facultatively polygynous (e.g. Johnsen & Lifjeld 1995), and females seem to defend their mates against intruding females at the territories by vocalising (S.A. Sæther & T. Amundsen, unpublished). So far, the results suggest no similar function related to coloration (T. Amundsen et al. unpublished).

Female ornaments and male mate choice

The bluethroat study described above is one of still only a handful of experiments aimed at testing the potential function of female showiness in mate choice of species with conventional sex roles. Hence, the current state of knowledge is limited. With one important exception (Jones & Hunter 1993), the few existing studies have been performed on species where the female ornament is a much weaker expression of the male ornament, like in the bluethroat. Even though such 'vestigial' female ornaments may well have been selected for, the strongest candidates for demonstrating functional female ornaments are probably species where the two sexes carry similar extravagant ornaments. This is the case with the crested auklet *Aethia cristatella*, a small Pacific alcid. Crested auklets live in dense colonies and both sexes display at common arenas. Jones and Hunter (1993) manipulated the crest size of mounts of both sexes, and recorded sexual displays from opposite-sex individuals to these mounts when placed at the mating arena. They showed that large-crested individuals of both sexes received significantly more displays, and concluded that, in crested auklets, mutual mate choice and sexual selection is taking place (Jones & Hunter 1993). It is as yet unknown what benefits male auklets may accrue from mating with large-crested females. Studies of other species (birds or other taxa) in which both sexes are equally ornamented should be highly rewarding, allowing us to assess whether the intriguing crested auklet results are typical for such situations.

Although still limited, more data exist for species where females carry a more or less reduced version of the male ornament. This is the case, for instance, in many passerine species and the order Passeriformes is the taxon having been subject to most research on the issue. The results provide no clear-cut picture. In aviary experiments, males have been shown to prefer a colourful female plumage in house finches *Carpodacus mexicanus* (Hill 1993) and bluethroats (Amundsen et al. 1997; see above), but did not exert significant colour preferences in pinyon jays *Gymnorhinus cyanocephalus* (Johnson 1988) or red-winged blackbirds (Muma & Weatherhead 1989). In zebra finches *Taeniopygia guttata*, males have been reported to prefer orange-billed females, in contrast to a conspecific female preference for red bills in males (Burley & Coopersmith 1987; but see Jennions 1998). Descriptive data from barn swallows *Hirundo rustica* suggested that long-tailed females might have a mating advantage (Møller 1993a, 1994), but this was not supported by a subsequent field experiment (Cuervo et al. 1996). Likewise, Tella et al. (1997) were unable to find any mating advantage (earlier breeding) for colourful over drab lesser kestrels *Falco naumanni* in a descriptive study. Although the results are mixed, I find it intriguing that the studies with probably the highest power (controlled aviary experiments with reasonable sample sizes) have tended to support the notion of functional female ornaments. However, several more studies are needed before any general conclusions can be drawn.

If males prefer ornamented females, is this because such females signal a high quality in some sense (e.g. Zahavi 1975; Hoelzer 1989)? While the understanding of the signalling content of male ornaments has improved dramatically during recent years (e.g. Andersson 1994; Johnstone 1995; Zahavi & Zahavi 1997; Olson & Owens 1998), extremely little is known about individual quality in relation to ornamentation in females. Hill (1993) examined various measures of female breeding performance that might correlate with female coloration in house finches, but found no such relationships. However, female epaulet colour seems to be condition-dependent in red-winged blackbirds (Johnsen et al. 1996). In bluethroats, female throat colour was related to body size (Amundsen et al. 1997), but, as stated above, appeared to be unrelated to parenting ability (Rohde et al. 1999, Smiseth & Amundsen, in press). However, Linville et al. (1998) recently reported evidence that colourful female northern cardinals *Cardinalis cardinalis* provide more food for their offspring than do drab mothers. Although this study did not experimentally control for male effects as was done by Smiseth and Amundsen, it is currently the strongest evidence present in favour of direct benefits to males from mating with ornamented females. Focussing on potential indirect benefits, Potti & Merino (1996) found that pied flycatcher *Ficedula hypoleuca* females with a white forehead patch had fewer parasites, but surprisingly found no similar relationship for males. A relationship between parasitism and female (and male) coloration has also been found in fishes. Skarstein & Folstad (1996) reported that bright red female arctic charr had fewer deleterious parasites than drabber females. Despite these few studies providing suggestive evidence in favour of both direct and indirect benefits, it is still largely unknown whether female showiness is generally reflective of any kind of individual quality.

Badges of status in female-female competition

Alternatively or additionally to a role in male mate choice, female ornaments may be selected through female-female competition over sexual or non-sexual resources (e.g. West-Eberhard 1983). Such a function of female showiness has received indicative support in field studies of dotterels *Charadrius morinellus* (Owens et al. 1994), capuchinbirds *Perissocephalus tricolor* (Trail 1990) and hummingbirds *Amazilia* spp. (Wolf 1969; Bleiweiss 1985). In dotterels, females are the more competitive (and ornamented) sex and compete for access to males on mating arenas. Similarly, females of capuchinbirds and certain hummingbird species frequently engage in aggressive interactions over mates (capuchinbirds; Trail 1990) or food (hummingbirds; e.g. Wolf 1969). Indeed, female-female competition has been suggested as a general explanation for monomorphism among many lekking taxa (Trail 1990). West-Eberhard (1983) has argued that, in species where both sexes participate in aggressive or territorial displays, conspicuous but sexually monomorphic appearances are unlikely to have evolved by sexual selection on males (with pleiotropic, or correlated, effects on females), but are more likely explained as a result of resource competition ('social selection', *sensu* West-Eberhard). According to West-Eberhard (1983), this may for instance be the case for birds like certain toucans (*Ramphastos* spp. and *Pteroglossus* spp.), colonial parrots (*Agapornis* spp.) and jays, as well as for various species of lizards, mammals and fish.

Only a few experimental studies have specifically been designed to test the idea that female showiness is selected through female-female competition. Again, results are not consistent and general conclusions are premature. In aviary experiments with pinyon jays, malar feather coloration was weakly related to dominance in competition for mates (Johnsen 1988). In contrast, Muma and Weatherhead (1989) did not find any significant relationship between aggression and epaulet colour in dummy-presentation experiments with captive red-winged blackbirds. Female convict cichlid fish *Cichlasoma nigrofasciatum* with a bright ventral coloration received more aggressive displays than drab ones (Beeching et al. 1998), but it is unclear whether this suggests that bright females are more, or less, dominant than drab ones. In my view, experiments recording aggressive responses by one party (bright or dull) to another party that is either dead (a dummy; Muma & Weatherhead 1989) or constrained in its response (separately enclosed; Beeching et al. 1998) cannot provide sufficient information to conclude which party is the dominant. In order to allow such conclusions, dominance should be inferred from the outcomes of repeated interactions between live individuals (as in Johnson's (1988) study).

Signals of dominance status may not only be relevant to intrasexual competition. In many species, males and females live in social assemblages during part of or the whole year, leading to both intrasexual and intersexual competition for limited (non-sexual) resources. Stutchbury (1994) tested the idea that 'male-like' female coloration in hooded warblers would reduce aggression from male competitors, but found no evidence to support this intersexual competition hypothesis.

I suggest that female-female competition should be a main focus of future studies related to the function of female ornamentation. There is now compelling evidence that female intrasexual competition is the main, although not sole (Langmore et al. 1996), context of female song among birds (e.g. Cooney & Cockburn 1995; Langmore & Davies 1997; Langmore 1998, this volume). This could well be true also for visual ornaments, but, as I have shown, this field of research is at present almost unexplored except for some suggestions based on descriptive work.

Asymmetry in female ornaments

Triggered by Møller's (1990, 1992) pioneering studies, a major body of sexual selection research in the 90s has been related to the geometry (symmetry or asymmetry) of sexual traits. Although it is questionable whether the symmetry of a given trait can be considered a signal *per se* (in the sense of having evolved for information transmission), there is now evidence suggesting that the degree of bilateral symmetry conveys information about genetic and/or phenotypic quality (reviewed by Møller & Swaddle 1997, but see Clarke 1998; Houle 1998). Ornaments have been suggested to be particularly sensitive to genetic or environmental stress interfering with the stability of development. Consistent with this idea, the degree of asymmetry is often particularly variable in ornamental traits (e.g. Møller & Höglund 1991; Manning & Chamberlain 1993; Møller 1993b, but see Balmford et al. 1993). Accordingly, many studies have been devoted to testing whether female animals prefer to mate with symmetrically ornamented males (e.g. Møller 1992; Swaddle & Cuthill 1994a; Bennett et al. 1996). Such preferences seem to be widespread, in birds and across the animal kingdom (e.g. Møller & Swaddle 1997; Møller & Thornhill 1998). Hence, ornament symmetry seems to play a role in female mate choice.

Very few studies have, however, considered whether males, and not only females, are sensitive to the symmetry of potential partners. If mutual mate choice is occurring, as may be the case when both sexes invest heavily in the young (Jones & Hunter 1993; Owens & Thompson 1994; Deutsch & Reynolds 1995; Johnstone et al. 1996), similar symmetry sensitivities might be expected in both sexes. Knowing whether one or both sexes discriminate between potential partners is highly important for understanding mate choice and its selective consequences. However, very little is known about male symmetry preferences. Except for a series of studies on human images (Grammer & Thornhill 1994; Langlois et al. 1994; Singh 1995; Swaddle & Cuthill 1995; Kowner 1996; Manning et al. 1997), some of which confound symmetry with 'averageness' (e.g. Langlois et al. 1994), there have only been a few non-manipulative attempts at studying male symmetry preferences. Almost all of these studies have focused on non-ornamental traits (Møller & Thornhill 1998). Møller and Zamora-Munoz (1997) argued, however, that the antennae of cerambycid beetles *Stenurella melanura* act as a sexual ornament, and reported that males of the species preferred the more symmetric female when given a choice between two females differing in antennal asymmetry. However, individuals with longer antennae were not preferred as mates and were not superior in male-male contests; hence, it is unclear whether the trait is really an ornament. Moreover, symmetry was not manipulated by Møller and Zamora-Munoz (1997) and it could thus not be conclusively determined whether males responded to asymmetry *per se*, or some correlated trait. To my knowledge, the only study manipulating ornamental asymmetry in wild animals has been conducted on bluethroats (Hansen et al. 1999). We arranged coloured leg bands (an artificial ornament) in symmetric and asymmetric combinations on females, and allowed males to choose between females wearing the two treatments. The males showed a significant preference for symmetric over asymmetric females (Hansen et al. 1999). In fact, the mean preference for symmetry among males was very similar to the female preference found in a previous study (Fiske & Amundsen 1997) (Fig. 1).

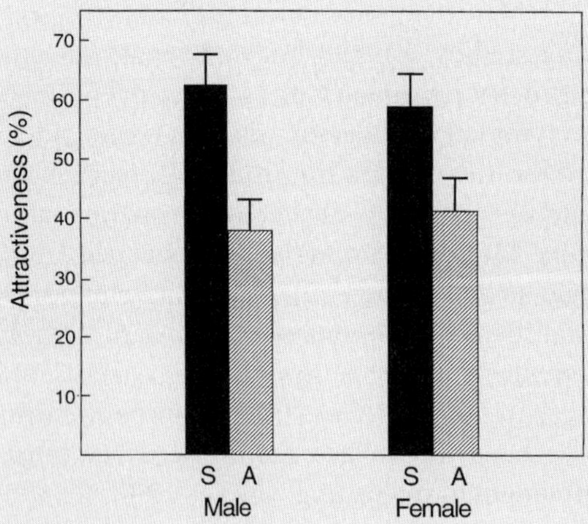

Figure 1. *Attractiveness of male and female bluethroats when wearing symmetric (S) and asymmetric (A) combinations of coloured leg bands. Attractiveness is measured as interest shown by members of the opposite sex, in terms of percent of time (+ 1 s.e.) spent in association with S and A individuals, respectively. Based on data from Fiske & Amundsen 1997 (males, N = 22) and Hansen et al. 1999 (females, N = 28).*

The bluethroat results suggest that the influence of symmetry on mate choice may be much more complex than if only one sex exerts preferences. There is an urgent need for more studies to test if male preferences for symmetry are as widespread as are those of females. A recent comparative study suggests that this is not the case (Møller & Thornhill 1998), but the analysis suffers from limited power in testing male preferences, due to few studies and lack of experimental manipulation of female traits. Experiments on a wide variety of species (and higher taxa) are needed before we can draw any firm conclusions on the issue. Our experiment using artificial ornaments with exaggerated asymmetry (Hansen et al. 1999) is a first step in the process; future experiments should aim at testing male (and female) sensitivity to treatments that better mimic natural ornaments (Swaddle & Cuthill 1994b, Rohde et al. 1997).

It is controversial whether mate preferences for symmetrical opposite-sex individuals have evolved as a means of acquiring partners of superior quality (Møller 1990). Alternatively, symmetry preferences may simply be an inherent property of information recognition systems, without any adaptive value (Enquist & Arak 1994; Johnstone 1994; Enquist & Johnstone 1997). In fact, the two hypotheses are not mutually exclusive, but may instead work in concert in shaping symmetry sensitivities. Whatever mechanism explains symmetry preferences, it has a profound influence on mate choice dynamics if both sexes, and not only females, exhibit mate choice based on the symmetry of potential partners.

Conclusions and prospects for the future: quentifying selection pressures and constraints on female showiness

It is now firmly established that female ornaments in some species are subject to sexual selection through male choice (e.g. Jones & Hunter 1993). There is also some evidence for female showiness being related to dominance (Johnsen 1988) or territory defence (e.g. Bleiweiss 1992) in female-female interactions. The frequent increases in female ornamentation without similar changes in the male, reported in a series of recent comparative studies (Irwin 1994; Peterson 1996; Burns 1998), strongly suggest that selection often acts directly on the females in favour of increased ornamentation. It remains to be seen how general such selection pressures are, and whether they apply primarily to cases with female ornaments similar to those of males, or also to species with 'vestigial male ornaments' occurring in females. Potentially, direct sexual selection on female traits could be quite widespread. It is a challenge for future theoretical work to make clear under what circumstances functional female ornaments should be expected, and for empirical work to test the predictions on a broad spectrum of organisms. Both theoretical and empirical approaches are needed to answer under which conditions one might expect selection due to male choice, and when female-female competition could be important. Comparative work should focus on the potential social and ecological correlates of female ornamentation, including sociality, territoriality, habitat, parental care and mating system.

There is also evidence that ornaments carry specific costs for female birds, at least among those species where the female alone incubates and/or broods the young (Martin & Badyaev 1996). Apparently, a high level of nest predation selects against conspicuous plumage in females, favouring a cryptic appearance. This kind of selection is probably the main reason for the loss of male-like plumage in females of many bird taxa, as demonstrated by comparative work. It cannot, however, explain the existence of highly variable female ornamentation seen in a variety of species (e.g. bluethroats).

In conclusion, females of many species are probably subject to simultaneous opposing selection pressures: predation favouring crypsis, and sexual selection favouring showiness. In species where males are ornamented, selection in favour of female crypsis may be constrained by genetic correlations (Lande 1980, 1987). Such correlations may not only result in non- or maladaptive female ornamentation, but could also provide a starting point for sexual selection on the trait in females. When mutual sexual selection is occurring, a genetic correlation between characters in the two sexes will in fact enforce the selection process.

The challenge for future researchers aiming to understand the evolution of female ornamentation is not only to demonstrate whether each of the various selection pressures, and constraints, are acting on the females, but also to assess the importance of each factor in shaping female appearances in different species. Birds have proved to be highly suitable model organisms for such studies and further research building on established knowledge is recommended. However, studies of female ornaments in other taxa (e.g. invertebrates, fish, mammals) are particularly needed if we are to approach a general understanding of the phenomenon. This is a formidable task, but one that, when completed, will fill one of last major gaps in our understanding of sexual selection and hence provide us with a more complete and realistic understanding of animal mate choice.

Acknowledgements

I am grateful to Helena Cronin, Susan Fitzpatrick, Elisabet Forsgren and Naomi Langmore for their constructive comments on a previous draft. I would also like to thank the participants of the Kongsvoll symposium, and the participants of the 'Female ornaments' roundtable at the IOC in Durban 1998, for stimulating discussions regarding the evolution of female showiness in birds and other animals. Frank Götmark provided unpublished information about chaffinches and European blackbirds. Finally, I want to thank the Royal Norwegian Society of Sciences and Letters' Foundation for providing an excellent opportunity to present and discuss issues related to signalling in females.

References

Amundsen, T., Forsgren, E. & Hansen, L. T. T. 1997. On the function of female ornamentation: male bluethroats prefer colourful females. *Proceedings of the Royal Society of London, Series B*, **264**, 1579-1586.

Andersson, M. 1994. *Sexual Selection.* Princeton, NJ: Princeton University Press.

Andersson, S. & Amundsen, T. 1997. Ultraviolet colour vision and ornamentation in bluethroats. *Proceedings of the Royal Society of London, Series B*, **264**, 1587-1591.

Balmford, A., Jones, I. L. & Thomas, A. L. R. 1993. On avian asymmetry: evidence of natural selection for symmetrical tails and wings in birds. *Proceedings of the Royal Society of London, Series B*, **252**, 245-251.

Beeching, S. C., Gross, S. H., Bretz, H. S. & Hariatis, E. 1998. Sexual dichromatism in convict cichlids: the ethological significance of female ventral coloration. *Animal Behaviour*, **56**, 1021-1026.

Bennett, A. T. D., Cuthill, I. C., Partridge, J. C. & Maier, E. J. 1996. Ultraviolet vision and mate choice in zebra finches. *Nature*, **380**, 433-435.

Berglund, A., Bisazza, A. & Pilastro, A. 1996. Armaments and ornaments: an evolutionary explanation of traits of dual utility. *Biological Journal of the Linnean Society*, **58**, 385-399.

Bleiweiss, R. 1985. Iridescent polychromatism in a female hummingbird: is it related to feeding strategies? *Auk*, **102**, 701-713.

Bleiweiss, R. 1992. Widespread polychromatism in female sunangel hummingbirds (*Heliangelus*: Trochilidae). *Biological Journal of the Linnean Society*, **45**, 291-314.

Burley, N. & Coopersmith, C. B. 1987. Bill color preferences of zebra finches. *Ethology*, **76**, 133-151.

Burns, K. J. 1998. A phylogenetic perspective on the evolution of sexual dichromatism in tanagers (Thraupidae): the role of female versus male plumage. *Evolution*, **52**, 1219-1224.

Clarke, G. M. 1998. Developmental stability and fitness: the evidence is not quite so clear. *American Naturalist*, **152**, 762-766.

Cooney, R. & Cockburn, A. 1995. Territorial defence is the major function of female song in the superb fairy-wren, *Malurus cyaneus*. *Animal Behaviour*, **49**, 1635-1647.

Cronin, H. 1991. *The Ant and the Peacock.* Cambridge: Cambridge University Press.

Cuervo, J. J., de Lope, F. & Møller, A. P. 1996. The function of long tails in female barn swallows (*Hirundo rustica*): an experimental study. *Behavioral Ecology*, **7**, 132-136.

Cunningham, E. & Birkhead, T. 1997. Female roles in perspective. *Trends in Ecology and Evolution*, **12**, 337-338.

Cunningham, E. J. A. & Birkhead, T. R. 1998. Sex roles and sexual selection. *Animal Behaviour*, **56**, 1311-1321.

Darwin, C. 1859. *On the Origin of Species*. London: John Murray.

Darwin, C. 1871. *The Descent of Man, and Selection in Relation to Sex*. London: John Murray.

Deutsch, J. C. & Reynolds, J. D. 1995. Design and sexual selection: the evolution of sex differences in mate choice. *Perspectives in Ethology,* **11**, 297-323.

Dewsbury, D. A. 1982. Ejaculate cost and male choice. *American Naturalist*, **119**, 601-610.

Enquist, M. & Arak, A. 1994. Symmetry, beauty and evolution. *Nature*, **372**, 169-172.

Enquist, M. & Johnstone, R. A. 1997. Generalization and the evolution of symmetry preferences. *Proceedings of the Royal Society of London, Series B*, **264**, 1345-1348.

Fisher, R. A. 1915. The evolution of sexual preference. *Eugenics Review*, **7**, 184-192.

Fisher, R. A. 1930. *The Genetical Theory of Natural Selection*. London: Oxford University Press.

Fiske, P. & Amundsen, T. 1997. Female bluethroats prefer males with symmetric colour bands. *Animal Behaviour*, **54**, 81-87.

Folstad, I. & Karter, A. J. 1992. Parasites, bright males and the immunocompetence handicap. *American Naturalist*, **139**, 603-622.

Gowaty, P.A. 1997. Principles of females' perspectives in avian behavioural ecology. *Journal of Avian Biology*, **28**, 95-102.

Grammer, K. & Thornhill, R. 1994. Human (*Homo sapiens*) facial attractiveness and sexual selection: the role of symmetry and averageness. *Journal of Comparative Psychology*, **108**, 233-242.

Halliday, T. R. & Arnold, S. J. 1987. Multiple mating by females: a perspective from quantitative genetics. *Animal Behaviour*, **35**, 939-941.

Hansen, L. T. T., Amundsen, T. & Forsgren, E. 1999. Symmetry - attractive not only to females. *Proceedings of the Royal Society of London, Series B*, **266**, 1235-1240.

Hill, G. E. 1993. Male mate choice and the evolution of female plumage coloration in the house finch. *Evolution*, **47**, 1515-1525.

Hoelzer, G. A. 1989. The good parent process of sexual selection. *Animal Behaviour*, **38**, 1067-1078.

Houle, D. 1998. High enthusiasm and low r-squared. *Evolution*, **52**, 1872-1876.

Houston, A. I. & Davies, N. B. 1985. The evolution of cooperation and life history in the dunnock *Prunella modularis*. In: *Behavioural Ecology: Ecological Consequences of Adaptive Behaviour* (Ed. by R. Sibly & R. Smith), pp. 471-487. Oxford: Blackwell.

Irwin, R. E. 1994. The evolution of plumage dichromatism in the New World blackbirds: social competition on female brightness? *American Naturalist*, **144**, 890-907.

Jennions, M. D. 1998. The effect of leg band symmetry on female-male associations in zebra finches. *Animal Behaviour*, **55**, 61-67.

Johnsen, A. & Lifjeld, J. T. 1995. Unattractive males guard their mates more closely: an experiment with bluethroats (Aves, Turdidae: *Luscinia s. svecica*). *Ethology*, **101**, 200-212.

Johnsen, T. S., Hengeveld, J. D., Blank, J. L., Yasukawa, K. & Nolan, V. 1996. Epaulet brightness and condition in female red-winged blackbirds. *Auk*, **113**, 356-362.

Johnson, K. 1988. Sexual selection in pinyon jays II: male choice and female-female competition. *Animal Behaviour*, **36**, 1048-1053.

Johnstone, R. A. 1994. Female preference for symmetrical males as a by-product of selection for mate recognition. *Nature*, **372**, 172-175.

Johnstone, R. A. 1995. Sexual selection, honest advertisement and the handicap principle: reviewing the evidence. *Biological Reviews*, **70**, 1-65.

Johnstone, R. A., Reynolds, J. D. & Deutsch, J. C. 1996. Mutual mate choice and sex differences in choosiness. *Evolution*, **50**, 1382-1391.

Jones, I. L. & Hunter, F. M. 1993. Mutual sexual selection in a monogamous seabird. *Nature*, **362**, 238-239.

Kirkpatrick, M. & Barton, N. H. 1997. The strength of indirect selection on female mating preferences. *Proceedings of the National Academy of Sciences, USA*, **94**, 1282-1286.

Kowner, R. 1996. Facial asymmetry and attractiveness judgment in developmental perpective. *Journal of Experimental Psychology: Human Perception and Performance*, **22**, 662-675.

Lande, R. 1980. Sexual dimorphism, sexual selection, and adaptation in polygenic characters. *Evolution*, **34**, 292-305.

Lande, R. 1987. Genetic correlations between the sexes in the evolution of sexual dimorphism and mating preferences. In: *Sexual Selection: Testing the Alternatives* (Ed. by J. W. Bradbury & M. B. Andersson), pp. 83-95. Chichester: Wiley.

Lande, R. & Arnold, S. J. 1985. Evolution of mating preference and sexual dimorphism. *Journal of Theoretical Biology*, **117**, 651-664.

Langlois, J. H., Roggman, L. A. & Musselman, L. 1994. What is average and what is not average about attractive faces? *Psychological Science*, **5**, 214-219.

Langmore, N. E. 1998. Functions of duet and solo songs of female birds. *Trends in Ecology and Evolution*, **13**, 136-140.

Langmore, N. E. & Davies, N. B. 1997. Female dunnocks use vocalizations to compete for males. *Animal Behaviour*, **53**, 881-890.

Langmore, N. E., Davies, N. B., Hatchwell, B. J. & Hartley, I. R. 1996. Female song attracts males in the alpine accentor *Prunella modularis*. *Proceedings of the Royal Society of London, Series B*, **263**, 141-146.

Lewontin, R. C. 1974. Darwin and Mendel - the materialist revolution. In: *The Heritage of Copernicus: Theories 'Pleasing to the Mind'* (Ed. by J. Neyman), pp. 166-183. Cambridge, MA: MIT Press.

Linville, S. U., Breitwisch, R. & Schilling, A. J. 1998. Plumage brightness as an indicator of parental care in northern cardinals. *Animal Behaviour*, **55**, 119-127.

Manning, J. T. & Chamberlain, A. T. 1993. Fluctuating asymmetry, sexual selection and canine teeth in primates. *Proceedings of the Royal Society of London, Series B*, **251**, 83-87.

Manning, J. T., Scutt, D., Whitehouse, G. H. & Leinster, S. J. 1997. Breast asymmetry and phenotypic quality in women. *Evolution and Human Behaviour*, **18**, 223-236.

Martin, T. E. & Badyaev, A. V. 1996. Sexual dichromatism in birds: importance of nest predation and nest location for females versus males. *Evolution*, **50**, 2454-2460.

Mayr, E. 1982. *The Growth of Biological Thought*. Cambridge, MA: Harvard University Press.

Muma, K. E. & Weatherhead, P. J. 1989. Male traits expressed in females: direct or indirect sexual selection? *Behavioral Ecology and Sociobiology*, **25**, 23-31.

Møller, A. P. 1990. Fluctuating asymmetry in male sexual ornaments may reliably reveal male quality. *Animal Behaviour*, **40**, 1185-1187.

Møller, A. P. 1992. Female swallow preference for symmetrical male sexual ornaments. *Nature*, **357**, 238-240.

Møller, A. P. 1993a. Sexual selection in the barn swallow (*Hirundo rustica*). III. Female tail ornaments. *Evolution*, **47**, 417-431.

Møller, A. P. 1993b. Patterns of fluctuating asymmetry in sexual ornaments predict female choice. *Journal of Evolutionary Biology*, **6**, 481-491.

Møller, A. P. 1994. *Sexual Selection and the Barn Swallow*. Oxford: Oxford University Press.

Møller, A. P. & Höglund, J. 1991. Patterns of fluctuating asymmetry in avian feather ornaments: implications for models of sexual selection. *Proceedings of the Royal Society of London, Series B*, **245**, 1-5.

Møller, A. P. & Swaddle, J. P. 1997. *Asymmetry, Developmental Stability, and Evolution*. Oxford: Oxford University Press.

Møller, A. P. & Thornhill, R. 1998. Bilateral symmetry and sexual selection: a meta-analysis. *American Naturalist*, **151**, 174-192.

Møller, A. P. & Zamora-Munoz, C. 1997. Antennal asymmetry and sexual selection in a cerambycid beetle. *Animal Behaviour*, **54**, 1509-1515.

Olson, V. A. & Owens, I. P. F. 1998. Costly sexual signals: are carotenoids rare, risky or required? *Trends in Ecology and Evolution*, **13**, 510-514.

Omland, K. E. 1997. Examining two standard assumptions of ancestral reconstructions: repeated loss of dichromatism in dabbling ducks (Anatini). *Evolution*, **51**, 1636-1646.

Owens, I. P. F., Burke, T. & Thompson, D. B. A. 1994. Extraordinary sex roles in the Eurasian dotterel: female mating arenas, female-female competition, and female mate choice. *American Naturalist*, **144**, 76-100.

Owens, I. P. F. & Thompson, D. B. A. 1994. Sex differences, sex ratios and sex roles. *Proceedings of the Royal Society of London, Series B*, **258**, 93-99.

Parker, G. A. 1983. Mate quality and mating decisions. In: *Mate Choice* (Ed. by P. Bateson), pp. 141-166. Cambridge: Cambridge University Press.

Parker, P.G. & Burley, N.T. (Eds) 1998. *Avian reproductive tactics: Female and Male Perspectives. Ornithological Monographs No. 49.* Washington: American Ornithologists' Union.

Peterson, A. T. 1996. Geographic variation in sexual dichromatism in birds. *Bulletin of the British Ornithologists' Club*, **116**, 156-172.

Potti, J. & Merino, S. 1996. Decreased levels of blood trypanosome infection correlate with female expression of a male secondary sexual trait: implications for sexual selection. *Proceedings of the Royal Society of London, Series B*, **263**, 1199-1204.

Price, T. & Birch, G. L. 1996. Repeated evolution of sexual color dimorphism in passerine birds. *Auk*, **113**, 842-848.

Rohde, P.A., Amundsen, T. & Fiske, P. 1997. Fluctuating asymmetry, mate choice and experimental designs. *Animal Behaviour,* **54**, 1030-1033.

Rohde, P. A., Johnsen, A. & Lifjeld, J. T. 1999. Female plumage coloration in the bluethroat: no evidence for an indicator of maternal quality. *Condor*, **101**, 96-104.

Rosenqvist, G. 1990. Male mate choice and female-female competition for mates in the pipefish *Nerophis ophidion*. *Animal Behaviour*, **39**, 1110-1115.

Rosenqvist, G. & Berglund, A. 1992. Is female sexual behaviour a neglected topic? *Trends in Ecology and Evolution*, **7**, 174-176.

Saino, N., Calza, S. & Møller, A. P. 1997. Immunocompetence of nestling barn swallows in relation to brood size and parental effort. *Journal of Animal Ecology*, **66**, 827-836.

Singh, D. 1995. Female health, attractiveness, and desirability for relationships: role of breast asymmetry and waist-to-hip ratio. *Ethology and Sociobiology*, **16**, 465-481.

Skarstein, F. & Folstad, I. 1996. Sexual dichromatism and the immunocompetence handicap: an observational approach using Arctic charr. *Oikos*, **76**, 359-367.

Smiseth, P. T. & Amundsen, T. In press. Does female plumage coloration signal parental quality? A male removal experiment with the bluethroat. *Behavioral Ecology and Sociobiology*.

Stutchbury, B. J. 1994. Competition for winter territories in a neotropical migrant - the role of age, sex and color. *Auk*, **111**, 63-69.

Stutchbury, B. J. & Howlett, J. S. 1995. Does male-like coloration of female hooded warblers increase nest predation? *Condor*, **97**, 559-564.

Swaddle, J. P. & Cuthill, I. C. 1994a. Preference for symmetric males by female zebra finches. *Nature*, **367**, 165-166.

Swaddle, J. P. & Cuthill, I. C. 1994b. Female zebra finches prefer males with symmetric chest plumage. *Proceedings of the Royal Society of London, Series B*, **258**, 267-271.

Swaddle, J. P. & Cuthill, I. C. 1995. Asymmetry and human facial attractiveness: symmetry may not always be beautiful. *Proceedings of the Royal Society of London, Series B*, **261**, 111-116.

Tella, J. L., Forero, M. G., Donazar, J. A. & Hiraldo, F. 1997. Is the expression of male traits in lesser kestrels related to sexual selection? *Ethology*, **103**, 72-81.

Trail, P. W. 1990. Why should lek-breeders be monomorphic? *Evolution*, **44**, 1837-1852.

Wallace, A. R. 1889. *Darwinism*. London: Macmillan.

West-Eberhard, M. J. 1983. Sexual selection, social competition, and speciation. *Quarterly Review of Biology*, **58**, 155-183.

Whittingham, L. A., Kirkconnell, A. & Ratcliffe, L. M. 1992. Differences in song and sexual dimorphism between Cuban and North American red-winged blackbirds (*Agelaius phoeniceus*). *Auk*, **109**, 928-933.

Wolf, L. L. 1969. Female territoriality in a tropical hummingbird. *Auk*, **86**, 490-504.

Wolf, L. L. 1975. Female territoriality in the purple-throated carib. *Auk*, **92**, 511-522.

Zahavi, A. 1975. Mate selection - a selection for a handicap. *Journal of Theoretical Biology*, **53**, 205-214.

Zahavi, A. & Zahavi, A. 1997. *The Handicap Principle*. Oxford: Oxford University Press.

IN: ESPMARK,Y., AMUNDSEN,T. & ROSENQVIST,G. (EDS.) 2000
ANIMAL SIGNALS:
SIGNALLING AND SIGNAL DESIGN IN ANIMAL COMMUNICATION, PP. 155-175
TAPIR ACADEMIC PRESS, TRONDHEIM, NORWAY

7

Is Fluctuating Asymmetry a Visual Signal?

JOHN P. SWADDLE

Centre for Behavioural Biology, School of Biological Sciences, University of Bristol, Woodland Road, Bristol BS8 1UG, UK

E-mail: john.swaddle@bristol.ac.uk

Abstract

Fluctuating asymmetries are the small, random deviation from perfect symmetry that occur during the development of otherwise symmetric traits and are a measure of developmental stability. It has frequently been claimed that fluctuating asymmetry is related to aspects of fitness, as asymmetry is an indicator of how well the genome can buffer developmental processes against perturbations from the environment. As asymmetry could be related to fitness, it has been further proposed that animals could use fluctuating asymmetry as a signal or cue to reveal fitness-related information to conspecifics (and perhaps heterospecifics). In this study I present a literature review which indicates that asymmetry is inconsistently related to both parameters. In some taxa, asymmetry can be related to fitness and be an effective signal; however, in other taxa fluctuating asymmetry is clearly not related to either aspects of fitness or signalling. In many cases, fluctuating asymmetry may be too small and inconspicuous to be detected as a visual signal and I discuss a number of symmetry perception parameters that may influence the receiver psychology of symmetry-signalling and hence influence the efficacy of any symmetry-signalling system. In particular I present a number of empirical results from visual perception experiments with European starlings Sturnus vulgaris, *which indicate that symmetry detection and discrimination is particularly error-prone, hence reducing the probability that fluctuating asymmetry could be used as an effective signal in this species. The general role of perceptual processes and their influence on signalling by fluctuating asymmetry is discussed.*

Fluctuating asymmetry is a measure of developmental instability. Specifically, fluctuating asymmetries are the small, random deviations from perfect bilateral symmetry that occur during the development of traits that we would normally describe as being symmetric (Ludwig 1932; Van Valen 1962). These asymmetries can arise for a number of reasons that can be conveniently grouped into two categories. First, there are a number of genetic factors that can increase levels of asymmetry. These include intense inbreeding, incorporation of mutant genes into the genome, recent episodes of hybridization that splits up previously co-adapted gene complexes, and perhaps extreme homozygosity in some taxa (Møller & Swaddle 1997). Second, there is a whole suite of environmental factors that can disrupt developmental processes to increase asymmetry. Such factors include temperatures that are too high or too low, nutritional stress (e.g. low food availability, or poor quality diet), high parasitic infection and chemical pollution (most commonly reported for aquatic species in polluted waterways) (Møller & Swaddle 1997). In summary, fluctuating asymmetry is a reflection of how well the genome can buffer developmental processes against environmental perturbations, and particular genotypes are better at buffering development than others.

Fluctuating asymmetry and fitness

As fluctuating asymmetry is an indicator of how well the genotype is coping with environmental conditions, researchers have investigated associations between asymmetry and fitness parameters to examine whether asymmetry could be a marker for fitness. In a number of cases, individuals with low asymmetry can have significant fitness advantages, and a recent meta-analysis has revealed that there is a relatively small, but significant, negative relationship between asymmetry and fitness measures (Leung & Forbes 1996). In this study I have conducted a further review of this literature, incorporating all of the studies cited by Leung and Forbes as well as ones not included and more recent references (excluding studies on human subjects). These studies are summarised in Table 1.

As can be seen from my literature review (reported on the basis of published species accounts, $N = 74$), there are inconsistent relationships between asymmetry and fitness measures (survival, fecundity, growth rates, body condition, parasitic infestation) among taxa. This is especially true when considering data from correlational and observational studies of asymmetry; only 52% of published correlational studies report the predicted negative relationship between asymmetry and fitness. The percentage of species accounts reporting the predicted negative relationship is much greater in studies where asymmetries have been induced experimentally (approximately 94% of cases). This is, perhaps, not surprising as many of the experiments reported have disrupted developmental stability through fairly gross environmental manipulation away from optimal developmental conditions. In sub-optimal conditions we should expect negative effects on fitness, as the optimality of developmental conditions is often determined by fitness effects.

Table 1. *Review of non-human species accounts of the relations between fluctuating asymmetry and aspects of fitness (survival, fecundity, growth rates, body condition, parasitic infestation). Study results are summarised as follows: -, individuals with low asymmetry have a fitness advantage.* No effect, *no relationship between asymmetry and fitness. +, individuals with greater asymmetry have a fitness advantage.* Correlations *refers to correlational studies.* Experiments *refers to studies in which trait asymmetry has been manipulated in some manner or the experiment was performed under standardised conditions. In many of these experiments, the asymmetry was induced by environmental disturbance. References are listed in the Appendix along with the number of species assessed in each study. If species numbers are not given, one species was studied. Only one study is included for each species account, but multiple citations are given if there are conflicting data for the same speies.*

	-	*No effect*	+	*N*
Correlations	30	22	5	57
Experiments	16	1	0	17
TOTAL	46	23	5	74

The most generalised conclusion that I can draw from this literature review is that fluctuating asymmetry can be a marker for fitness in some systems and that the relationship between fitness and asymmetry is taxon- and trait-specific. This is a conclusion that other reviewers have previously reached (Markow 1995; Palmer 1996; Clarke 1997, 1998).

Fluctuating asymmetry and visual signalling

As fluctuating asymmetry can be a marker for fitness, it is possible that morphological asymmetry could be used a direct visual signal or cue of fitness in some systems (Møller 1990). This prediction has generated great interest in studies of fluctuating asymmetry and developmental instability, particularly in relation to sexual selection. In the eight years since Møller (1990) first hypothesised that asymmetry of secondary sexual traits could be used as a cue in mate choice, there have been at least 194 non-human species accounts relating some aspect of visual signalling (mostly indirect measures) to morphological fluctuating asymmetry. I have summarised these studies in Table 2.

As can be seen from this large body of literature, there is a general lack of consensus as to whether asymmetry is used as a visual signal. A recent meta-analysis of the influence of fluctuating asymmetry on sexual selection processes suggests that symmetry does play an important role (Thornhill & Møller 1998), but the review I have performed here incorporates a broader range and number of studies and also focuses on the differences

Table 2. *Review table of number of non-human species accounts relating some aspect of visual signalling to magnitude of trait fluctuating asymmetry. Study results are summarised as follows: -, individuals with low asymmetry are preferred or are inferred to do better in a signalling context.* No effect, *no relationship between asymmetry and signalling parameters. +, individuals with greater asymmetry are preferred or are inferred to do better in a signalling context.* Correlations *refers to correlational studies.* Experiments *refers to studies in which trait asymmetry has been manipulated in some manner.* Natural range experiments *refers to experimental studies that have performed asymmetry manipulations within a reasonable morphological range set by natural data sets for that particular species and have not confounded manipulations with other variables, such as behaviour, size or signal intensity. Note the large proportion of correlational studies (approximately 92%) and the relative paucity of species that have been investigated under experimentally controlled conditions. References are listed in the Appendix along with the number of species assessed in each study. If species numbers are not given, one species was studied. Only one study is included for each species account, but multiple citations are given if there are conflicting data for the same species.*

	-	*No effect*	+	*N*
Correlations	58	113	3	174
Experiments	5	5	1	11
Natural range experiments	3	2	0	5
TOTAL	66	120	4	190

between correlational and experimental studies. The correlational studies summarised above can only provide indirect evidence for any relationship between asymmetry and signalling. It is probable that in many of the signalling systems studied, asymmetry is either correlated with another signalling feature or that the asymmetry itself is not directly assessed. Therefore, correlational studies will tend to over-estimate the occurrences in which asymmetry is used a signal. Nevertheless, the majority of species examples (approximately 65%) indicate that morphological asymmetry does not function as a visual signal.

To confirm a signalling function, experimental manipulation of signalling structures are needed. Despite the large number of such studies, I could find only five in which asymmetry has been manipulated independent of other signalling features (trait size, signal intensity, behaviour) and the resulting experimental asymmetries have fallen within a natural range of asymmetry. In these '*natural range*' experiments, three studies have indicated the predicted negative relationship (symmetric individuals are preferred) for zebra finches *Taeniopygia guttata* (Swaddle & Cuthill 1994a), swordtail fish *Xiphophorus cortezi* (Morris 1998), and model flowers attracting insect pollinators (Møller & Sorci 1998). However, the remaining two studies have demonstrated that signalling trait asymmetry in European starlings *Sturnus vulgaris* (Swaddle & Witter 1995) and chaffinches *Fringilla coelebs* (Jablonski & Matyjasiak 1997) do not influence intra-specific competitive encounters.

Although only five natural range experiments have been reported, it is interesting that the three which support the predicted negative relationship have examined signalling in the context of mate preferences and sexual selection. On the other hand, the two studies in which asymmetry has no effect have examined signalling in the context of social dominance. This could indicate that symmetry signalling is context-specific and is related to the relative benefit of accurate discrimination of the signal (Swaddle 1999). In cases of mate choice the benefit of getting a high quality mate may outweigh the cost of choosing a poor quality mate. The relative cost of inaccurately assessing individuals in social dominance interactions may be much smaller. Interestingly, where the benefits/costs of dominance interactions have been experimentally manipulated, symmetric individuals became relatively more dominant when the benefits of social dominance were increased (Swaddle & Witter 1994; Witter & Swaddle 1994).

Receiver psychology of symmetry-signalling

The preceding review of the symmetry-signalling literature indicates that there is extreme heterogeneity among the data reported. In some cases, asymmetry can be used as a signal, whereas in other cases asymmetry appears unimportant. Most of the evidence commonly cited (and the majority of studies included in Table 2) to support the notion that asymmetry can be used as a signal reports correlations between trait asymmetry (i.e. the signal) and some property of the signaller (Møller & Pomiankowski 1993; Watson & Thornhill 1994; Møller & Swaddle 1997). A vital element of any signalling system has been largely over-looked by researchers interested in assessing asymmetry signalling: the receiver's ability to detect, discriminate and respond to the asymmetry signal (receiver psychology) (Guilford & Dawkins 1991). Without reliable detection of the asymmetry signal and accurate discrimination among natural asymmetry levels it is hard to envisage that signalling by fluctuating asymmetry could occur (Swaddle 1999). A central issue is thus the perceptual ability of organisms to discriminate symmetry from asymmetry. For the remainder of this manuscript I will focus on symmetry perception and its role in symmetry-signalling, as this has seldom been discussed in the literature.

Visual perception of symmetry

Compared with our knowledge of human visual symmetry perception, we know little about how well animals detect and respond to symmetry. In the following section I will indicate features of the human visual system that may be relevant to how other animals perceive symmetry, and discuss the implications of this for symmetry signalling.

Human symmetry perception

Human visual symmetry perception has been studied for over a hundred years and a review of the processes involved is presented elsewhere (Swaddle 1999). Here, I will briefly discuss features of symmetry perception that could act to either amplify or attenuate a symmetry-signal.

(i) Most researchers agree that symmetry perception is a fundamental property of the human visual system and decoding of visual information (Wagemans et al. 1993; Tyler & Miller 1994; Osorio 1996).
(ii) Bilateral symmetry is more readily detected than rotational or translational symmetry (Wagemans 1996). This could imply that asymmetry differences may be easiest to detect in traits with bilateral symmetry.
(iii) Symmetry is easiest to detect about a vertical axis compared with symmetry about horizontal or diagonal axes (Palmer & Hemenway 1978). If this is similar for other animals, asymmetry differences should generally be easiest to discriminate in signalling structures with a vertical axis of symmetry.
(iv) Symmetry is most easily assessed in the pattern area closest to the axis of symmetry (Jenkins 1982). Therefore, symmetry-signalling may be more prevalent where traits span the axis of symmetry rather than left and right elements appearing a large distance apart on either sides of the body.
(v) Geometric patterns aid symmetry perception (Baylis & Driver 1994). Hence, asymmetry differences may be most easily assessed in complex traits composed of regular geometric shapes.
(vi) Contrast and colour asymmetry can influence the assessment of morphological symmetry (Zhang & Gerbino 1992). Therefore, when assessing asymmetry, contrast and colour differences may also be important. This is not only relevant to the organisms involved in the signalling system, but also to researchers quantifying asymmetry in scientific studies.

Non-human symmetry perception

There are a number of features of non-human visual systems that appear to parallel our knowledge of the human visual system. I have summarised these below.
(i) Symmetry perception appears widespread among many taxonomic groups. There is experimental evidence to indicate that species of bees, fish, birds and primates can perceive and respond to symmetry (Swaddle 1999).
(ii) Symmetry perception may be a fundamental property of the vertebrate visual system and a corollary of how vertebrates detect edges and lines within structures (Osorio 1996).
(iii) Bees can discriminate bilateral from rotational symmetry (Horridge 1996).
(iv) Pigeons *Columba livia* can learn a general symmetry preference (Delius & Habers 1978; Delius & Nowak 1982).

Investigations of receiver psychology of symmetry-signalling

As can be seen from the preceding section, we know rather little about how animals detect and respond to symmetry and asymmetry differences. Therefore, in the following section I will present four fundamental areas of research in which I feel we need to investigate symmetry perception further. These areas are supported with some empirical data that I have collected from studies of wild-caught European starlings. The methodological details and full data will be presented elsewhere (J. P. Swaddle & S. Pruett-Jones, unpublished).

Do organisms exhibit a general symmetry preference?

General symmetry preferences have been invoked as an explanation for some of the symmetry preferences observed in the experiments summarised earlier in this manuscript (Enquist & Arak 1994; Johnstone 1994; Swaddle & Cuthill 1994a, b). If organisms have a pre-disposed preference for symmetry (over asymmetry) or have acquired a symmetry preference in a specified context and can transfer that preference to another context, then signalling via fluctuating asymmetry may have evolved indirectly as a by-product of some other function. There is generally a lack of evidence that organisms display a strong predisposition toward symmetric objects, although there have been very few studies (Rensch 1958; Delius & Nowak 1982; Lehrer et al. 1995).

My experimental studies with starlings also indicate that birds do not show a spontaneous or pre-acquired preference for symmetric objects. By presenting symmetric and asymmetric pairs of monochromatic random dot patterns to starlings in a 2-key Skinner

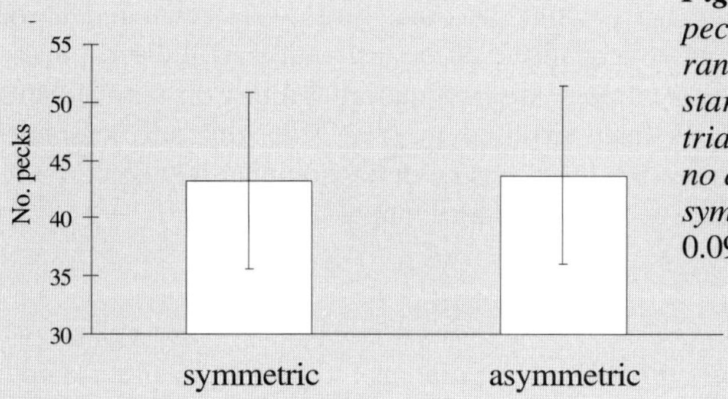

Figure 1. *Mean (± S.E.) number of pecks at symmetric and asymmetric random dot patterns presented to starlings ($N = 6$) over 10 one hour trials in a 2-key Skinner box. There is no difference in number of pecks on symmetric or asymmetric keys ($t = 0.09$, $df = 5$, $P = 0.93$).*

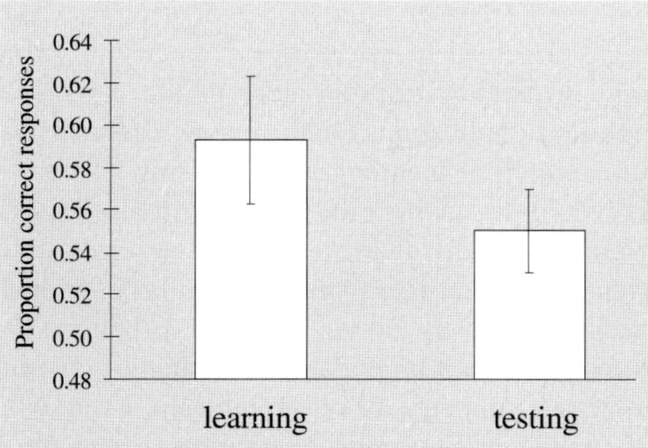

Figure 2. *Mean (± S.E.) proportion of 'correct' responses (i.e. correct discrimination of symmetry from asymmetry) during the last 10 learning trials (exposure to learning images) and the 10 testing trials (testing on novel images). There is no significant difference ($t = 1.51$, $df = 5$, $P = 0.19$), indicating that there was some transfer of the general ability to discriminate symmetry from asymmetry. However, performance with the novel images (testing) was not quite significantly different from random (50%) pecking at symmetric and asymmetric images ($t = 2.23$, $df = 5$, $P = 0.076$), indicating that the generalized symmetry discrimination was not strong.*

box, I could compare spontaneous preferences to peck at (choose) either symmetric or asymmetric images. Although birds showed a high degree of key-pecking, there was no relative preference for either symmetric or asymmetric patterns (Fig.1) (J. P. Swaddle & S. Pruett-Jones, unpublished).

Can organisms acquire generalised symmetry detection?

In a similar manner to the previous question, if organisms can acquire a symmetry preference and transfer that preference to another context, symmetry preferences in mate choice may have evolved through indirect mechanisms. This can be viewed as a 'generalised' symmetry preference and has been demonstrated to some degree in captive-bred pigeons (Delius & Habers 1978).

My experiments with wild-caught starlings show mixed evidence for the capacity to acquire a generalised symmetry preference. Over a series of 76 one hour learning trials, birds were trained by operant-conditioning (using food rewards) to discriminate between symmetric and asymmetric random dot patterns. Half of the birds received food rewards if they pecked at the symmetric images, while the other birds were rewarded if they pecked at the asymmetric images. After acquisition of the task (successful discrimination by symmetry) birds were tested for symmetry discrimination abilities on a novel set of images that they had not previously been exposed to. The birds did not completely transfer their pre-acquired preferences, but there was an indication that they could do this to some extent (Fig. 2) (J. P. Swaddle & S. Pruett-Jones, unpublished).

Is reception of an symmetry-signal error-prone?

Asymmetry discrimination may be a difficult perceptual task to perform, as indicated by the relatively low levels of performance shown in the starling experiments described above. Therefore, I postulate that there may be a relatively high degree of error associated with the perceptual and cognitive elements involved in detecting asymmetry, discriminating various levels of asymmetry and responding appropriately to the asymmetry signal. Errors of reception have two important implications (Johnstone & Grafen 1992). (i) The response of the receiver may not accurately reflect the true advertising level of the signal (i.e. individual asymmetry). This would de-stabilise the system and also weaken the intensity of selection against asymmetric individuals. (ii) Receiver-error could influence the signalling tactics of the signaller in that there may be little benefit for the signaller to invest in a high (very symmetric) and costly advertising level. Empirical data from my starling experiments indicates that there is a large degree of among-individual variation in ability to reliably discriminate symmetry from asymmetry, even when asymmetry is very large compared with asymmetries observed in nature (Fig. 3) (J. P. Swaddle & S. Pruett-Jones, unpublished).

Perceptual thresholds

Perhaps the most important issue is that of perceptual thresholds for asymmetry discrimination, i.e., what magnitude of asymmetry can be discriminated from symmetry?

There is very little data relevant to this question. Indirect evidence from a study on pigeons suggest that these birds are able to discriminate at best a 2% relative asymmetry (e.g. left side of trait is 2% larger than right side) from zero asymmetry (Schwabl & Delius 1984). Many of the asymmetries observed in nature may be too small to be discriminated from zero asymmetry. For example, mean trait asymmetry for signalling structures in birds have been reported as approximately 1.9 to 3.5% of trait size (Møller & Höglund 1991; Balmford et al. 1993).

Fluctuating asymmetry is characterised by a normal distribution of signed asymmetry scores (Palmer & Strobeck 1986), although it is predicted that distributions could be more leptokurtic. Receivers will select individuals on the basis of unsigned asymmetry scores; hence individuals will be chosen from a truncated half-normal distribution. It is, therefore, straightforward to see that mean relative asymmetry will have to be substantially greater than 2% for the receiver to be able to discriminate among the majority of individuals in the population based on signal asymmetry. This illustrates the importance of receiver psychology in assessing the signalling role of asymmetry and may help to explain discrepancies among studies in this highly disputed field of research. In species where the mean population relative asymmetry is low in relation to the ability to perceive asymmetry, symmetry-signalling is not likely to have evolved.

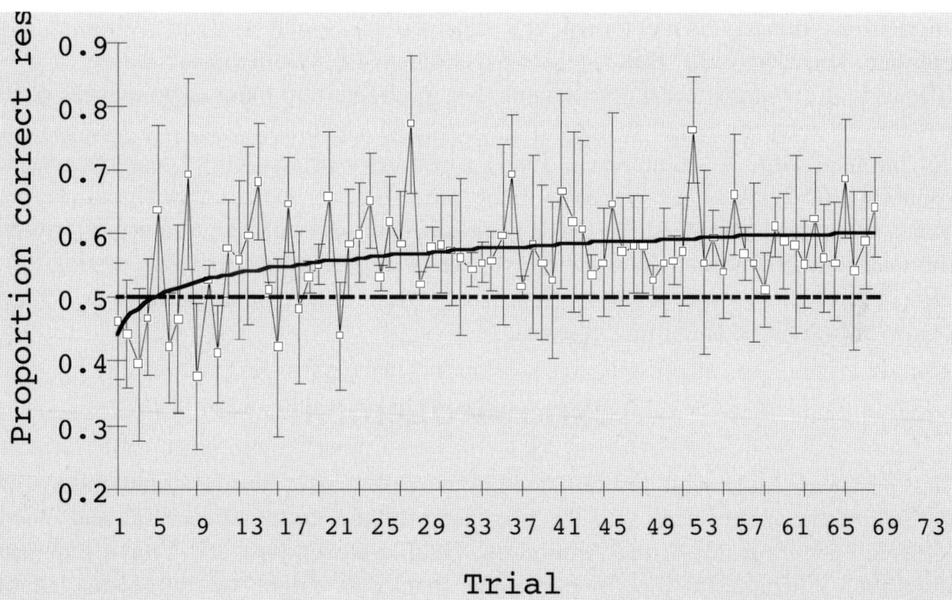

Figure 3. Mean (± S.E.) proportion of 'correct' discrimination of asymmetry from symmetry versus trial no. where birds received operant food rewards upon correct discrimination. The solid dark line is a fitted logarithmic regression line indicating a significant increase in the proportion of correct responses over time. The dashed line represents 50% preference, i.e., no preference for either correct or incorrect responses. The large error bars indicate among-individual variation in performance.

It also important to point out that all of the experiments in which the signalling role of asymmetry has been investigated have either studied traits with far greater asymmetries than 2% or manipulated asymmetries to extremely large degrees (commonly 20-30%). The smallest experimentally-induced asymmetry comes from a study of chest plumage asymmetry on mate preference in zebra finches, which resulted in a relative asymmetry of approximately 10% (Swaddle & Cuthill 1994a).

Conclusions

The question posed in the title of this chapter was 'Is fluctuating asymmetry a visual signal?' The answer to this question is not clear and I doubt if it is possible to generalise among signalling systems. There is clear evidence that asymmetry can influence some signalling systems, and symmetric individuals are sometimes preferred in mate choice situations. However, there is also compelling evidence to indicate that fluctuating asymmetry does not influence signalling systems in other situations.

Why is there variation in the importance of fluctuating asymmetry as a visual signal? As the relationship between fluctuating asymmetry and fitness parameters appears to be taxon- and trait-specific, it seems likely that symmetry-signalling should also be similarly specific to certain signalling systems. In addition, we know very little about how symmetry/asymmetry is perceived and detected by visual systems. Without a more complete knowledge of the receiver-psychology of symmetry-signalling it is very difficult to draw any general conclusions. It is probable that many asymmetries observed in nature will be too small to play a direct visual signalling role, and that asymmetry perception of large asymmetries is likely to be error-prone. These features reduce the probability of fluctuating asymmetry being an effective and reliable signal. Perceptual studies in humans also illustrate that signal trait features could act to amplify or interfere with an asymmetry signal, and the relative role of these features will vary among taxa and traits. Therefore, I urge researchers to study these perceptual parameters within the context of individual signalling systems.

Acknowledgements

I thank Staffan Andersson and an anonymous reviewer for helpful comments on earlier versions of the manuscript, and the organisers of the highly successful and enjoyable Kongsvoll Symposium: Trond Amundsen, Gunilla Rosenqvist and Yngve Espmark for doing such a wonderful job. In particular, thanks to Yngve for organising all of the subsequent manuscripts and pushing them through to publication. I was financially supported by the Royal Commission for the Exhibition of 1851, the Natural Environment Research Council (NERC grant GR9/03414) and by a Royal Society of London University Research Fellowship.

References

Agnew, P. & Koella, J. 1997. Virulence, parasite mode of transmission, and host fluctuating asymmetry. *Proceedings of the Royal Society of London, Series B,* **264**, 9-15.

Allen, G. R. & Simmons, L. W. 1996. Coercive mating, fluctuating asymmetry and male mating success in the dung fly *Sepsis cynipsea. Animal Behaviour*, **52**, 737-741.

Arcese, P. 1994. Harem size and horn symmetry in oribi. *Animal Behaviour*, **48**, 1485-1488.

Arnqvist, G., Thornhill, R. & Rowe, L. 1997. Evolution of animal genitalia: morphological correlates of fitness components in water striders. *Journal of Evolutionary Biology*, **10**, 613-640.

Bagchi, S. K., Sharma, V. P. & Gupta, P. K. 1989. Developmental instability in leaves of *Tectonia grandis. Silvae Genetica,* **38**, 1-6.

Balmford, A., Jones, I. L. & Thomas, A. L. R. 1993. On avian asymmetry: evidence of natural selection for symmetrical tails and wings in birds. *Proceedings of the Royal Society of London, Series B,* **252**, 245-251.

Baylis, G. C. & Driver, J. 1994. Parallel computation of symmetry but not repetition within single visual shapes. *Visual Cognition*, **1**, 377-400.

Beardmore, J. A., Dobzhansky, T. & Pavlovsky, O. A. 1960. An attempt to compare fitness of polymorphic and monomorphic experimental populations of *Drosophila pseudoobscura. Heredity*, **14**, 19-33.

Björklund, M. 1996. The effect of male presence on nestling growth and fluctuating asymmetry in the blue tit (*Parus caeruleus*). *Condor,* **98**, 172-175.

Bonn, A., Gasse, M., Rolff, J. & Martens, A. 1996. Increased fluctuating asymmetry in the damselfly *Coenagrion puella* is correlated with ectoparasitic water mites: implications for fluctuating asymmetry theory. *Oecologia,* **108**, 596-598.

Brault, A.-M. & de Oliveira, D. 1995. Seed number and an asymmetry index of 'McIntosh' apples. *Horticultural Science*, **30**, 40-46.

Brooks, R. & Caithness, N. 1995. Female choice in a feral guppy population: are there multiple cues? *Animal Behaviour*, **50**, 301-307.

Brown, W. D. 1997. Female remating and the intensity of female choice in black-horned tree crickets, *Oecanthus nigricornis. Behavioral Ecology*, **8**, 66-74.

Campbell, W. B. & Emlen, J. M. 1996. Developmental instability analysis of BKD-infected spring chinook salmon, *Oncorhynchus tshawytscha*, prior to seawater exposure. *Oikos,* **77**, 540-548.

Campbell, W. B., Emlen, J. M. & Hershberger, W. K. 1998. Thermally induced chronic developmental stress in coho salmon: integrating measures of mortality, early growth, and developmental instability. *Oikos*, **81**, 398-410.

Carrascal, L. M., Senar, J. C., Mozetich, I., Uribe, F. & Domenech, J. 1998. Interactions among environmental stress, body condition, nutritional status, and dominance in great tits. *Auk,* **115**, 727-738.

Chenuil, C. 1991. Asymetrie du nombre d'ecailles ventrales chez le lezard vivipare. M.Sc. Thesis, Department of Biology, Université de Paris-Sud, France.

Clarke, G. M. 1997. The genetic and molecular basis of developmental stability: the *Lucilia* story. *Trends in Ecology and Evolution*, **12**, 89-90.

Clarke, G. M. 1998. Developmental stability and fitness: the evidence is not quite so clear. *American Naturalist*, **152**, 762-766.

Córdoba-Aguilar, A. 1995. Fluctuating asymmetry in paired and unpaired damselfly males *Ischnura denticollis* (Burmeister)(Odonata: Coenagrionidae). *Journal of Ethology*, **13**, 129-132.

Danzmann, R. G., Ferguson, M. M., Allendorf, F. W. & Knudsen, K. L. 1986. Heterozygosity and developmental rate in a strain of rainbow trout (*Salmo gairdneri*). *Evolution,* **40**, 86-93.

David, P., Hingle, A., Grieg, D., Rutherford, A., Pomiankowski, A. & Fowler, K. 1998. Male sexual ornament size but not asymmetry reflects condition in stalk-eyed flies. *Proceedings of the Royal Society of London, Series B,* **265**, 2211-2216.

Delius, J. D. & Habers, G. 1978. Symmetry: can pigeons conceptualize it? *Behavioral Biology*, **22**, 336-342.

Delius, J. D. & Nowak, B. 1982. Visual symmetry recognition by pigeons. *Psychological Research*, **44**, 199-212.

Downhower, J. F., Blumer, L. S., Lejeune, P., Gaudin, P., Marconato, A. & Bisazza, A. 1990. Otolith asymmetry in *Cottus bairdi* and *C. gobio*. *Polskie Archiwum Hydrobiologii,* **37**, 209-220.

Dufour, K. W. & Weatherhead, P. J. 1998a. Bilateral symmetry and social dominance on captive male red-winged blackbirds. *Behavioral Ecology and Sociobiology*, **42**, 71-76.

Dufour, K. W. & Weatherhead, P. J. 1998b. Bilateral symmetry as an indicator of male quality in red-winged blackbirds: associations with measures of health, viability, and parental effort. *Behavioral Ecology*, **9**, 220-231.

Dunn, P. O. 1994. Extra-pair paternity in tree swallows: why do females mate with more than one male? *Behavioral Ecology and Sociobiology*, **35**, 273-281.

Enquist, M. & Arak, A. 1994. Symmetry, beauty and evolution. *Nature,* **372**, 169-172.

Eriksson, M. 1996. Consequences for plant reproduction of pollinator preference for symmetric flowers. Ph.D. Thesis, Department of Zoology, Uppsala University, Sweden.

Escós, J. M., Alados, C. L. & Emlen, J. M. 1997. The impact of grazing on plant fractal architecture and fitness of a Mediterranean shrub *Anthyllis cytisoides* L. *Functional Ecology*, **11**, 66-78.

Fiske, P. & Amundsen, T. 1997. Female bluethroats prefer males with symmetric colour bands. *Animal Behaviour*, **54**, 81-87.

Fiske, P., Kålås, J. A. & Sæther, S. A. 1994. Correlates of male mating success in the lekking great snipe (*Gallinago media*): results from a four-year study. *Behavioral Ecology*, **5**, 210-218.

Folstad, I., Arneberg, P. & Karter, A. J. 1996. Antlers and parasites. *Oecologia*, **105**, 556-558.

Forbes, M., Leung, B. & Schalk, G. 1996. Fluctuating asymmetry in *Coenagrion resolutum* damselflies in relation to size, age, and parasitism. *Odonatologica,* **26**, 9-16.

Forkman, B. & Corr, S. 1996. Influence of size and asymmetry on sexual characteristics in the rooster and hen on the number of eggs laid. *Applied Animal Behaviour Science*, **49**, 285-291.

Gest, T. R., Siegel, M. I. & Anistranski, J. 1986. The long bones of neonatal rats stressed by cold, heat, and noise exhibit increased fluctuating asymmetry. *Growth,* **50**, 385-389.

Guilford, T. & Dawkins, M. S. 1991. Receiver psychology and the evolution of animal signals. *Animal Behaviour*, **41**, 1-14.

Harvey, I. F. & Walsh, K. J. 1993. Fluctuating asymmetry and lifetime mating success are correlated in males of the damselfly *Coenagrion puella* (Odonata: Coenagrionidae). *Ecological Entomology*, **18**, 198-202.

Hoffmann, A. A., Hercus, M. & Dagher, H. 1998. Population dynamics of the *Wolbachia* infection causing cytoplasmic incompatibility in *Drosophila melanogaster. Genetics*, **148**, 221-231.

Hoikkala, A., Aspi, J. & Suvanto, L. 1998. Male courtship song frequency as an indicator of male genetic quality in an insect species, *Drosophila montana. Proceedings of the Royal Society of London, Series B,* **265**, 503-508.

Horridge, G.A. 1996. The honeybee (*Apis mellifera*) detects bilateral symmetry and discriminates its axis. *Journal of Insect Physiology*, **42**, 755-764.

Hoysak, D.J. & Ankney, C.D. 1996. Correlates of behavioural dominance in mallards and American black ducks. *Animal Behaviour*, **51**, 409-419.

Hunt, J. & Simmons, L. W. 1997. Patterns of fluctuating asymmetry in beetle horns: an experimental examination of the honest signalling hypothesis. *Behavioral Ecology and Sociobiology*, **41**, 109-114.

Jablonski, P. G. & Matyjasiak, P. 1997. Chaffinch (*Fringilla coelebs*) epaulette display depends on the degree of exposure but not symmetry of intruder's epaulettes. *Behaviour*, **134**, 1115-1121.

Jenkins, W. 1982. Redundancy in the perception of bilateral symmetry in dot textures. *Perception and Psychophysics*, **32**, 171-177.

Jennions, M. D. 1998a. The effect of leg band symmetry in female-male association in zebra finches. *Animal Behaviour*, **55**, 61-67.

Jennions, M. D. 1998b. Tibial coloration, fluctuating asymmetry and female choice behaviour in the damselfly *Platycypha caligata*. *Animal Behaviour*, **55**, 1517-1528.

Johnstone, R. A. 1994. Female preference for symmetrical males as a by-product of selection for mate recognition. *Nature*, **372**, 172-175.

Johnstone, R. A. & Grafen, A. 1992. Error-prone signalling. *Proceedings of the Royal Society of London, Series B*, **248**, 229-233.

Karino, K. 1997. Female mate preference for males having long and symmetric fins in the bower-holding cichlid *Cyathopharynx furcifer*. *Ethology*, **103**, 883-892.

Kimball, R. T., Ligon, J. D. & Merola-Zwartjes, M. 1997. Fluctuating asymmetry in red junglefowl. *Journal of Evolutionary Biology*, **10**, 441-457.

Kodric-Brown, A. 1997. Sexual selection, stabilizing selection and fluctuating asymmetry in two populations of pupfish (*Cyprinodon pecosensis*). *Biological Journal of the Linnean Society*, **62**, 553-566.

Lehrer, M., Horridge, G. A., Zhang, S. W. & Gadagkar, R. 1995. Shape vision in bees: innate preference for flower-like patterns. *Philosophical Transactions of the Royal Society of London, Series B*, **347**, 123-137.

Leung, B. & Forbes, M. R. 1996. Fluctuating asymmetry in relation to stress and fitness: Effects of trait type as revealed by meta-analysis. *Ecoscience,* **3**, 400-413.

Leung, B. & Forbes, M. R. 1997. Fluctuating asymmetry in relation to indices of quality and fitness in the damselfly, *Enallagma ebrium* (Hagen). *Oecologia,* **110**, 472-477.

Liggett, A. C., Harvey, I. F. & Manning, J. T. 1993. Fluctuating asymmetry in *Scatophaga stercoraria* L.: successful males are more symmetrical. *Animal Behaviour*, **45**, 1041-1043.

Ligon, J. D., Kimball, R. & Merola-Zwartjes, M. 1998. Mate choice by female red junglefowl: the issues of multiple ornaments and fluctuating asymmetry. *Animal Behaviour*, **55**, 41-50.

Ludwig, W. 1932. *Das Rechts-Links Problem im Tierreich und beim Menschen.* Berlin: Springer-Verlag.

Markow, T. A. 1987. Genetics and sensory basis of sexual selection in *Drosophila*. In: *Evolutionary Genetics and Invertebrate Behavior* (Ed. by M. D. Huettel), pp. 89-95. New York: Plenum.

Markow, T. A. 1995. Evolutionary ecology of developmental instability. *Annual Reviews of Entomology*, **40**, 105-120.

Markow, T. A. & Ricker, J. P. 1992. Male size, developmental stability, and mating success in natural populations of three *Drosophila* species. *Heredity*, **69**, 122-127.

Markow, T. A. & Sawka, S. 1992. Dynamics of male mating success in experimental groups of *Drosophila melanogaster* (Diptera: Drosophilidae). *Journal of Insect Behaviour*, **5**, 375-383.

Markowski, J. 1993. Fluctuating asymmetry as an indicator for differentiation among roe deer *Capreolus capreolus* populations. *Acta Theriologica*, **38**, (supplement 2), 19-31.

Markusson, E. & Folstad, I. 1997. Reindeer antlers: visual indicators of individual quality? *Oecologia*, **110**, 501-507.

McKenzie, J. A. & O'Farrell, K. 1993. Modification of developmental instability and fitness: malathion-resistance in the Australian sheep blowfly, *Lucilia cuprina*. *Genetica*, **89**, 67-76.

McLachlan, A. & Cant, M. 1995. Small males are more symmetrical: mating success in the midge *Chironomus plumosus* L. (Diptera: Chironomidae). *Animal Behaviour*, **50**, 841-846.

Møller, A. P. 1990. Fluctuating asymmetry in male sexual ornaments may reliably reveal male quality. *Animal Behaviour*, **40**, 1185-1187.

Møller, A. P. 1992. Patterns of fluctuating asymmetry in weapons: evidence for reliable signalling of quality in beetle horns and bird spurs. *Proceedings of the Royal Society of London, Series B*, **248**, 199-206.

Møller, A. P. 1993a. Female preference for apparently symmetrical male sexual ornaments in the barn swallow *Hirundo rustica*. *Behavioral Ecology and Sociobiology*, **32**, 371-376.

Møller, A. P. 1993b. Morphology and sexual selection in the barn swallow *Hirundo rustica* in Chernobyl, Ukraine. *Proceedings of the Royal Society of London, Series B*, **252**, 51-57.

Møller, A. P. 1994. Sexual selection in the barn swallow (*Hirundo rustica*). IV. Patterns of fluctuating asymmetry and selection against asymmetry. *Evolution*, **48**, 658-670.

Møller, A. P. 1995. Bumblebee preference for symmetrical flowers. *Proceedings of the National Academy of Sciences, USA*, **92**, 2288-2292.

Møller, A. P. 1996a. Developmental stability of flowers, embryo abortion, and developmental selection in plants. *Proceedings of the Royal Society of London, Series B*, **263**, 53-56.

Møller, A. P. 1996b. Sexual selection, viability selection, and developmental stability in domestic flies *Musca domestica*. *Evolution*, **50**, 746-752.

Møller, A. P., Cuervo, J. J., Soler, J. J. & Zamora-Muñoz, C. 1996. Horn asymmetry and fitness in gemsbok *Oryx g. gazella*. *Behavioral Ecology*, **7**, 247-253.

Møller, A. P. & Eriksson, M. 1995. Flower asymmetry and sexual selection in plants. *Oikos,* **73**, 15-22.

Møller, A. P. & Höglund, J. 1991. Patterns of fluctuating asymmetry in avian feather ornaments: implications for models of sexual selection. *Proceedings of the Royal Society of London, Series B,* **245**, 1-5.

Møller, A. P., Magnhagen, K., Ulfstrand, A. & Ulfstrand, S. 1995a. Phenotypic quality and moult in the barn swallow *Hirundo rustica*. *Behavioral Ecology*, **6**, 242-249.

Møller, A. P. & Pomiankowski, A. 1993. Fluctuating asymmetry and sexual selection. *Genetica,* **89**, 267-279.

Møller, A. P., Sanotra, G. S. & Vestergaard, K. S. 1995b. Developmental stability in relation to population density and breed of chickens *Gallus gallus*. *Poultry Science*, **74**, 1761-1771.

Møller, A. P. & Sorci, G. 1998. Insect preference for symmetrical artificial flowers. *Oecologia,* **114**, 37-42.

Møller, A. P. & Swaddle, J. P. 1997. *Asymmetry, Developmental Stability and Evolution.* Oxford: Oxford University Press.

Møller, A. P. & Zamora-Muñoz, C. 1997. Antennal asymmetry and sexual selection in a cerambycid beetle. *Animal Behaviour*, **54**, 1509-1515.

Morris, M. R. 1998. Female preference for trait symmetry in addition to trait size in swordtail fish. *Proceedings of the Royal Society of London, Series B,* **265**, 907-911.

Naugler, C. T. & Leech, S. M. 1994. Fluctuating asymmetry and survival ability in the forest tent caterpillar moth *Malacosoma disstria*: implications for pest management. *Entomologia Experimentalis et Applicata*, **70**, 295-298.

Norry, F. M., Vilardi, J. C. & Hasson, E. 1998. Sexual selection related to developmental stability in *Drosophila buzzatii*. *Hereditas*, **128**, 115-119.

Novak, J. M., Rhodes, O. E., Jr., Smith, M. H. & Chesser, R. K. 1993. Morphological asymmetry in mammals: genetics and homeostasis reconsidered. *Acta Theriologica*, **38**, (supplement 2, 7-18.

Oakes, E. J. & Barnard, P. 1994. Fluctuating asymmetry and mate choice in paradise whydahs, *Vidua paradisaea*: an experimental manipulation. *Animal Behaviour*, **48**, 937-943.

Omland, K. E. 1996. Female mallard preferences for multiple male ornaments. I. Natural variation. *Behavioral Ecology and Sociobiology*, **39**, 353-360.

Osorio, D. 1996. Symmetry detection by categorization of spatial phase, a model. *Proceedings of the Royal Society of London, Series B,* **263**, 105-110.

Otronen, M. 1998. Male asymmetry and postcopulatory sexual selection in the fly *Dryomyza anilis. Behavioral Ecology and Sociobiology*, **42**, 185-191.

Packer, C. & Pusey, A. E. 1993. Should a lion change its spots? *Nature,* **362**, 595.

Palmer, A. R. 1996. Waltzing with asymmetry. *Bioscience,* **46**, 518-532.

Palmer, A. R. & Strobeck, C. 1986. Fluctuating asymmetry: measurement, analysis, patterns. *Annual Review of Ecology and Systematics*, **17**, 391-421.

Palmer, S. E. & Hemenway, K. 1978. Orientation and symmetry: effects of multiple, rotational and near symmetries. *Journal of Experimental Psychology: Human Perception and Performance*, **4**, 691-702.

Pankakoski, E. 1985. Epigenetic asymmetry as an ecological indicator in muskrats. *Journal of Mammalogy*, **66**, 52-57.

Parker, L. T. & Leamy, L. 1991. Fluctuating asymmetry of morphometric characters in house mice: the effects of age, sex, and phenotypical extremeness in a randombred population. *Journal of Heredity*, **82**, 145-150.

Parsons, P.A. 1962. Maternal age and developmental variability. *Journal of Experimental Biology*, **39**, 251-260.

Polak, M. 1993. Parasites increase fluctuating asymmetry of male *Drosophila nigrospiracula:* implications for sexual selection. *Genetica,* **89**, 255-265.

Polak, M. 1997a. Ectoparasitism in mothers causes higher FA in their sons: implications for sexual selection. *American Naturalist*, **149**, 955-974.

Polak, M. 1997b. Parasites, fluctuating asymmetry, and sexual selection. In: *Parasites: Effects on Host Hormones and Behavior* (Ed. by N. Beckage), pp. 246-276. New York: Chapman and Hall.

Radesäter, T. & Halldórsdóttir, H. 1993. Fluctuating asymmetry and forceps size in earwigs, *Forficula auricularia. Animal Behaviour*, **5**, 626-628.

Reimchen, T. E. 1997. Parasitism of asymmetrical pelvic phenotypes in sticklebacks. *Canadian Journal of Zoology*, **75**, 2084-2094.

Rensch, B. 1958. Die Wirksamkeit ästhetischer Faktoren bei Wirbeltieren. *Zeitschrift für Tierpsychologie*, **15**, 447-461.

Rintamäki, P. T., Alatalo, R. V., Höglund, J. & Lundberg, A. 1997. Fluctuating asymmetry and copulation success in lekking black grouse. *Animal Behaviour*, **54**, 265-269.

Roldan, E. R. S., Cassinello, J., Abaigar, T. & Gomiendo, M. 1998. Inbreeding, fluctuating asymmetry, and ejaculate quality in an endangered ungulate. *Proceedings of the Royal Society of London, Series B,* **265**, 243-248.

Ryan, M. J., Warkentin, K. M., McClelland, B. E. & Wilczynski, W. 1995. Fluctuating asymmetries and advertisement call variation in the cricket frog, *Acris crepitans*. *Behavioral Ecology*, **6**, 124-131.

Saino, N. & Møller, A. P. 1994. Secondary sexual characters, parasites and testosterone in the barn swallow, *Hirundo rustica*. *Animal Behaviour*, **48**, 1325-1333.

Savalli, U. M. 1995. Notes on the natural history and mating system of the pin-tailed whydah *Vidua macroura*. *Ostrich*, **66**, 129-134.

Schall, J. J. 1996. Malarial parasites of lizards: diversity and ecology. *Advances in Parasitology*, **37**, 255-333.

Schandorff, S. 1997. Developmental stability and the harbour seal epizootic in 1988. *Annales Zoologici Fennici*, **34**, 167-175.

Schlüter, A., Parzefall, J. & Schlupp, I. 1998. Female preference for symmetrical vertical bars in male sailfin mollies. *Animal Behaviour*, **56**, 147-153.

Schwabl, U. & Delius, J. D. 1984. Visual bar length discrimination threshold in the pigeon. *Bird Behaviour*, **5**, 118-121.

Sheridan, L. & Pomiankowski, A. 1997. Female choice for spot asymmetry in the Trinidadian guppy. *Animal Behaviour*, **54**, 1523-1529.

Simmons, L. W. 1995. Correlates of male quality in the field cricket, *Gryllus campestris* L.: age, size, and symmetry determine pairing success in field populations. *Behavioral Ecology*, **6**, 376-381.

Somarakis, S., Kostikas, I., Peristeraki, N. & Tsimenides, N. 1997. Fluctuating asymmetry in the otoliths of larval anchovy *Engraulis encrasicolus* and the use of developmental instability as an indicator of condition in larval fish. *Marine Ecology Progress Series*, **151**, 191-203.

Stockley, P., Searle, J. B., Macdonald, D. W. & Jones, C. S. 1996. Correlates of reproductive success within alternative mating tactics of the common shrew. *Behavioral Ecology*, **7**, 334-340.

Suchentrunk, F. 1993. Variability of minor tooth traits and allozymic diversity in brown hare *Lepus europaeus* populations. *Acta Theriologica*, **38**, (supplement 2), 59-69.

Swaddle, J. P. 1996. Reproductive success and symmetry in zebra finches. *Animal Behaviour*, **51**, 203-210.

Swaddle, J. P. 1997. Developmental stability and predation success in an insect predator-prey system. *Behavioral Ecology*, **8**, 433-436.

Swaddle, J. P. 1999. Visual signalling by asymmetry: a review of perceptual processes. *Philosophical Transactions of the Royal Society of London, Series B*, **354,** 1383-1393.

Swaddle, J. P. & Cuthill, I. C. 1994a. Female zebra finches prefer males with symmetric chest plumage. *Proceedings of the Royal Society of London, Series B,* **258**, 267-271.

Swaddle, J. P. & Cuthill, I. C. 1994b. Preference for symmetric males by female zebra finches. *Nature,* **367**, 165-166.

Swaddle, J. P. & Witter, M. S. 1994. Food, feathers and fluctuating asymmetries. *Proceedings of the Royal Society of London, Series B,* **255**, 147-152.

Swaddle, J. P. & Witter, M. S. 1995. Chest plumage, dominance and fluctuating asymmetry in female starlings. *Proceedings of the Royal Society of London, Series B,* **260**, 219-223.

Thornhill, R. 1992a. Fluctuating asymmetry and the mating system of the Japanese scorpionfly, *Panorpa japonica. Animal Behaviour*, **44**, 867-879.

Thornhill, R. 1992b. Fluctuating asymmetry, interspecific aggression and male mating tactics in two species of Japanese scorpionflies. *Behavioral Ecology and Sociobiology*, **30**, 357-363.

Thornhill, R. & Møller, A. P. 1998. The relative importance of size and asymmetry in sexual selection. *Behavioral Ecology*, **9**, 546-561.

Thornhill, R. & Sauer, P. 1992. Genetic sire effects on the fighting ability of sons and daughters and mating success of sons in a scorpionfly. *Animal Behaviour*, **43**, 255-264.

Tomkins, J. L. & Simmons, L. W. 1995a. Dimorphisms and fluctuating asymmetry in the forceps of male earwigs. *Journal of Evolutionary Biology*, **9**, 753-770.

Tomkins, J. L. & Simmons, L. W. 1995b. Patterns of fluctuating asymmetry in earwig forceps: no evidence for reliable signalling. *Proceedings of the Royal Society of London, Series B,* **259**, 89-96.

Tyler, C. W. & Miller, R. H. 1994. Pattern identification by trajectory analysis in autocorrelation hyperspace. *Proceedings of the World Congress on Neural Networks,* **3**, 312-316.

Ueno, H. 1994. Fluctuating asymmetry in relation to two fitness components, adult longevity and male mating success in a ladybird beetle, *Harmonia axyridis* (Coleoptera: Coccinellidae). *Ecological Entomology*, **19**, 87-88.

Van Valen, L. 1962. A study of fluctuating asymmetry. *Evolution,* **16**, 125-162.

Veiga, J. P., Salvador, A., Martin, J. & López, P. 1997. Testosterone stress does not increase asymmetry of a hormonally mediated sexual ornament in a lizard. *Behavioral Ecology and Sociobiology*, **41**, 171-176.

Vøllestad, L. A. & Hindar, K. 1997. Developmental stability and environmental stress in *Salmo salar* (Atlantic salmon). *Heredity,* **78**, 215-222.

Vrijenhoek, R. C. 1994. Genetic diversity and fitness in small populations. In: *Conservation Genetics* (Ed. by V. Loeschcke, J. Tomiuk & S. K. Jain), pp. 37-53. Basel: Birkhäuser.

Wagemans, J. 1996. Detection of visual symmetries. In: *Human Symmetry Perception and its Computational Analysis* (Ed. by C. W. Tyler), pp. 25-48. Utrecht: VSP.

Wagemans, J., Van Gool, L., Swinnen, V. & Van Horebeek, J. 1993. Higher-order structure in regularity detection. *Vision Research*, **33**, 1067-1088.

Wardhaugh, K. G., McMahon, R. J., Axelsen, A., Rowland, M. W. & Wanjura, W. 1993. Effects of ivermectin residues in sheep dung on the development and survival of the bushfly, *Musca vetustissima* Walker, and a scarabine dung beetle, *Euoniticellus fulvus* Goese. *Veterinary Parasitology*, **48**, 139-157.

Watson, P. J. & Thornhill, R. 1994. Fluctuating asymmetry and sexual selection. *Trends Ecology and Evolution*, **9**, 21-25.

Wauters, L. A., Dhondt, A. A., Knothe, H. & Parkin, D. T. 1996. Fluctuating asymmetry and body size as indicators of stress in red squirrel populations in woodland fragments. *Journal of Applied Ecology*, **33**, 735-740.

Wiggins, D. A. 1997. Fluctuating asymmetry in *Colophospermum mopane* leaves and oviposition preference in an African silk moth *Imbrasia belina*. *Oikos,* **79**, 484-488.

Windig, J. J., Brakefield, P. M., Reitsma, N. & Wilson, J. G. M. 1994. Seasonal polyphenism in the wild: survey of wing patterns in five species of *Bicyclus* butterflies in Malawi. *Ecological Entomology*, **19**, 285-298.

Witter, M. S. & Swaddle, J. P. 1994. Fluctuating asymmetries, competition and dominance. *Proceedings of the Royal Society of London, Series B,* **256**, 299-303.

Wolf, S. D., Silk, W. K. & Plant, R. E. 1986. Quantitative patterns of leaf expansion: comparison of normal and malformed growth in *Vitis vinifera* cv Ruby Red. *American Journal of Botany*, **73**, 832-846.

Zakharov, V. M., Pankakoski, E., Sheftel, B.I., Peltonen, A. & Hanski, I. 1991. Developmental stability and population dynamics in the common shrew, *Sorex araneus*. *American Naturalist*, **138**, 797-810.

Zhang, L. & Gerbino, W. 1992. Symmetry in opposite-contrast dot patterns. *Perception,* **21** (supplement 2), 95.

Zvereva, E. L., Kozlov, M. V., Niemelä, P. & Haukioja, E. 1997. Delayed induced resistance and increase in leaf fluctuating asymmetry as responses of *Salix borealis* to insect herbivory. *Oecologia*, **109**, 368-373.

Zvereva, E. L., Kozlov, M. V. & Haukioja, E. 1997. Stress response of *Salix borealis* to pollution and defoliation. *Journal of Applied Ecology*, **34**, 1387-1396.

Appendix

Notes to Table 1.

Correlations - effect: Danzmann et al. 1986; Pankakoski 1985; Gest et al. 1986; Brault & de Oliveira 1995; Agnew & Koella 1995; Parsons 1962; Downhower et al. 1990; Chenuil 1991; Zakharov et al. 1991; Møller 1996b; Forbes et al. 1996; Møller et al. 1996; Stockley et al. 1996; Novak et al. 1993; Parker & Leamy 1991; Packer & Pusey 1993; Markowski 1993; Wolf et al. 1986; Zvereva et al. 1997; Bonn 1996; Escós et al. 1997; Reimchen 1997; Vrijenhoek 1994; Schall 1996; Saino & Møller 1994; Markusson & Folstad 1997; McKenzie & O'Farrell 1993; Somarakis et al. 1997; Roldan et al. 1998; Simmons 1995. *Correlations no effect*: Chenuil 1991; Markusson & Folstad 1997; Björklund 1996; Møller 1994; Arcese 1994; Leung & Forbes 1997; Windig et al. 1994; Dufour & Weatherhead 1998b; Wiggins 1997; Forbes et al. 1996; Hoffmann et al. 1998; Polak 1997b, 4 spp.; Schandorff 1997; Vøllestad & Hindar 1997; Arnqvist et al. 1997; Hoikkala et al. 1998; Kimball et al. 1997; Carrascal et al. 1998. *Correlations + effect*: Packer & Pusey 1993; Reimchen 1997; Dufour & Weatherhead 1998b; Suchentrunk 1993; Campbell & Emlen 1996. *Experiments - effect*: Bagchi et al. 1989; McKenzie & O'Farrell 1993; Møller et al. 1995b; Møller et al. 1995a; Møller 1996a; Swaddle & Witter 1994; Swaddle 1996; Swaddle 1997; Ueno 1994; Naugler & Leech 1994; Thornhill 1992a; Zvereva 1997; Polak 1993; Folstad et al. 1996; Campbell et al. 1998; Beardmore et al. 1960. *Experiments no effect*: Wardhaugh et al. 1993.

Notes to Table 2.

Correlations - effect: Møller & Eriksson 1995, 8 spp.; Eriksson 1996, 2 spp.; Harvey & Walsh 1993; Córdoba-Aguilar 1995; Radesäter & Halldórsdóttir 1993; Møller & Zamora-Muñoz 1997; Simmons 1995; Otronen 1998; Thornhill 1992b, 2 spp.; Thornhill 1992a; Thornhill & Sauer 1992; McLachlan & Cant 1995; Allen & Simmons 1996; Liggett et al. 1993; Markow 1987; Markow & Ricker 1992; Møller 1996b; Kodric-Brown 1997; Karino 1997; Sheridan & Pomiankowski 1997; Rintamäki et al. 1997; Forkman & Corr 1996; Hasegawa 1995; Borgia 1985; Møller 1993b; Wauters et al. 1996; Arcese 1994; Møller et al. 1996; Roldan et al. 1998; Møller & Höglund 1991, 3 spp.; Møller 1992, 9 spp.; Córdoba-Aguilar 1997; Malyon & Healy 1994; Tomkins & Simmons 1995b, 2 spp.; Balmford et al. 1993; Carrascal et al. 1998; Norry et al. 1998. *Correlations no effect*: Møller & Eriksson 1995, 2 spp.; Markow & Ricker 1992, 3 spp.; Kodric-Brown 1997; Rintamäki et al. 1997; Møller & Höglund 1991, 12 spp.; Møller 1992, 5 spp.; Tomkins & Simmons 1995b, 23 spp.; Balmford et al. 1993, 41 spp.; Jennions 1998b; Tomkins & Simmons 1995a, 5 spp.; Hunt & Simmons 1997; David et al. 1998; Arnqvist et al. 1997; Ueno 1994; Brown 1997; Hoikkala et al. 1998; Polak 1997a; Taylor, M. I., Turner, G. F., Robinson, R. L. & Stauffer, J. R., unpublished; Brooks & Caithness 1995; Ryan et al. 1995; Veiga et al. 1997; Omland 1996; Fiske et al. 1994; Dunn 1994; Savalli 1995; Dufour & Weatherhead 1998a; Hoysak & Ankney 1996; Stockley et al. 1996. *Correlations + effect*: Markow & Ricker 1992; Tomkins & Simmons 1995b; Balmford et al. 1993. *Experiments - effect*: Møller 1995; Møller 1993a; Fiske & Amundsen 1997; Swaddle & Cuthill 1994b; Schlüter et al. 1998. *Experiments no effect*: Jennions 1998b; Schlüter et al. 1998; Ligon et al. 1998; Jennions 1998a; Markow & Sawka 1992. *Experiments + effect*: Oakes & Barnard 1994. *Natural range experiments - effect*: Morris 1998; Møller & Sorci 1998; Swaddle & Cuthill 1994a. *Natural range experiments no effect*: Swaddle & Witter 1995; Jablonski & Matyjasiak 1997.

IN: ESPMARK,Y., AMUNDSEN,T. & ROSENQVIST,G. (EDS.) 2000
ANIMAL SIGNALS:
SIGNALLING AND SIGNAL DESIGN IN ANIMAL COMMUNICATION, PP. 177-193
TAPIR ACADEMIC PRESS, TRONDHEIM, NORWAY

8

Problems in Studying Receiver Biases and Their Effects on Signal Evolution

ALEXANDRA L. BASOLO

School of Biological Sciences, University of Nebraska-Lincoln, Lincoln, NE, USA

E-mail: basolo@niko.unl.edu

Abstract

Receiver biases can have important consequences for signal evolution. In this paper, the evolution of receiver biases is discussed, including the establishment of biases and the maintenance of biases. The utility of the comparative method in studying receiver biases is discussed, as are problems inherent to comparative studies. Both adaptive and non-adaptive origins for receiver biases are detailed. A discussion of how biases can play a role in both natural selection and sexual selection, including intrasexual selection and intersexual selection, is presented. The primary examples of cases in which receiver biases have been implicated as important to trait evolution in fish, frogs, spiders, water mites and crabs are reviewed. Problems in the study of preexisting receiver biases and their role in signal evolution are discussed and some possible solutions are presented.

In the past eight years, one model that explains the evolution of mating preferences and their influence on male trait evolution, the preexisting receiver bias model, has received increasing attention (reviewed by Endler & Basolo 1998). This model suggests that receiver biases can be established via mutation and genetic drift, or via selection in a context other than intersexual selection on the specific male trait under study. Such biases can exist at any level along the signal reception and processing continuum from stimulation of a peripheral sensory system to synthesis at the higher levels of integration and cognitive processing. Receiver biases can influence both the direction and rate of signal evolution. Preexisting receiver biases have been indicated as important in female mate choice in a wide range of animals, including spiders (Clark & Uetz 1992), water mites (Proctor 1991, 1992), fish (Basolo 1990b, 1995a, b), frogs (Ryan et al. 1990; Ryan & Rand 1993) and crabs (Christy 1995).

Preexisting receiver biases can be important in the evolution of traits via both natural selection and sexual selection. In terms of female mate choice, such biases can affect male signal evolution when male traits arise to which females are differentially sensitive; when such traits arise, they may convey a mating advantage and may thus be favored by sexual selection. The presence of a preexisting receiver bias may be sufficient by itself to cause the evolution of a male trait, or it can set the stage for other mechanisms of mate choice evolution. For instance, receiver biases can act in conjunction with Fisherian processes (Fisher 1958); females with the bias will mate with males with new traits that stimulate the bias, producing offspring that have both the bias and the trait. In this way, there will be coevolution and linkage between the bias and a preferred male trait, if both the bias and the male signal show heritable variation. Preexisting receiver biases can also work in conjunction with good genes mechanisms, and some direct benefits mechanisms (when the male trait has a genetic basis; Basolo 1996) of mate choice evolution to affect the evolution of male signals.

The utility of the comparative method in the study of preexisting receiver biases

The comparative method has proven indispensable in identifying receiver biases; by measuring preferences in extant taxa, the state of a bias in ancestral taxa can be hypothesized. For example, consider the hypothetical phylogeny represented in Figure 1 for taxa A, B and C. Taxa A and B share a trait and a preference favoring that trait, while taxon C has only the preference favoring the trait. Given that males in the outgroups do not possess the trait, from this phylogeny and using the principle of parsimony (Hennig 1966), we would infer that the preference was present in a common ancestor of A, B and C, whereas the trait arose in the common ancestor of A and B, after the divergence of taxon C. Thus, a receiver bias is present in the ancestor prior to the appearance of the male trait it favors; this conclusion is based on the assumption that the state of a bias in extant taxa can be used to infer the state of the bias at the branch points (nodes 1 and 2, Fig. 1).

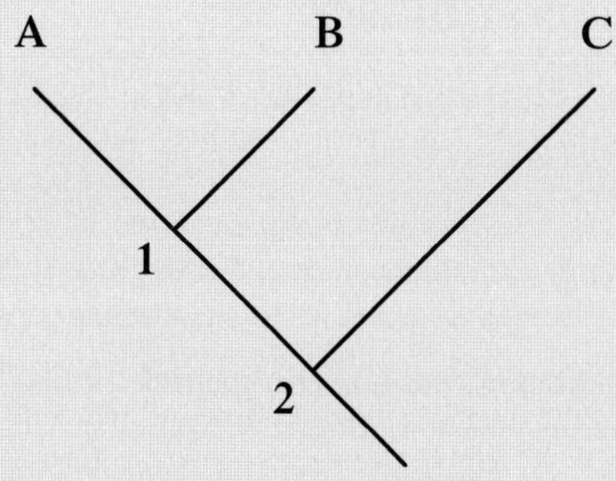

Figure 1. Hypothetical phylogeny for taxa A, B and C. Nodes 1 and 2 represent the branch point nodes.

The comparative method can also be used to investigate how receiver biases change over time. For instance, in the hypothetical example given for taxa A, B and C (Fig. 1), all three taxa share the presence of an ancestral bias. However, the strength of this preference could have changed after divergence of these taxa. That is, following divergence but prior to male trait appearance, sources of evolutionary change, such as genetic drift and selection in another context, could have resulted in a change in the strength of the bias (see the following section for sources of change due to selection). In addition, after the appearance of the male trait in taxa A and B, the receiver bias may act in conjunction with these sources of evolutionary change, as well as good genes mechanisms, direct benefits mechanisms, and Fisherian selection mechanisms; such mechanisms may alter the bias by either increasing or decreasing the strength of the resulting preference. Thus by using phylogenetic information, it is also possible to assess whether there has been change in a bias since its origin.

An example of the utility of the comparative method for investigating both the presence or absence of a receiver bias, as well change in the bias, has been demonstrated for a group of poeciliid fishes. Males in most species of swordtails of the genus *Xiphophorus* possess long, colored rays at the base of the caudal fin called a sword; swords are only found in this genus. Female green swordtails *X. helleri* have been found to prefer males with longer swords (Basolo 1990a), and this sword preference was found to be an increasing function of sword length (Basolo 1990b). Females in the proposed sister genus, *Priapella* (Rosen 1979; Meyer et al. 1994), which is swordless, were tested for the presence of a preference favoring a sword by surgically-attaching artificial swords to conspecific males. Female *P. olmecae* were found to prefer males with artificial swords to males with clear attachments of equal length (Basolo 1995b). The combined results from the preference tests suggest that the common ancestor of the unsworded *P. olmecae* and the sworded *X. helleri* possessed a bias for the sword that pre-dates the origin of the sword itself in *Xiphophorus*. Thus, by utilizing the comparative method, the hypothesis that a preexisting receiver bias has played a role in the evolution of a female preference for a male trait in poeciliid fishes is supported.

Evolution of receiver biases

When considering the evolution of preexisting receiver biases and their effects on male trait evolution, three factors should be considered: origin of the bias, changes in the bias after its origin and maintenance of the bias.

Establishment of biases

Preexisting biases can originate either in a non-adaptive context (Kirkpatrick 1987a, b; Basolo 1995a) or in an adaptive context (Kirkpatrick 1987a, b; Endler & McLellan 1988; Basolo 1990b, 1995a; Endler 1992; Dawkins & Guilford 1996). Causes of bias establishment in a nonadaptive context include pleiotropy (Kirkpatrick 1987b; Basolo 1990b), and mutation with genetic drift. In terms of pleiotropy, a bias that is favored in one context could have consequences for receiver responses in a different context. The establishment of such biases have been investigated with neural network models (Arak & Enquist 1993; Enquist & Arak 1993; also see review in Endler & Basolo 1998). For a bias that arises as a result of mutation, genetic drift could assure that the bias is not lost prior to playing a role in sexual selection. In the absence of selection, drift could ultimately lead to fixation of the receiver bias.

Preexisting receiver biases that have originated in an adaptive context could do so via either natural selection or sexual selection. Receiver biases established via natural selection could have an original context pertaining to foraging, prey detection, predator detection, habitat selection, parental care, etc. To illustrate how a receiver bias that arose in a natural selection context could result in a female mating preference, consider a hypothetical scenario (Fig. 2). In an ancestral bird species, females have a response bias for a particular leaf phenology and configuration because these leaves are important to female fitness in terms of survival of her young; if placed in the nest, these leaves may reduce the incidence of ectoparasites on young. This ancestral response bias can be likened to a genetically-based search image for a specific shape and configuration. This ancestral bias may subsequently affect the direction and nature of sexual selection such that males in descendent species that exhibit structures to which females are differentially sensitive may gain a mating advantage. The initial bias can favor different male traits; two possible outcomes of the ancestral bias are depicted in Figure 2. In one lineage, the initial receiver bias has resulted in a descendent species in which males have a head ornament. In the other lineage, the same initial bias has resulted in the evolution of a male tail ornament. This scenario illustrates how a receiver bias established in a maternal care context could subsequently result in a female preference, as well as subsequently bias the nature and direction of male trait evolution.

Intrasexual (female-female or male-male competition) selection could result in receiver biases that subsequently act in a mate choice context. For example, if females use a specific trait to identify female competitors, detection of females in the environment may evolve in both sexes as a result of female-female competition. Males may then respond to the same female trait when searching for potential mates. In terms of intersexual

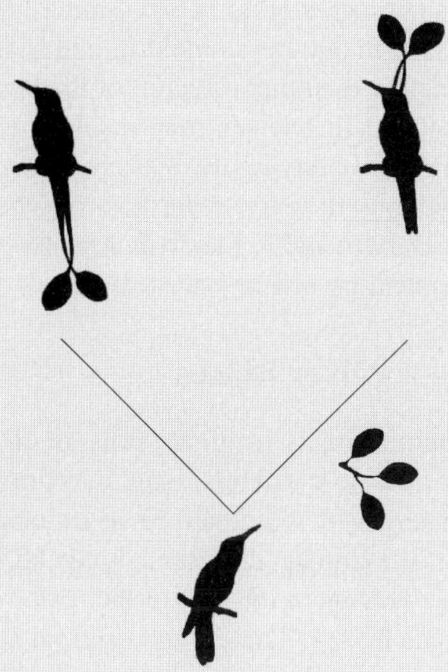

Figure 2. Hypothetical scenario for the establishment of a receiver bias in a maternal care context in an ancestral taxon and the subsequent effects of this bias on male trait evolution in two descendent lineages. (Figure created by W. E. Wagner, Jr.).

selection, several conditions could give rise to receiver biases. First, a preference for a female trait established in a male mate choice context could result in a preexisting bias in females that could favor a different trait in males. (The situation in which an initial male bias for a trait in females results in a female bias for the same trait in males would be a case of a correlated effect (Lande & Arnold 1985; Halliday & Arnold 1987; Basolo 1996)). Likewise, a preference for a trait in males established in a female mate choice context could result in a preexisting bias in males favoring a different trait in females. Second, a preference could be established in females for a specific male trait and subsequently bias female responses to an alternate male trait (and vice versa: male preference for female trait). For example, while *P. olmecae* females have a preexisting bias favoring sworded males, the bias may not be for the sword per se. That is, the bias could be for a single, or several components of the multi-component sword (Basolo 1996). Or the original intersexual selection context may have been a female preference for large males; if so, the sword could possibly increase perceived body size in the horizontal plane (Basolo 1990b; Ryan 1990). Other possibilities for a bias favoring a sword include an initial bias favoring coloration in males or enlarged fins in males. While this list of possible contexts is not exhaustive, it illustrates the point that the number of possible original contexts that could establish a bias favoring a sword is numerous.

Whether a bias is established in an adaptive or a nonadaptive context, the bias need not be sex-limited. If a bias arises as a consequence of mutation with subsequent genetic drift, or if it arises due to selection, the bias is expected to be shared between the sexes, unless there is a cost to carrying the bias to one of the sexes. The role such a receiver bias may play when a trait favored by the bias arises may be different for the sexes; that is, for males, the bias may play a role in intraspecific competition, while in females, the bias may affect mate choice. If a bias is the result of pleiotropy and the other effect(s) of the gene(s) have adaptive value in both sexes, the bias is not expected to be sex-limited. If the basis of a bias is the result of pleiotropy that is adaptive in another context in one sex, the bias may or may not be sex-limited, depending on whether the other effect(s) of the gene(s) are sex-specific. Biases originating in an adaptive context in one sex may result in either expressed or non-expressed biases in the other sex, if costs of having the bias are negligible for that sex. Lastly, if a bias was originally established in both sexes, regardless of whether it had an original function in both (correlated effect: Darwin 1871; Lande & Arnold 1985; Basolo 1996), it could later be co-opted in another context by one or both sexes.

Maintenance and change in receiver biases

After the establishment of a bias, it may be maintained in its original form, or it may change via either selection or genetic drift. Consider how post-establishment evolutionary processes may influence a female receiver bias. In terms of maintenance in its original form, biases established in an adaptive context, as previously discussed, may be constrained by that context, and thus there would be little change in the bias, either prior to or subsequent to the appearance of the male trait the bias favors. That is, the original selective context could act as opposing selection to limit change in the bias. Alternatively, if the direction of selection is the same in terms of both the original selective context and the bias in a new mate choice context, the bias may change in a positive direction. Biases established in a non-adaptive context may change due to genetic drift in the absence of counter-selection. Because a bias established in a non-adaptive context may be less likely to be constrained than a bias established in an adaptive context, a bias with a non-adaptive origin may be more labile and thus subject to a greater degree of change, both prior to male trait evolution and during male trait evolution. Likewise, if a bias is a result of past selection that is no longer operating, the bias may be much more labile than a bias maintained by selection presently operating in another context.

As discussed, once a male trait is favored by a preexisting bias, linkage disequilibrium between the preference and trait via Fisherian selection (Fisher 1958; Lande 1981; Kirkpatrick 1982) will cause the bias to increase in frequency until counter-selection acts against it. However, if an increase in the frequency of a bias as a result of the Fisherian process enhances the bias in its original adaptive function, the male trait may be carried past its optimum. In this case, one of two situations would occur; either the trait and preference would become disassociated, or the male trait may be maintained at a higher, maladaptive state. Positive selection via good genes and correlated effects mechanisms, as well as some direct benefits mechanisms, may affect bias evolution after the appearance of the male trait. A change in the environment resulting in strong negative selection on the trait could cause the bias to decrease to a low frequency or even be lost. If the bias exists as a result of non-adaptive processes, then we would not expect the bias to constrain the trait when negative selection is directed at the trait, and we would expect both the trait and the preference to decrease (if they are in linkage disequilibrium). Lastly, the direction of change in a bias, whether established in a non-adaptive or an adaptive context, can be affected by constraints due to allometry, constraints due to the existing sensory system, and developmental constraints.

Evidence for evolutionary change in receiver biases

Research on the preexisting female bias favoring the sword in poeciliid fishes suggests that receiver biases do evolve after their origin. The strength of the female preference for sword length in both *P. olmecae* and *X. helleri* has been examined by comparing the relationship between the amount of time females spent with conspecific males differing in sword length between the two species (Basolo 1998). Males in both species were randomly given artificial swords of varying length. For both species, there was a significant positive correlation between strength of female response and sword length; females more strongly preferred longer sworded males. However, controlling for the effect of male body size on female response, there was a significant difference between the preference functions for *P. olmecae* and *X. helleri*. When the slopes were compared, female *X. helleri* showed a significantly weaker preference based on sword length than female *P. olmecae*. There are several possible explanations for why the strength of the preference favoring a sword differs for these sister taxa. First, it may be that the sword preference is weaker in species in which the sword has evolved, i. e. the green swordtail *X. helleri* due, to costs associated with expression of the preference, the sword, or for both. Second, it may be that preference in the *P. olmecae* population tested may not be typical of other species possessing the primitive state for the male trait; it may be that other species of poeciliids with the preexisting receiver bias favoring a sword would show a weaker preference than *P. olmecae*. Third, the population of *X. helleri* that was tested may be neither typical of other populations of green swordtails, or other species of swordtails. It may be that other populations of *X. helleri* and other species of swordtails would show a preference equal to or stronger than that found in *P. olmecae*. Although it is unclear whether the preference in *X. helleri* or *P. olmecae* more closely resembles the ancestral state of the receiver bias favoring a sword, it is clear there has been change in the bias, either in the population of *X. helleri* or the population of *P. olmecae* tested. This conclusion is based on the assumption that the sword biases in these sister taxa are homologous.

Studying preexisting receiver biases

To explain why a receiver bias has evolved, we need to investigate the nature of the bias. A number of approaches can be employed to study the nature of receiver biases. The approaches that have been most successful in furthering our understanding of how pre-existing biases are established and how they affect male signal evolution are discussed in this section.

Investigating the sensory and neurobiological basis of the bias

Properties of a species sensory system can bias the direction of the evolution of signals (West-Eberhard 1984; Kirkpatrick 1987b; Endler & McLellan 1988; Ryan et al. 1990). For example, Ryan et al. (1990) examined the morphology of the basilar papillae in the peripheral auditory system of the frog *Physalaemus coloradorum* and the Túngara frog, *P. pustulosus*. They found that the closely related species share peak basilar papillae sensitivities at 2200 Hz. Phylogenetic evidence suggests that this is the ancestral condition for female tuning. Chucks at this frequency are produced by calling male *P. pustulosus*, but not by *P. coloradorum*. They concluded that the chuck in *P. pustulosus* evolved as a consequence of this bias in female tuning. In addition, they predicted that the receiver bias in *P. coloradorum* females could result in a female preference in this species for male calls with chucks. This hypothesis was subsequently tested by Ryan and Rand (data discussed by Kirkpatrick & Ryan 1991); they presented *P. coloradorum* females with a synthetic call that included a chuck. Although male *P. coloradorum* do not produce calls at the targeted frequencies, females preferred the call with a chuck to the call lacking a chuck, as predicted. Thus, by combining phylogenetic information with knowledge about the sensory basis of female hearing, the researchers were able to test the sensory basis of a preexisting receiver bias. Additional studies directed at understanding how receiver biases in different sensory and cognitive processes influence male signal evolution in a range of animals are needed before we can fully understand why such biases exist and how they influence male signal evolution.

Correlational evidence

Several studies have looked at the behavior of animals in contexts other than mate choice to investigate the potential origins of receiver biases that affect male signal evolution. This type of correlational approach has indicated that biases established in anti-predator contexts, as well as foraging contexts, have played a role in the establishment of receiver biases and the subsequent evolution of male behaviors.

Proctor (1991, 1992) has suggested that a response bias maintained in a foraging context in female water mites *Neumania papillator* has resulted in the evolution of a male signal. Male water mites vibrate their forelegs, which often causes nearby females to orient towards and clutch at the trembling male. These male vibrations are within the range of vibrations made by copepod prey. Females exhibit orienting and clutching behavior only in the presence of males or copepods, and females on restricted food diets orient towards and clutch at males more often. Proctor hypothesized that males capitalize on the predatory behavior of females orienting towards vibrations of potential prey to attract female attention, and, females recognize males as conspecific mates when they approached, allowing males the opportunity to mate. This hypothesis based on correlational evidence was supported by phylogenetic evidence suggesting that courtship behavior, e. g. trembling and the male net-stance posture, either arose independently twice after predatory behavior appeared, or co-evolved with predatory behavior (Proctor 1992). Proctor concluded that the first possibility, a female receiver bias that was established in a predatory context, supported the female bias hypothesis. She also concluded that the second possibility, co-evolution of male behavior and female predatory behavior, neither supported nor rejected the female bias hypothesis.

Clark and Uetz (1992) used video playback to test the hypothesis that the initial movement of a male was the basis of mate choice in the jumping spider *Maevia inclemens* and that this movement bias was a result of a receiver bias established in a predation context. In the jumping spider, an orientation response is elicited by movement of potential prey that stimulates the posterior lateral eyes. Orientation movement allows jumping spiders to shift the field of vision to the higher resolution of the anterior median eyes. Therefore, an initial female response to a moving male singles that male out as a potential mate. Thus, the correlational evidence suggests that a receiver bias in female jumping spiders favors a male behavior and that this same bias plays a role in female predatory behavior.

A third study suggests that a receiver bias established in an anti-predator context affects male mating success in the fiddler crab *Uca beebei* (Christy & Salmon 1991). One morph of male *U. beebei* constructs mud pillars which appear to attract females to their burrows. An anti-predator behavior exhibited by fiddler crabs is rapid orientation to cover, including structures that are perpendicular to the ground. Pillar construction is postulated to have evolved because males and females use pillars to find their burrows easily in order to avoid capture by predatory birds. Christy (1995) conducted a comparative study to test this hypothesis and found that females of five crab species that were investigated, including species in which males do not build pillars, differentially oriented towards burrows with pillars rather than burrows without pillars. He concluded that landmark orientation that evolved in an anti-predator context is a shared trait in *Uca*, and that this receiver bias may have affected the evolution of pillar construction by males in some *Uca* species.

These three studies suggest that receiver biases that affect mate choice can evolve due to their function in a natural selection context. For water mites and jumping spiders, correlational evidence suggest that biases have evolved in a prey capture context, and for fiddler crabs, a receiver bias evolved in a predator-avoidance context. These types of studies are only suggestive, however, as they are based on plausible scenarios and on present-day environmental conditions. They may thus not provide the strongest evidence for the factors favoring the origin and maintenance of the receiver biases. Additional evidence for the water mites is provided by phylogenetic information, which strengthens the conclusion that a preexisting receiver bias in females has played a role in male signal evolution. Although comparative information is provided for the fiddler crabs, phylogenetic information for both the fiddler crabs and the jumping spiders could determine whether the female preferences in fiddler crabs and jumping spiders reflect the ancestral or derived condition for each. That is, whether the female preferences evolved prior to the appearance of the male traits. However, these studies demonstrate that the study of extant species and present day environmental conditions may provide information useful in formulating hypotheses about the evolution of preexisting receiver biases and the traits they favor.

Sex-specificity of receiver biases

The state of a receiver bias can be investigated in a taxon to determine whether it is sex-specific or shared. Finding that a bias is shared by the sexes would suggest either that: (1) it arose in a non-adaptive context and was not selected against in either sex, (2) it arose in an adaptive context in one sex and was carried along in the other sex as a correlated consequence, (3) it was established in the same adaptive context in both sexes, or (4) it was established in different adaptive contexts in the two sexes. Finding that a bias is not presently shared by the sexes would indicate that: (1) it originated in one sex via either selection or mutation with drift, without a correlated response in the opposite sex, or (2) it arose in both sexes via either selection or mutation with drift, but was maintained in only one sex. Both situation (1) and (2) would suggest that there is a net cost to possessing the receiver bias in one sex, but not in the other. Comparisons across taxa may shed light on the original selective context in which a receiver bias was established. For example, consider the situation in which a receiver bias was established as a sex-limited trait, arising in females, as depicted in Figure 3. First, we would test females in species D to establish that sexual selection via female choice is present in the species possessing the male trait (species D). If a female preference of the trait is present in D, then we would test for the presence of the preference in species C, the sister species which lacks the male trait, to determine if a bias favoring the male trait arose prior to the trait. If species C possesses the receiver bias favoring the male trait, we would then test additional outgroups to determine when the receiver bias arose. Here, females in species A lack the bias, while females in species B possess the bias. We would conclude that the bias arose in the common ancestor of the B and the C/D group, after the divergence of species A. Knowing the state of the bias in females, we can then test whether males also possess the bias in the four species, even though females lack the trait. Here, males in species C and D possess a receiver bias for conspecific females with the trait present in males in species D, while species A and B do not. From these results, we would conclude that the bias arose in females initially, and subsequently appeared in males. If males in species B are found to possess the receiver bias, then we would conclude that the bias originated in both sexes. Such a comparative approach could also be used to test whether the receiver bias has played a role in the evolution of the male trait in a male-male competition context.

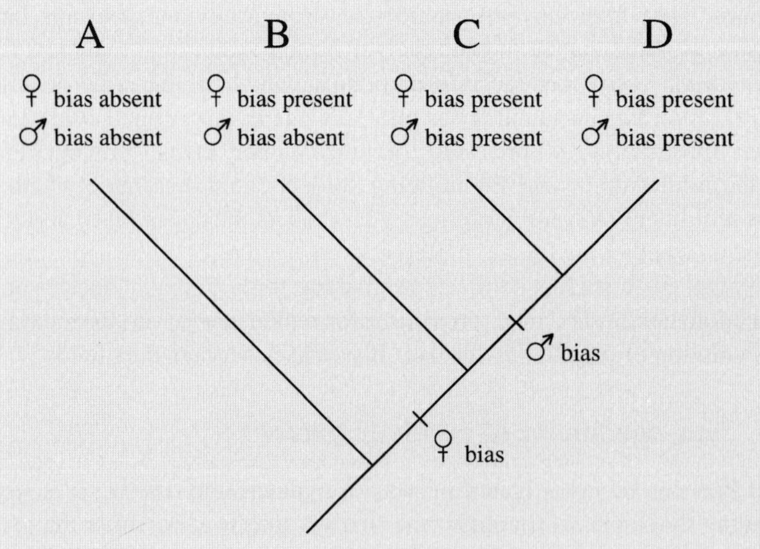

Figure 3. *Hypothetical phylogeny for the state of a female receiver bias and a male trait favored by the bias for taxa A, B, C and D.*

Is it reasonable to assume that intersexual differences exist (e. g. species B, Fig. 3) that can result in different responses to the same signal? Evidence for the hooded warbler

Wilsonia citrina indicates that males and females respond differently to the angle of stripes, and that this difference between the sexes has resulted from differences in habitat usage. Morton (1990) tested warblers on stripes at vertical angles to the floor versus stripes at oblique angles to the floor; females oriented toward the oblique stripes while males oriented towards the vertical stripes. He suggested that this difference between the sexes was a result of sex-specific preferences for wintering habitat. Female hooded warblers prefer nonbreeding territories that differ from breeding habitats, while males establish winter territories in habitat similar to the habit used during the breeding season. In winter, males prefer taller vegetation, as is found in secondary growth and forest habitat, while females defend shrub or field habitat. Thus, Morton suggests that the response of males to vertical stripes reflects a preference for components of a forest, e. g. vertical vegetative elements such as tree trunks, while the response of females to oblique stripes reflect their habitat preference for shrubby plants, e. g. oblique vegetative elements. Morton concluded that the difference between the sexes in orientation towards stripes was 'largely innate'. Thus, it is not unreasonable to assume that there can be differences between the sexes in their responses to stimuli, and that such differences can be the source of intersexual differences in the state of receiver biases.

Problems with studying receiver biases and some possible solutions

The study of receiver biases requires inferences about evolutionary history. For example, when considering a male trait that evolves as a consequence of female bias, inferences must be made about both the trait and bias. It also requires some inferences about historical patterns of selection. As a result, there are some methodological problems associated with establishing the presence of receiver biases and with examining the evolutionary consequences of such biases. There are also problems associated with determining why biases have evolved. Some of the problems inherent to studying receiver biases that have not been discussed in the previous sections, as well as some suggestions on addressing specific problems, are presented in this section.

Phylogenies and receiver biases

To test hypotheses about evolutionary history, a robust phylogenetic hypothesis is essential. However, for many taxa, robust phylogenies are not available. Even when phylogenies are available, there may be multiple conflicting hypotheses about the taxa of interest. For example, there are several phylogenetic hypotheses for the evolutionary relationships within *Xiphophorus*. In two of the proposed phylogenies the swordtails are a monophyletic group (Rosen 1979; Haas 1993). In a third proposed phylogeny the swordtails are not a monophyletic group; instead, the unsworded platyfish are proposed to have evolved from a sworded ancestor and to have secondarily lost the sword (Meyer et al. 1994). A fourth proposed phylogeny is not rooted, and thus difficult to interpret (Borowsky et al. 1995). If the swordtails constitute a monophyletic group, parsimony would suggest that a bias favoring the sword evolved in the ancestor of platyfish and swordtails, with the sword evolving after the divergence of the swordtails (Basolo 1990b). In contrast, if the platyfish have secondarily lost the sword, the bias in platyfish may represent a retained primitive preference, or 'residual' preference (e.g. Ryan & Wagner 1987; Basolo 1995a) rather than a preexisting bias established in another context that subsequently favored the evolution of the sword.

A potential resolution of the problem of conflicting phylogenies is to examine taxa outside of the group for which phylogenetic information is conflicting. For example, as discussed earlier, female *P. olmecae* and *X. helleri* appear to share a bias favoring a sword, while phylogenetic evidence suggests that the sword did not arise until after the divergence of *Priapella* and *Xiphophorus*. However, this type of approach has potential only if the bias evolved prior to the divergence of the group in which the trait is present and other closely-related groups. If the bias evolved within the unresolved clade in which the trait arose, better resolution of the phylogenetic relationships within the clade possessing the trait will be necessary.

Homology and inferring the ancestral state of receiver biases

If a robust phylogeny is available, it is necessary to infer ancestral character states in order to infer when the bias and the trait evolved. The principle of parsimony is one of the primary methods used to infer ancestral character states. However, as pointed out by Frumhoff and Reeve (1994), the use of this method is only possible if the rate of evolution for these traits is slow. More recently, a maximum likelihood method has been proposed by Schluter et al. (1997). They found that when changes between discrete character states are rare, maximum likelihood methods give results that are similar to those obtained using parsimony. But when changes between discrete characters states are frequent, inferences about ancestral character states are highly uncertain. As a result, parsimony estimates may not be appropriate, depending upon the group under study. Instead, it may be better to obtain probabilistic estimates using maximum likelihood, particularly if there is reason to believe that evolutionary changes between character states are common.

The reconstruction of ancestral character states requires information on both trait and bias homology. A trait can be present in two taxa because it was inherited from a common ancestor (homology) or because it evolved independently in the two taxa (homoplasy). In terms of receiver biases, it is possible that numerous biases favor the same signal. Studies on the peripheral sensory system, as well as at higher levels of signal processing within receivers could reveal receiver biases, as has been suggested in crickets (Römer 1993) and can provide evidence that similar responses among taxa to a signal results from a shared bias. However, because the neurobiological basis of receiver biases are not often known, it is can be difficult to test whether two biases are homologous.

There are indirect methods that can potentially be used to assess the homology of biases in different taxa. For example, we might predict that homologous biases will have similar consequences. If we manipulate a trait that appears to be favored by an ancestral bias, females in taxa that share the putative homologous bias should respond similarly to these trait variants. If females do not respond similarly, we might conclude that the biases are not homologous.

Indirect methods that compare the consequences of putative homologous biases may provide suggestive evidence about the homology of biases. However, similarity of consequence does not necessarily mean that the biases were inherited from a common ancestor, and dissimilarity of consequences does not necessarily mean that the biases were not inherited from a common ancestor. First, non-homologous biases may have similar consequences. If so, we might mistakenly infer that a bias was present in the ancestor of two taxa when in fact different biases evolved independently in the two taxa. In poeciliids, there is some evidence that sword biases may not be homologous. In a poeciliid distantly related to *Xiphophorus*, *Brachyrhaphis rhabdophora*, females prefer males with swords. However, they do not prefer longer swords to shorter swords (A. L. Basolo, unpublished), as do female *Xiphophorus* and *Priapella*. This difference might suggest that the bias in *B. rhabdophora* has a different neurobiological basis than that in the other two taxa. Alternatively, the biases may share the same origin, but there may have been evolutionary changes in the biases in the different lineages. Second, homologous biases may favor different traits in different lineages. This may result from the modification of an ancestral bias in one or more lineages or from identical biases favoring different traits in different lineages. For example, a bias that favors larger caudal fins may have the consequence of favoring mutations that increase caudal fin area, caudal fin length, or any elaboration of the caudal fin that increases either area or length. Because different mutations for new male traits may arise in different lineages, the same bias can favor what appear to be very different traits.

Testing the homology of male traits can also present problems. For example, Meyer (1997) has suggested that 'sword genes' may be widespread in poeciliid fishes. Some swordless *Xiphophorus* species will express protrusions when treated with hormones (Gordon et. al 1943; Zander & Dzwillo 1969), and several poeciliids other than *Xiphophorus* have protrusions extending from the base of the caudal fin (see discussion in Basolo 1995a). If the ancestor of poeciliids possessed a sword that was then lost in some lineages, this would affect our interpretation of the relative origin of the trait and the bias. However, few poeciliids have such protrusions, and similar traits often evolve independently in different taxa, e. g. a sword-like structure, including elongation of ventral caudal fin rays with a caudal margin brown-black stripe, topped by a white-orange stripe, topped by red coloration, is found in the distantly-related rivulid *Pterolebias xiphophorus* (Thomerson & Taphorn 1992). Even if the protrusions have a similar genetic basis in all poeciliids with protrusions, and even if the genes that allow protrusions are ubiquitous in poeciliids, this by itself does not constitute evidence for homology of the protrusions. The expression of genes important in caudal fin development could be affected by derived changes in the timing of hormone activity, or by changes in the regulatory genes, to produce protrusions without the ancestor of poeciliids having possessed a protrusion. Clearly, however, the issue of bias and trait homology are important in the study of receiver biases and their evolutionary consequences must be given careful consideration.

Multiple receiver biases and composite signals

When phylogenetic information indicates the role of receiver biases in the evolution of a male signal, there are additional questions to consider. Many signals have more than one component, and individual components could act to enhance the overall effectiveness of the signal. For example, the sword in green swordtails is a composite trait consisting of elongation of a specific set of lower caudal fin rays and a tri-layer of coloration: two outer black stripes and an inner strip of orange and/or green. The female preference for the sword in green swordtails could result from a single receiver bias, or may be the result of multiple biases, each acting on different components of the sword. Thus, when considering a complex signal, one must consider whether there is a single bias or whether there are multiple biases that have acted in signal evolution. In addition, one must consider whether all of the components of the signal are under intersexual selection via the receiver biases, or whether some of the components were established in a context other than intrasexual selection via receiver biases. By testing female preferences for single components and for different combinations of components of the composite trait, one may be able to distinguish between these different possibilities.

Final comments

Despite the problems involved in testing the preexisting receiver bias model for the evolution of mating preferences and preferred traits, we know that biases arising from a number of potential sources are present and that they can influence the evolution of mating signals. Thus, despite these inherent problems, we need to continue to study receiver biases if we want to understand why particular mating signals evolve, as well as how receiver biases set the stage for other processes that then affect the subsequent evolution of mating preferences and signals used in sexual selection.

Acknowledgements

I thank A. Aspbury, K. Delaney, A. Olivedo, W. E. Wagner, Jr. and two anonymous reviewers for comments on early manuscript versions. Support was provided by NSF-RPG grant IBN9509211 and NSF IBN9629318 during manuscript preparation.

References

Arak, A. & Enquist, M. 1993. Hidden preferences and the evolution of signals. *Philosophical Transactions of the Royal Society of London, Series B*, **340**, 207-213.

Basolo, A. L. 1990a. Female preference for male sword length in the green swordtail, *Xiphophorus helleri* (Pisces, Poeciliidae). *Animal Behaviour*, **40**, 331-338.

Basolo, A. L. 1990b. Female preference predates the evolution of the sword in swordtail fish. *Science*, **250**, 808-810.

Basolo, A. L. 1995a. A further examination of a pre-existing bias favouring a sword in the genus *Xiphophorus*. *Animal Behaviour*, **50**, 365-375.

Basolo, A. L. 1995b. Phylogenetic evidence for the role of a pre-existing bias in sexual selection. *Proceedings of the Royal Society of London, Series B*, **259**, 307-311.

Basolo, A. L. 1996. The phylogenetic distribution of a female preference. *Systematic Biology*, **45**, 290-307.

Basolo, A. L. 1998. Evolutionary change in a receiver bias: a comparison of female preference functions. *Proceedings of the Royal Society of London, Series B*, **265**, 2223-2228.

Borowsky, R. L ., McClelland, R ., Cheng, R. & Welsh, J. 1995. Arbitrarily primed DNA fingerprinting for phylogenetic reconstruction in vertebrates: the *Xiphophorus* model. *Molecular Biology and Evolution*, **12**, 1022-1032.

Christy, J. H. 1995. Mimicry, mate choice, and the sensory trap hypothesis. *American Naturalist*, **146**, 171-181.

Christy, J. H. & Salmon, M. 1991. Comparative studies of reproductive behavior in mantis shrimps and fiddler crabs. *American Zoologist*, **31**, 329-337.

Clark, D. L. & Uetz, G. W. 1992. Morph-independent mate selection in a dimorphic jumping spider: demonstration of movement bias in female choice using video-controlled courtship behaviour. *Animal Behaviour*, **43**, 247-254.

Darwin, C. 1871. *The Descent of Man, and Selection in Relation to Sex.* London: J. Murray.

Dawkins, M. S. & Guilford, T. 1996. Sensory bias and the adaptiveness of female choice. *American Naturalist*, **148**, 937-942.

Endler, J. A. 1992. Signals, signal conditions, and the direction of evolution. *American Naturalist*, **139**, S125-S153.

Endler, J. A. & Basolo, A. L. 1998. Sensory ecology, receiver biases and sexual selection. *Trends in Ecology and Evolution*, **13**, 415-420.

Endler, J. A. & McLellan, T. 1988. The processes of evolution: towards a newer synthesis. *Annual Review of Ecology and Systematics*, **19**, 395-421.

Enquist, M. & Arak, A. 1993. Selection of exaggerated male traits by female aesthetic senses. *Nature*, **361**, 446-448.

Fisher, R. A. 1958. *The Genetical Theory of Natural Selection*. 2nd edn. New York: Dover.

Frumhoff, P. C. & Reeve, H. K. 1994. Using phylogenies to test hypotheses of adaptation: a critique of some current proposals. *Evolution*, **48**, 172-180.

Gordon, M ., Cohen, H. & Nigrelli, R. F. 1943. A hormone-produced taxonomic character in *Platypoecilus maculatus* diagnostic of wild *Platypoecilus xiphidium*. *American Naturalist*, **77**, 569-572.

Haas, V. 1993. *Xiphophorus* phylogeny, reviewed on the basis of the courtship behavior. In: *Trends in Ichthyology* (Ed. by J. H. Schroder, J. Bauer & M. Schartl), pp. 279-288. London: Blackwell.

Halliday, T. & Arnold, S. A. 1987. Multiple mating by females: a perspective from quantitative genetics. *Animal Behaviour*, **35**, 939-941.

Hennig, W. 1966. *Phylogenetic Systematics*. Urbana, IL: University of Illinois Press.

Kirkpatrick, M. 1982. Sexual selection and the evolution of female choice. *Evolution*, **36**, 1-12.

Kirkpatrick, M. 1987a. The evolutionary forces acting on mating preferences in polygynous animals. In: *Sexual Selection: Testing the Alternatives* (Ed. by J. W. Bradbury & M. B. Andersson), pp. 67-82. New York: Wiley.

Kirkpatrick, M. 1987b. Sexual selection by female choice in polygynous animals. *Annual Review of Ecology and Systematics*, **18**, 43-70.

Kirkpatrick, M. & Ryan, M. J. 1991. The evolution of mating preferences and the paradox of the lek. *Nature*, **350**, 33-38.

Lande, R. M. 1981. Models of speciation by sexual selection on polygenic traits. *Proceedings of the National Academy of Sciences, USA*, **78**, 3721-3725.

Lande, R. M. & Arnold, S. J. 1985. Evolution of mating preferences and sexual dimorphism. *Journal of Theoretical Biology*, **117**, 651-664.

Meyer, A. 1997. The evolution of sexually selected traits in male swordtail fishes (*Xiphophorus*, Poeciliidae). *Heredity*, **79**, 329-337.

Meyer, A., Morrissey, J. & Schartl, M. 1994. Molecular phylogeny of fishes of the genus *Xiphophorus* suggests repeated evolution of a sexually selected trait. *Nature*, **368**, 539-542.

Morton, E. S. 1990. Habitat segregation by sex in the hooded warbler: experiments on proximate causation and discussion of its evolution. *American Naturalist*, **135**, 319-333.

Proctor, H. C. 1991. Courtship in the water mite *Neumania papillator*: males capitalize on female adaptation for predation. *Animal Behaviour*, **42**, 589-598.

Proctor, H. C. 1992. Sensory exploitation and the evolution of male mating behaviour: a cladistic test using water mites (Acari, Parasitengona). *Animal Behaviour*, **44**, 745-752.

Römer, H. 1993. Environmental and biological constraints for the evolution of long-range signalling and hearing in acoustic insects. *Philosophical Transactions of the Royal Society of London, Series B*, **340**, 179-185.

Rosen, D. E. 1979. Fishes from the uplands and intermontane basins of Guatemala, revisionary studies and comparative geography. *Bulletin of the American Museum of Natural History*, **162**, 267-376.

Ryan, M. J. 1990. Sexual selection, sensory systems, and sensory exploitation. *Oxford Surveys in Evolutionary Biology*, **7**, 157-195.

Ryan, M. J. & Wagner, W. E. Jr. 1987. Asymmetries in mating preferences between species: female swordtails prefer heterospecific males. *Science*, **236**, 595-597.

Ryan, M. J., Fox, J. H., Wilczynski, W. & Rand, S. A. 1990. Sexual selection for sensory exploitation in the frog *Physalaemus pustulosus*. *Nature*, **34**, 66-67.

Ryan, M. J. & Rand, A. S. 1993. Sexual selection and signal evolution: the ghost of biases past. *Philosophical Transactions of the Royal Society of London, Series B*, **340**, 187-195.

Schluter, D., Price, T., Mooers, A. & Ludwig, D. 1997. Likelihood of ancestor states in adaptive radiation. *Evolution*, **51**, 1699-1711.

Thomerson, J. E. & Taphorn, D. C. 1992. Two new annual killifishes from Amazonas Territory, Venezuela (Cyprinodontiformes, Rivulidae). *Ichthyological Exploration of Freshwaters*, **3**, 377-384.

West-Eberhard, M. J. 1984. Sexual selection, competitive communication, and species-specific signals in insects. In: *Insect Communication* (Ed. by T. Lewis), pp. 283-324. Toronto: Academic Press.

Zander, C. D. & Dzwillo, M. 1969. Untersuchungen zur Entwicklung und Vererbung des Caudalfortsatzes der *Xiphophorus*-Arten (Pisces). *Zeitschrift für Wissenschaftlichen Zoologie*, **178**, 276-315.

IN: ESPMARK,Y., AMUNDSEN,T. & ROSENQVIST,G. (EDS.) 2000
ANIMAL SIGNALS:
SIGNALLING AND SIGNAL DESIGN IN ANIMAL COMMUNICATION, PP. 195-207
TAPIR ACADEMIC PRESS, TRONDHEIM, NORWAY

9

Environmental Dependence of Genetic Indicator Mechanisms

BEN C. SHELDON

Department of Animal Ecology, Evolutionary Biology Centre, Uppsala University, Norbyvägen 18d, S-752 36 Uppsala, Sweden

Present address: Edward Grey Institute, Department of Zoology, South Parks Road, Oxford OX1 3PS, UK

E-mail: ben.sheldon@zoology.oxford.ac.uk

Abstract

The last five years have produced several clear demonstrations in support of genetic indicator mechanisms in sexual selection. The key finding, that the phenotypic expression of secondary sexual characters predicts the signaller's contribution of genes influencing offspring fitness has been shown in at least six species of bird, one amphibian and six arthropods. These results have been obtained using a variety of experimental designs in both natural and laboratory populations. Given that such relationships exist, one next step is to understand how consistent such relationships are across environments, and what consequences any environmental dependence has for selection on traits in signallers and preferences in receivers. I describe some recent work on collared flycatchers Ficedula albicollis *which demonstrates the existence of genotype x environment interactions affecting the relationship between male phenotype and measures of offspring performance. Little is known of the underlying causes of such relationships, yet consideration of some likely pathways linking father's phenotype to offspring performance suggest that in many cases these pathways can be expected to show environmental dependence. The existence of such interactions has a number of potential consequences for sexual selection.*

Of numerous models of the evolution of extravagant epigamic traits, one class could be argued to have particularly captured the attention of empirically oriented biologists. The idea that large secondary sexual traits are signals of high genetic quality (hence genetic indicator mechanism) can be traced back at least as far as Fisher's 1915 paper (Fisher 1915; see Andersson 1994). Resuscitated by the influence of Zahavi's (1975) Handicap Principle, and increasing interest in the evolution of female choice, the idea initially came under attack from evolutionary geneticists who argued that theory suggested that it was unlikely that there was sufficient genetic variance for fitness present in natural populations to account for such evolutionary trajectories (reviewed in Andersson 1994). It has always been possible to suggest alternative reasons for females preferring more ornamented males, even in cases where at first sight genes are all that seem likely to be important (e.g. Reynolds & Gross 1990; Kirkpatrick & Ryan 1991). As a consequence, the evidence required for empiricists to convince doubters has been difficult to obtain. However, increased attention to quantitative genetics and experimental design have solved these problems. There are now a reasonable number of studies from a wide variety of taxa, employing a range of experimental designs and approaches, supporting indicator models of sexual selection. I briefly review these studies. Consideration of the possible pathways linking a male's phenotype to the performance of his offspring via purely genetic effects suggests that the environment in which genes are expressed may often affect their expression. I describe some recent work on the collared flycatcher *Ficedula albicollis* which suggests the importance of genotype x environment interactions in sexual selection for genetic quality. These findings, and complementary evidence from other systems, suggest that more attention should be paid to the evolutionary consequences of such interactions. I discuss two main implications: (1) The maintenance of genetic variation for traits related to fitness. (2) Variable selection on female choice.

Empirical support for indicator models

At the heart of indicator models is the assumption that there should exist a genetic correlation between a male sexual display trait and fitness of offspring. Simply measuring the correlation between male phenotype and offspring fitness is something that is very challenging for free-ranging organisms, and an additional problem with obtaining empirical evidence in support of indicator models is that many things other than genes could be responsible for such relationships. There are two main classes of confounding variables, which can be said to result from (1) correlations between factors influencing fitness (that is, attractive males may provide not only genes that increase fitness of offspring, but also have other beneficial effects, such as providing high quality territories or parental care), and (2) correlations between parents (that is, that attractive males may mate with more fecund females, or females that produce offspring of higher quality: c.f. Kirkpatrick et al. 1990). Empirical studies supporting indicator models have attempted to control for these problems in a number of ways (Table 1), either by (i) statistical means, where the influence of potentially confounding variables is removed, by (ii) exploiting systems where such effects are unlikely to be of importance or by (iii) employing some form of breeding design, or selection experiment, where individuals are mated randomly, or the effect of sires is nested within that of dams. Table 1 summarises these studies.

Table 1. *Summary of published studies offering support for genetic indicator mechanisms operating in sexual selection. The criteria for inclusion were that the study presented evidence that some fitness component of offspring was correlated with the size of a secondary sexual character known (or reasonably inferred) to be subject to directional mate preference, and that major confounding variables were controlled in some way.*

Study organism	Control for		Offspring fitness component affected	Source
	(i) Correlated fitness components	(ii) Intra-parental correlations		
Insects				
Cockroach *Nauphoeta cinerea*	Breeding design	Breeding design	Developmental rate	Moore 1994
Wax moth *Achroia grisella*	Breeding design	Breeding design	Developmental rate	Jia & Greenfield 1997
Stalk-eyed fly *Cyrtodiopsis* spp.	Selection experiment	Selection experiment	Offspring viability[1]	Wilkinson et al. 1998
Drosophila	Breeding design	Breeding design	Offspring viability	Hoikkala et al. 1998
Spiders				
Wolf spider *Hygrolycosa rubrofasciata*	Selection experiment	Selection experiment	Offspring viability	Alatalo et al. 1998
Sierra dome spider *Neriene litigiosa*	Breeding design	Statistical	Growth rate	Watson 1998
Amphibians				
Gray tree frog *Hyla versicolor*	Breeding design	Breeding design	Growth and developmental rate	Welch et al. 1998
Birds				
Great tit *Parus major*	Cross-fostering	None	Survival	Norris 1993
Peacock *Pavo pavo*	Breeding design	Breeding design	Growth and survival[2]	Petrie 1994
Barn swallow *Hirundo rustica*	Statistical	Statistical	Longevity	Møller 1994
Great reed warbler *Acrocephalus arundinaceus*	Statistical	Statistical	Survival[2]	Hasselquist et al. 1996
Ring-necked pheasant *Phasianus colchicus*	Genotype specific[3]	Genotype specific[3]	Survival	von Schantz et al. 1996
Collared flycatcher *Ficedula albicollis*	Breeding design	Sires nested within dams	Fledging condition	Sheldon et al. 1997

1. Via meiotic-drive suppression
2. Survival to first breeding
3. Association between specific alleles at MHC loci and both survival and ornamentation

Parental investment decisions should depend, among other things, on the likely reproductive value of offspring to parents. In the case of investment decisions made by mothers, one factor that can potentially affect the optimal level of investment in offspring (and hence their fitness) is the quality of mate that a female has obtained. For example, a female breeding with a male on a territory that provides lots of food for offspring, or is particularly safe from predators, may invest relatively more in offspring than a female breeding under less favourable conditions. This argument can also be extended to situations where the reproductive value of offspring is determined in part by the genes that they have inherited from their father. Such differential allocation models (Burley 1986) have received some empirical support, particularly from studies of birds (Burley 1986; deLope & Møller 1993; Petrie & Williams 1993; reviewed in Møller & Thornhill 1998). It has been argued by some authors (e.g. Møller & Thornhill 1998) that the existence of differential allocation makes the interpretation of empirical studies offering support for indicator models problematic, because any difference in the performance of offspring with different sires could be explained as resulting from differential allocation. I believe that this argument is mistaken for two, related, reasons. First, in many of the cases listed in Table 1, the only possible source of a difference in reproductive value of offspring is paternally inherited genes, in which case the existence of differential allocation in response to genetic differences between sires actually supports indicator models. Second, the idea that fitness of offspring could somehow be separated into a component due to the action of paternal genes alone, and a component due to differential maternal investment in response to genetically-based differences in reproductive value of offspring, implies an unnecessarily restrictive definition of the phenotypic effects of genes. In these situations I would argue that differential maternal investment should be considered to be part of the extended phenotype (Dawkins 1982) of the offspring.

As already mentioned, fitness is very difficult to measure in natural populations. Consequently, all studies that can be said to support indicator models have instead measured something else, which is assumed to be related to fitness. Purists may object that, in that case, the possibility of negative correlations between the measured fitness component and other fitness components has not been ruled out, so that the overall relationship between male phenotype and offspring fitness has not been determined. Two arguments suggest that this objection has little weight. First, it is difficult to understand why there should be any relationships of this nature at all unless they have evolved through sexual selection. Second, in most cases the effects of fathers on offspring have been observed to act early in development, and are often related to ameliorating substantial rates of juvenile mortality (Table 1; this no doubt in part reflects ease of study of juvenile mortality as opposed to other fitness components measured later in life). Because natural selection acting early in life-histories has more force than selection acting late in life (for simple demographic reasons), counter-vailing selection would need to be strong in order to counteract the observed effects. Thus, although it may be argued that no single completely convincing empirical study exists, taken as a whole the evidence in Table 1 does provide a convincing demonstration that indicator mechanisms operate frequently in natural populations.

Case study: the collared flycatcher

The population of collared flycatchers breeding on Gotland, Sweden has been the subject of intensive study as regards its life-history since 1980 (see Gustafsson 1989; Pärt & Gustafsson 1989 for some general details). In recent years a substantial amount of work has been directed at studying the operation of sexual selection in this population, particularly as it relates to the signalling function of conspicuous patches of white feathering on males. Using a combination of observational and experimental approaches in the field, together with molecular genetic work to determine parentage and the sex of nestlings, we have shown that a wide range of selective mechanisms act on variation in the size of the white forehead patch of males, and that numerous processes contribute to variation among individuals in the size of the patch. Experimentally staged contests between males reveal that forehead patch area is the best phenotypic predictor of the outcome of disputes over territory ownership (Pärt & Qvarnström 1997), and experimental manipulations of forehead patch size suggest that males have to trade-off effort spent in male-male competition with effort spent in paternal care (Qvarnström 1998). Manipulations of paternal care (by changing brood size) reveal that this trait is condition-dependent, since males given more young to rear return in the following year with smaller patches, whereas males given fewer young return with larger patches (Gustafsson et al. 1995). Although these experiments reveal that a large proportion of variation in badge size can be explained by variation in individual condition, cross-fostering experiments suggest that the trait also has moderate heritability (= 0.44: Sheldon et al. 1997). Consistent with a genetic component to variation in this trait, females mated to large patched males skew the sex ratio of their offspring in favour of sons, whereas females mated to small patched males skew the sex ratio of their offspring in favour of daughters (Ellegren et al. 1996). Analysis of paternity of nestlings using microsatellite markers has shown that small patched males are more likely to lose paternity of offspring, and that large patched males are more likely to be the fathers of these extra-pair sired offspring (Sheldon & Ellegren 1999).

We have used the occurrence of extra-pair paternity as a means to ask whether the white forehead patch of male collared flycatchers acts as a signal of genes influencing the fitness of their offspring. The rationale behind this approach was that since offspring within mixed paternity broods share the same rearing environment and mother, any differences between them can only arise as a result of the influence of differences in the genes that they inherit from their fathers. Comparisons between maternal half-siblings should thus be a sensitive means to detect genetic effects of fathers on offspring, while being free from the problem of confounding variables discussed above. In the first year in which we employed this approach, we found a positive correlation between the difference in the size of the forehead patches of the two sires within a brood, and the difference in fledging condition of their offspring (Sheldon et al. 1997). In this species, fledging condition (defined as the residual from a regression of body mass at fledging on tarsus length) has been shown to predict the likelihood of recruitment of nestlings under a wide range of conditions (Lindén et al. 1992). The result that we obtained is consistent with the existence of a positive genetic correlation between male badge size and fledging condition of nestlings, and therefore provides support for the operation of a genetic indicator mechanism in this species, involving male forehead patch size.

In the first year in which we employed the technique of comparisons among maternal half-siblings, we relied on naturally occurring broods of mixed paternity to provide us with material for our comparisons. This is an inefficient technique, because only about one third of natural broods are multiply sired (Sheldon & Ellegren 1999). More recently we have been employing maternal half-sibs produced experimentally, by removing mated males for two days, shortly before the female of the pair begins egg-laying. These experiments increase the proportion of multiply sired broods to about 75% (Sheldon & Ellegren 1998), and also facilitate the identification of extra-pair sires, since the usual outcome of the experiments is that the removed male is temporarily replaced by an immediate territorial neighbour (Sheldon et al. 1999). In these experiments, we also cross-foster all experimental broods with non-experimental broods at the egg stage, so that all offspring are reared by foster parents. This prevents the possibility, however unlikely, that males or females direct parental care preferentially to their own genetic offspring within broods of half-siblings.

When we repeated our earlier analysis of the relationship between the forehead patch area of sires and the fledging condition of half-sibs, we were surprised to find a relationship of opposite sign to that in the previous year; the interaction between year and forehead patch size on nestling fledging condition was highly significant (Fig. 1a). Hence it appears that the information about genes affecting offspring viability signalled by a male's forehead patch size may depend on the environment in which the offspring of the male will grow. Clearly, it can be dangerous to suggest post-hoc explanations for such contrasting patterns, but one such explanation is suggested by the fact that the two years differed greatly in terms of the conditions experienced by nestling flycatchers (Fig. 1b). The year in which the relationship was positive, 1994, was one in which fledging success was relatively good, and mean fledging mass relatively high. That in which the relationship was negative, 1996, was a year with poor fledging success and an unusually low mean

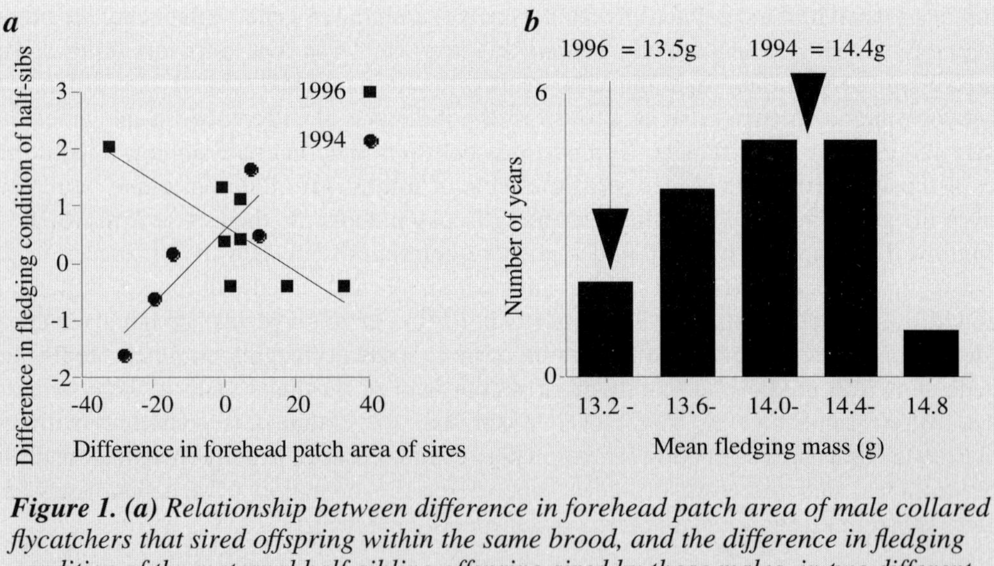

***Figure 1.* (*a*) *Relationship between difference in forehead patch area of male collared flycatchers that sired offspring within the same brood, and the difference in fledging condition of the maternal half-sibling offspring sired by those males, in two different years. The interaction between year and the difference in forehead patch size on the difference in fledging condition is statistically significant* ($F_{1,9} = 21.76$, $P < 0.001$). **(*b*)** *Distribution of annual mean fledging mass of collared flycatchers breeding on Gotland between 1980 and 1998.*

fledging mass (Fig. 1b). One could tentatively suggest, therefore, that the reason for the difference in the relationships between years lies in variation in food availability for nestlings. Other possibilities cannot be discounted, however. For example, females in the first year (1994, that with a positive relationship) were allowed to control which males they copulated with, while those in the second year were forced by the experiment to increase the rate of extra-pair paternity in their broods (Sheldon & Ellegren 1998). It might be that epistatic interactions between male and female alleles had differing effects in the two years. Experiments are currently underway to test the hypothesis that food availability is the critical factor directly.

An alternative way to study the existence of genotype x environment interactions involving indicator traits is to compare the covariance of father's and son's phenotypes under a variety of environmental conditions. Here, one asks whether a particular genotype (judged from the phenotype) will have a consistent effect in all environments. This approach has been taken by Qvarnström (1999) in the Gotland collared flycatcher population. The large population, high natal philopatry and the long-term nature of the study have resulted in an unusually extensive data set with which to investigate the heritability of a secondary sexual character currently under selection (c.f. Alatalo et al. 1997): over 400 'families' are available for analysis, many of which had been the subject of experimental manipulations. Numerous analyses conducted by Qvarnström (1999) yield a consistent heritability of forehead patch size of between 0.25 and 0.50 (corresponding with the estimate of 0.44 based on cross-fostering experiments), but under some conditions the covariance between fathers and sons is much reduced. Two such effects concern (i) experimental manipulations of the rearing environment of nestlings (by changing brood size) and (ii) maternal age. Males that recruited from experimentally increased broods resembled their fathers less than did males which recruited from experimentally decreased broods (this was not simply a result of an increased environmental component of variance, since the phenotypic variances did not differ across groups). Similarly, while there was a significant resemblance between fathers and sons when the mother of the nestling was either one or two years of age, this resemblance was no longer present for females aged three years or older. Previous work, in a different context, has suggested that maternal age has a negative effect on the condition of daughters in the collared flycatcher (Schluter & Gustafsson 1993). In addition to the simple maternal age effect on the heritability of forehead patch size, there was a significant interaction between maternal age and father's forehead patch size, indicating that the effect of maternal age was dependent on the forehead patch size of the father. Both of these results provide good evidence that the rearing environment of a nestling affects the way in which genes linked to those affecting forehead patch size act: hence genotype x environment interactions. Two other recent studies in quite different systems have also found that the relationship between the male secondary sexual character and offspring performance is dependent upon environment (Jia & Greenfield 1997; Welch et al. 1998). Given that there is at least some evidence that genotype x environment interactions affect expression of genetic indicator mechanisms, I now discuss, first, how likely it is that such interactions are common, and second, what consequences they may have.

Likelihood of genotype x environment interactions

Although there are a small number of studies which suggest that the operation of genetic indicator mechanisms shows environmental dependence, relatively few studies have explicitly sought this evidence. Nevertheless, one could ask how general such relationships might be. Figure 2 presents a schematic representation of some of the factors that can potentially contribute to variation in offspring viability. There are at least three ways in which genotype x environment interactions (where environment is defined in the broadest sense possible) can affect the operation of genetic indicator mechanisms (arrows labelled *I* in Fig. 2). Given the large number of loci likely to be involved in producing relationships between quantitative secondary sexual characters and fitness-related traits, there is potential for a very large number of epistatic interactions between loci (I_1). Thus, the effect of a particular paternally inherited allele may depend upon interactions with maternally inherited alleles at other loci. Although there are no empirical examples of epistatic interactions between loci influencing the operation of indicator mechanisms, there are examples for other quantitative traits. Lynch & Walsh (1998) argue that epistasis is likely to be of considerable importance in determining both the expression of quantitative traits and levels of additive genetic variance. In contrast, there are numerous empirical demonstrations that the phenotypic effects of genes depend on characteristics of the abiotic environment in which they are expressed (I_2; see Lynch & Walsh 1998). Coulson et al. (1998) presented evidence from an analysis of the association between genotypes at genetic marker loci and measures of fitness in red deer *Cervus elaphus* to

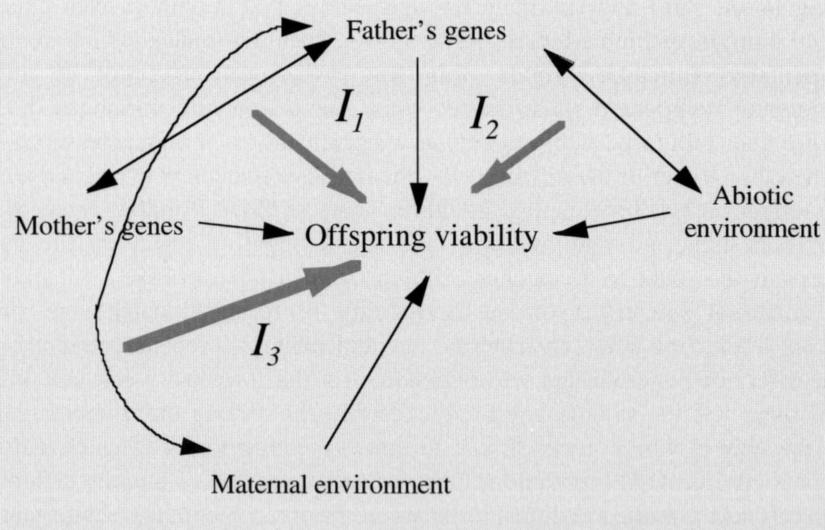

Figure 2. *Schematic representation of potential effects on offspring viability in a species where a genetic indicator mechanism can operate. Broad grey arrows indicate three classes of genotype x environment interaction: I_1 is that due to epistatic interactions between maternal and paternal genes, I_2 is that due to interactions between paternal genes and the abiotic environment, and I_3 that due to interactions between paternal genes and the maternal environment.*

suggest that the frequency of genotype x environment (G x E) interactions of this kind may be underestimated. Most of the evidence from our recent work on collared flycatchers, together with that of Jia & Greenfield (1997) and Welch et al. (1998) concerning G x E interactions concerns interactions with abiotic environmental factors.

The final class of G x E interaction (I_3) concerns interactions between paternally inherited genes and the maternally derived environment in which offspring develop. The occurrence of this sort of interaction is suggested by Qvarnström's (1999) work showing the dependence of the heritability of forehead patch size in collared flycatchers on maternal age. In theory, there are numerous ways in which interactions between maternal environments and offspring genotypes could contribute to phenotypic variance (Rossiter 1996), although documenting these effects, and measuring their contribution to phenotypic variance is likely to be a very difficult task (Lacey 1998).

These arguments suggest that the three main classes of interactions identified in Figure 2, which can potentially affect the way in which indicator mechanisms operate, may be rather common in natural systems. It is clear that this is an area at which more empirical work should be directed.

Implications

If the phenotypic effect of genes signalled by male secondary sexual traits commonly depends on the environment in which those genes are expressed, as argued above, I believe this has a number of implications for future empirical and theoretical work on indicator mechanisms. Two implications are methodological. First, that the environment in which a study is done may have an important influence on the results obtained, and indeed affect the interpretation of them. In particular, although laboratory models offer the possibility of much greater control of confounding variables than field-based systems, they also suffer from the possibility that organisms are either too well buffered from environmental variation, or exposed to environments different from those in which they evolved, or both. Second, field studies should be replicated under a range of environmental conditions, over several different years if possible.

Two of the implications are rather more interesting. First, G x E interactions for traits related to fitness can easily cause the temporary reversal of selection gradients on male secondary sexual traits, on the genes determining condition that underlie them, and also on genes determining female choice rules. If selection fluctuates in direction and intensity over time or space, then such fluctuations can have the effect of increasing the amount of genetic variance for the traits subject to fluctuating selection present in the population (e.g. Chesson & Warner 1981; Ellner & Hairston 1994; Sasaki & Ellner 1997). The ability to find empirical support in favour of indicator models suggests that there is non-negligible genetic variance for fitness present in natural populations: the challenge then becomes to explain what mechanisms maintain the variance, and G x E interactions provide one potential mechanism. A second interesting implication of frequent G x E interactions affecting indicator mechanisms is that female choice rules might potentially be expected to evolve in order to take account of these interactions. If the key environmental factors are very predictable, then one outcome might be that female choice will be phenotypically plastic: females may exercise different choices in different environments. Alternatively, if the environment is rather unpredictable, then one might expect females to 'bet-hedge', perhaps showing intermediate levels of preference. In either case, the existence of such behavioural flexibility would add another layer of complexity to the interpretation of data from female choice experiments, and might have interesting consequences for the coevolution of male traits and female preferences.

To conclude, in this paper I have argued that, while recent empirical work strongly supports the idea that male secondary sexual characters can act as signals of genetic quality, recent work also hints at greater levels of complexity involving such signalling systems. In particular, genotype x environment interactions can have the effect of making the message conveyed by a particular signal environmentally dependent: the same signal may mean different things in different environments. This is an area ripe for more empirical work, and with implications for a number of different areas of sexual selection theory.

Acknowledgements

I am grateful to Marlene Zuk and an anonymous referee for constructive comments. My work is supported by a grant from the Swedish Natural Science Research Council.

References

Alatalo, R. V., Kotiaho, J., Mappes, J. & Parri, S. 1998. Mate choice for offspring performance: major benefits or minor costs? *Proceedings of the Royal Society of London, Series B*, **265**, 2297-2301.

Alatalo, R. V., Mappes, J. & Elgar, M. A. 1997. Heritabilities and paradigm shifts. *Nature*, **385**, 402-403.

Andersson, M. 1994. *Sexual Selection.* Princeton: Princeton University Press.

Burley, N. 1986. Sexual selection for aesthetic traits in species with biparental care. *American Naturalist*, **127**, 415-445.

Chesson, P. L. & Warner, R. R. 1981. Environmental variability promotes co-existence in lottery competitive systems. *American Naturalist*, **117**, 923-943.

Coulson, T. N., Albon, S. D., Pemberton, J. M., Slate, J., Guinness, F. E. & Clutton-Brock, T. H. 1998. Genotype x environment interactions in winter survival in red deer. *Journal of Animal Ecology*, **67**, 434-445.

Dawkins, R. 1982. *The Extended Phenotype*. Oxford: Oxford University Press.

Ellegren, H., Gustafsson, L. & Sheldon, B. C. 1996. Sex ratio adjustment in relation to paternal attractiveness in a wild bird population. *Proceedings of the National Academy of Sciences, USA*, **93**, 11723-11728.

Ellner, S. & Hairston, N. G. Jr. 1994. Role of overlapping generations in maintaining genetic variation in a fluctuating environment. *American Naturalist*, **143**, 403-417.

Fisher, R. A. 1915. The evolution of sexual preference. *Eugenics Review*, **7**, 184-192.

Gustafsson, L. 1989. Collared flycatcher. In: *Lifetime Reproduction in Birds* (Ed. by I. Newton), pp. 75-88. London: Academic Press.

Gustafsson, L., Qvarnström, A. & Sheldon, B. C. 1995. Trade-offs between life-history traits and a secondary sexual character in male collared flycatchers. *Nature*, **375**, 311-313.

Hasselquist, D., Bensch, S. & von Schantz, T. 1996. Correlation between male song repertoire, extra-pair paternity and offspring survival in the great reed warbler. *Nature*, **381**, 229-232.

Hoikkala, A., Aspi, J. & Suvanto, L. 1998. Male coutship song frequency as an indicator of male genetic quality in an insect species. *Proceedings of the Royal Society of London, Series B*, **265**, 503-508.

Jia, F.-Y. & Greenfield, M. D. 1997. When are good genes good? Variable out-comes of female choice in wax moths. *Proceedings of the Royal Society of London, Series B*, **264**, 1057-1063.

Kirkpatrick, M., Price, T. & Arnold, S. J. 1990. The Darwin-Fisher theory of sexual selection in monogamous birds. *Evolution*, **44**, 180-193.

Kirkpatrick, M. & Ryan, M. J. 1991. The evolution of mating preferences and the paradox of the lek. *Nature*, **350**, 33-38.

Lacey, E. P. 1998. What is an adaptive environmentally induced parental effect? In: *Maternal Effects as Adaptations* (Ed. by T.A. Mousseau & C.W. Fox), pp. 54-66. New York: Oxford University Press.

Lindén. M., Gustafsson, L. & Pärt, T. 1992. Selection on fledging mass in collared flycatchers and great tits. *Ecology*, **73**, 336-343.

de Lope, F. & Møller, A. P. 1993. Female reproductive effort depends on the degree of ornamentation of their mates. *Evolution*, **47**, 1152-1160.

Lynch, M. & Walsh, B. 1998. *Genetics and Analysis of Quantitative Traits.* Sunderland, MA: Sinauer.

Møller, A.P. 1994. Male ornament size as a reliable cue to enhanced offspring viability in the barn swallow. *Proceedings of the National Academy of Sciences, USA*, **91**, 6929-6932.

Møller, A.P. & Thornhill, R. 1998. Male parental care, differential parental investment by females and sexual selection. *Animal Behaviour*, **55**, 1507-1515.

Moore, A. J. 1994. Genetic evidence for the 'good genes' process of sexual selection. *Behavioral Ecology and Sociobiology*, **35**, 235-241.

Norris, K. 1993. Heritable variation in a plumage indicator of viability in male great tits *Parus major*. *Nature*, **362**, 537-539.

Pärt, T. & Gustafsson, L. 1989. Breeding dispersal in the collared flycatcher: possible causes and reproductive consequences. *Journal of Animal Ecology*, **58**, 307-322.

Pärt, T. & Qvarnström, A. 1997. Badge size in collared flycatchers predicts outcome of male competition over territories. *Animal Behaviour*, **54**, 893-899.

Petrie, M. 1994. Improved growth and survival of offspring of peacocks with more elaborate trains. *Nature*, **371**, 598-599.

Petrie, M. & Williams, A. 1993. Peahens lay more eggs for peacocks with larger trains. *Proceedings of the Royal Society of London, Series B*, **251**, 127-131.

Qvarnström, A. 1998. *Sexual Selection in the Collared Flycatcher: a Life-History Perspective*. PhD thesis, Uppsala University, Sweden.

Qvarnström, A. 1999. Genotype-by-environment interactions in the determination of the size of a secondary sexual character in the collared flycatcher (*Ficedula albicollis*). *Evolution*, **53**, 1564-1572.

Reynolds, J. D. & Gross, M. R. 1990. Costs and benefits of female mate choice: is there a lek paradox? *American Naturalist*, **136**, 230-243.

Rossiter, M. C. 1996. Incidence and consequences of inherited environmentally induced parental effects. *Annual Review of Ecology and Systematics*, **27**, 451-476.

Sasaki, A. & Ellner, S. 1997. Quantitative genetic variance maintained by fluctuating selection with overlapping generations: variance components and covariances. *Evolution*, **51**, 682-696.

Schluter, D. & Gustafsson, L. 1993. Maternal inheritance of condition and clutch size in the collared flycatcher. *Evolution*, **47**, 658-667.

Sheldon, B. C., Davidson, P. & Lindgren, G. 1999. Mate replacement in experimentally widowed collared flycatchers: determinants and outcomes. *Behavioral Ecology and Sociobiology*, **46**, 141-148.

Sheldon, B. C. & Ellegren, H. 1998. Paternal effort related to experimentally manipulated paternity of male collared flycatchers. *Proceedings of the Royal Society of London, Series B*, **265**, 1737-1742.

Sheldon, B. C. & Ellegren, H. 1999. Sexual selection resulting from extra-pair paternity in collared flycatchers. *Animal Behaviour*, 57, 285-298.

Sheldon, B. C., Merilä, J., Qvarnström, A., Gustafsson, L. & Ellegren, H. 1997. Paternal genetic contribution to offspring condition predicted by size of male sexual ornament. *Proceedings of the Royal Society of London, Series B*, **264**, 297-302.

von Schantz, T., Wittzell, H., Göransson, G., Grahn, M. & Persson, K. 1996. MHC genotype and male ornamentation: genetic evidence for the Hamilton-Zuk model. *Proceedings of the Royal Society of London, Series B*, **263**, 265-271

Watson, P.J. 1998. Multi-male mating and female choice increase offspring growth in the spider *Neriene litigiosa* (Linyphiidae). *Animal Behaviour*, **55**, 387-403.

Welch, A. M., Semlitsch, R. D. & Gerhardt, H. C. 1998. Call duration as an indicator of genetic quality in male gray tree frogs. *Science*, **280**, 1928-1930.

Wilkinson, G. S., Presgraves, D. C. & Crymes, L. 1998. Male eye span in stalk-eyed flies indicates genetic quality by meiotic drive suppression. *Nature*, **391**, 276-279.

Zahavi, A. 1975. Mate selection - a selection for handicap. *Journal of Theoretical Biology*, **53**, 205-214.

IN: ESPMARK,Y., AMUNDSEN,T. & ROSENQVIST,G. (EDS.) 2000
ANIMAL SIGNALS:
SIGNALLING AND SIGNAL DESIGN IN ANIMAL COMMUNICATION, PP. 209-236
TAPIR ACADEMIC PRESS, TRONDHEIM, NORWAY

10

Signal Function, Signal Efficiency and the Evolution of Anoline Lizard Dewlap Color

LEO J. FLEISHMAN

Department of Biological Sciences, Union College, Schenectady, New York 12308, USA

E-mail: fleishman@union.edu

Abstract

Animal signals should evolve features which allow them to efficiently stimulate the sensory system of intended receivers. The specific design features that will make a signal efficient depend on (1) the function served by the signal, (2) response properties of the sensory system relevant to this function and (3) the signaling environment. I explored the potential influence of these factors on the evolution of the color of Anolis *lizard dewlaps. I used computer modeling to examine the impact of dewlap color on efficiency for different combinations of signal environment, receiver visual system and hypothesized signal function. I compared predictions based on these models to the dewlap colors observed in seventeen species of anoles. The hypothesis that dewlap color functions to increase the probability of detection of the displaying animal was supported. The hypothesis that dewlap color patterns are designed to optimize the ability of conspecifics to make subtle discriminations between conspecifics was not supported. A third hypothesized function – that dewlaps serve a role in species recognition – was not tested directly, but seemed consistent with the observed dewlap color patterns. Selection for making moving dewlaps easy for conspecifics to detect under different sets of habitat light conditions, possibly combined with selection for dewlap color patterns which differ from those of sympatric congeners and other objects in the environment, may explain the evolution of the naturally- observed among-species diversity in anoline dewlap color patterns.*

Animal signals can perform a variety of different functions. The term 'efficacy' refers to the effectiveness with which a particular function is carried out by a particular signal component. In general we expect signals to evolve in the direction of increased efficacy. The effectiveness with which a signal carries out a function depends, at least to some extent, on the efficiency with which it stimulates the sensory system of the receiver (Guilford & Dawkins 1993; Johnstone 1998). The specific physical design features which give a signal component the greatest efficiency will depend on (1) the function (or functions) of the signal (or signal component), (2) the environmental conditions (including noise, transmission, etc.) in which the signal is typically performed and detected, and (3) the response properties of the sensory system of the receiver associated with the particular perceptual task associated with a particular function (Guilford & Dawkins 1993). In this paper I examine the way in which these three factors may interact to influence the evolution of signal design. I attempt this with two aims in mind. First, I wish to see if among-species differences in environmental conditions and/or sensory response properties can account for the naturally-observed diversity in signal design. Second, I wish to indirectly test hypotheses about the natural function of a signal component, by seeing if the design features predicted for a particular function are similar to those found in nature.

In this paper I examine the evolution of color of the dewlaps of anoline lizards. The dewlap is a colorful, expandable throat fan used by male anoles in a variety of display contexts. In the most frequent use of the dewlap, males patrol their territories and display spontaneously from conspicuous perches using stereotyped motion patterns of the head and body in combination with opening and closing of the dewlap (Fleishman 1992). The dewlaps of different species vary in color over a wide range. Here I explore the role that selection for efficient visual-system stimulation may have had in the evolution of this diversity.

My approach is to use computer modeling to attempt to identify optimal dewlap colors (i.e., the colors which provides the greatest signal efficiency) under a variety of conditions. The computer models were carried out using different combinations of the three different factors which influence efficiency: signal function, receiver sensory system and signaling environment. I tested two hypothetical signal functions (I also discuss a third). I tested two distinctly different visual systems found in different anoline species. I then compared the predictions of the computer models to dewlap spectral patterns of seventeen species of *Anolis*. The seventeen species are listed and described in Table 1.

Sensory system response

Retinal anatomy and physiology

The retinas of a number of species of anoles have been examined, and in all cases there are four classes of cones. Each cone class has a particular visual pigment and a particular type of colored oil droplet (Fleishman et al. 1993). Light must pass through the oil droplet in order to reach the visual pigment, so that the oil droplet acts as an optical filter. Therefore the spectral response (i.e., the magnitude of response as a function of

Table 1. *Overview of seventeen species of* Anolis *lizards.*

			DEWLAP CHARACTERISTICS		
Species	**Origin**	**Habitat light**	**Desaturation*** **(Y_{MW}/Y_{LW})**	**Short wavelength amplitude*** **(Y_{SW}/Y_{LW})**	**λ_c****
gundlachi	Puerto Rico	full shade	0. 45	0.98	544
evermanni	Puerto Rico	full shade	0. 13	0.28	522
krugi	Puerto Rico	partial shade	0. 36	0.95	532
cristatellus	Puerto Rico	partial shade	0. 59	1.23	515
stratulus	Puerto Rico	partial shade	0. 22	0.33	543
cooki	Puerto Rico	no shade	0. 16	0.40	556
pulchellus	Puerto Rico	no shade	0. 20	0.60	586
garmani	Jamaica	full shade	0. 10	0.55	530
lineatopus	Jamaica	partial shade	0. 20	0.46	545
grahami	Jamaica	partial shade	0. 08	0.36	548
opalinus	Jamaica	partial shade	0. 19	0.47	550
*mestrei(red)****	Cuba	full shade	0. 17	0.25	595
mestrei(white)	Cuba	full shade	0. 91	1.17	560
allogus	Cuba	full shade	0. 47	0.20	520
homolechis	Cuba	partial shade	0. 83	0.83	488
sagrei	Cuba	no shade	0. 17	0.24	575
carolinensis	N. America	partial shade	0. 56	1 15	577

* See Figures 5 and 6 and text for explanation of these terms.

** λ_c= cut-on wavelength. See Figures 5 and 6 and text for explanation.

*** The dewlap of *A. mestrei* has two different patches of color (red and white). Each of these patches are indicated seperately here.

wavelength) of the cone depends on the spectral transmission properties of the oil droplet as well as spectral absorption properties of the pigment. Spectral response functions for each class of cone in the retina of *Anolis cristatellus* are shown in Figure 1. Dr. Ellis Loew has measured absorption functions of cones and oil droplets for many anoline species (using microspectrophotometry), including all of the species shown in Table 1, except for three of the Cuban species. Except for *A. carolinensis*, all species which have been measured have four classes of cones nearly identical to those shown in Figure 1A (E.R. Loew, unpublished). The three Cuban species, which have not been measured, are closely related to *A. sagrei* (J.B. Losos, personal communication) which fits the common pattern, and for this paper I assume that they have the same visual system as the others.

The exceptional case is *A. carolinensis*, whose cones have the characteristics shown in Figure 1B (based on Provencio et al. 1992 and E.R. Loew, personal communication). The overall pattern is similar to that of the other species, but the wavelength of peak sensitivity of some cone classes is shifted to longer values.

Another important aspect of visual-system response is the additive response of the cones, or the overall spectral sensitivity. Spectral sensitivity has been determined for the Puerto Rican species in Table 1 (Fleishman et al. 1997), as well as for *A. sagrei* and *A. carolinensis* (L. Fleishman & M. Frye, unpublished), using electroretinographic flicker photometry (rate of flicker = 6 Hz). The spectral sensitivity for wavelengths longer than 450 nm are shown in Figure 1A and B (the reason for limiting this to wavelengths longer than 450 nm is explained below). The spectral sensitivity function is, to a large degree, predictable from the cone absorption functions, and I am therefore confident that the response pattern of *A. cristatellus* as shown in Figure 1A is typical for all of the species in Table 1, except for *A. carolinensis*, whose pattern of response is shown in Figure 1B.

Signal function and sensory response

An animal visual system is a complex structure based on many interacting components, which can perform a number of different tasks. Different visual tasks draw on different combinations of components in different ways (e.g. Goldsmith 1990; Neumeyer 1998).

The ability to discriminate between two colors (i.e., two different spectral distributions) is based on a comparison of the relative stimulation of the different classes of cones in the retina that arise in response to each of the two stimuli. Thus the brain compares the magnitude of response of the different classes of cones.

There are other visual tasks which rely on the sensation of brightness, which arises from the added response of different classes of photoreceptors, and is directly related to overall spectral sensitivity. For example, brightness contrast between a moving object and its background appears to be the most important factor in detection of motion (Schaerer & Neumeyer 1996; Persons et al. 1999), although contrast in spectral properties between a moving object and its background also contribute to a smaller degree in some cases (Przyrembel et al. 1995; Persons et al. 1999).

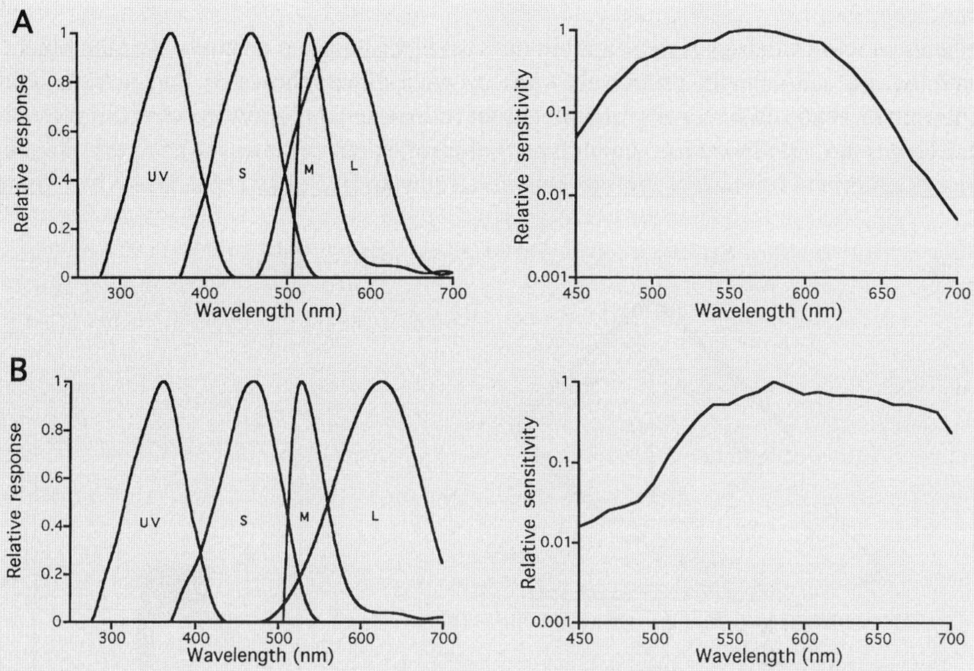

Figure 1. *(**A**) The plot on the left shows the spectral response of each visual photoreceptor class for* Anolis cristatellus. *The functions are estimates based on microspectrophotometry carried out by E. R. Loew (personal communication). The pigment absorption of each photoreceptor class was multiplied by the relative spectral transmission of its associated oil droplet to calculate the response function for each class of photoreceptor. The righthand plot shows the overall spectral sensitivity of* A. cristatellus *from 450-700 nm, based on electroretinographic flicker photometry. See Fleishman et al. (1997) for details. (**B**) The lefthand plot shows the spectral response function of each visual photoreceptor class for* Anolis carolinensis, *estimated based on microspectrophotometry carried out by E. R. Loew (personal communication). The pigment absorption of each photoreceptor class was multiplied by the relative spectral transmission of its associated oil droplet. The righthand plot shows the overall spectral sensitivity (from 450-700 nm) for* A. carolinensis, *which was estimated using electroretinographic flicker photometry (M. Frye & L. Fleishman, unpublished). In the graphs of (**A**) and (**B**), each photoreceptor class is identified by the region of the spectrum of the peak sensitivity: UV (ultraviolet), S (short wavelength), M (middle wavelangth), and L (long wavelength). Methods are described in Fleishman et al. (1997).*

Visual response of anoles to dewlap-like stimuli

In the analyses described below I attempted to determine optimal dewlap color patterns for two of the three hypothesized signal functions: detection and discrimination. I now describe the relationship between signal design and visual-system response relevant to these two functions.

Detection

When a moving stimulus is presented in the visual periphery of an anole, and the object is detected, the animal will reflexively shift its gaze directly towards the moving object (Fleishman 1986). In a recently completed set of experiments (Persons et al., in press; M. H. Persons & L. J. Fleishman, unpublished), this reflex response was used to test the influence of color and brightness contrast between a dewlap-like object (i.e., a moving colored

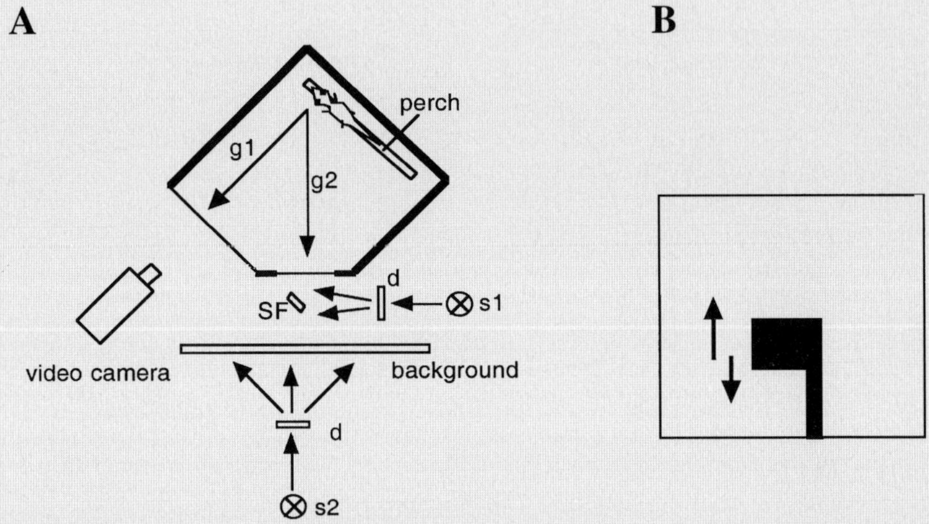

Figure 2. *(**A**) A schematic diagram illustrating the methods used to determine the effect of stimulus/background contrast on detection probability of a small moving stimulus flag. The lizard sits on the perch. A noise from behind a blind causes the lizard to gaze monocularly directly towards the video camera (*g1*). The stimulus flag (*SF*) is then moved up and down a single time. If the lizard notices the movement it shifts its gaze towards it (*g2*). The lizard views the stimulus flag through a small square opening on the front wall of the enclosure. The front face of the stimulus flag is illuminated by the light source (*s1*) after it is filtered and diffused (*d*). The stimulus flag (*SF*) is viewed against a diffuse transluscent background of white paper, which is irradiated from behind with light from source (*s2*) passed through diffusers and filters (*d*). By varying the spectral quality and intensity of the stimulus flag and its background a wide variety of brightness contrast and color contrasts between stimulus and background were tested. Each combination was presented to approximately 50 lizards, and the probability of response was determined as a function of the stimulus/background contrast. (**B**) A 'lizard's eye view' of the stimulus flag and its background is shown. The stimulus flag has a white reflective surface which is illuminated with a light source (*s1*). The arrows indicate the direction of motion of the stimulus flag. Experiments were carried out using two different patterns of motion. In the first set of experiments the stimulus flag was in constant view, and was moved up then down in a trapezoidal motion pattern. In the second set of experiments the flag was initially kept out of view, and then moved up into view, then down out of view in the same motion pattern as in the first experiments. Results were the same for both patterns of motion. Complete results will be published in M.H. Persons & L.J. Fleishman (unpublished).*

stimulus flag of approximately the size of a dewlap) and its background in determining the probability of detection of the moving stimulus. The set-up used in these experiments is illustrated in Figure 2. In each experimental trial a lizard was placed on the perch, and its attention (and gaze) drawn directly forward. A small stimulus flag which was in the lizard's visual periphery was then moved up and down one time. If the lizard noticed the moving stimulus it would shift its gaze toward the flag. The spectral quality and intensity of the stimulus and background were systematically varied, and the probability of response was determined as a function of stimulus and background qualities.

It was found that the probability of an inattentive anole directing its gaze towards a small moving stimulus flag (i.e., similar to a dewlap) depended on (1) the brightness contrast between the stimulus and the background and (2) the spectral contrast between stimulus and background. Brightness contrast was the more important of the two factors. It was established empirically that the probability of detection could be accurately predicted with the following equation:

$$(\text{Probability of detection})\ \text{x}100 = (0.43\ \text{x}\ \text{ABS(B.C.)})+(63\ \text{x}\ \text{C.C.}) + 14$$

$$(\text{if C.C.}>0.3,\ \text{then C.C.}=0.3)$$

In this equation ABS(B.C.) equals the absolute value of percent brightness contrast between background and stimulus, with percent contrast defined as $(B_s-B_b)/(B_s+B_b)$x100. B_s = stimulus brightness, and B_b = background brightness. Brightness is equal to the spectral sensitivity function (determined electroretinographically) multiplied by the stimulus (or background) spectrum measured from 450 -700 nm. It is limited to this range of wavelengths because it was determined, in the behavioral experiments, that shorter wavelengths do not to contribute to brightness for this detection task (M. H. Persons & L. J. Fleishman, unbuplished).

In addition to the effect of brightness contrast, it was found that the extent to which the stimulus differed from the background in spectral quality also influenced the probability of detection. In order to quantify this difference we calculated a value we called color contrast (C.C.), which is an estimate of how differently two spectral patterns stimulate the anoline color vision system. Color perception is determined by the relative degree of stimulation of different classes of cones. One way to estimate the response of a color vision system is to plot the relative extent to which a color stimulates each cone class on a multi-dimensional set of axes. One can then plot two different colors and calculate the vector distance between them. We used this approach, and used this vector distance as an estimate of color contrast.

In this case we assumed that for a pure white stimulus (i.e., one with equal quanta at all wavelengths) the output of each of the four classes of cones is equal. For a given stimulus spectrum the stimulation of each cone class was determined by multiplying the spectrum by the cone spectral response function. This value was then corrected for the area under the cone response function (so that the condition of equal output for all cones with a white stimulus holds). The values for all four cone classes were then totaled. The relative stimulation of each cone was then calculated as the fraction for that cone over the total for all four. The spectrum could then be described with four coordinates, each representing the relative stimulation of one cone class. We then plot these four coordinates as a point in a four-dimensional space in which each axis is the relative stimulation of one cone class (each axis has a value from 0 to 1). In order to compare two colors, we plot each in this four-dimensional space, and calculate the vector distance between the two points.

Assuming that the sensation of color arises from the ratio of the degree of stimulation of each of the different cone classes, our color contrast term is an estimate of how differently the color vision system responds to two different spectra. This estimate of color contrast is based only on the spectral shape of each stimulus and is independent of the total intensity of the stimuli.

We do not know how the anoline visual system weights the inputs from the different cone classes, so here we assume that each class has equal weight. Although our estimate of color contrast may not accurately reflect the perceived difference between two color patterns as perceived by an anole, we found that our calculated color contrast was sufficient to allow us to accurately predict the effects of color contrast between moving stimulus and background in our behavioral experiments.

Discrimination

The behavioral experiments described above tested the probability that a moving stimulus presented in the visual periphery would be detected. A visual-system must perform a rather different task when the problem is to discriminate between two similar stimuli both of which are clearly visible. If an anole is going to distinguish among two or more conspecifics based on the appearance of the dewlap, the color (i.e., the shape of the radiance spectrum) is likely to be a much more important cue than the brightness (i.e. perceived intensity) of the dewlap. This is the case because brightness changes dramatically with habitat light conditions, while color tends to remain fairly constant (especially if the animal visual system has some color constancy mechanism as is found in other vertebrates). Therefore, I assume for this analysis that information about the animal (e.g. genetic quality, health) would be encoded in the spectral quality (i.e., color) of the dewlap, independent of its brightness. In order to use such information an animal must be able to make fine distinctions between two dewlaps which differ slightly in color.

There is no empirical data available on color discrimination in anoles. Here I make the simplifying assumption that the ability to discriminate two colors is directly related to the vector distance between the two stimuli when they are plotted in the four-dimensional color space defined in the previous section. In other words, I assume that the magnitude of the color contrast between two dewlaps is directly related to the ability of an anole to distinguish between them.

Signal environment

An anoline dewlap is a relatively thin flat structure consisting of skin which is stretched open and closed by movement of the hyoid bone. When light strikes the dewlap it will either reflect off, be absorbed by, or transmit through the dewlap. Both transmission and reflection occur in a diffuse manner. The three major factors influencing the appearance and/or detectability of a dewlap are (1) the spectral quality and intensity of the light striking the dewlap from the back and from the front (i.e., the side facing towards a viewer, or the side facing away from a viewer), (2) the extent to which the dewlap reflects, absorbs and transmits light as a function of wavelength, and (3) the background against which the dewlap is viewed. These factors are illustrated in Figure 3. All of the

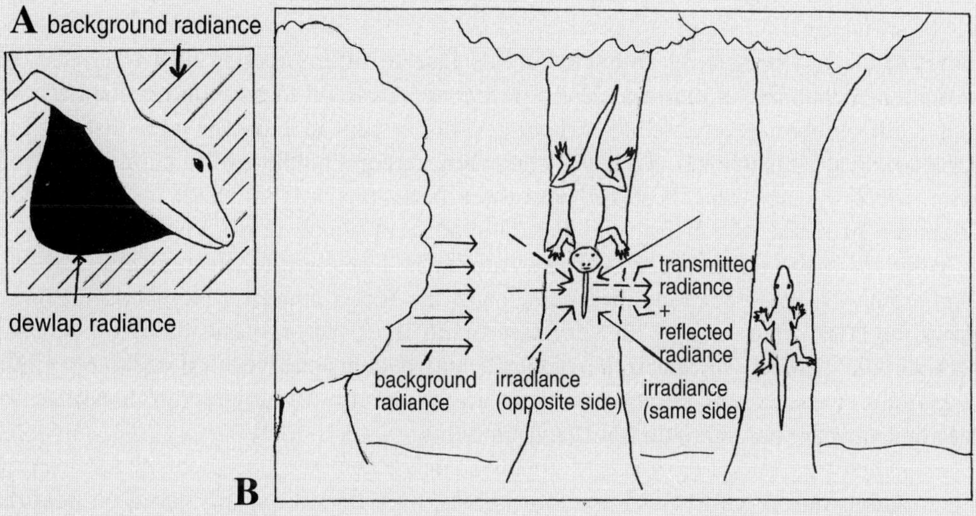

***Figure 3. (A)** A conspecific viewing a male anole with the dewlap extended sees the dewlap radiance contrasted against the radiance of the vegetation forming the background. **(B)** illustrates the factors influencing the visibility of the expanded dewlap of the male lizard in the center of the picture when viewed by the smaller lizard at the right. When expanded, the dewlap forms a thin flat surface which both reflects and transmits light in a diffuse manner. Irradiance is defined as all of the light striking a flat surface originating from a full hemisphere centered on the surface. In this case the dewlap is the surface. There are two irradiance functions that are relevent: the light striking the dewlap from the same side as the viewer (the front), and the light striking the dewlap from the side opposite the viewer (the back). Light emanating from a small finite solid angle is defined as radiance. Radiance is used to characterize the light emanating from a small area of a surface or source. The dewlap radiance depends on the habitat irradiance striking it from the back and the extent to which the dewlap transmits this light, and the irradiance striking the dewlap from the front, and the extent to which it reflects this light. When radiance or irradiance are described as a function of wavelength, the terms spectral radiance and spectral irradiance are used.*

light striking a flat surface (arriving from a hemisphere with the flat surface as its base) is known as *irradiance.* The light emanating from a small finite solid angle of a surface (or source) is know as *radiance.* We use irradiance to describe the intensity of light striking the surface of the dewlap, and radiance to describe the intensity of light coming from the dewlap or from the background against which it is viewed. The units of irradiance are μmol m^{-2} s^{-1} and units of radiance are μmol m^{-2} s^{-1} sr^{-1}, where 1 μmol = 6.02 x 10^{17} quanta, and the steridian (sr) is a unit of solid angle. If irradiance or radiance are plotted per unit wavelength, they are referred to as spectral irradiance and spectral radiance, respectively. The units of spectral radiance are, for example, μmol m^{-2} s^{-1} sr^{-1} nm^{-1}. As will be described below, the radiance of a dewlap can be calculated if we know the spectral irradiance striking the surface of the dewlap from each side. In order to determine the contrast between the dewlap and its background we must also know the spectral radiance of the natural background (typically green vegetation, tree bark or sky).

Anoles are found in heavily vegetated areas. Due to differences in heat tolerance, and locomotory capability among species, different species tend to be found in habitats with distinct differences in vegetation structure. This results in habitats with distinct light characteristics. Fleishman et al. (1997) characterized light habitats for a number of Puerto Rican *Anolis* species, and these fell into three basic types. 'Full shade' is found in the understory of moist or wet forests. 'Partial shade', in which sky is visible but trees are available for some shade, occurs in dry forests, at forest edges, in managed forests in which some canopy trees have been cut, or in the forest canopy. 'No shade' occurs in grassy fields or other areas with very few or no trees. These basic habitat types have characteristic patterns of spectral irradiance and background spectral radiance. Endler (1993) reviewed habitat light conditions in a variety of geographic localities and identified similar characteristic light environments.

In the models described below I use four habitat light conditions, all based on field data collected in Puerto Rico. I use the average spectral irradiance and spectral radiance for the habitat of a full shade species (*A. gundlachi*), for a partial shade species (*A. krugi*), and for a no shade species (*A. pulchellus*). The data for these habitats were collected by observing male anoles for 10 minute periods. Light was then measured at the location the anole was last observed. If the anole displayed in this time, a measurement was taken from the display location. For each measurement an irradiance detector (which measures all light from a complete hemisphere) was placed at the location where the male's dewlap would have been, and oriented parallel to the ground (i.e., pointed so that the center of its hemisphere of reception is pointed parallel to the ground), in order to measure the light irradiating the dewlap. This was done to the right and to the left. Next a radiance detector (which measures light over a small cone of solid angle) was used to measure the spectral radiance of the background in each direction. Typically this detector was pointed parallel to the ground. These data were averaged for twenty individuals for each species. No difference was found between display and non-display locations in any of the habitats, so these were combined.

In addition to the three conditions described above, a fourth habitat condition was used for one of the models. In this case irradiance values from the 'no shade' habitat were used. However, in order to measure background radiance, the radiance detector was oriented upward at angle of 45 degrees. Since the radiance detector samples from only a small solid angle, this typically measures the radiance of the sky. In species from 'no shade' habitats males tend to display from perches which emerge from the ground vegetation, while females and non-displaying males are often near the ground. In such cases the dewlap would be viewed at an upward angle, typically with sky as background. Thus our fourth condition, referred to as 'radiance angle up', combines the 'no shade' irradiance with the radiance measured at an upward angle.

I have measured habitat light for a fairly large number of anoline species. Although there are gradations from one habitat to the other, they all can generally be assigned to one of the three basic habitat light conditions used in this study.

All of the field data collected for this study was measured under sunny skies, with less than one third of the sky covered with clouds.

Dewlap color patterns

Animal color patterns are typically specified in terms of their percent reflectance, which is the amount of light reflected at each wavelength divided by the light reflected (using the same light source) from a white standard (nearly 100% reflective at all wavelengths). However, in this study I needed to calculate the spectral radiance of dewlaps in quantal units. For this reason I measured dewlap color patterns in a slightly different way.

The dewlaps of all the species from Table 1 were measured in my laboratory using methods described in Figure 4. Each dewlap was held open and oriented with one side facing a spectrally broad diffuse light source (300 W xenon arc lamp passed through a diffuser). A black light proof curtain was placed behind the measurement point. First the irradiance of light striking the dewlap was measured. I then measured dewlap radiance; first for reflected light by pointing the radiance detector towards the front of the dewlap (the side facing the light), then for transmitted light by pointing the radiance detector at the back of the dewlap (the side opposite the light source). Then, for every wavelength from 350-700 nm I divided the radiance by the irradiance to get an 'irradiance-to-radiance conversion factor'. These values allow me to calculate spectral radiance of the dewlap for any given spectral irradiance function, simply by multiplying the irradiance at each wavelength by the irradiance-to-radiance conversion factor.

Figure 5 gives two examples of dewlap spectral radiance. Both transmitted and reflected radiance values are shown. Nearly all dewlaps have a similar pattern: there is a short wavelength region (SW) below 400 nm, and a middle wavelength region (MW), the amplitudes of which vary among species. Then there is a long wavelength region (LW) of relatively high, constant amplitude continuing to the long wavelength limits of visual sensitivity. The shift from the MW region to the higher amplitude LW region occurs

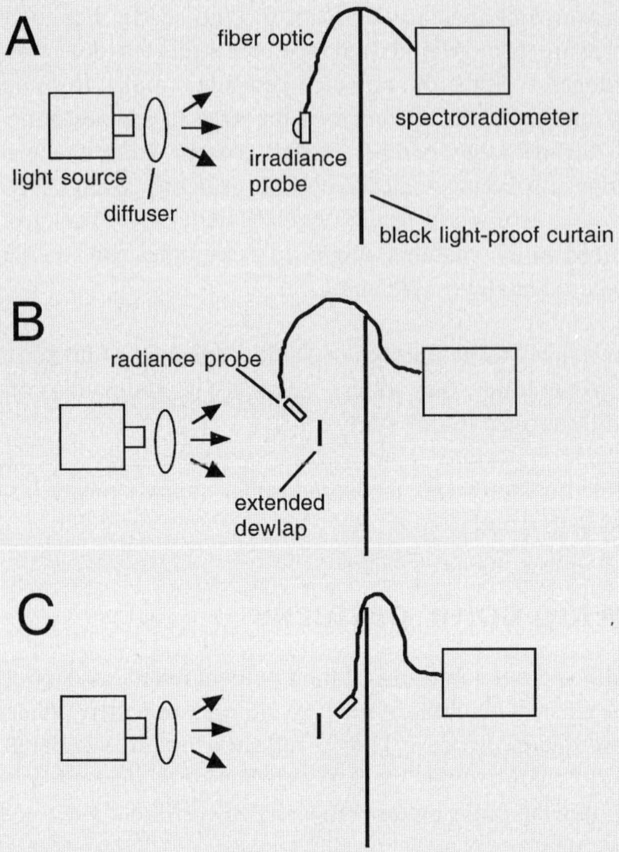

Figure 4. *In order to be able to calculate the spectral radiance of a dewlap under field conditions the illustrated measurements were made in the lab. (**A**) The spectral irradiance of a spectrally broad light source (300 W xenon arc lamp) passed through a diffuser is measured at a fixed location using a cosine-corrected irradiance probe directed toward the source (which measures all light arriving from a solid hemisphere). The irradiance detector serves as the input to the fiber optic spectroradiometer. (**B**) The irradiance probe is replaced with a radiance probe input to the spectroradiometer. The radiance probe measures the light from the dewlap surface over a small solid angle. Each lizard's dewlap is held fully extended with its flat surface directed towards the light source, and it is placed at the same location where the irradiance probe was placed in the previous step. The radiance detector is positioned at a 45 degree angle to the dewlap surface and measures spectral radiance of light reflected from the dewlap. After this measurement is completed the intensity value (in quanta) for radiance at each wavelength is divided by the irradiance (in quanta) at that wavelength to produce a radiance-to-irradiance conversion factor for each wavelength. This factor can be multiplied by any known irradiance value (at each wavelength) to calculate the reflected radiance of the dewlap under any known spectral irradiance. (**C**) The radiance probe is then repositioned behind the dewlap in order to measure the spectral radiance of light passing through the dewlap. As with reflected light, a radiance-irradiance conversion factor is calculated for transmitted radiance for each wavelength. Based on these data one can calculate total dewlap spectral radiance if one knows the spectral irradiance striking the dewlap from the front and the back.*

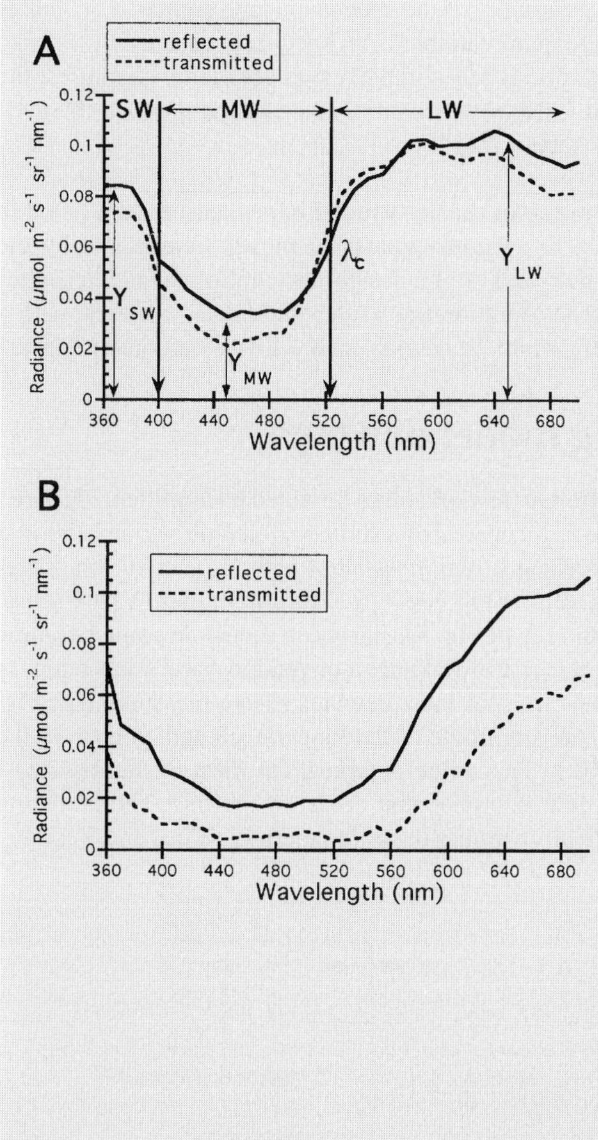

Figure 5. *Examples of dewlap radiance are shown for two species: (**A**)* Anolis krugi *and (**B**)* Anolis pulchellus. *Transmitted and reflected radiance are shown separately for each curve. In this case the irradiant light source is assumed to be an ideal white light with equal quantal intensity at all wavelengths (360-700 nm) equal to 1.0 μmol m^{-2} s^{-1} nm^{-1}. Since the irradiance source is assumed to have a value of 1.0 at each wavelength, this plot can also be thought of as a plot of the radiance-to-irradiance conversion factors vs. wavelength for each of these dewlaps. The dewlaps were measured as described in Figure 4. In (**A**) the terms used to describe dewlaps in this paper are illustrated. The dewlap is divided into three spectral regions: SW (short wavelength), MW (middle wavelength), and LW (long wavelength). The amplitudes at these wavelengths are characterized by the variables Y_{SW}, Y_{MW} and Y_{LW}, which are the amplitudes at 360, 450 and 650 nm, respectively. For comparison purposes the dewlaps are described in terms of two ratios: the desaturation ratio (or just 'desaturation') = Y_{MW}/Y_{LW}, and the UV ratio = Y_{SW}/Y_{LW}. The wavelength at which amplitude is half way between Y_{MW} and Y_{LW} is defined as the cut-on wavelength = λ_c.*

fairly quickly. The wavelength at which there is a shift from the MW region to the LW region is referred to as the 'cut-on wavelength', or λ_c. For descriptive purposes I defined three terms: Y_{SW} (short wavelength amplitude) = the amplitude at 360 nm, Y_{MW} (middle wavelength amplitude) = amplitude at 450 nm, and Y_{LW} (long wavelength amplitude) = the amplitude at 650 nm. λ_c is defined as the wavelength at which the amplitude is half way between Y_{MW} and Y_{LW}.

Although there is some variation, Y_{LW} for most of the species is quite similar. The values of Y_{MW}, Y_{SW}, and the position of λ_c are quite variable from species to species. In order to facilitate comparison among different species I also identify two additional quantities: the desaturation ratio = Y_{MW}/Y_{LW} (which I also refer to simply as 'desaturation' in the text) and, to quantify UV radiance, the UV ratio = Y_{SW}/Y_{LW}.

The characteristics of the dewlaps of different species which I have measured are listed in Table 1. The appearance of a dewlap to a human viewer is mostly determined by the position of λ_c (which determines its hue) and on the desaturation ratio which determines how white it appears. The higher the desaturation the whiter the dewlap appears, with a value of 1.0 being essentially perfectly white. A dewlap with a low desaturation appears strongly colored.

Creating model dewlaps

In order to model the efficiency of different dewlap colors I created a simplified, idealized dewlap color pattern, shown in Figure 6. It consists of a short wavelength region (250-400 nm) of amplitude Y_{SW}, a middle wavelength region of amplitude Y_{MW}, a cut-on wavelength λ_c, and a long wavelength region of amplitude Y_{LW}. The amplitudes (Y) are specified in units of radiance-irradiance ratio (as in Fig. 5). For these model dewlaps I assume that the transmitted and reflected radiance-to-irradiance conversion values are equal at each wavelength. This is approximately true for most dewlaps I have measured (see Fig. 5). In all of the modeling I have kept the amplitude of the long wavelength portion of the dewlap (Y_{LW}) constant at a value of 0.1. This value is typical for most of the dewlaps I have measured, although there are a few examples that deviate from this. The modeling described below consists of systematically varying the parameters λ_c, Y_{MW}, and Y_{SW}.

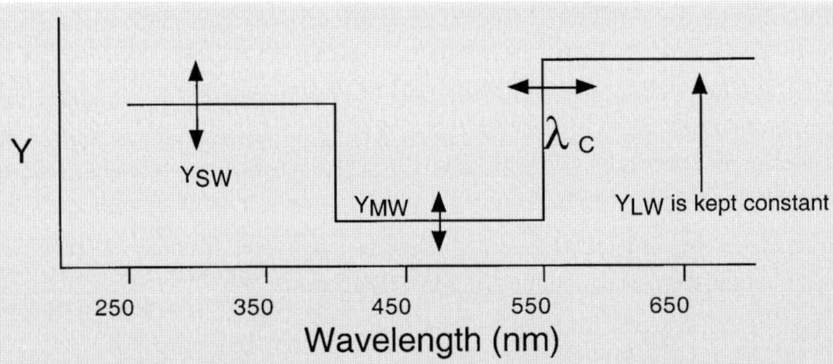

Figure 6. *The simplified dewlap used in the computer models. Amplitude (Y) values were created in units of radiance/irradiance per nm (i.e., in terms of radiance to irradiance conversion factor). These values are made equal for transmitted and reflected light. The value for Y_{LW} was kept the same (and constant) for all of the models at a value which was a typical average for most of the dewlaps we measured. A value of radiance/irradiance = 0.1 was used. Y_{SW}, Y_{MW} and λ_c were systematically manipulated in the computer models. In order to facilitate comparison to natural dewlaps the model dewlap is described quantitatively in terms of the desaturation ratio Y_{MW}/Y_{LW} and the UV ratio Y_{SW}/Y_{LW}.*

Signal function

The actual function (or functions) of anoline dewlap color is not known. Here I consider three hypotheses. I have named these hypotheses in terms of the task each requires the visual system to perform. The most widely stated hypothesis is that the dewlap color pattern allows conspecifics to recognize the bearer as a conspecific (Rand & Williams 1970; Williams & Rand 1977; Losos 1985; Macedonia & Stamps 1994; Losos & Chu 1998). I refer to this hypothesis as 'recognition'. Color may serve to help anoles distinguish conspecifics from heterospecifics as well as to distinguish anoles from other distractions in the environment. The second hypothesis is that the color acts in combination with display motion patterns to increase the probability that the displaying animal will be seen by a conspecific (Fleishman 1992). I refer to this hypothesis as 'detection'. The third hypothesis is that the color provides information about the health and/or genetic fitness of the displaying individual (Johnstone 1995). From a sensory point of view, this is equivalent to hypothesizing that the color allows viewers to discriminate among different individuals of the same species. I therefore refer to this hypothesis as 'discrimination'. In this paper I will present computer models only for the 'detection' and 'discrimination' functions. I will also briefly discuss 'recognition'.

Modeling the efficiency of different dewlap color patterns for the hypothesized function of detection

In order for a signal to be useful it must be seen by its intended receiver. Male anoles move about their territories and display spontaneously every few minutes. While the specific purpose of the display is not known, it probably attracts females and/or keeps other males away. In either case to be effective it must be seen from some distance away, and seen by potentially inattentive individuals. Fleishman (1986) showed that different patterns of motion are differentially effective at eliciting a viewer's attention. Here I examine the relative effectiveness of different color patterns when used in a moving display.

Running the model for detection: Effect of λ_c, desaturation, and SW amplitude

At the beginning of the model run a set of habitat light values were selected. The program was started by setting Y_{MW} to a certain value. The value of λ_c was then systematically shifted from 400 nm to 700 nm in 4 nm steps. For each value the spectral radiance of the dewlap was calculated. This was done by multiplying habitat irradiance times the model dewlap's irradiance-to-radiance conversion factor. The resulting value was doubled because dewlaps reflect and transmit light in roughly equal amounts, so that irradiance striking the dewlap from each side must be accounted for. After dewlap spectral radiance was calculated it was compared to the average habitat background spectral radiance. Brightness contrast and color contrast were calculated in the manner described above, and from these values a probability of detection was determined according to the behaviorally derived equation listed above, which relates detection probability of a moving object to its color and brightness contrast with the background. After completing these calculations for each λ_c, the value of Y_{MW} was altered to create a new desaturation ratio and the entire process was repeated. These steps were carried out for six desaturation ratios (Y_{MW}/Y_{LW}): 0.2, 0.3, 0.5, 0.7, and 0.8. The entire model was carried out for the four different habitat conditions (full shade, partial shade, no shade, and no shade with radiance measured at an upward angle) using the *A. cristatellus* visual system (which is typical of nearly all species). It was then carried out for the *A. carolinensis* visual system using the partial shade habitat.

Results for the model of detection

The outcome of the model for the full shade habitat is summarized in Figure 7. In Figure 7A detection probability versus λ_c is shown for three levels of desaturation. The optimal solution appears to be a dewlap with very low desaturation with λ_c = 650 nm or more. Such a dewlap would appear essentially black to an anole (i.e., the color is too dark to be perceived as a color). It is effective because it is much darker than the background and therefore has a very high brightness contrast with the background. The next best color pattern is a highly desaturated dewlap of almost any λ_c. This is essentially a white dewlap. Finally, it is worth noting that a dewlap with very low desaturation and a λ_c of 500-525 nm is also quite effective. In Figure 7B the two parameters which determine detection probability are plotted versus λ_c for the lowest desaturation example. It is clear that for any λ_c shorter than 520 nm the dewlap is brighter than the background (a positive number indicates dewlap brighter than background). At longer λ_c values brightness drops, and the effectiveness of the long wavelength cut-on (650 nm) is because the dewlap is darker than the background. Note that detection probability is a function of the absolute value of

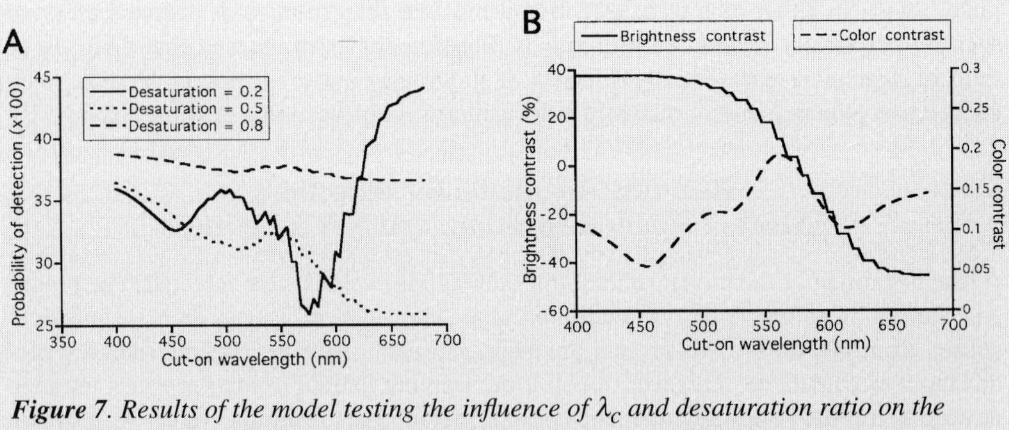

***Figure 7.** Results of the model testing the influence of λ_c and desaturation ratio on the probability of detection for a full shade habitat. In (**A**) the probability of detection is plotted versus wavelength for three levels of desaturation. In (**B**) the variables which determine response probability – brightness contrast and color contrast – are plotted seperately for the case where desaturation = 0.2.*

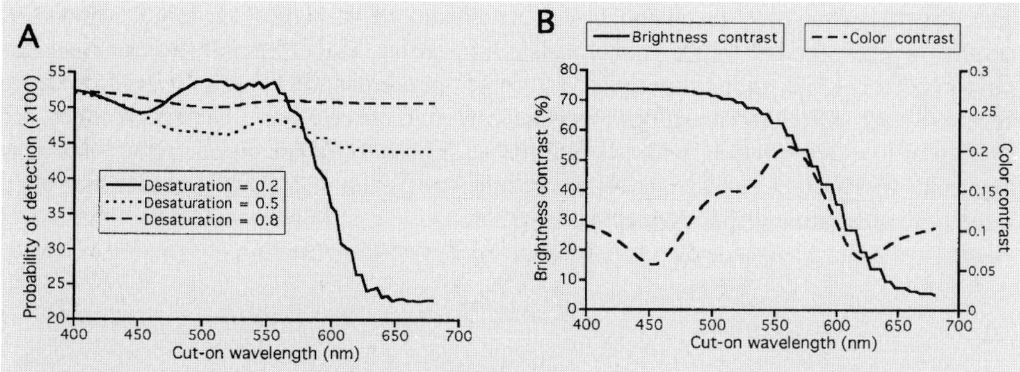

***Figure 8**. As in Fig. 7, for a partial shade habitat.*

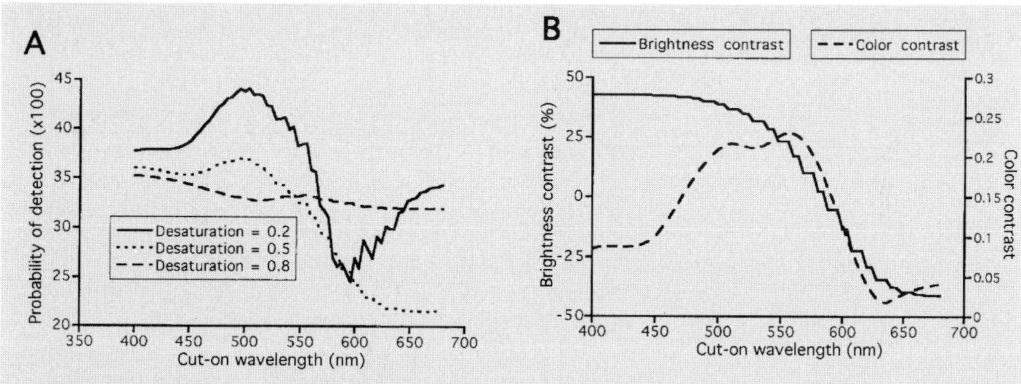

***Figure 9**. As in Fig. 7, for an unshaded habitat.*

brightness contrast (between dewlap and background), which means that high contrast (and therefore high probability of detection) can be achieved with the dewlap being either lighter or darker than the background.

The results for partial shade habitat light are summarized in Figure 8. Here it can be seen that the most effective color pattern is a dewlap with low desaturation and a λ_C anywhere from 500 to 570 nm. A highly desaturated (whitish) dewlap with almost any λ_C also works very well.

The results for no shade are shown in Figure 9. Here the most effective dewlap color pattern is one with low desaturation (i.e., very colored) with λ_C in the 500-570 nm range. However, as was mentioned above, the species from these habitats are also likely to be viewed from below with some regularity. In Figure 10 the effect of measuring background radiance with the radiance detector pointed at an upward angle is explored. In this case a dewlap with low desaturation and with a λ_C of 560 nm or greater was most effective. As can be seen in Figure 10B this is effective because such a dewlap is much darker than background radiance (i.e., the blue) sky, creating a high brightness contrast.

The effect of changing the visual system is illustrated in Figure 11, which shows the results of this model for *A. carolinensis* in partial shade habitat. Recall that the photoreceptor absorption functions of some photoreceptors, and overall spectral sensitivity are shifted toward longer wavelengths in *A. carolinensis*. The best solution is a dewlap of low desaturation, with the optimal λ_c shifted to longer wavelengths relative to the result in Figure 8 (which shows the results for partial shade habitat with the other, more typical, anoline visual system).

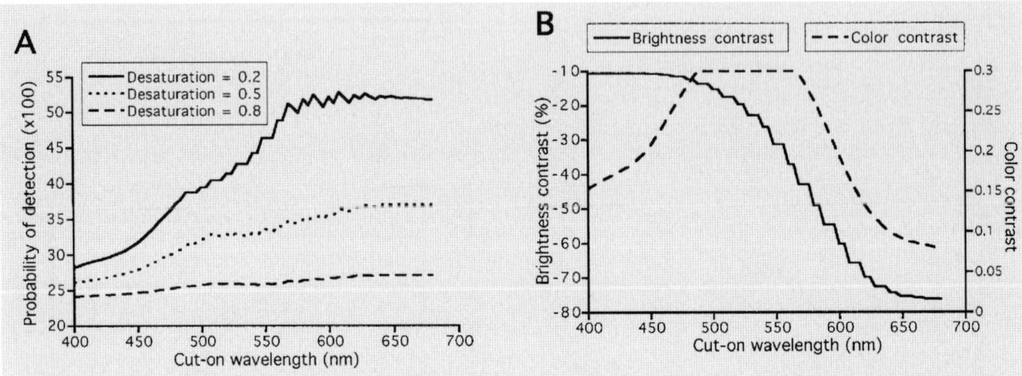

Figure 10. As in Fig. 7 for the unshaded habitat. In this case background radiance was measured with the radiance detector pointed at an upward angle of 45 degrees.

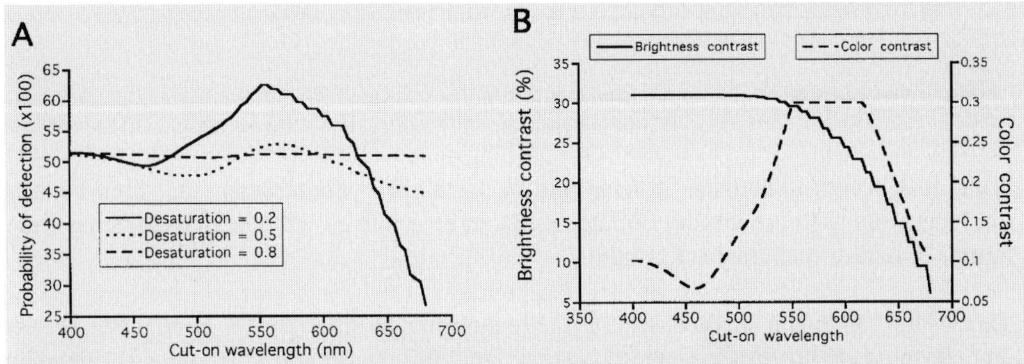

Figure 11. As in Fig. 7, for A. carolinensis *and partial shade habitat light conditions.*

The effect of short wavelength amplitude on detection probability

In the next runs of the computer model I chose a few parameters for λ_c and desaturation, and for each I systematically varied the amplitude of the short wavelength portion (Y_{SW}). The results are illustrated for various dewlaps and different habitat conditions in Figure 12. In full shade conditions (Fig. 12A), increasing short wavelength reflectance increased detection probability to a modest degree. Under partial shade conditions (Fig. 12B), high short wavelength reflectance added modestly to the detection probability of a low desaturation dewlap with a λ_c of 570 nm, or a high desaturation dewlap with a λ_c of 520 nm. In the case of the low desaturation dewlap with a λ_c of 520 nm, elevated short wavelength reflectance reduced the predicted detection probability slightly. For the unshaded habitat (Fig. 12C) addition of short wavelength reflectance added to detection probability for radiance measured parallel to the ground, although only at very high Y_{SW} amplitude. For viewing against the sky (angled up) increasing short wavelength amplitude reduced the predicted detection probability. Finally, for *Anolis carolinensis* the addition of short wavelength dewlap radiance had almost no effect.

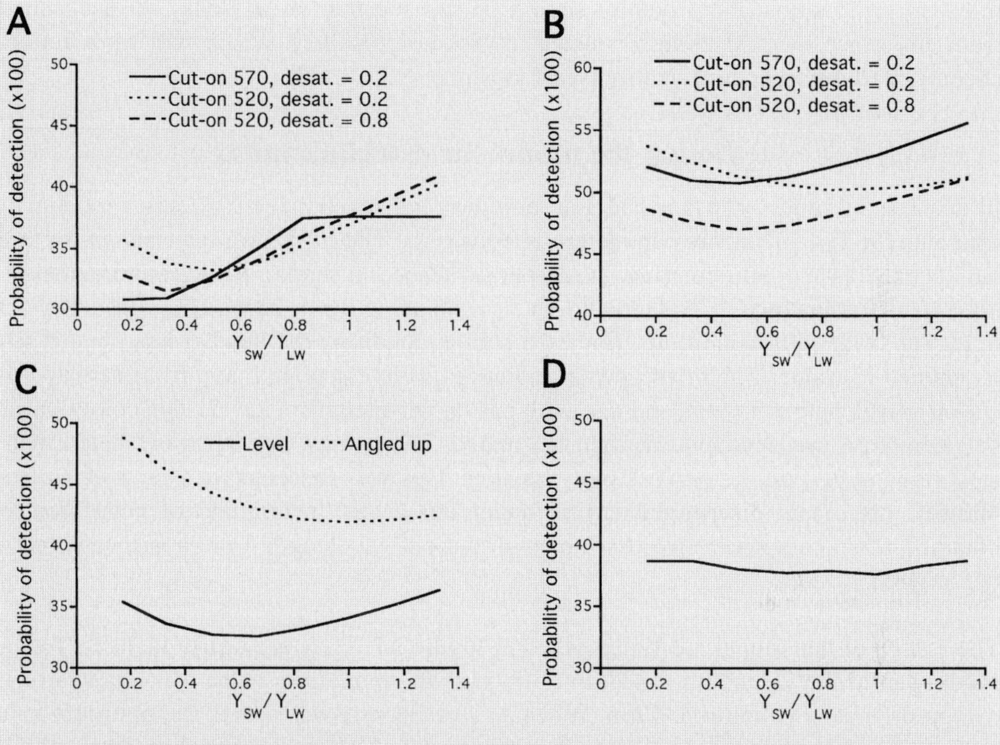

Figure 12. *Summary of the effects of altering the amplitude of the short wavelength (SW) portion of the model dewlap. Amplitude of the short wavelength portion is given as Y_{SW}/Y_{LW}. Examples are shown of (**A**) full shade, (**B**) partial shade and (**C**) no shade habitats. In (**D**) the results of the model carried out on* A. carolinensis *in partial shade are shown. In (**A**) and (**B**), the model was carried out on three dewlaps: (1) λ_c = 520 nm, desaturation ratio (Y_{MW}/Y_{LW}) = 0.2, (2) λ_c = 520 nm, desaturation ratio = 0.8 and (3) λ_c = 570 nm, desaturation = 0.2. In (**C**) only one dewlap (λ_c = 570 nm, desaturation = 0.2) was modeled under each of the two conditions: (1) background radiance measured with the detector pointed parallel to the ground, and (2) background radiance measured with the detector pointed at an upward angle of 45 degrees. In (**D**) the model was carried out on one dewlap (λ_c = 570 nm, desaturation = 0.2).*

Modeling for the hypothesized function of discrimination

If the dewlap color is an honest indicator of quality it is important that conspecifics be able to easily discriminate slight variations in dewlap spectral pattern. There is no empirical data available on color discrimination by anoles. For this model I made the assumption that the ability to discriminate two similar color patterns was related to the color contrast between them, or the vector distance between each dewlap spectrum plotted in four-dimensional color space. The spectral pattern which most efficiently transfers information about quality should be the one that is very easy to discriminate from any other spectral pattern, since it is this color pattern which will most reliably encode small differences in quality (see Johnstone 1995, 1998).

Running the model for discrimination

In this case all model dewlaps had a desaturation value = 0.2 and Y_{SW} was kept equal to the value for Y_{MW}. The only parameter varied was λ_c. The model was carried out for three habitat irradiance conditions (full shade, partial shade, no shade). Background radiance is not a factor in this model. The model was also run using the *A. carolinensis* visual system in partial shade light conditions. For each habitat condition the dewlap was created with its defined λ_c value. Then two dewlap radiances were calculated: one for a dewlap with cut-on wavelength = $\lambda_c + 2$, and one with cut-on wavelength = $\lambda_c - 2$. The color contrast between these two dewlaps was then calculated. Then a new dewlap was created with a new value of λ_c and the process was repeated. This was repeated for values of λ_c from 400-680 nm. Thus discriminability is quantified as the magnitude of color contrast obtained when comparing two dewlaps with cut-on wavelength 4 nm apart, centered on the original λ_c value.

The results of this model are summarized in Figure 13. They are nearly identical for full shade, partial shade, and no shade. In every case the maximum value was a dewlap with λ_c approximately equal to 535 nm. When *A. carolinensis* was tested, the optimum value shifted to a slightly longer wavelength of 544 nm. I also ran the model on dewlaps with higher desaturation values. Increased desaturation did not affect the shape of the resulting curves, but reduced the color contrast for all values of λ_c.

Hypothesized function of recognition

In order to use a color to facilitate rapid recognition it is necessary to create a color pattern that is (1) easily visible and (2) as different as possible from other objects in the environment. There are various approaches to trying to determine whether this function has influenced the evolution of dewlap color. One approach is to examine the color patterns of groups of sympatric anoline species to see if the most similar species in terms of microhabitat and body type are the most different in dewlap color pattern. This approach has been used (Rand & Williams 1970; Williams & Rand 1977), although

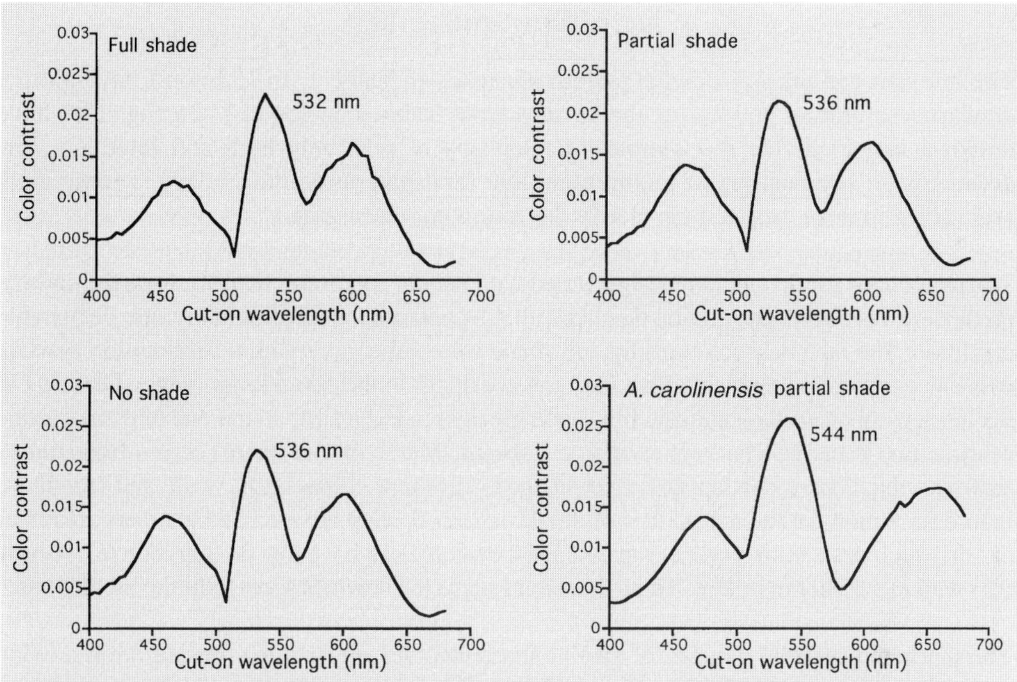

Figure 13. The results of the model testing dewlap discriminability as a function of its spectral composition. For each cut-on wavelength (=λ_c) value on the x-axis, the color contrast was calculated between a dewlap with cut-on = (λ_c + 2) and one with cut-on wavelength = (λ_c - 2). Three different habitat irradiance spectra were tested (first three plots). The fourth plot shows the model results for A. carolinensis *under partial shade conditions. The number within each graph indicates the wavelength at which peak discrimination from similar spectra is possible.*

dewlap colors were assessed visually by humans, and it would be instructive to repeat this approach using the anoline visual system. Another approach would be to sample as many objects in the environment as possible, including sympatric congeners, and attempt to determine what spectral patterns are least likely to be found in the habitat. In order to facilitate rapid recognition a dewlap color pattern should be an uncommon pattern in the environment.

Discussion

I have modeled optimal color patterns for two different functions. I will now compare my predictions with naturally occurring dewlap color patterns.

Detection probability

The relevant variables of dewlap pattern are listed in Table 1. In addition I have plotted dewlap desaturation and λ_c for the species from Table 1 in Fig. 14. On Fig. 14 I have drawn a line to divide the sample into dewlaps of relatively high and relatively low desaturation. The position of the dividing line (at desaturation ratio = 0.3) is completely arbitrary, and done only to help clarify the following discussion.

The predictions for the optimal design varied with habitat condition. For full shade the optimal prediction was a low desaturation dewlap with $\lambda_c = 650$ nm. No species in my sample met that condition. The next best solutions for full shade were either a very high desaturation dewlap, or a low desaturation dewlap with a cut-on wavelength in the 500-525 nm range. Glancing at the black symbols (which indicate full shade species) in Figure 14, all but one of them fits one of these two patterns. The lone exception (labeled 'Mes2' on the figure) comes from *Anolis mestrei*, which has a dewlap with two distinctly different colors on it: white and red. If we ignore for a moment the red portion of the *A. mestrei* dewlap we see that the others are either of fairly high desaturation (greater than 0.3, for example), or have low desaturation (e.g. below 0.3) with λ_c at 530 nm or less. Thus, in a broad sense the dewlaps seem to fit the predictions.

The predicted optimal pattern for partial shade was for either a low desaturation dewlap with a λ_c in the range 500-570 nm, or else a dewlap with very high desaturation of any λ_c.

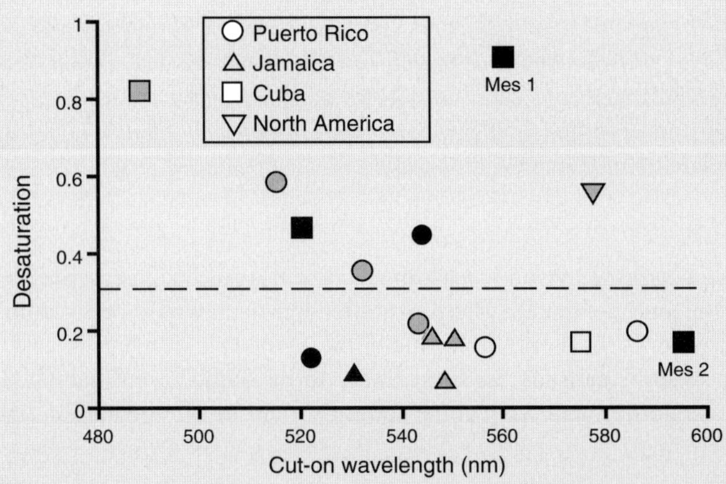

Figure 14. *A summary of the dewlap characteristics of the species which appear in Table 1. The filled symbols indicate species from full shade habitat. The gray symbols are partial shade species. The open symbols indicate species from unshaded habitat. One of the species (*A. mestrei *from Cuba) has a dewlap that consists of two distinctly different colors, and one measurement for each of these colors appears on the graph. The two values for* A. mestrei *are labeled 'Mes1' and 'Mes2'. The single North American species is* A. carolinensis. *Desaturation and cut-on wavelength (=λ_c) are as defined in the text. The dashed line indicates an arbitrarily chosen value of desaturation = 0.3, which is used to clarify the discussion in the text.*

Partial shade species are indicated by the gray symbols in Figure 14. We find a cluster of four low desaturation dewlaps with λ_c = ca. 550 nm, and four dewlaps with fairly high desaturation (greater than 0.3) spread over a wide range of λ_c values. The general prediction of having either a dewlap with high desaturation, or low desaturation with a λ_c between 500-570 nm seems to hold up.

For the no shade habitats the dewlaps need to be a compromise between the two different viewing conditions. The dewlap may be viewed against vegetation (i.e., the radiance measured with the detector level to the ground) or against the sky (measured with the radiance detector pointed at an upward angle). A λ_c of approximately 560-570 nm might be expected since it works well in both conditions (Figs 9 and 10). For both radiance conditions (i.e., measured parallel to the ground, or measured at an upward angle) a low desaturation dewlap is predicted. The three species in my sample (white symbols in Fig. 14) from no shade habitat have low desaturation and their λ_c values are of longer wavelength than most of the other species, ranging from 550-586 nm.

Finally the prediction for *A. carolinensis* is for a low desaturation dewlap with λ_c in the region of 560-580 nm. In reality the dewlap has a fairly high desaturation, but it does have a fairly long λ_c value (577 nm).

In summary, the predictions from the computer models for optimal detection probability are not badly out of line with what is found in reality. The predictions for each condition were rather broad. Nevertheless there are some overall patterns for different signaling environments which are consistent with the model predictions. For example, as one moves from full shade to partial shade to no shade there is (both predicted and observed) a general shift to longer wavelengths of the value of λ_c, and the no shade species have lower desaturation values than the other species overall.

One gets the impression, overall, that selection for detection is imposing some constraints, rather than driving dewlap design to particular optima. Comparing Figures 7-11 to Figure 14 it is generally true that the patterns which would produce very low detection probabilities are rare. However, within each set of habitat light conditions there are a variety of possible color patterns that work well, and we do find diversity in dewlap color among the species from a given habitat type.

It is difficult to make a case for short wavelength reflectance being selected for by the need for high detection probability. For example, in full shade, high UV reflectance would increase detection probability. In most instances (see Table 1) it is actually quite low for full shade species. High UV reflectance is also predicted for partial shade species. It is, in reality, quite high in some species and quite low in others. In general there is little match between the pattern of where high UV reflectance is found and where it is predicted by this model. My feeling is that UV reflectance, in general, has a relatively modest impact on detection probability, and its presence or absence in the dewlap probably relates to the function of recognition discussed below.

It is worth pointing out that all habitat measurements were made under clear blue skies. While our field data has indicated relatively little change in light conditions (measured in the way described here) in going from sunny to cloudy skies (Fleishman et al. 1997) for irradiance, and for radiance measured parallel to the ground, one might expect a more dramatic change for the case in which radiance was measured with the detector pointed upward toward the sky. For this case there should be much less UV in the background radiance under cloudy conditions and this might alter the predictions for optimal values for UV in the dewlaps of no shade species. In particular, high UV radiance in the dewlap should substantially increase color contrast under cloudy sky conditions.

Discrimination

Little or no case can be made that there has been strong selection on dewlap color for ease of discrimination. The model for discriminability makes a clear prediction in every habitat condition. In every case a λ_C of ca. 535 nm is predicted. Figure 14 reveals that dewlaps occur with a variety of different λ_C values with no evidence of clustering near 535 nm. Moreover high desaturation dewlaps have a lower discriminability at all λ_C values, and yet desaturated dewlaps are quite common. The hypothesis that dewlap colors are designed to allow the visual system to easily discriminate among individuals with small color variations is not supported.

Recognition

The analyses so far suggest that the need for detectability acts as a constraint on dewlap design. Yet there is a great deal of variability of design within these constraints. I have not considered the function of recognition in any great detail, yet it seems likely to be an important function. This hypothesis could be tested with a phylogenetic analysis in which sympatric and allopatric congeners are compared for similarity, but this test has not yet been carried out. It would also be interesting to see if dewlap spectra are a fairly rare subset of natural object colors in the habitat – since conspecifics need to be distinguished from other objects in the environment as well as congeners.

At this time I can only offer anecdotal support for the recognition function by comparing certain sympatric species pairs which are very similar in body form and habitat preference. In southwestern Puerto Rico for example, *A. cristatellus* and *A. cooki* are nearly identical in appearance and found in the same habitat. Looking at Table 1, we see that they differ by a great deal in desaturation and in short wavelength radiance. Another very similar species pair in Puerto Rico are the grass anoles *A. pulchellus* and *A. krugi*. Although *krugi* has much greater preference for shade, at higher elevations they are often found in the same fields, with *pulchellus* tending to stay out in the sun, and *krugi* in the shade. *A. pulchellus* has a highly saturated orange/red dewlap, while that of *krugi* is fairly desatured yellow. In Cuba, *A. mestrei* and *A. allogus* are found in very similar locales. *A. allogus* has a very desaturated yellowish dewlap. *A. mestrei* has a dewlap which is half dark red and half white. One could speculate that the red portion is critical for distinguishing it from *allogus*, while the white portion improves its detectability. Thus the hypothesized function of recognition seems plausible.

A number of studies have attempted to assess the role of dewlap color in species recognition by experimentally manipulating dewlap color with colored paint (Greenberg & Noble 1944; Crews 1975; Sigmund 1983; Losos 1985) or with the use of video images (Macedonia & Stamps 1994). The results have been mixed (reviewed in Tokarz 1995), with some studies finding no role for dewlap color in species recognition and others finding strong support for such a function. Unfortunately, in all of these experiments the appearance of the colors of the manipulated dewlaps were assessed solely by human viewers. Such studies must be viewed somewhat sceptically since the anoline visual system is quite different from the human visual system. A particular difficulty is that many anoline dewlaps possess ultraviolet components which the anoles can see (Fleishman et al. 1993) and humans cannot. Thus the experimental color may not have appeared to the anoles as they did to the experimenters, and even more problematic, the control colors may actually not have matched natural colors from the lizards' point of view. Further experimentation will be required to assess the importance of recognition as a selective force on anoline dewlap color, taking the anoline visual system into account.

Another piece of circumstantial evidence which suggests that species recognition may be an important function of dewlap color is the presence of high ultraviolet reflectance and transmittance in some dewlaps. This UV component does not seem to enhance the probability that dewlaps will be detected by an inattentive viewer in most habitat light conditions. However, since green vegetation absorbs UV, colors which have an ultraviolet component will tend to be quite rare in the habitat, and a dewlap which includes an ultraviolet component may be easy to recognize quickly.

Conclusion

In summary, it appears that color patterns in anoles may result from selection for efficiency in more than one capacity. I have modeled dewlap efficiency assuming different functions. I have found little support for the hypothesis that dewlap color is selected for by the need for discrimination among slight color variants. The hypothesis that the need to make the moving dewlap easily detectable has been a selective force – or at least a constraint – is supported. Moreover the color patterns which result in high efficiency for this function vary in a predictable way with habitat condition and with visual system design, which suggests that selection for efficiency in different habitats can help to explain signal diversity among closely related species. This supports the hypothesis of 'sensory drive' (Endler & McLellan 1988; Endler 1992; Marchetti 1993) as an important evolutionary force in the generation of signal diversity.

It seems doubtful, however, that the need for detectability is the only selective force. Selection for detectability in signals may only be acting as a broad constraint. Within the limits of this constraint there is some evidence that sympatric anoles tend to diverge in their dewlap color patterns.

There are, no doubt, other influences on the evolution of dewlap color patterns which I have not considered, such as counter-selection from visually hunting predators, as well as physiological or biochemical constraints on what patterns may be formed. Moreover, my models are based on simplified idealized dewlaps which do not exhibit all of the complexity of natural dewlap color patterns. Nevertheless, the approach of focusing on the need to stimulate the visual system efficiently, in a manner that is appropriate to the function of the signal, has provided considerable insight into some of the important factors contributing to the evolution of dewlap color.

Acknowledgments

Special thanks to M. Persons, E. Loew and M. Frye for allowing me to include their unpublished results in this paper. I also thank J. Losos, M. Leal and L. Rodriguez-Schetino for assistance in the field, and L. Almstead for assistance with the computer programming. J. Endler and G.-P. Sætre made suggestions which led to improvements of the manuscript. This research was supported by National Science Foundation grant IBN-9307019 and Whitehall Foundation grant S93-03.

References

Crews, D. 1975. Effects of different components of male courtship behaviour on environmentally induced ovarian recrudescence and mating preferences in the lizard, *Anolis carolinensis*. *Animal Behaviour*, **23**, 349-356.

Endler, J. A. 1992. Signals, signal conditions, and the direction of evolution. *American Naturalist*, **139** (supplement), s125-s153.

Endler, J. A. 1993. The color of light in forests and its implications. *Ecological Monographs*, **63**, 1-27.

Endler, J. A. & McLellan, T. 1988. The process of evolution: toward a newer synthesis. *Annual Review of Ecology and Systematics*, **19**, 395-421.

Fleishman, L. J. 1986. Motion detection in the presence and absence of background motion in an *Anolis* lizard. *Journal of Comparative Physiology, Series A*, **159**, 711-720.

Fleishman, L. J. 1992. The influence of the sensory system and the environment on motion patterns in the visual displays of anoline lizards and other vertebrates. *American Naturalist*, **139** (supplement), s36-s61.

Fleishman, L. J., Loew, E. R. & Leal, M. 1993. Ultraviolet vision in lizards. *Nature*, **365**, 397.

Fleishman, L. J., Bowman, M., Saunders, D., Miller, W. E., Rury, M. J. & Loew, E. R. 1997. The visual ecology of Puerto Rican anoline lizards: habitat light and spectral sensitivity. *Journal of Comparative Physiology, Series A*, **181**, 446-460.

Goldsmith, T. H. 1990. Optimization, constraint, and history in the evolution of eyes. *Quarterly Review of Biology*, **65**, 281-322.

Greenberg, B. & Noble, G. K. 1944. Social behavior of the American chameleon (*Anolis carolinensis,* Voigt). *Physiological Zoology*, **17**, 392-439.

Guilford, T. & Dawkins, M. S. 1993. Receiver psychology and the design of animal signals. *Trends in Neurosciences*, **16**, 430-436.

Johnstone, R. A. 1995. Sexual selection, honest advertisement and the handicap principle: reviewing the evidence. *Biological Reviews*, **70**, 1-65.

Johnstone, R. A. 1998. Efficacy and honesty in communication between relatives. *American Naturalist*, **152**, 45-58.

Losos, J. B. 1985. An experimental demonstration of the species recognition role of *Anolis* dewlap color. *Copeia*, **1985**, 905-910.

Losos, J. B. & Chu, L.-R. 1998. Examination of factors potentially affecting dewlap size in Caribbean anoles. *Copeia*, **1998**, 430-438.

Macedonia, J. M. & Stamps, J. A. 1994. Species recognition in *Anolis grahami* (Sauria, Iguanidae): evidence from responses to video playbacks of conspecific and heterospecific displays. *Ethology*, **98**, 246-264.

Marchetti, K. 1993. Dark habitats and bright birds illustrate the role of the environment in species divergence. *Nature*, **362**,149-152.

Neumeyer, C. 1998. Color vision in lower vertebrates. In: *Color Vision: Perspectives from Different Disciplines* (Ed. by W. G. K. Backhaus, R. Kliegl & J. S. Werner), pp. 149-162. Berlin: Walter de Gruyter.

Persons, M. H., Fleishman, L. J., Frye, M. A. & Stimphil, M. E. 1999. Sensory response patterns and the evolution of visual-signal design in anoline lizards. *Journal of Comparative Physiology, Series A*, **184**, 585-607.

Przyrembel, C., Keller, B. & Neumeyer, C. 1995. Trichromatic color vision in the salamander (*Salamandra salamandra*). *Journal of Comparative Physiology, Series A*, **176**, 575-586.

Provencio, I., Loew, E. R. & Foster, R. G. 1992. Vitamin A_2-based visual pigments in fully terrestrial vertebrates. *Vision Research*, **32**, 2201-2208.

Rand, A. S. & Williams, E. E. 1970. An estimation of redundancy and information content of anole dewlaps. *American Naturalist*, **104**, 99-103.

Schaerer, S. & Neumeyer, C. 1996. Motion detection in goldfish investigated with the optimotor response is "color blind". *Vision Research*, **36**, 4025-4034.

Sigmund, W. R. 1983. Female preference for *Anolis carolinensis* males as a function of dewlap color and background coloration. *Journal of Herpetology*, **17**, 137-143.

Tokarz, R. R. 1995. Mate choice in lizards: a review. *Herpetological Monographs*, **9**, 17-40.

Williams, E. E. & Rand, A. S. 1977. Species recognition, dewlap function, and faunal size. *American Zoologist*, **17**, 261-270.

IN: ESPMARK,Y., AMUNDSEN,T. & ROSENQVIST,G. (EDS.) 2000
ANIMAL SIGNALS:
SIGNALLING AND SIGNAL DESIGN IN ANIMAL COMMUNICATION, PP. 237-257
TAPIR ACADEMIC PRESS, TRONDHEIM, NORWAY

11

Sexual Signals and Speciation

GLENN-PETER SÆTRE

Department of Biology, Division of Zoology, University of Oslo, P.O. Box 1050 Blindern, N-0316 Oslo, Norway

E-mail: g.p.satre@bio.uio.no

Abstract

Speciation is the set of evolutionary processes that lead to reproductive isolation of different populations. Empirical evidence shows that sexual signals are important in producing prezygotic barriers to gene exchange between members of different populations. Speciation may result when sexual signals diverge in isolated, allopatric populations as a response to different ecological conditions or as a result of processes of sexual selection. Stability of female mate preferences compared to other traits determines the relative importance of sexual selection in speciation, but little is known about this subject. Different models of sexual selection differ in this regard and Fisherian runaway selection on arbitrary traits is the most powerful in predicting rapid speciation. Sexual signals may also diverge in sympatry as a response to natural selection against unfit hybrids between differentiated populations (i.e., reinforcement). Theoretical models suggest that speciation by reinforcement may not be a likely outcome of interactions between differentiated populations except under some rather restricted conditions. However, reinforcement may set the stage for sexual selection to take new directions in populations. The joint effects of reinforcement and self-reinforcing sexual selection may cause speciation more easily. It is not known to what extent sympatric interactions between differentiated populations in general would tend to speed up divergence of sexual signals and speciation. However, in some organisms, such a relationship apparently exists. Finally, sexual selection could in theory induce ecological speciation. Sexual selection has the potential to move populations around in an adaptive landscape because it maximises mating success and not survival. As a result, sexual selection could move a population away from an ecological peak in the adaptive landscape, whereafter natural selection would reinforce the evolutionary trend and move the population to an alternative ecological peak.

Speciation can be defined as the set of evolutionary processes that leads to reproductive isolation between members of different populations. According to the biological species concept, speciation is completed and two species have been formed when members of two populations no longer can, or otherwise do not, produce fertile offspring of mixed ancestry when (actually or potentially) living within the same geographical locality (Mayr 1942). Reproductive isolation may result from prezygotic barriers to gene exchange, of which positive assortative mating will be considered here, and/or from postzygotic barriers (e.g. hybrid sterility). The present paper is concerned with the significance of sexual signals in processes of speciation. By sexual signals, I mean characters that evolve by means of sexual selection within populations. In particular, I will be concerned with secondary sexual characters favoured by mate choice and the corresponding mate preferences. A prezygotic barrier to gene exchange by sexual signalling has developed if the courted individuals (usually females) do not respond to the sexual signals of courting individuals (usually males) of the other species (e.g. Sætre et al. 1997a), or if the courting individuals do not display to members of the other species. The combination of sexual signals and responses to the signals within a species has been called the specific mate recognition system of a species (Paterson 1985), but below I will refer to these as sexual signals in short. Although emphasis here is placed on mate choice, characters favoured by intrasexual selection (e.g. male-male competition for mating opportunity) may also be relevant in processes of speciation.

I will mainly discuss three processes of speciation in which sexual signals are, or may be of importance. First, sexual signals may be important in producing prezygotic barriers to gene exchange during allopatric speciation. For instance, in two geographically separated populations, female choice on male secondary sexual traits may take different routes, that is, females select for different male traits or different expression of a trait in different populations. If, after some time, the populations experience secondary contact, the males will be differentiated in secondary sexual characters and the females will be differentiated in mate preferences. As a result, there will be positive assortative mating in sympatry. The second process is that natural selection, due to incomplete postzygotic barriers to gene exchange, would favour a reinforcement of reproductive isolation by increasing prezygotic isolation (Dobzhansky 1940, 1941). That is, positive assortative mating would increase in frequency or magnitude as a response to natural selection against the production of unfit hybrids, eventually completing the speciation process. This could be accomplished by divergence of secondary sexual characters and corresponding mate preferences. Finally, it is possible that evolution of sexual signals can induce ecological speciation. To exemplify this process I will use Sewall Wright's metaphor of adaptive landscapes and borrow from parts of his shifting balance theory of evolution (Wright 1968-1978). In short, because sexual selection may oppose survival selection, the former could in theory move a population away from an ecological peak in the adaptive landscape whereafter natural selection would reinforce the evolutionary trend and move the population towards another ecological peak.

Allopatric divergence of sexual signals

Speciation can result as a by-product whenever sexual signals diverge in isolated allopatric populations (Fig. 1). Sexual signals may diverge in isolated population because natural selection would tend to shape the signal according to local environmental or ecological conditions, or they may diverge as a result of processes of sexual selection. Allopatric speciation may, however, also occur because of ecological selection. That is, isolated populations become adapted to local ecological conditions and reproductive isolation is a side effect of the adaptive divergence (e.g. Grant 1986; Schluter & Nagel 1995). In many cases it is difficult to separate the relative importance of these processes. For instance, closely related species that differ in sexual signals may have diverged ecologically. After establishment of secondary contact, sexual signals may then have diverged to help species recognition to avoid reproductive wastage (e.g. Butlin 1987). It is also possible that populations can diverge even when there is some gene flow between them (e.g. Turner & Burrows 1995; Smith et al. 1997).

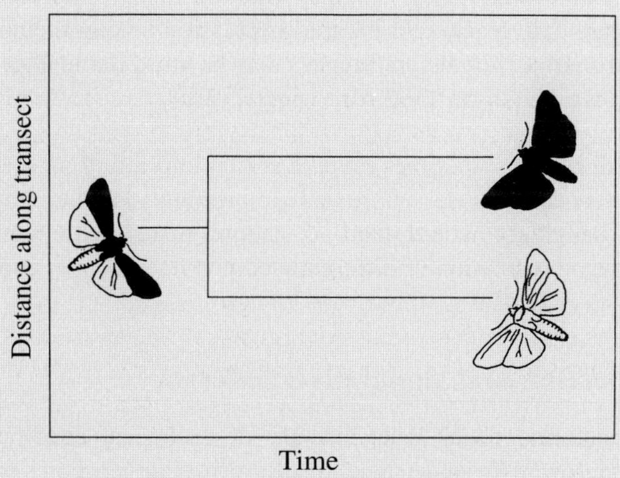

Figure 1. *A hypothetical example of allopatric speciation by divergence of sexual signals in populations that become isolated. Reproductive isolation is produced by prezygotic barriers to gene exchange caused by differentiated mate preferences and expressions of secondary sexual traits in the two populations.*

Ecological conditions and signal divergence

Both sexual signals and responses to the signals are subjected to a variety of selection pressures arising both from the need to achieve successful mating and several other aspects of the life and environment of the organism (Endler 1992). Ecological differences may affect signal evolution. It has been demonstrated that habitat structure and climatic conditions affect signal transmission in sound signals of birds and primates and that geographical variation in sound signals within a species is correlated with habitat (e.g. Chappuis 1971; Morton 1975; Hunter & Krebs 1979). For instance, in open habitats, great tits *Parus major* have songs with higher maximum frequency, more rapidly repeated notes, and a wider range of frequencies than those of forest types (Hunter & Krebs 1979). Likewise, other receivers than potential mates, such as predators and parasites, may receive sexual signals. Selection pressure from natural enemies is likely to affect signal evolution. Different allopatric populations may be subjected to different levels, or different species of such enemies, and may diverge in sexual signals as a result. Hence, geographically separated populations can diverge in sexual signals because natural selection affects the signal according to local ecological and environmental conditions. This divergence may be sufficient to provide a complete barrier to gene exchange between the populations.

However, for prezygotic barriers to develop, not only the shape of the signal must diverge, but also the corresponding mate preference. For instance, natural selection could reduce the expression of a male trait in the more risky environment when there are differences in levels of predation among habitats. Females preferring less conspicuous males need not accompany this, and if so, prezygotic isolation would not increase. Likewise, different colours may be conspicuous in habitats with different light regimes. A female preference for conspicuously coloured males would then result in divergence of male colour in the different habitats. Again, prezygotic isolation need not result, because in secondary contact, females of the differentiated populations would prefer the same males (that is, the most conspicuously coloured ones in that particular habitat). Divergence in mate preferences of the choosing sex may thus be more important in producing prezygotic isolation than a divergence in the trait expression among members of the sex that is being chosen. Moreover, female preferences may be more stable than the expression of the male traits (but see Magurran 1998 for a special case).

On the other hand, mate recognition may sometimes involve sexual imprinting on sexual signals (Price 1998). For instance, because of sexual imprinting on the sexual characteristics of their father, daughters would tend to choose mates with similar characteristics. In that case, prezygotic isolation of differentiated populations may evolve even without evolutionary changes of female preferences.

Sexual selection and signal divergence

Divergence of sexual signals may also occur independently of ecological conditions. Understanding sexual selection is relevant for understanding allopatric divergence in sexual signals. In particular, how stable are female mate preferences in this process? This is an important question. Knowledge about the relative stability of sexual signals vs. traits associated with ecological adaptations (and those affecting hybrid fertility) can provide information about the relative importance of ecological speciation vs. speciation by divergence of sexual signals. Comparative evidence suggests that male secondary sexual characteristics often diverge at a faster rate than other morphological traits (e.g. Joseph & Moritz 1993; Arnquist 1998). Similar evidence for mate preferences does not exist, and as argued above, there may not necessarily be any correspondence between divergence of these traits.

According to some models of sexual selection, the cues favoured by mate choice may be arbitrary and hence, likely to diverge in isolated populations independently of ecological factors. These include the suggestion that females may prefer more conspicuous, novel, or even 'supernormal' traits because such traits enable females to find mates faster (e.g. Hinde 1970; West-Eberhard 1979). Arbitrary traits may cause the evolution of different female preferences and male traits in different populations, for instance by a Fisherian process (Fisher 1930; Lande 1981; Iwasa & Pomiankowski 1995; Pomiankowski & Iwasa 1998).

Models of Fisherian runaway sexual selection suggest that this process may easily cause speciation. Lande (1982) showed that such sexual selection may cause speciation by premating isolation even in a continuous line of populations (a cline), because sexual selection may take different routes in different parts of the species' geographical distribution when there is geographically variable viability selection on male secondary sexual characters. Cyclic evolution in which the preference and the male character evolve in alternating directions over generations is one potential result when there is an ecological cost associated with female choice. This may cause divergence in mate preferences and male traits in different populations (Iwasa & Pomiankowski 1995; Pomiankowski & Iwasa 1998). If there is any genetic variation in female responses to several different male characters, different characters or combination of characters may evolve depending on initial genetic conditions, and again differentiation of sexual signals could be the result (Pomiankowski & Iwasa 1993). Thus, runaway sexual selection can follow different routes in different populations. I think, however, that some of these models over estimate the power of runaway selection in causing divergence in mate preferences in natural populations. This is because they assume that escalation of preferences is equally likely to occur in the positive (e.g. for longer tail) as in the negative (e.g. for shorter tail) direction. Apparently, however, sexual selection usually escalates male traits (and thus, presumably female preferences) in the positive direction: brighter, longer, louder etc. (Andersson 1994). Other models of sexual selection may not be equally powerful in causing divergence of sexual selection (Iwasa & Pomiankowski 1995).

Ryan (1990, 1994) and Basolo (1990, 1995) suggested that females might have sensory biases residing in the organisation of the sense organs and the nervous system toward certain stimuli. Characters favoured because of a sensory bias are also in a sense arbitrary and may thus be unstable. In some cases, however, a sensory bias may represent a constraint on signal evolution. Closely related populations may have similar sensory biases and as a result, their sexual signals (or rather mate preferences) would not diverge (see also Basolo 1990).

Sexual signals closely linked to fitness, such as those assumed in indicator models of sexual selection (Zahavi 1977; Hamilton & Zuk 1982; Pomiankowski 1988) are probably more stable than those assumed in runaway models of sexual selection on arbitrary traits (e.g. Iwasa & Pomiankowski 1995). However, even such indicator traits can be relatively unstable in some cases. For instance, a specific indicator trait may after some time lose its information content concerning viability and be replaced by another indicator trait (e.g. Hasson 1990). Alternatively, different indicator traits may differ in information content and in detectability. A less indicative male trait can displace a more indicative one if its detectability is sufficiently high. Differences between habitats in detectability of male traits can thus trigger a change in the female preference (Schluter & Price 1993).

Stability of sexual signals

Stability of mate preferences has rarely been tested in natural populations. Hill (1994) found that female preference for male colour ornaments in the housefinch *Carpodacus mexicanus* did not vary geographically although the males differed in coloration between different localities. Hence, in this case it seems that female preferences are more stable than the expression of the male trait (see also Price & Burley 1994; Basolo 1995; Møller 1995). Future work focusing on stability of sexual signals, and in particular the stability of mate preferences, may be important for our understanding of speciation.

Stability of sexual signals may also depend upon the potential for changing the signal. Rate of speciation may depend upon sensory abilities and potential for variation in the sexual signals. In some taxa, rate of speciation appears to be related to potential for variation in such traits. For example, the most species rich lineages of anurans have inner ears that can detect a wider range of frequencies than other lineages, and hence, mating calls can diverge over a wider range of frequencies, making speciation more likely (Ryan 1986, 1990). Likewise, bird taxa with advanced syringes (oscines, furnari and tyranni), capable of producing a higher variety of sounds than other bird taxa, are the most species rich ones (Raikow 1986, 1987).

Empirical evidence for allopatric divergence of sexual signals

Numerous lines of evidence suggest that sexual selection is important in speciation. Taxa in which sexual selection appears to be intense and diverse often have high species richness. Some famous examples are the large number of African cichlid species. The species are distinguishable almost exclusively by male colour pattern and females have been shown to choose mates on this basis (Greenwood 1974; Dominey 1984; McKaye et al. 1984). The large number of *Drosophila* species in the Hawaiian archipelago, with conspicuous differences mainly in mating systems and sexual signals, provides another example (e.g. Carson 1978; Templeton 1979; Kaneshiro & Boake 1987; Hoy et al. 1988; Ahern & Templeton 1989; Carson et al. 1989). Groups of birds that are highly dimorphic in secondary sexual characters, such as hummingbirds (Trochilidae), pheasants (Phasianidae), ducks (Anatidae) and birds of paradise (Paradisaeidae) also typically contain many species. In some cases it has been shown that species within these groups of animals are genetically very similar and that the species are interfertile although they rarely hybridise in the wild. These findings suggest that the speciation events may occur rapidly by sexual selection in these species, and that sexual selection provides more important barriers to gene exchange than postzygotic barriers. Comparative studies on species diversity in relation to intensity of sexual selection (measured as degree of sexual dimorphism or type of mating system) support the suggestion that sexual selection promotes speciation (Barraclough et al. 1995; Mitra et al. 1996; Grant & Grant 1997; Møller & Cuervo 1998).

Evidence as presented above suggests that sexual selection is important in speciation. However, these data do not necessarily support the allopatric divergence hypothesis. Sympatric character displacement or reinforcement are alternative explanations that often cannot be ruled out. Also, some models suggest that sexual selection may cause speciation even without or with only limited geographical isolation (e.g. Lande 1982; Turner & Burrows 1995). More direct evidence for allopatric divergence exists in some cases, however. For instance, species-specific courtship displays and assortative mating in the laboratory of allopatric, congeneric species has been demonstrated in a few cases (e.g. Pinto 1980; Nevo & Capranica 1985). The problem of separating the relative roles of speciation by sexual selection from ecological divergence is also noticeable. Differentiation of sexual signals could sometimes be a secondary process, e.g. a response to selection against unfit hybrids adapted to different ecological conditions. Possibly, both sexual and ecological divergence are important processes and the relative importance of each may vary in different cases.

To conclude this section, sexual selection is most probably highly important in causing divergence in sexual signals, thereby producing prezygotic barriers to gene exchange. It is likely that geographically separated populations would diverge in sexual signals by processes of sexual selection and by natural selection acting differentially on the signals in different ecological environments. However, most evidence is of an indirect nature and it has rarely been demonstrated conclusively that speciation has been accomplished by allopatric divergence of sexual signals. Finally, a general problem is that studies on sexual selection have mainly focused on variation in secondary sexual characters rather than variation in mate preferences. As argued above, mate preferences and the traits they affect may not diverge in parallel. Thus, future studies focusing on the choosing sex may provide important insight into processes of speciation.

Sympatric divergence of sexual signals - reinforcement

When two differentiated populations come into secondary contact they may have diverged to such an extent that offspring between them would have reduced, but still non-zero fitness. It is frequently observed that offspring of such mixed ancestries have reduced fertility and/or viability (see Harrison 1990 for a review). Yet, the sexual signals of the two populations may be sufficiently similar for heterospecific mating to occur. The theory of reinforcement suggests, in such cases, that natural selection against the production of unfit hybrids would reinforce prezygotic isolation eventually completing the speciation process that started in allopatry (Dobzhansky 1940, 1941). This process could favour any character that increases the probability of positive assortative mating (e.g. differentiation in timing of breeding and habitat separation), but here I am concerned with the potential for a sympatric divergence in sexual signals that would facilitate species recognition (Fig. 2). This effect of reinforcement is known as reproductive character displacement (Brown & Wilson 1956; Grant 1972). Theoretically, such divergence of sexual signals is not problematic if hybrids have zero fitness, i.e., when the postzygotic barrier is complete. In this case, no recombination could possibly occur between the two species and natural selection against reproductive wastage could favour a divergence of the sexual signals of the two species that would lead to perfect, positive assortative mating. Butlin (1987) restricted the use of the term 'reproductive character displacement' to the case of complete postzygotic isolation to contrast it with the theoretically more difficult case when gene flow between species is occurring (i.e., incomplete postzygotic barriers). However, I will use the traditional definition in which reproductive character displacement would be the result of both complete and incomplete postzygotic isolation. Why is incomplete postzygotic isolation problematic? Moreover, is it a real problem or only related to the assumption of the models? Let us look at the theory.

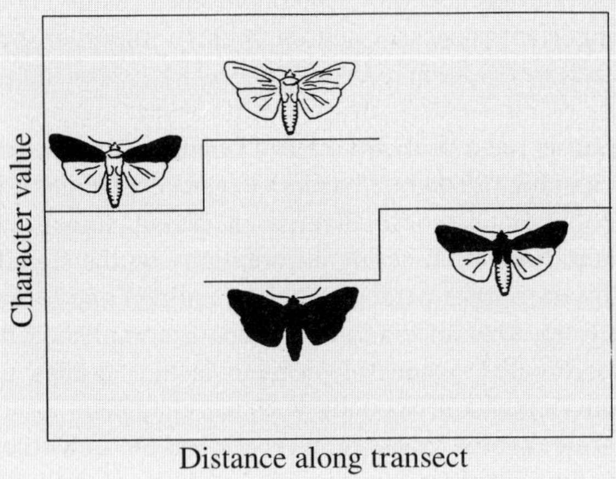

Figure 2. A hypothetical example of reproductive character displacement. In sympatry, natural selection against the production of unfit hybrids favours a divergence of sexual signals that would help species recognition. This would lead to prezygotic isolation being increased.

Theoretical objections against reinforcement

Most models of reinforcement envisage a situation in which at least one locus yields low fitness when heterozygous (i.e., hybrid disadvantage), and then envisage the eventual spread of another gene (or genes) that promotes prezygotic isolation in two semispecies (e.g. Crosby 1970; Caisse & Antonovics 1978; Felsenstein 1981; Spencer et al. 1986; Liou & Price 1994). In particular, Felsenstein (1981) and Sanderson (1989) stressed the phenomenon that recombination would oppose selection for alleles that confer prezygotic isolation. Thus, because hybrids by definition are at least partially fertile, recombination would tend to break down the association between alleles affecting hybrid fitness and alleles promoting prezygotic isolation. As a result, isolation-alleles may become randomly combined with the alleles that reduce the fitness of hybrids and the isolation-alleles will no longer confer any selective advantage to the bearer. The opposing effect of recombination becomes progressively stronger the higher the fitness of the hybrids and the higher the frequency of heterospecific mating (i.e., more opportunity for recombination to break the association between fitness and assortative mating). Similarly, reinforcement is a more likely outcome if: (1) the alleles for prezygotic isolation and those affecting hybrid fitness are tightly linked, (2) if selection against hybrids is strong, and (3) some assortative mating occurs even at the formation of the contact zone. The problem of recombination demonstrates that reinforcement need not result in a contact zone even though hybrids have reduced fitness and appropriate genes causing assortative mating are present in the population(s). However, there is considerable disagreement on how restricted the conditions for reinforcement to evolve are likely to be. Knowledge about the genetic basis for hybrid inviability and infertility, and also the genetic basis for the genes promoting prezygotic isolation is clearly needed to resolve this controversy (Civetta & Singh 1998).

The second main objection against reinforcement is the effect of gene flow from respective allopatric populations (Butlin 1987, 1989; Sanderson 1989; Servedio & Kirkpatrick 1997). Gene flow opposes reinforcement because the selection for increased assortative mating occurs only where the two populations meet. As a result of selection against hybridisation, which is likely to be of the hard type (i.e., reducing population growth), mean fitness would tend to be lower in sympatry than in allopatry. Hence, influx of genes from allopatry would dominate and would tend to dilute the effect of reinforcement. As a result, a likely outcome when two differentiated populations meet is a relatively stable and narrow contact zone maintained by selection against hybridisation within the contact zone, and gene flow from allopatry. Indeed, many natural hybrid zones appear to be maintained in this way (e.g. Harrison 1990).

Finally, several authors have pointed out that selection against hybridisation would tend to be weakened during time in a contact zone. This is for three reasons. First, selection for modifiers that reduce hybrid disadvantage may occur in parallel with selection for assortative mating. Second, backcrossing will produce a hybrid zone in which the individuals eventually are various backcrossed genotypes that may have higher fitness than first-generation hybrids. Third, as an isolation-allele increases in frequency, so that cross-mating is reduced but not eliminated, selection for further reinforcement will become lower because unfit hybrids will arise less often. Hence, full reproductive isolation may not be likely to be achieved (Butlin 1987).

All these factors limit the potential for reinforcement to evolve. It is unlikely that natural selection against hybridisation alone would be sufficient for producing complete prezygotic isolation except under some rather restricted conditions (e.g. hybrid fitness close to zero and limited gene flow from allopatric populations).

Sexual selection may reinforce reinforcement

Two studies, one modelling reinforcement (Liou & Price 1994), and one modelling a scenario relevant for reinforcement (Lande 1982), suggest that the joint effects of sexual selection and natural selection against hybridisation increase the probability that prezygotic isolation would evolve. That is, the range of conditions in which prezygotic isolation will develop is broader according to these models than in models only considering natural selection on an isolation-allele. I find it very likely that processes of sexual selection will interplay with reinforcement, because the cues used in mate recognition are those preferred by females and affecting male secondary sexual characters. Hence, in a situation where postzygotic (and prezygotic) barriers to gene exchange are incomplete among populations that come into secondary contact, the appropriate scenario may be to consider selection mainly on the mate-choosing sex (e.g. females). If there is variation between females in mate preferences and some variant is less likely to mate with a male of the other species than another variant, natural selection would favour the former variant (i.e. reinforcement). The corresponding male character will diverge from those of the other species, as a response to selection on females. Reinforcement may then set the stage for sexual selection to take a different route in one or both of the species within the contact zone. Sexual selection theory suggests that this process may be self-rein-forcing because females are selecting not only for the male trait, but also, indirectly through a genetic correlation between the male trait and the preference, for more intense mating preferences (Fisher 1930; Lande 1981, 1982; Pomiankowski & Iwasa 1998). Hence, even though the initial benefit of avoiding hybridisation decreases over time, as hybridisation becomes more rare, self-reinforcing sexual selection could nevertheless complete the speciation process. Moreover, in this case the selection for divergence of sexual signals needs not be restricted to the contact zone but could spread from sympatry into allopatry. This would reduce the problem of dilution effects of gene flow from allopatry.

Indicator-models of sexual selection (e.g. Pomiankowski 1988) may be relevant for understanding the initial stages of reproductive character displacement in zones of secondary contact, viz. the more species-specific male traits would be viability signals parallel to those signalling high genetic quality in such models (Sætre et al. 1997b). As hybridisation becomes more rare, the effect of viability selection will be reduced, but self-reinforcing sexual selection (i.e., the runaway process) could nevertheless complete speciation. As in the case of allopatric divergence, knowledge about processes of sexual selection is essential for speciation theory also in the case of sympatric divergence.

Evidence for character displacement and reinforcement

Several reviews exist on the empirical evidence for reinforcement occurring or not occurring in hybrid zones (e.g. Barton & Hewitt 1985, 1989; Butlin 1987; Harrison 1990, 1993). The conclusions of these are that evidence for reinforcement is absent or equivocal in the majority of hybrid zones examined. This result may suggest that reinforcement is not a common outcome of interactions in hybrid zones. On the other hand, current hybrid zones may be over-represented by those in which the necessary conditions for reinforcement to evolve are not met (see above). For instance, hybridisation would have ceased to occur in instances in which reinforcement has completed the speciation process. These cases would not be included in the comparisons above. Moreover, because reinforcement may be more likely to occur when linked to sexual selection, traces of reproductive character displacement would tend to disappear as the divergent phenotypes spread into allopatry. Hence, lack of evidence for ongoing reinforcement in current hybrid zones does not necessarily imply that it is an unimportant process of speciation. There are, however, some rather convincing examples of reinforcement in the literature (e.g. Waage 1979; Wasserman & Koepfer 1979; Coyne & Orr 1989; Johannesson et al. 1993, 1995; Noor 1995; Sætre et al. 1997b). I will use my own study as an example.

In the European *Ficedula* flycatchers there is evidence for reproductive character displacement, particularly in male plumage coloration (Sætre et al. 1997b; Fig. 3), but perhaps also in male song (Wallin 1986; Sætre 1998; G.-P. Sætre, personal observations). In various allopatric populations the average male has a black-and-white plumage and there is relatively little variation in male colour between the different populations. However, in Central and Eastern Europe two flycatcher forms live, mainly in sympatry, that differ considerably in plumage colour: the brown form of the pied flycatcher *Ficedula hypoleuca* and the collared flycatcher *F. albicollis*. Males of the former species are dull brown, whereas the latter males are black with larger patches of white on the forehead, neck and wings than any allopatric flycatcher has. Phylogenetic reconstruction based on mitochondrial DNA suggests that the allopatric phenotype is the ancestral state, suggesting that the sympatric forms have diverged phenotypically in sympatry (Sætre et al. 1997b). Field data from the hybrid zone shows that hybridisation occurs, but at a much lower frequency than expected from random mating. The postzygotic barrier against gene exchange is significant but not complete, viz. hatching success of first-generation hybrids is much lower than that of 'pure' individuals. Hence, conditions for reinforcement to occur are present and assortative mating predominates, suggesting that reinforcement may operate. In mate choice trials in aviaries we showed that divergent, typically sympatric, colours facilitate species recognition (Sætre et al. 1997b). Females of both species only chose conspecific males when choosing between a con- and a heterospecific male

Figure 3. *Character displacement in flycatchers. Allopatric populations have similar plumage colours, whereas the two sympatric forms in Central and Eastern Europe are highly differentiated in colour.*

with divergent, sympatric colours. However, when females chose between con- and heterospecific males painted to resemble allopatric flycatchers a significant proportion of the females chose the heterospecific male. Finally, also in aviary experiments, we demonstrated that the sympatric colour divergence probably is driven by female mate choice. In sympatry, female pied flycatchers prefer brown to black conspecific males, whereas female collared flycatchers prefer the conspecific male with the larger white patches (i.e., character states that would increase the difference between the males of the two species; Sætre et al. 1997b). Interestingly, mate preferences of female pied flycatchers are opposite in allopatry (Sætre et al. 1994).

A further interesting aspect of the flycatcher species complex is that there exist other hybrid zones with a completely different structure, i.e., the island hybrid zones of the Baltic Isles (e.g. Alatalo et al. 1990, 1994). On these islands there is less evidence for reproductive character displacement, viz. male pied flycatchers are mainly black, similar to those in surrounding allopatric areas. It is possible that the difference between the hybrid zones is due to differences in patterns of gene flow. That is, on the islands gene flow may be more asymmetric than on the continent; collared flycatchers being isolated on the island whereas there is influx of pied flycatchers from the surrounding mainland (Fig. 4). Theory suggests that asymmetric gene flow limits the potential for reinforcement evolution (Servedio & Kirkpatrick 1997). Consistent with the suggestion that reinforcement may operate to a lesser extent on the island hybrid zones, hybrids there are more abundant and more fertile than on the continent (Sætre et al. 1999). The latter may be explained by more extensive backcrossing occurring on the islands, breaking down the hybrid disadvantage. Hence, hybrid zones of flycatchers may exemplify both a clear case of reinforcement (i.e., the continental, clinal hybrid zone), and a case in which the conditions for reinforcement possibly are not met (i.e., on the islands). Further studies

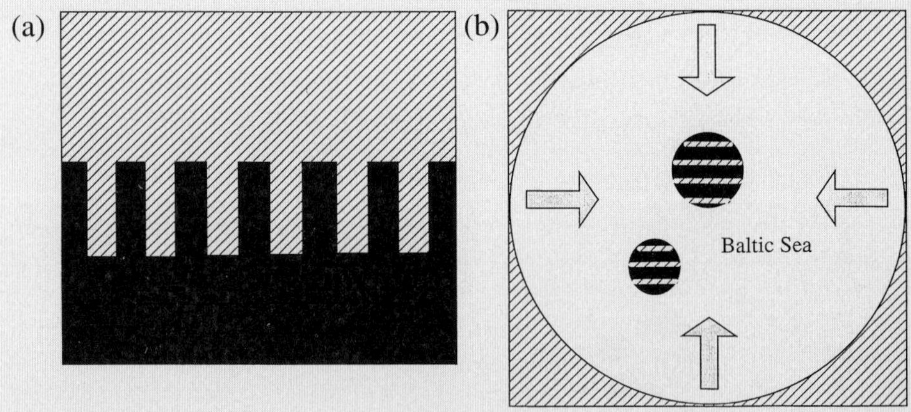

Figure 4. *Structure of (**a**) the clinal hybrid zone in Central and Eastern Europe and (**b**) the island hybrid zones of the Baltic Isles. Black area: geographical distribution of the collared flycatcher, hatched area: distribution of the pied flycatcher. Patterns of distribution suggest that gene flow would be more asymmetric on the islands. Arrows indicate the main direction of gene flow of pied flycatchers, i.e. from the mainland to the Baltic Isles (small circles).*

making use of molecular markers may shed light on the differences between these hybrid zones. If indeed gene flow can be shown to prevent reinforcement on the islands, it would be a remarkable demonstration of the sensitivity of reinforcement to opposing factors. Besides case studies, such as the one cited above, one comparative study provides important evidence for reinforcement (Coyne & Orr 1989). These authors compared prezygotic and postzygotic isolation between pairs of *Drosophila* species. Some of the species pairs live sympatrically in at least part of their geographic range, while others are allopatric. Forty-two independent pairs of species, for which degree of prezygotic and postzygotic isolation, genetic distance and geographic distribution had been measured, were compared. Both prezygotic and postzygotic isolation increased with genetic distance between the species in the pairs. However, most interestingly, prezygotic isolation was found to be higher in sympatric than in allopatric species. No such difference was apparent for postzygotic isolation. The latter result suggests that selection acts in sympatry to reinforce the prezygotic isolation of closely related species (see also Civetta & Singh 1998).

In conclusion, reinforcement of prezygotic isolation remains a controversial theory in evolutionary biology. Population genetic and quantitative genetical models suggest that reinforcement can evolve under some conditions, but the various models vary concerning how restricted these conditions are. Reproductive character displacement and prezygotic isolation will evolve more easily when natural selection against unfit hybrids works in concert with self-reinforcing sexual selection. Empirical evidence for reinforcement exists but is apparently absent in many zones of secondary contact in which it could be expected to be found. Hence, it is still not known how common this mode of speciation is in nature. Both the case study and the comparative study cited above support the reinforcement hypothesis, but are also consistent with a joint effect of reinforcement and sexual selection. Moreover, they show that sympatric interactions between differentiated populations at least sometimes may speed up a divergence of sexual signals and speciation.

Sexual signals and ecological speciation

Natural selection moves populations to local peaks in the adaptive landscape (Wright 1932). According to this metaphor of adaptive landscapes, populations may be in an equilibrium state at an adaptive peak, unable to cross the valley and reach another, and perhaps higher peak by natural selection alone. This is because natural selection can only increase fitness, and a peak shift would have to involve a phase of decreasing fitness. A peak shift may be accomplished if the gene frequencies of the population drift away from equilibrium by stochastic factors (e.g. Wright 1968-1978; Lande 1976). However, sexual selection may have a similar potential to interplay with natural selection in the evolution of adaptations and in ecological speciation. Darwin (1871) invoked the theory of sexual selection to explain evolution of sexually dimorphic characters that seemed to hinder survival of the possessor, the resolution being that such characters promote mating success. Due to the dichotomy between the effect of sexual selection on mating success and effects of natural selection on survivorship, the possibility that sexual selection might

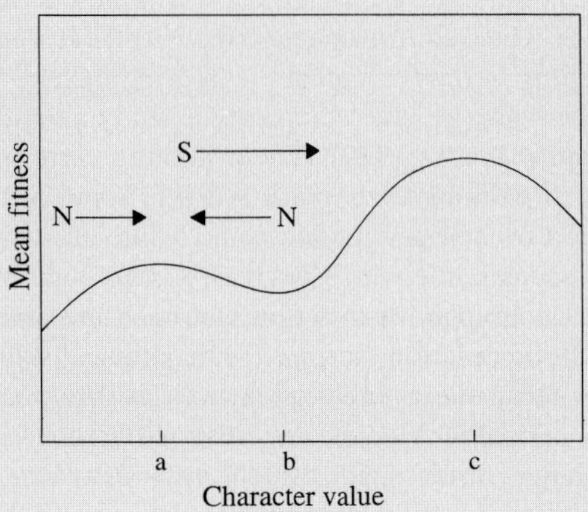

Figure 5. *Hypothetical fitness surface in terms of probability of survival, with respect to a character influenced by natural (N) and sexual selection (S). See text for further explanation.*

also be involved in the evolution of traits that enhance survivorship has often been neglected. I will consider one hypothetical scenario for ecological speciation by sexual selection (Fig. 5).

Consider a fitness surface in terms of probability of survival for one character (body size or whatever you like). Stabilising natural selection maintains the population mean at a local fitness peak (point a in Fig. 5). Then we introduce a force of sexual selection acting on, say males, driving the trait away from the local optimum under natural selection. According to sexual selection theory (e.g. Lande 1980) the homologous trait in females may evolve almost in parallel with the male trait during some period because of a genetic correlation between the sexes, whereafter a phase of sexual dimorphism may evolve gradually. The important point is that the evolution of the male trait away from the local optimum does not need continue to reduce male fitness as it evolves away from the original population mean. Sexually selected traits, such as large body size and weapons, have sometimes been shown to decrease male fitness (Andersson 1994). However, at some threshold value (point b in Fig. 5) the fitness of males may eventually increase. This may be for several reasons; for instance, sexual selection on increased male body size may eventually lead to a situation in which a previously unexplored food resource can be utilised by the males. It may also be that at some point in its evolution a costly sexually selected trait becomes useful in avoiding predation. For instance, the enlarged claw of the male fiddler crab *Uca pugilator* seems to make males less vulnerable to avian predators than females (Bildstein et al. 1989). In such cases natural selection would reinforce the effect of sexual selection, accelerating the evolution of the trait away from the origin. If there is still a substantial genetic correlation between the sexes, the homologous trait in females would drift again, along with selection on the male trait, possibly beyond the threshold value and thus, the population may reach a new adaptive optimum (point c in Fig. 5). Otherwise, males and females may reach different optima depending on the strength of the selective forces operating and the magnitude of genetic correlation between the sexes. Other scenarios yielding the same result are also possible. A simpler case would be if sexual selection acts in the same direction on both males and females.

Lande and Kirkpatrick (1988) have treated a scenario rather similar to that above in a formal model. They found that female choice, based on ecologically important characters, could initiate such a sudden shift into a new ecological niche. Pre- and postzygotic isolation may be produced rapidly, as well as ecological separation between populations that can then co-exist in the same area as distinct species. This mode of speciation is therefore theoretically possible. However, I have not seen any discussion of its potential relevance in nature. Hence, to my knowledge, no data supports (nor contradicts) the hypothesis and it becomes difficult to discuss its importance. I think it is a wonderful idea and I hope biologists eventually will address it. Evidence exists that sexual selection may result in evolution away from an ecological optimum (Andersson 1994). The important question is therefore, how likely is it that sexually selected traits would become ecologically adaptive? I have already mentioned body size and weapons as potential candidates, and other examples exist too. For instance, Grant (1986) showed that Darwin's finches *Geospiza* spp. choose mates on general morphology such as beak size. Even ornaments and ritual displays may at some stage become adaptive. I have no problems imagining that avian flight could initially have evolved by sexual selection. Indeed, contemporary flightless birds and those with poor flight abilities make frequent use of their wings in sexual displays. Female preferences for flapping and aerial displays in a Mesozoic pro-bird could, so to speak, have lifted the avian ancestor into the air.

Conclusions

Undoubtedly, sexual signals are important in producing prezygotic barriers to gene exchange, and hence, in speciation. Developments in the field of sexual selection are relevant for speciation theory, because the characters used in species recognition are those favoured by sexual selection. For instance, studies investigating geographical variation of sexual signals and the factors causing the variation would be important for our understanding of speciation by allopatric divergence of sexual signals. Likewise, studies addressing the effects of both genetics and behaviour in hybrid zones are important for understanding the complex dynamics of such zones and the potential importance of speciation by reinforcement. At least three still unanswered questions have been identified in this review. First, how stable are female mate preferences? Second, under what circumstances would female preferences tend to diverge (e.g. in allopatry by natural selection or by sexual selection, or as a response to sympatric interactions between differentiated populations)? The answers to these two closely related questions would provide information about the relative role of sexual selection and ecological adaptation in speciation, and the relative importance of allopatric and sympatric divergence of sexual signals. Finally, how likely is it that a sexually selected character would become ecologically adaptive? This is probably the key question as to whether sexual selection may be important in causing ecological speciation.

Acknowledgements

I would like to thank Alexandra Basolo and Innes Cuthill for comments on the manuscript. Financial support was received from the Norwegian Research Council.

References

Ahearn, J. N. & Templeton, A. R. 1989. Interspecific hybrids of *Drosophila heteroneura* and *D. silvestris*. I. Courtship success. *Evolution,* **43**, 347-361.

Alatalo, R. V., Eriksson, D., Gustafsson, L. & Sundberg, A. 1990. Hybridization between pied and collared flycatchers - sexual selection and speciation theory. *Journal of Evolutionary Biology*, **3**, 375-389.

Alatalo, R. V., Gustafsson, L. & Lundberg, A. 1994. Male coloration and species recognition in sympatric flycatchers. *Proceedings of the Royal Society of London, Series B*, **256**, 113-118.

Andersson, M. 1994. *Sexual Selection.* Princeton, NJ: Princeton University Press.

Arnqvist, G. 1998. Comparative evidence for the evolution of genitalia by sexual selection. *Nature*, **393**, 784-786.

Barraclough, T. G., Harvey, P. H. & Nee, S. 1995. Sexual selection and taxonomic diversity in passerine birds. *Proceedings of the Royal Society of London, Series B*, **259**, 211-215.

Barton, N. H. & Hewitt, G. M. 1985. Analysis of hybrid zones. *Annual Review of Ecology and Systematics,* **16**, 113-148.

Barton, N. H. & Hewitt, G. M. 1989. Adaptation, speciation and hybrid zones. *Nature*, **341**, 497-503.

Basolo, A. L. 1990. Female preferences predates the evolution of the sword in swordtail fish. *Science*, **250**, 808-810.

Basolo, A. L. 1995. Phylogenetic evidence for the role of a pre-existing bias in sexual selection. *Proceedings of the Royal Society of London, Series B*, **259**, 307-311.

Bildstein, K. L., McDowell, S. G. & Brisbin, I. L. 1989. Consequences of sexual dimorphism in sand fiddler crabs, *Uca pugilator*: differential vulnerability to avian predation. *Animal Behaviour*, **37**, 133-139.

Brown, W. L. Jr. & Wilson, E. O. 1956. Character displacement. *Systematic Zoology*, **5**, 49-64.

Butlin, R. 1987. Speciation by reinforcement. *Trends in Ecology and Evolution*, **2**, 8-13.

Butlin, R. 1989. Reinforcement of premating isolation. In: *Speciation and its Consequences* (Ed. by D. Otte & J.A. Endler), pp. 158-179. Sunderland, MA: Sinauer.

Cade, W. H. 1979. The evolution of alternative male reproductive strategies in field crickets. In: *Sexual Selection and Reproductive Competition in Insects* (Ed. by M. Blum & N.A. Blum), pp. 343-379. London: Academic Press.

Caisse, M. & Antonovics, J. 1978. Evolution in closely adjacent plant populations. IX. Evolution of reproductive isolation in clinal populations. *Heredity*, **40**, 371-384.

Carson, H. L. 1978. Speciation and sexual selection in Hawaiian *Drosophila*. In: *Ecological Genetics: The Interface* (Ed. by P.F. Brussard), pp. 93-107. New York: Springer Verlag.

Carson, H. L., Kaneshiro, K. Y. & Val, F. C. 1989. Natural hybridisation between the sympatric Hawaiian species *Drosophila silvestris* and *Drosophila oneura*. *Evolution*, **43**, 190-203.

Chappuis, C. 1971. Un example de l'influence du milieu sur les èmissions vocales des oiseaux: l'evolution des chants en forêt èquitoriale. *La Terre et la Vie*, **25**, 183-202.

Civetta, A. & Singh, R. S. 1998. Sex-related genes, directional sexual selection, and speciation. *Molecular Biology and Evolution*, **15**, 901-909.

Coyne, J. A. & Orr H. A. 1989. Patterns of speciation in *Drosophila*. *Evolution*, **43**, 362-381.

Crosby, J. L. 1970. The evolution of genetic discontinuity: computer models of the selection of barriers to interbreeding between subspecies. *Heredity*, **25**, 253-297.

Darwin, C. 1871. *The Descent of Man, and Selection in Relation to Sex*. London: John Murray.

Dobzhansky, T. 1940. Speciation as a stage in evolutionary divergence. *American Naturalist*, **74**, 312-321.

Dobzhansky, T. 1941. *Genetics and the Origin of Species*. New York: Columbia University Press.

Dominey, W. J. 1984. Effect of sexual selection and life history on speciation: species flocks in African cichlids and Hawaiian *Drosophila*. In: *Evolution of Fish Species Flocks* (Ed. by A.A. Echelle & I. Kornfield), pp. 231-249. Orono: University of Maine at Orono Press.

Endler, J. A. 1992. Signals, signal condition, and the direction of evolution. *American Naturalist*, **139**, 125-153.

Felsenstein, J. 1981. Scepticism toward Santa Rosalia or why are there so few kinds of animals. *Evolution,* **35**, 124-138.

Fisher, R. A. 1930. *The Genetical Theory of Natural Selection*. Oxford: Clarendon Press.

Grant, P. R. 1972. Convergent and divergent character displacement. *Biological Journal of the Linnean Society*, **4**, 39-68.

Grant, P. R. 1986. *Ecology and Evolution of Darwin's Finches*. Princeton, NJ: Princeton University Press.

Grant, P. R. & Grant, B. R. 1997. Genetics and the origin of bird species. *Proceedings of the National Academy of Sciences, USA*, **94**, 7768-7775.

Greenwood, P. J. 1974. The cichlid fishes of Lake Victoria, East Africa: the biology and evolution of a species flock. *Bulletin of the British Museum of Natural History (Zoology),* **Suppl. 6**, 1-134.

Hamilton, W. D. & Zuk, M. 1982. Heritable true fitness and bright birds: a role for parasites? *Science*, **218**, 384-387.

Harrison, R. G. 1990. Hybrid zones: windows on evolutionary process. *Oxford Surveys in Evolutionary Biology*, **7**, 69-128.

Harrison, R. G. (Ed.) 1993. *Hybrid Zones and the Evolutionary Process*. New York: Oxford University Press.

Hasson, O. 1990. The role of amplifiers in sexual selection: an integration of the amplifying and the Fisherian mechanisms. *Evolutionary Ecology*, **4**, 277-289.

Hill, G. E. 1994. Geographic variation in male ornamentation and female mate preferences in the house finch: A comparative test of models of sexual selection. *Behavioural Ecology*, **5**, 64-73.

Hinde, R. A. 1970. *Animal Behaviour: A Synthesis of Ethology and Comparative Psychology,* 2nd edn. New York: McGraw-Hill.

Hoy, R. R., Hoikkala, A. & Kaneshiro, K. 1988. Hawaiian courtship songs: evolutionary innovation in communication signals of *Drosophila*. *Science*, **240**, 217-219.

Hunter, M. L. & Krebs, J. R. 1979. Geographical variation in the song of the great tit (*Parus major*) in relation to ecological factors. *Journal of Animal Ecology*, **48**, 759-785.

Iwasa, Y. & Pomiankowski, A. 1995. Continual change in mate preferences. *Nature*, **377**, 420-422.

Johannesson, K., Johannesson, B. & Rolàn-Alvarez, E. 1993. Morphological differentiation and genetic cohesiveness over a microenvironmental gradient in the marine snail *Littorina saxatilis*. *Evolution*, **47**, 1770-1787.

Johannesson, K., Rolàn-Alvarez, E. & Ekendahl, A. 1995. Incipient reproductive isolation between two sympatric morphs of the intertidal snail *Littorina saxatilis*. *Evolution*, **49**, 1180-1190.

Joseph, L. & Moritz, C. 1993. Phylogeny and historical aspects of the ecology of eastern Australian scrubwrens *Sericornis* spp. - evidence from mitochondrial DNA. *Molecular Ecology*, **2**, 161-170.

Kaneshiro, K. Y. & Boake, C. R. B. 1987. Sexual selection and speciation: issues raised by Hawaiian *Drosophila*. *Trends in Ecology and Evolution*, **2**, 297-212.

Lande, R. 1976. Natural selection and random genetic drift in phenotypic evolution. *Evolution*, **30**, 314-334.

Lande, R. 1980. Sexual dimorphism, sexual selection and adaptation in polygenic characters. *Evolution*, **34**, 292-305.

Lande, R. 1981. Models of speciation by sexual selection on polygenic traits. *Proceedings of the National Academy of Sciences, USA*, **78**, 3721-3725.

Lande, R. 1982. Rapid origin of sexual isolation and character divergence in a cline. *Evolution*, **36**, 213-223.

Lande, R. & Kirkpatrick, M. 1988. Ecological speciation by sexual selection. *Journal of Theoretical Biology*, **133**, 85-98.

Liou, L. W. & Price, T. D. 1994. Speciation by reinforcement of premating isolation. *Evolution*, **48**, 1451-1459.

Magurran, A. E. 1998. Population differentiation without speciation. *Philosophical Transactions of the Royal Society of London, Series B*, **353**, 275-286.

Mayr, E. 1942. *Systematics and the Origin of Species*. New York: Columbia University Press.

McKaye, K. R., Kocher, T., Reinthal, P., Harrison, R. & Kornfield, I. 1984. Genetic evidence for allopatric and sympatric differentiation among color morphs of a Lake Malawi cichlid fish. *Evolution*, **38**, 215-219.

Mitra, S., Landel, H. & Pruett-Jones, S. 1996. Species richness covaries with mating system in birds. *Auk*, **113**, 544-551.

Møller, A. P. 1995. Sexual selection in the barn swallow (*Hirundo rustica*). 5. Geographic variation in ornament size. *Journal of Evolutionary Biology*, **8**, 3-19.

Møller, A. P. & Cuervo, J. J. 1998. Speciation and feather ornamentation in birds. *Evolution*, **52**, 859-869.

Morton, E. S. 1975. Ecological sources of selection on avian sounds. *American Naturalist*, **109**, 17-34.

Nevo, E. & Capranica, R. R. 1985. Evolutionary origin of ethological reproductive isolation in cricket frogs, *Acris. Evolutionary Biology*, **19**, 147-214.

Noor, M. A. 1995. Speciation driven by natural selection in *Drosophila. Nature*, **375**, 674-675.

Paterson, H. E. H. 1985. The recognition concept of species. In: *Species and Speciation* (Ed. by E.S. Vrba), pp. 21-29. Pretoria: Transvaal Museum Monograph No. 4.

Pinto, J. D. 1980. Behavior and taxonomy of the *Epicauta maculata* group (Coleoptera: Meloidae). *University of California Publications in Entomology*, **89**, 1-111.

Pomiankowski, A. 1988. The evolution of female mate preferences for male genetic quality. *Oxford Surveys in Evolutionary Biology*, **5**, 136-184.

Pomiankowski, A. & Iwasa, Y. 1993. Evolution of multiple sexual preferences by Fisher's runaway process of sexual selection. *Proceedings of the Royal Society of London, Series B*, **253**, 173-181.

Pomiankowski, A. & Iwasa, Y. 1998. Runaway ornament diversity caused by Fisherian sexual selection. *Proceedings of the National Academy of Sciences, USA*, **95**, 5106-5111.

Price, D.K. & Burley, N. T. 1994. Constraints on the evolution of attractive traits: selection in male and female zebra finches. *American Naturalist*, **144**, 908-934.

Price, T. 1998. Sexual selection and natural selection in bird speciation. *Philosophical Transactions of the Royal Society of London, Series B*, **353**, 251-260.

Raikow, R. J. 1986. Why are there so many kinds of passerine birds? *Systematic Zoology*, **35**, 255-259.

Raikow, R. J. 1987. Hindlimb myology and evolution of the old world suboscine passerine birds (Acanthisittidae, Pittidae, Philepittidae, Eurylaimidae). *Ornithological Monographs*, **41**, 1-81.

Ryan, M. J. 1986. Neuroanatomy influences speciation rates among anurans. *Proceedings of the National Academy of Sciences, USA*, **83**, 1379-1382.

Ryan, M. J. 1990. Sexual selection, sensory systems, and sensory exploitation. *Oxford Surveys in Evolutionary Biology*, **7**, 157-195.

Ryan, M. J. 1994. Mechanisms underlying sexual selection. In: *Behavioral Mechanisms in Evolutionary Ecology* (Ed. by L.A. Real), pp. 190-215. Chicago: University of Chicago Press.

Sætre, G.-P., Dale, S. & Slagsvold, T. 1994. Female pied flycatchers prefer brightly coloured males. *Animal Behaviour*, **48**, 1407-1416.

Sætre, G.-P., Kràl, M. & Bures, S. 1997a. Differential species recognition abilities of males and females in a flycatcher hybrid zone. *Journal of Avian Biology*, **28**, 259-263.

Sætre, G.-P., Kràl, M., Bures, S. & Ims, R. A. 1999. Dynamics of a clinal hybrid zone and a comparison with island hybrid zones of flycatchers. *Journal of Zoology,* **247**, 53-64.

Sætre, G.-P., Moum, T., Bures, S., Kràl, M., Adamjan, M. & Moreno, J. 1997b. A sexually selected character displacement in flycatchers reinforces premating isolation. *Nature*, **387**, 589-592.

Sanderson, N. 1989. Can gene flow prevent reinforcement? *Evolution*, **43**, 1223-1235.

Schluter, D. & Nagel, L. M. 1995. Parallel speciation by natural selection. *American Naturalist*, **146,** 292-301.

Schluter, D. & Price, T. 1993. Honesty, perception and population divergence in sexually selected traits. *Proceedings of the Royal Society of London, Series B*, **253**, 117-122.

Servedio, M. R. & Kirkpatrick, M. 1997. The effects of gene flow on reinforcement. *Evolution*, **51**, 1764-1772.

Smith, T. B., Wayne, R. K., Girman, D. J. & Bruford, M. W. 1997. A role for ecotones in generating rainforest biodiversity. *Science*, **276**, 1855-1857.

Spencer, H. G., McArdle, B. H. & Lambert, D. J. 1986. A theoretical investigation of speciation by reinforcement. *American Naturalist*, **128**, 241-262.

Templeton, A. R. 1979. Once again, why 300 species of Hawaiian *Drosophila*? *Evolution*, **33**, 513-517.

Turner, G. F. & Burrows, M. T. 1995. A model of sympatric speciation by sexual selection. *Proceedings of the Royal Society of London, Series B*, **260**, 287-292.

Waage, J. K. 1979. Reproductive character displacement in *Calopteryx* (Odonata: Calopterygidae). *Evolution*, **33**, 104-116.

Wallin, L. 1986. Divergent character displacement in the song of two allo-species: the pied flycatcher *Ficedula hypoleuca*, and the collared flycatcher *Ficedula albicollis*. *Ibis*, **128**, 251-259.

Wasserman, M. & Koepfer, H. R. 1977. Character displacement for sexual isolation between *Drosophila mojavensis* and *Drosophila arizonensis*. *Evolution*, **31**, 812-823.

West-Eberhard, M. J. 1979. Sexual selection, social competition, and evolution. *Proceedings of the American Philosophical Society*, **123**, 222-234.

Wright, S. 1932. The roles of mutation, inbreeding, crossbreeding, and selection in evolution. *Proceedings of the VIth International Congress of Genetics,* **1**, 356-366.

Wright, S. 1968-1978. *Evolution and the Genetics of Populations*. 4 vols. Chicago: University of Chicago Press.

Zahavi, A. 1977. The cost of honesty (further remarks on the handicap principle). *Journal of Theoretical Biology*, **67**, 603-605.

IN: ESPMARK,Y., AMUNDSEN,T. & ROSENQVIST,G. (EDS.) 2000
ANIMAL SIGNALS:
SIGNALLING AND SIGNAL DESIGN IN ANIMAL COMMUNICATION, PP. 259-276
TAPIR ACADEMIC PRESS, TRONDHEIM, NORWAY

12

Prowess and the Resolution of Animal Fights

FELICITY A. HUNTINGFORD*, ALAN C. TAYLOR, LYNNE U. SNEDDON & FRANCIS C. NEAT

Fish Biology Group, Division of Environmental and Evolutionary Biology, IBLS, Graham Kerr Building, University of Glasgow, Glasgow G12 8QQ, UK

**E-mail: f.a.huntingford@bio.gla.ac.uk*

Abstract

In this article we discuss recent research on signalling during animal fights, concentrating on our own experimental subjects (portunid crabs and cichlid fish), but also reviewing related studies on other animal groups. We first consider the interacting effects of body size and resource value as determinants of fighting behaviour, reviewing several recent studies in which larger size does not increase the chances of victory and identifying general circumstances in which large size may be of neutral value or even detrimental. Next, we discuss the energetic effects of aggression, documenting major effects of fighting on the metabolic physiology and on levels of nutrient reserves, examining the extent to which these constitute genuine fitness costs of fighting. We then review some studies of the detailed behavioural exchanges that occur during fights (especially rapid colour changes) and discuss whether fighting animals exchange signals that accurately reflect their aggressive motivation. Finally, these examples are used to draw some general conclusions about signalling during animal fights.

Because fights in which one animal simply overpowers another are extremely uncommon, most instances of aggression involve more-or-less complex exchanges during which the appearance and/or behaviour of one animal influences what its opponent does (Huntingford & Turner 1987). In other words: fights involve communication, or the exchange of signals. Therefore, research into the reasons why animals fight as they do is relevant to the literature on signalling. In this paper we present some recent work on the behavioural biology of aggression and use this material to highlight areas of interest or confusion that warrant further discussion.

Fighting in crabs and cichlid fish

The data presented in this paper mainly come from studies of fights staged in the laboratory between pairs of male cichlid fish and pairs of male portunid crabs, both of which have proved very good models for the study of animal fighting. In the case of the cichlid species used, *Tilapia zillii*, we studied adult males fighting over breeding territories. Fights are long (more than an hour) and complex, and show a classic pattern of escalation (Archer 1988) from displays (lateral and frontal) through mouth wrestling to a phase of intense circling and biting (Neat et al. 1998a). Our experiments on crabs involve males fighting over food or over mates. Fights are shorter than those seen in cichlid fishes (lasting a few minutes), but they are also complex, involving a range of actions of different degrees of intensity (including many involving the major chelipeds). Unusually, fights between crabs show no gradual escalation, with actions involving intense contact and possible injury occurring at all stages (Huntingford et al. 1995; Sneddon et al. 1997). In both cichlids and crabs, fights end abruptly as a result of a behavioural decision by the loser, rather than because the winner physically overpowers the loser or because they simply run out of energy. Work on these two species has produced results on three related issues that are relevant to the subject of signalling during fights. These are:

1. The interaction between body size (and other aspects of fighting ability) and resource value in determining the content and outcome of animal fights.
2. The costs of fighting and how these fall on individuals of different fighting ability.
3. The signalling of aggressive motivation during fights.

Body size, resource holding potential and resource value

Crabs

Using carapace width as an index of body size, when crabs are fighting over food the larger individual of a pair has a significant advantage (Fig. 1a; Smith et al. 1994). A closer examination of the relationship between size and victory shows, that in shore crabs *Carcinus maenas,* the main determinant of success is not body size *per se*, but the size of the major cheliped, which is positively but imperfectly related to carapace width. Thus, when crabs vary in both body size and claw size, the size of the major claw is the main predictor of victory (Sneddon et al. 1997). Interestingly, crabs with relatively larger claws use them more actively in displays (Fig. 1b), suggesting that they are aware of their relative advantage in this respect well before a fight is over, and adjust their behaviour accordingly (Sneddon et al. 1997).

Thus, in fights over food, various traits related to physical strength determine which crab wins. However, this is not the case in fights over females, which are long and fierce and are equally likely to be won by the smaller animal (Fig. 1a; Smith et al. 1994). Fights won by the smaller crab are particularly long and fierce (Fig. 1c), indicating that small males can defeat larger opponents through persistent aggression when motivated to do so by a valuable resource.

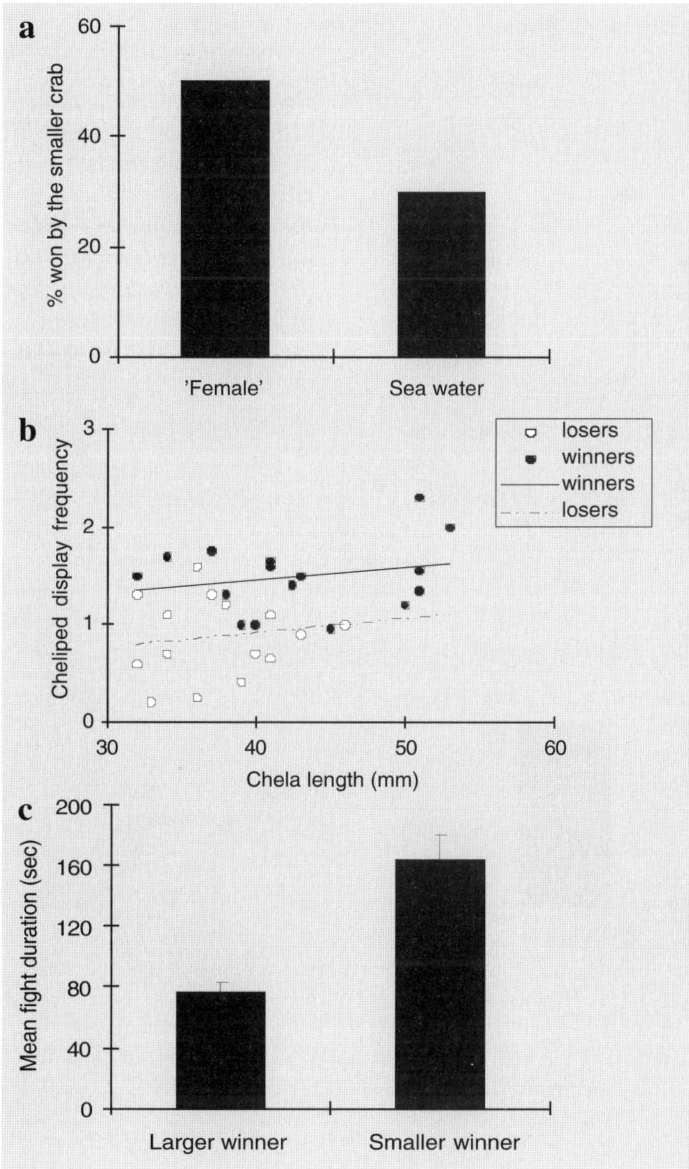

Figure 1. ***(a)*** *The proportion of fights won by the smaller of two male velvet swimming crabs* (Necora puber) *when fighting in the presence ('female') and in the absence (seawater) of pheromone from a receptive female (after Smith et al. 1994).* ***(b)*** *The frequency of cheliped display in shore crabs with different cheliped sizes (after Sneddon et al. 1997).* ***(c)*** *The duration of fights between male swimming crabs won by the larger and the smaller male in the presence of pheromone from a receptive female (after Smith et al. 1994).*

Cichlids

Many published studies on fighting in cichlid fish have shown that the larger of two opponents is likely to win fights and that, during fights, opponents assess their relative size (e.g. Enquist & Jacobsson 1986; Turner & Huntingford 1986). By contrast, in male *T. zillii* fighting over breeding territories, even with weight differences of up to 40%, no size advantage was detected (Neat et al. 1998a). Detailed analysis of fight sequences showed that relative size is assessed during mouth wrestling, with the larger of two males initiating progressively more bouts. In addition, relative size did influence behaviour in the final phase of the fight, when smaller winners showed a higher rate of biting (Fig. 2a). However, relative size did not determine the outcome of the fight. Instead, the fish that won was usually the one with the larger gonads (Fig. 2b; Neat et al. 1998a).

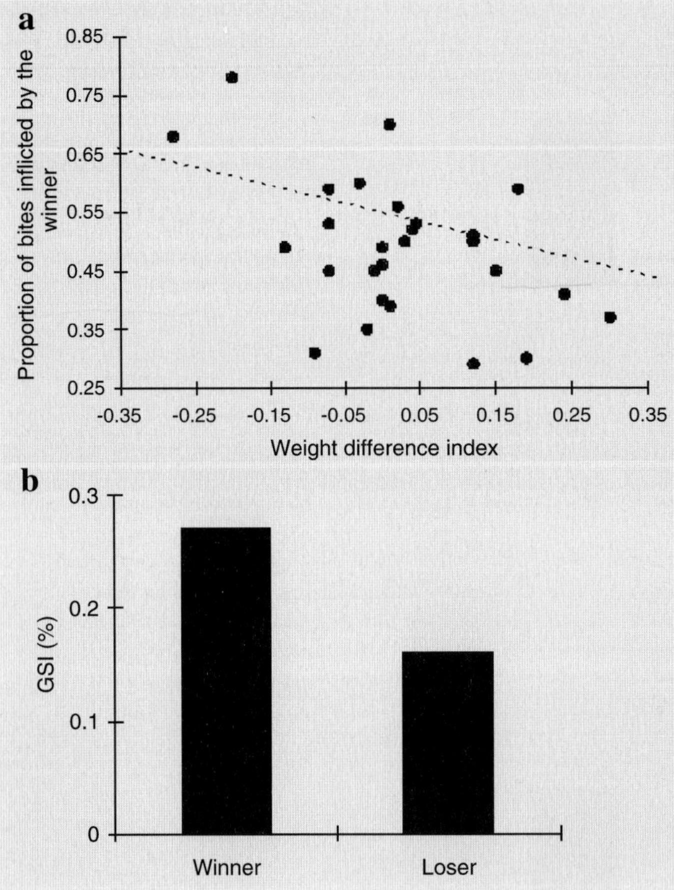

***Figure 2. (a)** Bite rate in relation to size differential in fights between male* T. zillii. *Negative values for the weight difference index indicates smaller winners. **(b)** Mean gonado-somatic index (GSI) in winners and losers (after Neat et al. 1998a).*

Size and victory in other species

The literature on information exchange during fights provides a number of examples in which relative size is not an important determinant of the outcome of fights. There are two kinds of explanation for the absence of a size advantage during fights.

Firstly, there are cases in which large size does not actually give an animal a competitive advantage, or at least is not the major determinant of fighting ability. For example, in midges *Chironomus plumosus*, small males are better at aerobatic manoeuvres in mating swarms, and this determines mating success (Neems et al. 1990). It is possible that something like this is going on in fights between *T. zillii* males, especially during the circling phase, which requires rapid and complex movement in three dimensions. Male damselflies *Calopteryx maculata* engage in long aerial fights over mating territories; in pairwise fights, winners are not larger than losers, but have larger lipid reserves (Fig. 3; Marden & Waage 1990). This result suggests that victory in damselfly fights is determined by how much energy in the form of fat stores each male has to fuel the energetic combat.

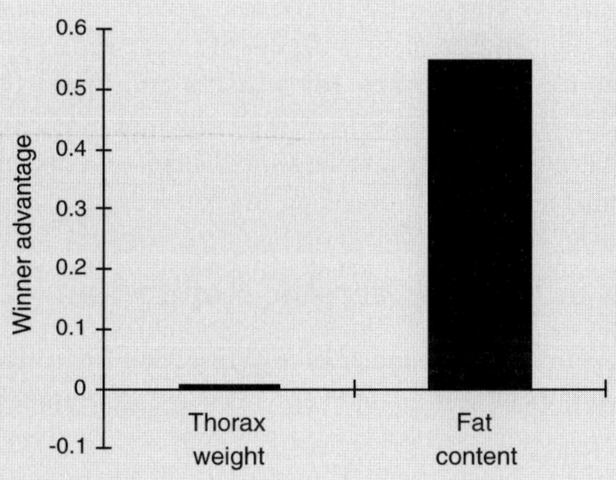

***Figure 3.** Mean difference in thorax weight and in fat reserves in males damselflies that won and lost fights over breeding territories (after Marden & Waage 1990).*

Secondly, there are cases where body size does indeed confer an advantage during fights, but smaller individuals can counteract this by fighting fiercely if the stakes are sufficiently high. This would be the case if there are immediate asymmetries in the value of a disputed resource, as depicted by Enquist & Leimar (1987) in their sequential assessment model incorporating variation in resource value. For example, during the breeding season, male elephants *Loxodonta africana* come in and out of musth asynchronously as androgen levels rise and fall. In non-musth males, larger animals are dominant, but musth overrides size as a determinant of victory; fights between evenly sized males in musth are extremely intense (Poole 1989). Something like this may be going on in *T. zillii*, which are known to undergo cycles of gonadal activity (Siddiqui 1979), on the untested assumption that having large gonads somehow enhances the value of a breeding territory.

Alternatively, there may be size-dependent asymmetries between large and small males with respect to the value of the future, as depicted in Enquist & Leimar's model (1990) of the evolution of fatal fighting and Grafen's (1987) model in which divisive asymmetries lead to the 'desperado effect'. Such an effect may also explain the fact that smaller males were able to win approximately half the fights in *T. zillii* (Neat et al. 1998a). Our subjects had been housed from hatching in large groups with a strongly polarised dominance hierarchy in which dominant males gained more food and grew larger. Hence, small males were likely to have a very low estimate of their ability to appropriate resources in the competitive environment they inhabited (i.e., a low expected value for the future), so when they acquired a breeding territory, they defended it very persistently.

Costs of fighting

Costs and the evolution of honest signals

The idea that signalling has costs, and is also differentially costly for competitors of different quality, is very important in theories about honest signals (e.g. Grafen 1990). It seems reasonable to suggest that this idea is correct, but there are few direct empirical data on the costs of signalling. During our studies of how fights are resolved, we have collected some information on possible costs of fighting for winners and losers that is potentially relevant to this debate. The risk of injury is a well documented cost of escalated fighting (see Huntingford & Turner 1986 for a review), but here we concentrate on studies of the energetic costs of fighting.

Energetic consequences of fighting: aerobic respiration

In an attempt to characterise the energetic consequences of fighting, we have tracked aerobic and anaerobic respiration as well as levels of nutrient reserves in winners and losers of different kinds of fight.

In velvet swimming crabs, ventillatory activity (a good correlate of oxygen consumption) increased dramatically during fights; on a fine scale, this is related to the performance of specific aggressive acts. Oxygen consumption only returns to resting levels after a period of an hour or so, taking somewhat longer in losers after a long fight (Smith & Taylor 1993).

Hack's detailed study (1997a) of the energetic costs of fighting in male house crickets *Acheta domesticus* showed that peaks in oxygen consumption coincide with peaks in activity and that signals and other actions of different intensity (e.g. low-intensity acts such stridulation versus high intensity actions such as wrestling) involve different levels of oxygen consumption (Fig. 4a). Actions that are more costly in this sense are used less frequently (Fig. 4a). Winners expended more energy than losers, especially during long fights (Fig. 4b), because they made use of more costly actions.

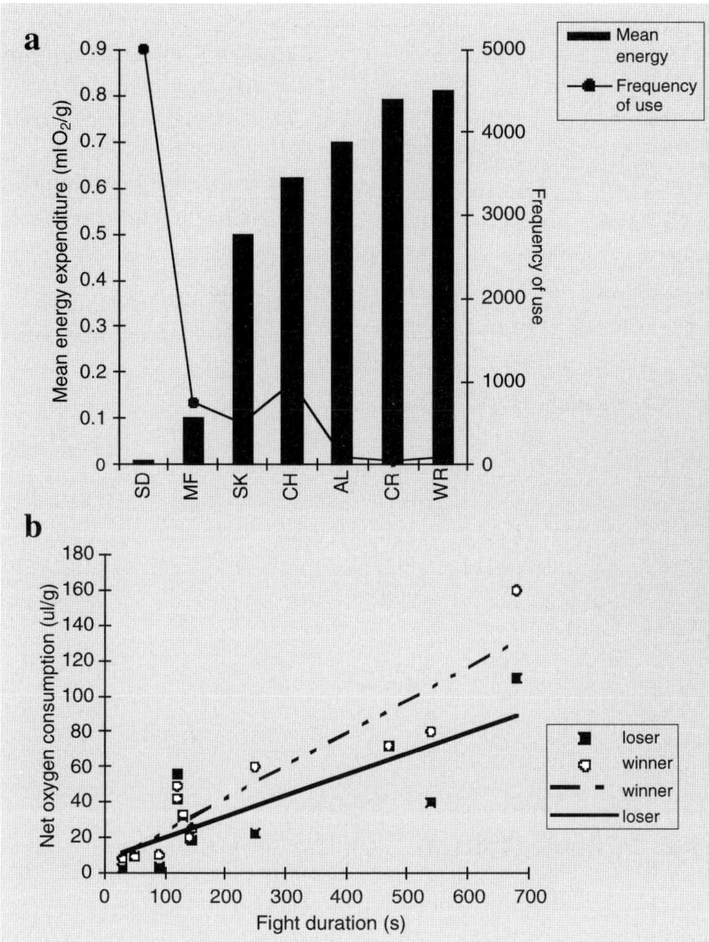

*Figure 4. (**a**) Oxygen consumption for different aggressive actions in male house crickets and their frequency of use during fights. Action key: SD (stridulate), MF (mandible flare), SK (shake), CH (charge), AL (antennae lash), CR (cerci raise), WR (wrestle). (**b**) Net oxygen consumption for fights of different duration in winners and losers (after Hack 1997a).*

Energetic consequences of fighting: anaerobic respiration

During vigorous exercise, anaerobic respiration can lead to accumulation of lactate in the muscles and in the circulatory system and this may constrain further activity. In seeking factors that might be involved in the resolution of fights, we have looked at lactate concentrations after fights in winners and losers. We found significant lactate accumulation in the muscles of cichlids after a fight (Fig. 5a), especially in the losers of long fights (Neat et al. 1998b). On the assumption that lactate accumulation has consequences for fitness (see below), then long fights are more costly and the cost falls more heavily on the loser than the winner.

In male shore crabs, fights get shorter (and the probability that the larger crab will win gets higher) as oxygen tensions decrease. At the same time, concentrations of lactate in the haemolymph at the end of the fight increased dramatically in both winners and losers (Fig. 6a; Sneddon et al. 1997). In this case, any cost arising from lactate accumulation falls equally on winners and losers, to a degree that is highly dependent on the environment in which fights occur.

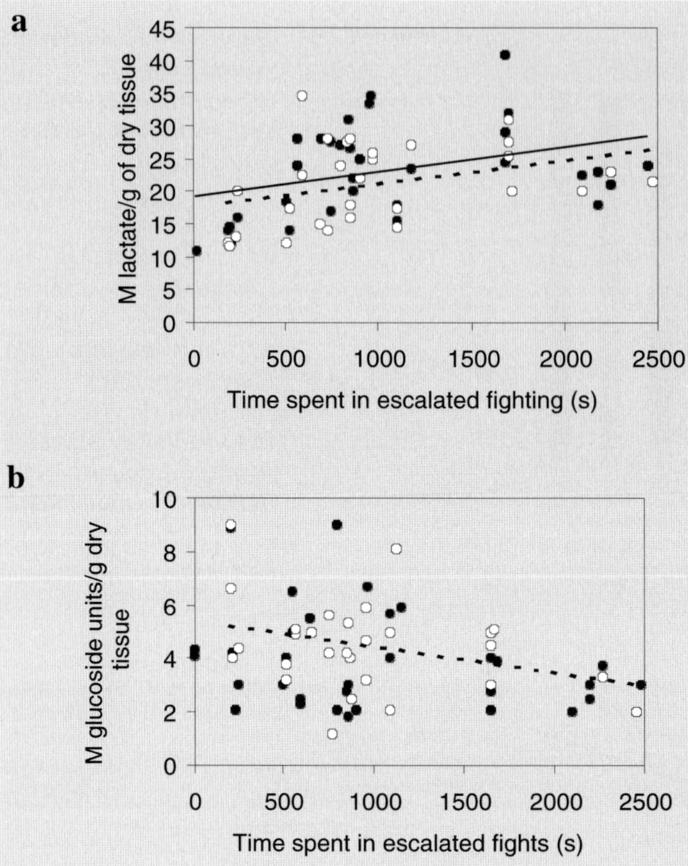

Figure 5. *Concentrations of **(a)** lactate and **(b)** total sugar in the muscles of winning and losing male* T. zillii *in relation to fight duration (after Neat et al. 1998b). (Solid circles and solid line: losers; open circle and dotted line: winners.)*

Depletion of nutrient reserves during fights

In both crabs and cichlids we found significant depletion of energy reserves after fights, as glycogen is mobilised for energy production. Total sugar concentrations in the muscles of male cichlids increase with fight length, with levels being significantly higher in losers (Fig. 5b; Neat et al. 1998b). In shore crabs, haemolymph glucose concentrations increased after fights, especially under hypoxic conditions (Fig. 6b; Sneddon et al. 1998).

Respiratory consequences and the costs of fighting

In the light of these results, fights in crabs and cichlids can reasonably be conceptualised at partial energetic wars of attrition in which opponents do not fight until they completely run out of energy reserves or until their muscles have seized up, but have some combined threshold level of lactate accumulation plus glycogen depletion at which they give up and pull out of the fight. Certainly, Marden and Waage (1990) use the term energetic war of attrition to describe the long fights between male damselflies.

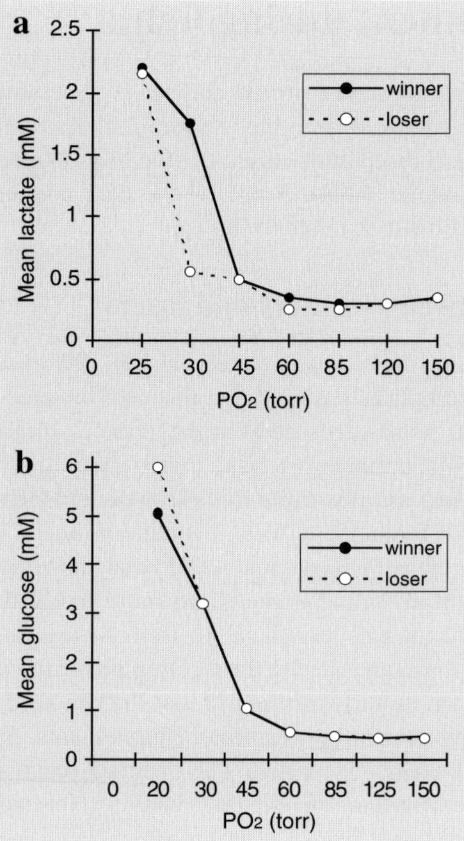

Figure 6. *Concentrations of **(a)** lactate and **(b)** glucose in the haemolymph of winning and losing shore crabs after fights staged at different oxygen tensions (after Sneddon et al. 1997).*

It is not clear to what extent any of these energetic consequences represent real fitness costs. In cichlids, comparison with the literature on the effects of exercise in other species of fish suggests that fights are energetically equivalent to vigorous exercise (see Neat et al. 1998b). Direct measurement of lactate and glucose concentrations in shore crabs that have been exercised on a treadmill until they raise their legs and cease to walk show that, in some circumstances at least, the physiological consequences of fighting are comparable to those of exhaustive exercise (Sneddon et al. 1998).

Hack (1997a) estimated that, in crickets, the cumulative energetic costs of fighting may represent a non-trivial component of the total energy budget ('a single spermatophore is equivalent to 19 escalated fights', *op. cit.*). He suggests that this will have negative consequences for fitness since depleted energy reserves may impair performance in subsequent fights (e.g. Marden & Waage 1990). Subsequent fighting ability was not significantly reduced in swimming crabs after fights over food (Thorpe et al. 1995), but impaired performance may well occur as a result of the longer and more intense fights over females, especially if these take place at reduced oxygen tensions. In addition, Hack (1997a) suggested that foraging to replenish the energy expended while fighting may incur predation costs. Further studies of such effects are needed before the increasingly well documented energetic consequences of fighting can be interpreted as a fitness cost.

Signalling and assessment during fights

There is an extensive theoretical and empirical literature on the subject of assessment during fights, concerning, for example, what aspects of an opponent should be assessed and how signals indicative of fighting ability and motivation might evolve (e.g. Maynard Smith 1982, 1994; Enquist 1985; Hurd & Enquist 1998). Some of the data collected during our studies of fighting in crabs and cichlid fishes relate to this topic.

Behavioural differences between winners and losers

In his study of the energetic costs of fighting in male house crickets, Hack (1997a) was able to obtain estimates of the oxygen consumption required for the performance of particular actions even though his two crickets were necessarily in the same respiration chamber. This was possible because during fights, opponents tend to match each other act for act. This presumably makes it easier to assess various traits in their opponents. Even so, it is often possible to detect behavioural differences between eventual winners and losers well before the fight ends. In swimming crabs (Thorpe et al. 1995) and shore crabs (Sneddon et al. 1997), from early on in a fight, eventual winners are more likely than eventual winners to move towards their opponent and less likely to move away. In *T. zillii*, towards the end of the mouth wrestling phase and during the circling phase the fish engaged in mouth locking, which is distinct from mouth wrestling in that the fish hold on to each others lips with closed mouths, thereby preventing respiratory movements. This action lasts for a variable time before one fish shakes itself free. The eventual loser was much more likely to break off from mouth locking than the eventual winner (Neat et al. 1998a).

Colour changes during fights

Like many cichlids, *T. zillii* show complex colour changes during fights (Hulscher-Emeis 1992). In both the eventual winner and the eventual loser, the red colouration on the belly becomes more intense as fights progress (Fig. 7). The proportion of fish with darkened eyes also increase, but this change takes place more rapidly in the eventual winner (Neat 1996).

Rapid colour changes during fights (on a scale ranging from seconds to minutes) have been described for a number of ectotherms. Such changes can be under neural or hormonal control and are often associated with specific behavioural states. For example, male cuttlefish *Sepia officinalis* show different intensities of a particular facial colour pattern (the 'intense zebra display') and the degree of darkening during the initial stages of agonistic encounters is related to the probability that the individual concerned will persist in escalated encounters (Adamo & Hanlon 1996). Exactly how such complex signalling functions during the resolution of fights is not clear.

The cichlid *Nannacara anomala* shows a number of independently varying colour patterns during fights, two of which are dark vertical and medial lines. The appearance of

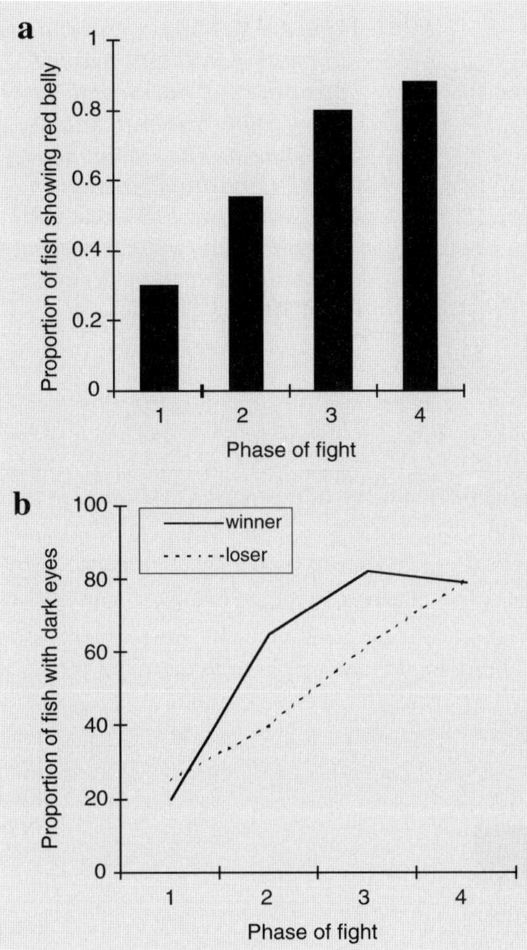

*Figure 7. The proportion of males showing (**a**) red belly colouration and (**b**) dark eyes in successive stages of fights in* T. zillii *(after Neat 1996).*

these colour patterns relates to the performance of certain acts that are known to be involved in assessment of relative size (Fig. 8; Hurd 1997). Thus, tail beating is particularly likely to occur just after the medial line appears and mouth wrestling is more common when both (or at least one) fish has darkened vertical bars. Hurd speculates that these colour patterns serve to facilitate the transmission of assessment information within the fight.

Taking an example from another cichlid species, firemouths *Cichlasoma meeki* are named after the red, carotenoid based pigment around their mouth, which is displayed during fights when the fish flare their gills. Under white light, fish fed on a high carotenoid diet (and so with brighter red colouration) win more fights than do opponents on a low carotenoid diet; when size is equal, the carotenoid advantage is 70%. This effect disappears under green light, suggesting that it is an effect of colour *per se* and not some other consequence of the experimental diet (Evans & Norris 1996). Interestingly, the rate at which the gills are flared tracks the strength of the visual cue. Speculating, it is as if the

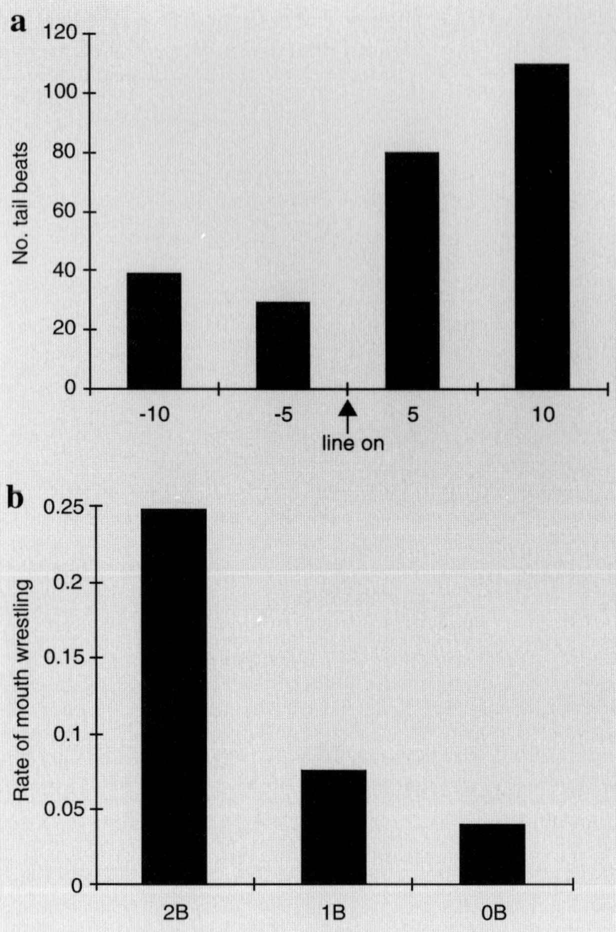

Figure 8. ***(a)*** *The occurrence of tail beating in relation to the appearance of the medial line during fights between male* N. anomala. ***(b)*** *The rate of mouth wrestling during fight sequences in which two (*2B*), one (*1B*) or neither (*0B*) male is showing the vertical bar pattern (after Hurd 1997).*

fish get some feedback from the strength of the visual cues generated by flaring their gills and use this signal more when the effect is impressive. This may be what is going on in shore crabs, when males with relatively large claws display these more than those with small claws (Sneddon et al. 1997), and in sand lizards *Lacerta agilis*, in which males whose green throat patch (a badge of status) had been artificially enlarged not only win more, but also initiate more contests (Olsson 1994).

Finally, green anole lizards *Anolis carolinensis* have a dark post-orbital eyespot, the pigment in which darkens in 90% of fights, under the influence of plasma noradrenaline. During staged pairwise fights, the eyespot darkens more quickly in the animal that eventually wins, and this is associated with higher noradrenaline levels in the winner at the end of the fight (Summers & Greenberg 1994). These authors suggest that rapid activation of the adrenergic system gives an animal an advantage in an aggressive encounter, through various physiological changes including increased blood pressure and heart rate and elevated blood sugar levels.

These instances of rapid colour changes during fights raise a number of issues concerning the nature of signalling during fights, including the perennial question of why fighting animals should signal how fiercely they are prepared to fight and the idea that they have a co-operative element, serving to co-ordinate mutually beneficial assessment.

Assessment of physiological asymmetries during fights

The greater probability of breaking off mouth locking by the eventual loser of fights between male *T. zillii* (Neat et al. 1998b) gives rise to the speculation that the ability to sustain this behaviour (which involves suppressed gill ventilation, personal observation) is unavoidably linked either to current oxygen debt or to the threshold oxygen debt that the fish are prepared to incur. Persistence in mouth locking might therefore provide information about an opponent's probability of persisting in the fight. The literature provides examples of cases in which some aspect of energetic physiology not only determines the outcome of fights, but is also assessed as the fights proceed. For example, in damselflies, the fatter of two males is more likely to win a territorial contest (Fig. 3) and there is a negative relationship between contest duration and energetic asymmetry (Fig. 9), indicating that males assess their relative energy reserves during fights.

In an experimental study of the effects of body temperature on fighting over sun spots in male wood butterflies *Pararge aegeria*, 90% of fights were won by the male with the highest body temperature (Stult & Willner 1998). The authors suggest (partly on the basis of the fact that fights were shorter when the males had an experimentally induced temperature asymmetry), that a rival's body temperature may be assessed during spiral fights, either directly or via wing beat rate.

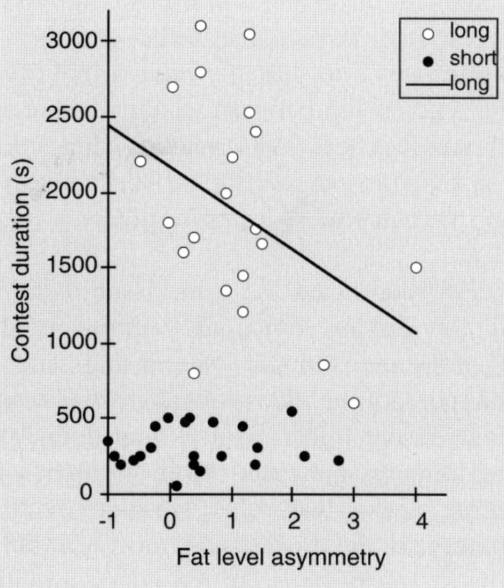

Figure 9. *Contest duration in relation to asymmetry in fat levels in male damselflies after long (open circles) and short (closed circles) fights (after Marden & Rollins 1994).*

Finally, Hack (1997b) showed that in house crickets, the cumulative energetic cost incurred by both combatants during the whole fight increased with decreasing size asymmetry, suggesting that behaviour during fights is adjusted according to an assessment of size asymmetry. In addition, however, cumulative energetic cost increased with decreasing degree of asymmetry between males in the amount of energy expended in the first two bouts of the fight (Fig. 10), which is unrelated to mass. This result indicates that some aspect of the capacity (or willingness) of an opponent to engage in energetically expensive actions is also being assessed.

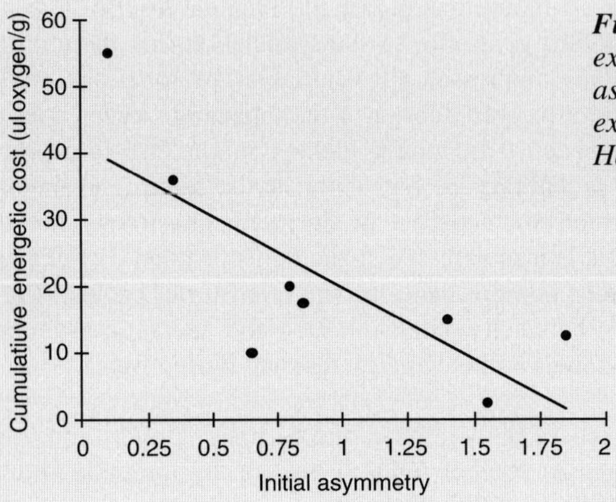

***Figure 10.** Cumulative energy expenditure in relation to asymmetry in initial energy expenditure in house crickets (after Hack 1997b).*

These studies suggest that various components of the respiratory physiology of fighting animals are influential in determining who wins a fight, and that relevant differences in state are assessed during fights. The assumption (supported by empirical data in some cases, e.g. Hack 1997b) is that differences in energetic investment in fights relate to the value placed by each opponent on the disputed resource. To examine further exactly how a difference in the value placed on a resource translates into strategic behavioural decisions during fights, it is necessary to explore the relationship between resource value and an animals' motivational state. Maynard Smith and Riechert (1984) made such a link in their model of the way in which relative body size and resource value influence levels of aggression and fear, and so interact to determine the outcome of fights in spiders.

In a recent empirical study, Elwood et al. (1998) addressed the same issue using an approach that offers considerable potential for linking functional studies of the determinants of fighting behaviour and studies of the motivational systems that generate strategic responses. These authors developed an index of the motivational state of hermit crabs that is independent of ongoing behaviour during fights, namely how long such behaviour is interrupted in response to a novel stimulus presented while the fight is in process. Crabs that have more to gain from taking over a rival's shell are more likely to win than those with less to gain, are more aggressive in the process and, most relevant in the current context, showed a markedly shorter startle response to the novel stimulus.

Asymmetries in size do not effect this index of aggressive motivation, perhaps because size does not give an unambiguous advantage in the circumstances of these experiments; small crabs defending large shells are able to take refuge far inside these shells and are difficult for a larger opponent to dislodge.

Concluding comments

The studies described in this paper raise a number of issues concerning signalling during fights that may warrant further consideration:

1. Animals such as crabs use aggressive actions with different levels of cost, yet there is no gradual escalation during a fight. Various components of fighting ability and of resource value are important in the resolution of fights, but it is difficult to accommodate fights with no pattern of gradual escalation within current models such as the sequential assessment game (Enquist & Leimar 1983, 1987).
2. A number of studies have demonstrated significant energetic consequences that are dependent on the duration and intensity of fighting. Information on their medium term effects on subsequent scope for action and ultimately on the fitness of the animals concerned are needed to determine whether these consequences constitute real costs.
3. Several cases have been identified in which fighting animals apparently adjust performance of displays to the quality of the signal they provide. The part that such an association might play in the dynamics of fight resolution warrants further consideration.
4. A number of examples have been described in which changes in colour patterns during fights reflect immediate behavioural state and probability of persistence during escalated fighting. It would be useful to explore the functional and causal bases of such changes in body colour (and possibly sound and smell) during fights.
5. The clear distinction made in the early game theory models of animal fights between resource holding potential (RHP), which should be signalled, and motivation to fight, which should not, has become blurred. Strength consequent on large size, the ability to injure due to large and/or strong claws (as in crabs) and agility in 3-dimensional combat (as in midges) clearly fall within the original concept of RHP, because they relate to more-or-less permanent physical properties that confer a mechanical advantage during fights. Possession of the energy reserves needed for long, athletic fights (as in damselflies) can also be seen as a component of RHP, since the building up of reserves takes place on a much longer time scale than that of the fight. However, a temporary increase in physiological capacity induced during a fight by secretion of catecholamines, as in anole lizards, or a decision to use more rather than less costly actions, as in crickets, are surely different, although Hack (1997b) discusses his results in terms of fighting ability ('the relative energetic cost incurred by combatants may reliably signal relative energy reserves and contribute to the active assessment of fighting ability', *op. cit.*). This kind of determinant of what happens during fights depends on temporary behavioural states, that determine the effort animals put going into the use of fighting equipment. It would be useful to clarify the point at which a fighting advantage due to higher RHP ends and one due to motivational difference begins.
6. Finally, although assessed levels of relative RHP and relative resource value must exert their influences on behavioural decisions via an effect on the motivational state

of fighting animals, little is known about the mechanisms by which these functionally important dimensions map onto motivational state. An investigation of this issue would be of value, either tracking motivational shifts as assessment proceeds, using behavioural tools such as colour patterns or the startle technique developed by Elwood et al. (1998), or by tracking changes in physiological state.

Acknowledgements

We would like to thank The Royal Norwegian Society of Sciences and Letters, The Foundation, for organising and funding the conference at which this material was presented, Brian Wisenden and Mats Grahn for comments on an earlier version of this manuscript and Jennifer Andrew for help with preparation of the manuscript and figures.

References

Adamo, S. A. & Hanlon, R. T. 1996. Do cuttlefish (Cephalopoda) signal their intentions to conspecifics during agonistic encounters. *Animal Behaviour*, **52**, 73-81.

Archer, J. 1988. *The Behavioural Biology of Aggression.* Cambridge: Cambridge University Press.

Elwood, R. W., Wood, K. E., Gallagher, M. B. & Dick, J. T. A. 1998. Probing motivational state during agonistic encounters in animals. *Nature*, **393**, 66-68.

Enquist, M. 1985. Communication during aggressive interactions with particular reference to variation in choice of behaviour. *Animal Behaviour*, **33**, 1152-1161.

Enquist, M. & Jakobsson, S. 1986. Decision making and assessment in the fighting behaviour of *Nannacara anomala* (Cichlidae, Pisces). *Ethology*, **72**, 143-153.

Enquist, M. & Leimar, O. 1983. Evolution of fighting behaviour: decision rules and assessment of relative strength. *Journal of Theoretical Biology*, **102**, 387-410.

Enquist, M. & Leimar, O. 1987. Evolution of fighting behaviour: the effect of variation in resource value. *Journal of Theoretical Biology*, **127**, 187-205.

Enquist, M. & Leimar, O. 1990. The evolution of fatal fighting. *Animal Behaviour*, **39**, 1-9.

Evans, M. R. & Norris, K. 1996. The importance of carotenoids in signalling during aggressive interactions between male firemouth cichlids (*Cichlasoma meeki*). *Behavioral Ecology*, **7**, 1-6.

Grafen, A. 1987. The logic of divisively asymmetric contests: respect for ownership and the desperado effect. *Animal Behaviour*, **35**, 462-467.

Grafen, A. 1990. Biological signals as handicaps. *Journal of Theoretical Biology*, **144**, 517-546.

Hack, M. A. 1997a. The energetic costs of fighting in the house cricket, *Acheta domesticus* (L.). *Behavioral Ecology*, **8**, 28-36.

Hack, M. A. 1997b. Assessment strategies in the contests of male crickets, *Acheta domesticus* (L.). *Animal Behaviour*, **53**, 733-747.

Hulscher-Emeis, T. M. 1992. The variable colour patterns of *Tilapia zillii* (Cichlidae): integrating ethology, chromatophore regulation and the physiology of stress. *Netherlands Journal of Zoology*, **42**, 525-560.

Huntingford, F. A. & Turner, A. K. 1987. *Animal Conflict.* London: Chapman & Hall.

Huntingford, F. A., Taylor, A. C., Smith, I. P. & Thorpe, K. E. 1995. Behavioural and physiological studies of aggression in swimming crabs. *Journal of Experimental Marine Biology and Ecology*, **193,** 21-39.

Hurd, P. L. 1997. Is signalling of fighting ability costlier for weaker individuals? *Journal of Theoretical Biology*, **184**, 83-88.

Hurd, P. L. & Enquist, M. 1998. Conventional signalling in aggressive interactions: the importance of temporal structure. *Journal of Theoretical Biology*, **192**, 197-211.

Marden, J. H. & Rollins, R. A. 1994. Assessment of energy reserves by damselflies engaged in aerial contests for mating territories. *Animal Behaviour*, **48**, 1023-1030.

Marden, J. H. & Waage, J. K. 1990. Escalated damselfly territorial contests are energetic wars of attrition. *Animal Behaviour*, **39**, 954-959.

Maynard Smith, J. 1982. *Evolution and the Theory of Games.* Cambridge: Cambridge University Press.

Maynard Smith, J. 1994. Must reliable signals always be costly? *Animal Behaviour*, **47**, 1115-1120.

Maynard Smith, J. & Riechert, S. E. 1984. A conflicting-tendency model of spider agonistic behaviour - hybrid-pure population line comparisons. *Animal Behaviour*, **32**, 564-578.

Neat, F. C. 1996. Behavioural and physiological studies of fighting in male *Tilapia zillii* (Cichlidae). PhD Thesis, University of Glasgow, UK.

Neat, F. C., Huntingford, F. A. & Beveridge, M. M. C. 1998a. Fighting and assessment in male cichlid fish: the effects of asymmetries in gonadal state and body size. *Animal Behaviour*, **55**, 883-891.

Neat, F. C., Taylor, A. C. & Huntingford, F. A. 1998b. Proximate costs of fighting in male cichlid fish: the role of injuries and energy metabolism. *Animal Behaviour*, **55**, 875-882.

Neems, R. M., McLachlan, A. J. & Chambers, R. 1990. Body size and lifetime mating success of male midges (Diptera: Chironomidae). *Animal Behaviour*, **40**, 648-652.

Olsson, M. 1994. Nuptial colouration in the sand lizard, *Lacerta agilis*: an intra-sexually selected cue to fighting ability. *Animal Behaviour*, **48**, 607-613.

Poole, J. H. 1989. Announcing intent: the aggressive state of musth in African elephants. *Animal Behaviour*, **37**, 140-152.

Siddiqui, A. Q. 1979. Reproductive biology of *Tilapia zillii* (Gervais) in Lake Naivasha, Kenya. *Environmental Biology of Fishes*, **4**, 257-262.

Smith, I. P. & Taylor, A. C. 1993. The energetic cost of agonistic behaviour in the velvet swimming crab, *Necora* (L = *Liocarcinus*) *puber* (L.). *Animal Behaviour*, **45**, 375-391.

Smith, I. P., Huntingford, F. A., Atkinson, R. J. A. & Taylor, A. C. 1994. Strategic decisions during agonistic behaviour in the velvet swimming crab *Necora puber* (L.). *Animal Behaviour*, **47**, 885-894.

Sneddon, L. U., Huntingford, F. A. & Taylor, A. C. 1997. Impact of an ecological factor on the costs of resource acquisition: fighting and metabolic physiology of crabs. *Functional Ecology*, **12**, 808-815.

Sneddon, L. U., Huntingford, F. A. & Taylor, A. C. 1998. Influence of resource value on the agonistic behaviour of the shore crab, *Carcinus maenas* (L.). *Behavioral Physiology*, **30**, 225-237.

Stutt, A. D. & Willmer, P. 1998. Territorial defence in speckled wood butterflies: do the hottest males always win? *Animal Behaviour*, **55**, 1341-1347.

Summers, C. H. & Greenberg, N. 1994. Somatic correlates of adrenergic activity during aggression in the lizard, *Anolis carolinensis. Hormones and Behaviour*, **28**, 29-40.

Thorpe, K. E., Taylor, A. C. & Huntingford, F. A. 1995. How costly is fighting? Physiological effects of sustained exercise and fighting in swimming crabs, *Necora puber* (L.) (Brachyura, Portunidae). *Animal Behaviour*, **50**, 1657-1666.

Turner, G. F. & Huntingford, F. A. 1986. A problem for game theory analysis: assessment and intention in male mouthbrooder contests. *Animal Behaviour*, **34**, 961-970.

IN: ESPMARK,Y., AMUNDSEN,T. & ROSENQVIST,G. (EDS.) 2000
ANIMAL SIGNALS:
SIGNALLING AND SIGNAL DESIGN IN ANIMAL COMMUNICATION, PP. 277-300
TAPIR ACADEMIC PRESS, TRONDHEIM, NORWAY

13

Handicap, Index, and Conventional Signal Elements of Bird Song

SANDRA L. VEHRENCAMP

Biology 0116, University of California at San Diego, La Jolla, CA 92093, USA

Present address: Laboratory of Ornithology, Cornell University, Ithaca NY 14850, USA

E-mail: slv8@cornell.edu

Abstract

Signals given in social contexts where sender and receiver disagree about the optimal receiver response must be stabilized by costs or constraints on senders that guarantee signal honesty. The appropriate type of cost is associated with both the design and information content of the signal. Five classes of signals can be distinguished in theory, each associated with a particular type of cost, design, and information content. Index signals are physically or physiologically constrained to provide honest information, quality and general handicaps impose receiver-independent signal production costs, vulnerability handicaps place the sender in a position of greater risk of attack by the receiver, and conventional signals are stabilized by a receiver retaliation rule. The utility of this classification scheme is explored with a survey of the variable parameters of bird song. Features of song such as song rate, song-type switching rate, repertoire size, song sharing, song matching, and others are examined for differential production costs, likelihood of receiver retaliation, association with specific contexts, signal form, and possible information content. The signal parameters used in male-sender/female-receiver contexts tend to be either quality handicap or index signals, whereas signal parameters used in male-sender /male-receiver contexts can be either conventional, index, or general handicap signals. The same signal parameter is often used in different ways in male and female receiver contexts. Although a few song parameters are difficult to assign to one signal category, most parameters fit optimally into one of the five signal classes.

Considering only communication among conspecific individuals, signals are actions or structures that have been selected to provide information to receivers about sender attributes or environmental contexts. Senders encode this information by linking distinctive signals or signal variants to specific contexts. Receivers use the signaled information about the current context, along with non-signal cues, to decide what action to take. For a signal to evolve, both the sender and receiver must benefit, on average, from the receiver's response. In some signaling contexts, such as social integration, group movement, and alarm signals, the interests of the sender and receiver are similar and senders are selected to provide accurately encoded, honest information. In other contexts, sender and receiver may have conflicting interests over what the receiver's response should be. Conflict among conspecifics is greatest in agonistic contexts where two rivals both desire a non-divisible resource and benefit only if the other retreats. Conflict can also occur in the mate attraction context, where males are generally always eager to mate but females prefer to mate only with high-quality males. In signaling exchanges with conflicting sender-receiver interests, there will always be an incentive for senders to lie, exaggerate, or mislead in order to induce the receiver to choose the response preferred by the sender.

During the past decade, a variety of game theoretic models of signal evolution have attempted to identify the conditions under which signal reliability can be maintained (reviewed in Bradbury & Vehrencamp 1998). These models depict a range of signal functions and find that different sets of conditions are required to achieve an evolutionarily stable signaling system. However, all come to the conclusion that stable, reliable signals can only evolve in conflicting-interest contexts when there is some type of cost or constraint on senders that makes cheating a less optimal strategy than revealing the truth. Classifying signals by the type of cost or stabilizing process is a useful exercise for two reasons (Hasson 1994, 1997; Guilford & Dawkins 1995; Maynard Smith & Harper 1995). First, the way in which the costs are paid is a function of the form and design of the signal. Second, and more controversial, is the idea that the nature of the cost and the form of the signal determine the type of information the signal can relay and the social function(s) the signal can serve. The classification schemes mentioned above do not incorporate all of the different cost types identified in the game theory models, and they tend to emphasize mate attraction signals over other types of signals. The expanded classification scheme presented here emphasizes agonistic threat signals, in which a wider variety of costs have been shown to be important. Understanding the diversity, design, and information content of signals is a major goal of current studies of animal communication. Research effort focused on distinguishing among alternative types of stabilizing costs is critical for achieving this goal.

A cost-based classification scheme

The costs that maintain signal reliability can be divided into two general categories: receiver-dependent and receiver-independent costs (Adams & Mesterton-Gibbons 1995; Guilford & Dawkins 1995). When the stabilizing cost is independent of the target receiver's response, it means that senders signal at levels reliably indicating their underlying quality because it is too costly or difficult to do otherwise. This leads to selection for an optimal signal design determined by the type of information being conveyed. When the stabilizing cost is dependent on receiver responses, senders only pay a cost a certain fraction of the time. The probability of paying a cost may depend on the mismatch between the sender's actual quality and the receiver's interpretation of the sender's quality. Bluffing can therefore occur at low levels at the ESS. Furthermore, the form of the display is less closely linked to the information content of the signal, and signal production costs can be low. Both of these general cost categories can be further subdivided in ways that refine the predicted relationship between signal form and information content. A distinction can be made between two forms of receiver-independent costs: physical or physiological constraints, and signal production costs. Two different receiver-dependent costs can also be distinguished: vulnerability to receiver attack arising from the form of the display, and a receiver retaliation or punishment rule. Classifying signals into these four cost categories is very similar to the scheme proposed by Hurd (1997a). The way in which each of these costs has been invoked for stabilizing signaling systems is described below, along with a discussion of the effects each type of cost imposes on signal design. These signal classification features are summarized in Table 1.

***Table 1**. Classification of signals used in sender-receiver conflict contexts based on the type of cost that maintains signal honesty. Associations with signal design and information content are also indicated.*

COST	SIGNAL TYPE	SIGNAL DESIGN	INFORMATION
Physiological and physical constraints	INDEX	Discrete or graded displays, form linked to sender attributes	Body size, strength, age, natal area, pointing
Signal production, time, predation	QUALITY HANDICAP	Graded display, intensity correlated with sender quality	Fighting ability, stamina, health, condition
Signal production, time, predation	GENERAL HANDICAP	Graded display, intensity correlated with sender need	Motivation, need, resource valuation
Vulnerability to receiver attack	VULNERABILITY HANDICAP	Display performance places sender at risk of receiver attack	Willingness to take risks, intentions
Receiver retaliation	CONVENTIONAL	Arbitrary form, antithetical discrete or graded signal	Motivation, willingness to escalate, or fighting ability

Communication signals can convey different kinds of information to receivers, but for the purposes of this discussion a distinction is made between information about intrinsic sender quality and information about current sender intentions or needs. Intrinsic quality includes sender traits that vary between individuals and within individuals on a longer-term time scale, such as age, body size, fighting ability, foraging ability, health, and condition. Intentions and needs are a function of the benefit the sender stands to gain if the receiver responds in the manner preferred by the sender. These sender characteristics usually vary on a shorter-term time scale and affect the sender's motivation and willingness to escalate or take risks. Most of the game models described below assume or predict the transmission of a particular type of information; the presumed information content for each signal category is also included in Table 1.

Physical constraints

One type of receiver-independent process that stabilizes signal reliability is physical constraint or a physiological ceiling that invariably forces the signal to reveal honest information. Signals stabilized by this mechanism are called index signals, assessment signals, or unbluffable signals (Maynard Smith & Harper 1988, 1995; Maynard Smith 1991). The sequential assessment game model assumes the use of repeated index signals for assessing opponent differences in fighting ability (Enquist & Leimar 1983). The form of an index signal is absolutely linked to the information transmitted by the signal, and such signals by definition convey information about intrinsic sender attributes. Examples of index signals given in agonistic contexts include displays that reveal body size and fighting ability such as vocalization frequency in frogs, tail-beating and mouth-wrestling in fish, and height of scratches on trees by mammalian carnivores (Davies & Halliday 1978; Jakobsson et al. 1979; Ramer et al. 1983; Robertson 1986; Thapar 1986). Signals that are obligatorily linked with aging processes or health may be used by females to select high-quality males. Pointing displays that indicate direction of gazing or attention are also index signals (Hasson 1997). Many additional examples can be found in Zahavi & Zahavi (1997).

Production costs

Signal production costs are the type of cost first envisioned by Zahavi (1975) as necessary for maintaining signal honesty. Production costs include investment in signaling organs and ornaments, energetic expenditure during display, the risk of attracting predators, and time lost. In other words, they are not affected by the presence or response of a conspecific receiver. In order for production costs to stabilize signal honesty, signals must be sufficiently expensive, and the costs for senders must increase as benefits increase. In addition, either the costs or the benefits must be differentially scaled for senders of different states (Parker 1979). Cost-scaled and benefit-scaled models lead to very different kinds of signals (Johnstone 1997).

Cost scaling means that the cost curves increase at different rates for senders of different abilities, conditions, or qualities (Nur & Hasson 1984; Grafen 1990). The cost of signaling is higher than it needs to be for simple signal efficacy reasons as a result of prior coevolution between sender and receiver. Costly signals force senders into a trade-off between signal intensity and some component of fitness such as survivorship. High-quality individuals can expend more for signaling than low-quality individuals without suffering higher mortality. If receivers benefit from responding differentially to high- and low-quality senders, they will be selected to pay attention to those costly signals that most effectively help them distinguish among senders. At the ESS, there is a positive correlation between signal intensity and intrinsic sender quality. Honesty is maintained by the high cost that poorer competitors would pay to exaggerate their signal intensity. Such signals are called quality handicap signals.

The quality handicap model makes specific predictions about the form and information content of signals. Since signal intensity is an indicator of sender quality, the signal itself must be graded, i.e., vary continuously along some perceivable axis. Furthermore, the cost of signal production must be linked to the specific type of quality information that receivers are interested in obtaining, i.e., the signal must 'use up' the specific quality being advertised. For example, if receivers require information about stamina, the signal should be energetically costly. Quality handicaps can provide information to female receivers about the health, vigor, foraging efficiency, or parental abilities of potential male mates. Well-documented examples are tail length in barn swallows *Hirundo rustica* (Møller et al. 1998) and color-patch hue and size in house finches *Carpodacus mexicanus* (Hill 1990, 1991, 1996). Quality handicaps can also provide information to same-sex competitor receivers about the stamina and fighting ability of the sender, as suggested for roaring in red deer *Cervus elaphus* (Clutton Brock & Albon 1979) and chasing contests in damselflies *Odonata* (Marden & Waage 1990; Marden & Rollins 1994).

When benefits are scaled, senders differ in the benefits they expect to obtain from the receiver's response. All senders must still pay a general cost that increases with increasing signal intensity (Godfray 1991; Maynard Smith 1991; Johnstone & Grafen 1992). At the ESS, senders with greater potential benefits have a higher optimal signal intensity than senders with lower benefits. Since the benefit of signaling varies as a function of resource valuation or need, the information transmitted by the sender's relative signal intensity is motivation, need, or the value of winning a contest. An example of such a signal is the begging display of nestling birds, where hungrier nestlings display with greater vigor or intensity (Mondloch 1995). The display itself does not need to be very costly if there is another cost, such as predator attraction, that also increases with signal intensity. This type of signal has been called a general handicap (Hasson 1997).

Vulnerability to receiver attack

Many signals do not appear to be costly to produce, but the execution of the signal necessarily places the sender in a position of vulnerability to attack and injury by the receiver. For example, the performance of a display may require that a sender physically move closer to a receiver opponent. The closer the sender is to a receiver, the more effective his display is, but the greater the risk the sender takes (Zahavi 1977). This type of cost is therefore receiver-dependent, and is only likely to be relevant in the context of agonistic interactions. This idea is described in models by Enquist et al. (1985) and Adams & Mesterton-Gibbons (1995). The models share many features with the quality handicap model, in that the actual performance of the signal determines the cost (risk), but in this case the receiver is responsible for imposing the cost. Signals stabilized by this mechanism have been called vulnerability handicaps.

As in the quality handicap model, the form of the display is linked to the information revealed by the signal. The information conveyed is assumed to involve intentions, aggressive motivation, or willingness to take risks. Vulnerability handicap signals are often discrete visual movement displays and postures. Good examples are the threat displays of goldfinches *Carduelis tristis* and little blue penguins *Eudyptula minor*, where different sets of threat displays are given at different distances from the opponent (Popp 1987; Waas 1990, 1991). Threats that move the sender closer to the opponent are both more effective in deterring the opponent and more likely to result in receiver attack than threats that move the sender away from the opponent. Other possible examples of vulnerability signals include short-range signals such as olfactory or low-amplitude vocal signals that can only be perceived when delivered close to the receiver. Attack intention movements are also effective only at close range. This type of signal form contains a tactical element, in that the sender may place himself in a more effective position to attack. Finally, displays that expose a vulnerable part of the body should also be considered vulnerability handicaps.

Receiver retaliation

Many signals are neither costly to produce nor risky to execute. Such signals appear to be associated with specific contexts by arbitrary convention, and they have therefore been called conventional signals (Sense 2 of Guilford & Dawkins 1995). Conventional signals can be stable without costs if there is no conflict between sender and receiver, i.e., they agree about the optimal receiver response (Davies 1981; Maynard Smith 1994). Conventional signals do occur in conflicting-interest situations, however, and in this case there must be some type of stabilizing cost. A receiver retaliation or punishment rule is the stabilizing factor here. Conventional signal game models in the agonistic context have been developed and expanded by Enquist (1985, Model 1), Johnstone & Norris (1993), Hurd (1997b), and Hurd & Enquist (1998). A critical feature of these models is the presence of two opponents who each know their own willingness or ability to fight, but not that of the opponent. Both individuals give one of two alternative signals, A or B, and then base their responses on the joint result. The evolutionarily stable decision policy is: 'Use signal A when strong or highly motivated and signal B when weak or not motivated, then attack opponents who signal an equal strength, flee from opponents who signal a stronger state, and pause-attack opponents who signal a weaker state.' A policy such as this one implies that competitive senders frequently test each other, especially those competitors giving equal-strength signals. The meaning of the alternative signals is therefore established by convention, not by differential production costs. A prediction of the model is that one could exchange the meaning of signals A and B (i.e., the coding rule) and still maintain an honest signaling system. Honesty is maintained by the high retaliation cost of signaling a strong state when actually weak, i.e., other truly strong competitors will attack such cheaters. Although conventional signals are most likely to occur in intrasexual agonistic contexts, it has recently been suggested that they could also occur in intersexual contexts if females retaliate against dishonest males by divorcing them (Viljugrein 1997).

Since there is no obligate link between signal form and information content, the design of conventional signals can be arbitrary. Signals are likely to consist of either a set of two alternative discrete signals or a graded signal that varies along one easily distinguished axis. Conventional signals do not need to be excessively costly to produce, nor need the alternative signal variants differ in production costs, but the weak versus strong signals should be easily distinguishable and thus antithetical in form (Hurd et al. 1995). Agonistic conventional signals can convey information about either motivation or fighting ability. Color badges of status appear to be good examples of graded conventional signals, where the size of the color patch is correlated with dominance status and condition (Rohwer 1982; Studd & Robertson 1985; Møller 1987; Maynard Smith & Harper 1988; Lemel & Wallin 1993; Qvarnström 1997). Olfactory conventional signals can also provide information about aggressive intentions (Poole 1989). Conventional auditory signals and visual displays can vary over shorter time periods than plumage color signals and are therefore more likely to be used to convey information about short-term changes in motivation or intentions.

Assigning signal categories to the variable parameters of bird song

The songs of birds, especially the oscines, are complex signals that typically vary along several structural, temporal, and/or geographic axes, depending on the species. Some of the variable parameters include song-type repertoire size, song-type sharing and matching patterns, switching rate, song rate, synchronous switching, song overlap, duration, and amplitude. These variable parameters presumably encode different kinds of information to receivers. A major reason for the complexity of oscine song is certainly the fact that it simultaneously serves two very different social functions, mate attraction and territorial defense (Searcy & Andersson 1986; Kroodsma & Byers 1991; Smith 1991). In the mate attraction context, female receivers are attempting to choose the mate that will maximize their fitness. They are therefore expected to pay attention to signal attributes that provide information about aspects of male quality such as age, parental ability, and/or territory quality. In the territory defense context, rival males are the primary receivers. As receivers, males need information about the identity and location of rival males, and as senders they need to both broadcast their presence and ownership status to surrounding neighbors and sometimes direct their threats to specific rivals. In addition, males engage in countersinging contests to assess each other's fighting ability, motivation to fight, and likelihood of attack versus withdrawal. The questions I attempt to address here are: Can these variable song parameters be assigned to one of the five signal categories based on the signaling context and type of cost, and does this classification exercise help us understand the types of information encoded in signals? A related goal is to evaluate whether the existing models are sufficient to describe all of the signal parameters, or whether additional signal categories or modifications are needed.

Assigning a signal or signal parameter to one of the five signal categories involves assessing the differential signal costs or benefits, the specific contexts, and the nature of receiver responses and benefits for the variants of the parameter. Although all of these are difficult tasks, it should be easier to quantify and compare the costs, contexts, and responses for avian auditory signals than for olfactory and visual signals because the displays are detectable to humans, temporally discrete, and relatively easy to describe quantitatively. In addition, vocal signals can be experimentally simulated with playback experiments, and receiver responses to the signal can be quantified. To qualify as an index signal, a signal parameter must be constrained by some type of physical or physiological process so it is impossible to exaggerate or cheat, and variants should not be associated with differential signal production costs. To qualify as a quality or general handicap signal, a signal parameter would minimally have to vary in its signal production costs. Furthermore, the more costly variants must be associated with some aspect of sender and/or receiver benefits such as attraction of mates, higher reproductive success, or repulsion of rivals. To distinguish between quality and general handicaps, one must determine whether signal costs are associated with long-term quality or instead with short-term motivation levels. Vulnerability handicap signals must possess a form that restricts their effectiveness to close receiver distances. It must be the case that that a strong threat signal leads to both a higher probability of winning and a higher risk of receiver attack as a consequence of the form of the signal. Finally, to qualify as a conventional signal, there must *not* be any differential signal production cost associated with parameter variants, but there should be an association between parameter variants and aggressive stimulus levels, and an association between parameter variants and the subsequent tendencies of the sender and receiver to attack versus retreat. More precisely, when a sender signals a strong threat, it should be more likely to subsequently approach and attack the opponent, and it should be at a greater risk of receiving an attack if the opponent also signals a strong threat. Comparative studies can be used to help distinguish between conventional signals and handicap/index signals: if the various species using a particular signal parameter employ different (opposite) coding rules (linkage of signal variants with contexts), the parameter is more likely to be an arbitrary conventional signal, whereas handicap and index signal parameters are more likely to show consistent coding rules across species.

In the survey that follows, some of the more commonly measured song parameters in birds are described. I then summarize the studies that found that variations in a particular parameter were associated with different social contexts, such as different receiver gender or different vocal contest escalation levels. Information about production costs, receiver responses, and benefits to senders and receivers is sometimes available. Complete information for assigning a song parameter to a given signal type (and excluding the other four) is rarely available, so the survey identifies the most likely signal classification given the limited data. The goal of this review is to highlight the specific types of data we need to collect to classify signals.

Song rate

Song rate is a parameter that birds can vary, even with a single song type. Song rate clearly involves a change in energy expenditure. In the Carolina wren *Thryothorus ludovicianus*, sage grouse *Centrocercus urophasianus*, and several frog species, oxygen consumption has been shown to increase rapidly with increased song rate (Bucher et al. 1982; Taigen & Wells 1985; Vehrencamp et al. 1989; Eberhardt 1994). This parameter therefore has the potential to serve as a quality or general handicap signal. In lek-mating sage grouse, males increase their display rate when females approach them, and females prefer to mate with males having higher long-term and short-term display rates (Vehrencamp et al. 1989; Gibson 1996). The most vigorous males expend twice as much energy per day as males that do not display. Males that display at high rates appear to be able to maintain their body condition (weight/tarsus length) at a lean but constant level throughout the display season. They also travel farther from the lek each day to forage. In contrast, males that display very little show a decrease in condition during the season, and also forage closer to the lek. Lower-quality males therefore appear to pay more for a given level of display. This evidence is consistent with the notion that better foragers can afford to display at a higher rate, and that females who prefer such males obtain better-quality mates.

In territorial passerines, song rate can also provide information to females about resource abundance on a male's territory. Studies of the pied flycatcher *Ficedula hypoleuca*, willow warbler *Phylloscopus trochilus*, and Ipswich sparrow *Passerculus sandwichensis princeps* have shown that song rate and time spent singing are positively correlated with territory quality, and that females prefer males with high song rates (Gottlander 1987; Radesäter et al. 1987; Reid 1987; Alatalo et al. 1990; Nyström 1997). In two cases, food supplement experiments demonstrated a direct effect of resource levels on foraging rate and song rate. Studies on several species with a female preference for high song rates (stonechat *Saxicola torquata*, willow tit *Parus montanus*, and willow warbler) have also found associations between male song rate and other aspects of male quality such as male condition, survival rate, and paternal care of nests and offspring (Greig-Smith 1982; Welling et al. 1997; Nyström 1997; Saino et al. 1997). These effects could also be a consequence of variation in territory quality instead of intrinsic male differences. Stronger evidence for intrinsic effects was found in chickadees *Parus atricapillus*, where dawn song output was correlated with winter flock dominance (Otter et al. 1996). All of the above studies argue that prolonged and repetitive singing is a costly signaling behavior (but see Gaunt et al. 1996 for an alternative view). Song rate is therefore most likely a quality handicap signal used by females to assess male and/or territory quality.

Elevated song rates are associated with escalated countersinging interactions between males in the song sparrow *Melospiza melodia*, great tit *Parus major*, bobolink *Dolichonyx oryzivorus*, red-winged blackbird *Agelaius phoeniceus*, and dunnock *Prunella modularis* (Kramer & Lemon 1983; Weary et al. 1988; Searcy & Yasukawa 1990; Capp & Searcy 1991; Langmore 1997; Bower 1999). In song sparrows, males sing at increased rates when presented with song playback, and even higher rates with song playback plus a mounted male visual stimulus (Kramer et al. 1985). Using a microphone array recording system, Bower (1999) showed that challenger song sparrow males dramatically increase their song rate from about 3 to 6 songs/min in the minutes prior to a physical fight. Defenders also increase their song rate, but their rate is typically half that of the challenger at any point in time prior to the fight. (Spontaneous song rates in this species range from 1 to 4 songs/min) It is difficult to argue that a few minutes of elevated singing (in contrast to hours of higher display rates in mate-attracting birds) results in a significant trade-off with body condition or survivorship, or that the defender (a territory owner) is not fully capable of a higher song rate. Bower (1999) found no difference in song rate between challengers who won versus lost, or between defenders who won versus lost a particular encounter. However, territorial takeovers consist of a series of encounters than may continue for several hours, and changes in song rate during this period could be informative. The challenger appears to be signaling his short-term willingness to fight, not his fighting ability per se. There are two possible interpretations for this signal parameter. It could be a conventional signal, with the receiver-imposed cost of fighting being the primary stabilizing factor. This classification implies that the cost differential for song rate in an agonistic context is small and unimportant. Hurd (1997b) shows that if a small cost is added to one of the signals in a conventional signaling system stabilized by a retaliation cost, the more costly signal could be assigned to either the stronger or weaker opponent without destabilizing the system. Alternatively, a high song rate just prior to a fight could impose a significant cost. Song rate could therefore be a general handicap signal of motivation stabilized by production costs plus the cost of a fight. Additional support for this second assignment comes from the fact that the coding rule of high song rate for high motivation is widespread and thus does not seem to be arbitrary.

Repertoire size

Repertoire size, measured as either number of song types or syllable types, varies from one to several hundred in different oscine species. In some species repertoire size seems to be infinitely large, but this is often the result of a fixed repertoire of song elements that are recombined in many different ways (e.g. sedge warbler *Acrocephalus schoenobaenus*, Catchpole 1976). Within a species, repertoire size usually varies among males. In essentially all species with repertoires greater than one that have been examined, females exhibit a directional preference for males with larger repertoire sizes (Searcy 1992; Searcy & Yasukawa 1996). Considerable effort has been spent recently to determine what benefits females might receive from selecting large-repertoire males. The most commonly found correlate of repertoire size is male age (Catchpole & Slater 1995). In the mockingbird *Mimus polyglottos*, red-winged blackbird, western meadowlark *Sturnella neglecta*, starling *Sturnus vulgaris*, pied flycatcher, sedge and great reed warblers *Acrocephalus arundinaceus*, and New World warblers, this correlation occurs because males continue to learn new song types after their first year (Howard 1974; Yasukawa et al. 1980; Catchpole 1986; Cozens & Sealy 1986; Eens et al. 1991; Horn et al. 1993; Lampe & Espmark 1994; Lemon et al. 1994; Aweida 1995; Birkhead et al. 1997). In song sparrows and great tits, the correlation is a consequence of the selective attrition of small-repertoire males (Lambrechts & Dhondt 1987; Hiebert et al. 1989). Older males usually have higher quality or larger territories, which provides a direct benefit to female reproduction. In addition, older males have proven their ability to survive, and can pass this ability on to their offspring (Hasselquist et al. 1996). Possible mechanisms for the correlation between repertoire size and male quality include variation in brain size for storing song types and nutritional condition of the bird during learning (Catchpole 1996; Nowicki et al. 1998). Repertoire size thus appears to be a physiologically constrained index signal of male age, condition, or survivability.

Switching rate

Switching rate is a parameter that can only be employed by species with more than one song type. In several repeat mode species (species that typically repeat a given song type several times before switching to the next type), switching rate has been found to increase in increasingly agonistic contexts, e.g., meadowlarks, Carolina wren, cardinal *Cardinalis cardinalis*, plain and tufted titmice *Baeolophus inornatus* and *B. bicolor*, chestnut-sided warbler *Dendroica pensylvania*, and song sparrow (Lemon 1968; Dixon 1969; Lein 1978; Falls & D'Agincourt 1982; D'Agincourt & Falls 1983; Schroeder & Wiley 1983; Kramer et al. 1985; Simpson 1985; Falls et al. 1990; Horn & Falls 1991). In the red-winged blackbird, dunnock, and banded wren *Thryothorus pleurostictus* the opposite pattern occurs: switching rate decreases during escalated encounters with neighboring or intruding males (Searcy & Yasukawa 1990; Langmore 1997; Molles & Vehrencamp 1999). Two of the repeat-mode singers mentioned above, cardinals and song sparrows, also decrease their switching rate if they enter into a bout of type-matching (Lemon 1968; Nielsen & Vehrencamp 1995). Variations in switching rate do not appear to involve a change in energetic cost, predation risk, or time lost, although Lambrechts and Dhondt (1988) have argued that more repetitive singing exhausts syringeal muscles and is more difficult to sustain at high rates and amplitudes in the great tit. On the other hand, a receiver retaliation cost in the form of a stronger approach to the sender (speaker) was found in two experimental studies of switching in the song sparrow (Nielsen & Vehrencamp 1995; Searcy et al. 1995). The minimal evidence (so far) of an obvious differential energetic cost, evidence for a retaliation cost, and opposite coding rules in different species, all suggest that switching rate is a conventional signal of aggressive intentions or motivation in competitive male contexts.

Switching rate is also used by many species in a mate attraction context. A survey of the direction of change in both switching rate and song rate in oscines with available data reveals at least four different sets of rules. These patterns are summarized in Table 2. In song sparrows, meadowlarks, and cardinals, males increase song rate and switching rate in both male and female contexts (D'Agincourt & Falls 1983; Ritchison 1988; Horn & Falls 1991). Banded wrens increase song and switching rates for female receivers and decrease both parameters in male contexts (Molles & Vehrencamp 1999). Red-winged blackbirds and dunnocks increase switching rate for female receivers while decreasing switching rate and increasing song rate for male receivers (Searcy & Yasukawa 1990; Langmore 1997). The only rule used by all of these species is to increase switching rate in the mate attraction context. As mentioned above, females in numerous species with song-type repertoires have been shown to prefer males with larger repertoire sizes (Searcy 1992). By increasing their switching rate, males make their repertoire size more apparent

Table 2. *Direction of change in song rate and switching rate from control periods or spontaneous singing when territorial males are presented with a female receiver stimulus and a male receiver stimulus.*

SPECIES	FEMALE CONTEXT		MALE CONTEXT	
	Song rate	Switching rate	Song rate	Switching rate
Song sparrow Cardinal Meadowlark	↑	↑	↑	↑
Banded wren	↑	↑	↓	↓
Red-winged blackbird Dunnock	↓	↑	↑	↓

to receivers (Horn & Falls 1991). This song parameter could therefore be operating as an amplifier of repertoire size (Hasson 1990). Alternatively, switching rate could be exploiting a sensory bias in female receivers (Searcy 1992). In species that increase switching rate in both male and female contexts, this song parameter has been interpreted as a conventional signal indicating general readiness to interact (Falls & D'Agincourt 1982).

Song-type and element sharing

In many repertoire species, males share some of their song types with their neighbors. Sharing level is typically measured by the index $S = 2N_s/(R_1 + R_2)$, where N_s is the number of song types shared between two males and R_1 and R_2 are their respective repertoire sizes. Sharing is very high in some species (cardinal, several titmice, several wrens; Lemon 1968; Dixon 1969; Verner 1975; Schroeder & Wiley 1983; Morton 1987; Catchpole & Rowell 1993), intermediate in others (meadowlark, great tit, New World warblers, chaffinch *Fringilla coelebs*, European blackbird *Turdus merula*, nightingale *Luscinia megarhynchos*, western populations of the song sparrow; Wolffgramm 1979; Krebs & Kroodsma 1980; Hultsch & Todt 1981; Falls 1984; Cozens & Sealy 1986; Spector 1992; Lemon et al. 1994; Byers 1995; Beecher et al. 1996), and rare in others (sedge wren *Cistothorus platensis*, eastern song sparrow; Kroodsma & Verner 1978; Hughes et al. 1998). Significant song-type sharing can arise from two different learning strategies: age-restricted learning plus a sedentary population structure, and open-ended learning (Slater 1989). The type of information provided by song sharing and the classification of this song element are quite different for these two learning mechanisms. With age-restricted learning, young birds memorize the song types of adult tutors in their natal or floater area during the few months after hatching and subsequently obtain territories close to these tutors (Kroodsma & Verner 1978; Beecher et al. 1994; Ewert & Kroodsma 1994; Nordby et al. 1999). Males that obtain territories close to their tutors have a high sharing index, whereas males that disperse farther have a lower chance of sharing song types with their neighbors. Sharing level could therefore serve as an unbluffable index of a male's dispersal distance from his natal or tutor area. If males always attempt to obtain territories near their tutors and only disperse farther if they are unable to do so, then sharing level could also indicate some aspect of male fighting ability. Several studies have shown that females prefer local males over immigrants or more distantly recorded conspecific songs (King et al. 1980; Baker et al. 1982; Balaban 1988; Searcy et al. 1997; Bensch et al. 1998). With the open-ended learning strategy, males older than one year of age can modify existing songs or learn new songs from neighbors, thereby adjusting the fraction of shared song types to the optimum level. Examples include the great tit, American redstart *Setophaga ruticilla* and probably other New World warblers, indigo bunting *Passerina cyanea*, village indigobird *Vidua chalybeata*, tufted titmouse, and saddleback *Philesturnus carunculatus* (Jenkins 1977; Payne 1981, 1985; McGregor & Krebs 1982, 1989; Lemon et al. 1994). Not only does this learning strategy result in a more uniform level of sharing among pairs of adjacent males, but it also prevents a male's repertoire from reflecting his natal origin. Receivers are therefore unlikely to obtain useful information about a male's quality or fighting ability from his sharing level, and the most likely function of sharing is song matching during male-male interactions.

Song matching

Whether learning is age-restricted or open-ended, possession of shared song types enables males to use different kinds of song-matching strategies during countersinging interactions. If neighbors share some, but not all, of their song types (partial repertoire overlap), they can opt to *type match* (sing the same song type as) a countersinging rival, *repertoire match* (sing a shared but not recently sung song type), or sing an *unshared song type*. In the cardinal, great tit, song sparrow, chaffinch, meadowlark, European blackbird, and several wrens, type matching appears to be an aggressive signal whereas repertoire matching and non-matching are associated with a lower probability of escalating (Hinde 1958; Lemon 1974; Kroodsma 1979; Krebs et al. 1981; Todt 1981; Falls 1984; McGregor et al. 1992; Beecher et al. 1996; J. Burt, unpublished; L. E. Molles & S. L. Vehrencamp, unpublished). Interactive playback experiments on several of these species have demonstrated that receivers retaliate (approach the speaker) more strongly to type-matching stimuli. Some species avoid matching playback song stimuli (western meadowlark, corn bunting *Milaria calandra*, wood thrush *Hylocichla mustelina*; Falls & Krebs 1975; McGregor 1986; Whitney 1991), whereas others increase their tendency to match (cardinal, tufted titmouse, hill mynah *Gracula religiosa*, banded wren; Lemon 1968; Bertram 1970; Schroeder & Wiley 1983; L. E. Molles & S. L. Vehrencamp, unpublished). Dominant song frequency (kHz) is another variable parameter that can be matched or not matched (Horn et al. 1992). It seems unlikely that singing a matching versus non-matching song type entails differential production costs in birds, although frequency variants could be associated with differential costs or constraints (Lambrechts 1996). Song-type matching strategies are therefore most likely to be conventional signals. Is there comparative evidence for variation in coding rules (i.e., the linkage between signal form and specific context)? Most of the species listed above that use type-matching as a strong threat have moderate repertoire sizes in which the probability of type-matching by chance is reasonably low. Non-matching is therefore the most likely default signal, and it makes sense that a matching strategy should be reserved for the rare aggressive occasions. A 'uniqueness' factor therefore predisposes the coding rule toward one direction.

Conclusions

Table 3 summarizes the assignments of the variable song parameters discussed above. Assignments have been placed in separate columns for female and male receiver contexts, since they are often different for the two sexes. Most parameters seemed to fit one of the categories well, but a few were either difficult to assign or differed depending on the species. In the female receiver context, most song parameters are either quality handicap or index signals. The strongest cases have been made for song rate as a quality handicap signal and for repertoire size as an index signal of male age. Females obtain the benefits of good-condition, healthy, and experienced males and/or resource-rich territories by paying attention to these two signal parameters. Another index signal females may use is the element or song-type sharing level of males relative to their neighbors, which could indicate a male's population of origin or his ability to acquire a territory close to his natal or tutor area if learning is age-restricted. By contrast, song parameters used in male receiver contexts tended to be either conventional or index signals. Conventional signals include switching rate, song matching strategies, and possibly song rate. Song rate could also be a general handicap signal. As with female receivers, repertoire size and song/element sharing level appear to operate as indices of male age and natal origin for male receivers.

Table 3. *Summary of signal classification for the variable parameters of bird song in both intersexual (female receiver) and intrasexual (male receiver) contexts. Q = quality handicap, G = general handicap, C = conventional, I = index, and A = amplifier.*

Song parameter	Tentative signal type	
	Intersexual	Intrasexual
Song rate	Q	G or C
Repertoire size	I	I
Switching rate	A	C
Song and element sharing	I	I
Song matching	—	C

Switching rate, repertoire size, and song rate were assigned to different signal categories for male and female receiver contexts. Such differences underscore the conclusion of Qvarnström and Forsgren (1998) that the features used by females to select mates are not necessarily those used to settle male-male conflicts. Switching rate in the female context is classified as an amplifier, since male senders in essentially all repertoire species increase their switching rates in the female receiver context as a way of maximizing apparent repertoire size. Song-type switching in the male receiver context is used either as a conventional signal of aggressive intentions or, in the case of synchronous switching, as a pointing index signal. Repertoire size also seems to provide different information to male and female receivers, and males in many species use their repertoires in different ways, or have separate repertoires, for the two contexts (Spector 1992; Wiley et al. 1994; Byers 1996). Males maximize the perception of their repertoire size when singing to females.

When singing to males, many species actually reduce their song-type diversity, and instead use specific song types or temporal patterns of song delivery (e.g., song matching, synchronous switching, song overlap). In species with song-type sharing the number of shared song types may be a more important determinant of a male's ability to defend his territory than repertoire size per se.

Song rate in the female receiver context could be clearly assigned to the quality handicap category, since males sing for long periods of time to attract females and their ability to sustain high rates depends on their condition or territorial resources. However, this parameter was more difficult to assign in the male receiver context. Where song rate is used in a graded fashion to indicate variable aggressive stimulus levels or intentions, a higher rate is associated with increasing aggressive intentions. The cost of the subsequent fight is more likely to be the stabilizing cost than the higher signal production cost, supporting a conventional signal classification. On the other hand, the higher production cost just before a fight could be more significant than we realize and the parameter may be a general handicap indicating motivation. In the only study that has quantified the cost of alternative threat displays and their strategic use during contests, Hack (1997) found that the winners of male-male cricket *Acheta domesticus* encounters used more costly displays and expended more energy during the fight. However, in this system, more costly acts are those that place the sender closer to his opponent, so tactical position strategies are playing a role. This system fits the vulnerability handicap model as well as the general handicap model.

A final generalization that emerges from this survey is that conventional signal parameters do not always meet the prediction of arbitrary or exchangeable coding rules. At least one apparently conventional parameter with no evidence of differential production cost and reasonable evidence of a retaliation cost, song matching, exhibits the same coding rule across many species. It was argued that a type match was unlikely to occur by chance and should therefore serve as the strong threat signal, compared to non-matching which is very likely to occur by chance. The evidence from comparative studies therefore does not help discriminate between conventional and handicap or index signals. Additional modeling may help to clarify the situations in which conventional signals are not exchangeable.

Acknowledgements

Rufus Johnstone, Peter McGregor, Peter Slater, Pete Hurd, John Bower, Laura Molles, Marc Dantzker, and Jack Bradbury provided fruitful discussions and useful comments on earlier drafts of this chapter. I would also like to thank the symposium organizers, Yngve Espmark, Trond Amundsen, and Gunilla Rosenqvist, for creating a stimulating and friendly environment in which to discuss the latest developments in the study of animal communication.

References

Adams, E. & Mesterton-Gibbons, M. 1995. The cost of threat displays and the stability of deceptive communication. *Journal of Theoretical Biology,* **175**, 405-421.

Alatalo, R. V., Glynn, D. & Lundberg, A. 1990. Singing rate and female attraction in the pied flycatcher: An experiment. *Animal Behaviour,* **39**, 601-603.

Aweida, M. K. 1995. Repertoires, territory size, and mate attraction in western meadowlarks. *Condor,* **97**, 1080-1083.

Baker, M. C., Spitler-Nabors, K. J. & Bradley, D. C. 1982. The response of female mountain white-crowned sparrows to songs from their natal dialect and an alien dialect. *Behavioral Ecology and Sociobiology,* **10**, 175-179.

Balaban, E. 1988. Cultural and genetic variation in swamp sparrows (*Melospiza georgiana*). II. Behavioral salience of geographic song variants. *Behaviour,* **105**, 292-322.

Beecher, M. D., Campbell, S. E. & Stoddard, P. K. 1994. Correlation of song learning and territory establishment strategies in the Song Sparrow. *Proceedings of the National Academy of Sciences, USA*, **91**, 1450-1454.

Beecher, M., Stoddard, P., Campbell, S. & Horning, C. 1996. Repertoire matching between neighbouring song sparrows. *Animal Behaviour,* **51**, 917-923.

Bensch, S., Hasselquist, D., Nielsen, B. & Hansson, B. 1998. Higher fitness for philopatric than for immigrant males in a semi-isolated population of great reed warblers. *Evolution,* **52**, 877-883.

Bertram, B. 1970. The vocal behaviour of the Indian hill mynah, *Gracula religiosa. Animal Behaviour Monographs*, **3,** 79-192.

Birkhead, T. R., Buchanan, K. L., DeVoogd, T. J., Pellatt, E. J., Szekely, T. & Catchpole, C. K. 1997. Song, sperm quality and testes asymmetry in the sedge warbler *Acrocephalus schoenobaenus. Animal Behaviour,* **53**, 965-971.

Bower, J. L. 1999. Acoustic interactions during naturally occurring territorial conflict in a song sparrow (*Melospiza melodia*) neighborhood. Ph.D. thesis, Cornell University, New York, USA.

Bradbury, J. W. & Vehrencamp, S. L. 1998. *The Principles of Animal Communication*. Sunderland, MA: Sinauer.

Bucher, T. L., Ryan, M. J. & Bartholomew, G. A. 1982. Oxygen consumption during resting, calling, and nest building in the frog *Physalaemus pustulosus. Physiological Zoology*, **55**, 10-22.

Byers, B. E. 1995. Song types, repertoires and song variability in a population of chestnut-sided warblers. *Condor,* **97**, 390-401.

Byers, B. E. 1996. Messages encoded in the songs of chestnut-sided warblers. *Animal Behaviour,* **52**, 691-705.

Capp, M. S. & Searcy, W. A. 1991. Acoustical communication of aggressive intentions by territorial male bobolinks. *Behavioral Ecology,* **2**, 319-326.

Catchpole, C. K. 1976. Temporal and sequential organization of song in the sedge warbler (*Acrocephalus schoenobaenus*). *Behaviour,* **59**, 226-246.

Catchpole, C. K. 1982. The evolution of bird sounds in relation to mating and spacing behavior. In: *Acoustic Communication in Birds* (Ed. by D. E. Kroodsma & E. H. Miller), pp. 297-319. New York: Academic Press.

Catchpole, C. K. 1986. Song repertoires and reproductive success in the great reed warbler, *Acrocephalus arundinaceus. Behavioral Ecology and Sociobiology,* **19**, 439-445.

Catchpole, C. K. 1996. Song and female choice: good genes and big brains? *Trends in Ecology and Evolution,* **11**, 358-360.

Catchpole, C. K. & Rowell, A. 1993. Song sharing and local dialects in a population of the European wren *Troglodytes troglodytes. Behaviour,* **125**, 67-78.

Catchpole, C. K. & Slater, P. J. B. 1995. *Bird Song: Biological Themes and Variations*. Cambridge: Cambridge University Press.

Clutton-Brock, T. H. & Albon, S. D. 1979. The roaring of red deer and the evolution of honest advertisement. *Behaviour,* **69**, 145-170.

Cosens, S. E. & Sealy, S. G. 1986. Age-related variation in song repertoire size and repertoire sharing of yellow warblers (*Dendroica petechia*). *Canadian Journal of Zoology,* **64**, 1926-1929.

D'Agincourt, L. G. & Falls, J. B. 1983. Variation of repertoire use in the eastern meadowlark, *Sturnella magna. Canadian Journal of Zoology,* **61**, 1086-93.

Davies, N. 1981. Calling as an ownership convention on pied wagtail territories. *Animal Behaviour,* **29**, 529-534.

Davies, N. B. & Halliday, T. R. 1978. Deep croaks and fighting assessment in toads *Bufo bufo. Nature,* **274**, 683-85.

Dixon, K. L. 1969. Patterns of singing in a population of the plain titmouse. *Condor,* **71**, 91-101.

Eberhardt, L. S. 1994. Oxygen consumption during singing by male Carolina wrens (*Thryothorus ludovicianus*). *Auk,* **111**, 124-130.

Eens, M., Pinxten, R. & Verheyen, R. F. 1991. Male song as a cue for mate choice in the European starling. *Behaviour,* **116**, 210-238.

Enquist, M. 1985. Communication during aggressive interactions with particular reference to variation in choice of behaviour. *Animal Behaviour,* **33**, 1152-1161.

Enquist, M. & Leimar, O. 1983. Evolution of fighting behaviour: decision rules and assessment of relative strength. *Journal of Theoretical Biology,* **127**, 387-410.

Enquist, M., Plane, E. & Röed, J. 1985. Aggressive communication in fulmars (*Fulmarus glacialis*) competing for food. *Animal Behaviour,* **33**, 1007-1020.

Ewert, D. N. & Kroodsma, D. E. 1994. Song sharing and repertoires among migratory and resident rufous-sided towhees. *Condor,* **96**, 190-196.

Falls, J. B. 1984. Song matching in western meadowlarks. *Canadian Journal of Zoology,* **63**, 2520-2524.

Falls, J. B. & Krebs, J. R. 1975. Sequences of songs in repertoires of western meadowlarks (*Sturnella neglecta*). *Canadian Journal of Zoology,* **53,** 1165-1178.

Falls, J. B. & D'Agincourt, L. G. 1982. Why do meadowlarks switch song types? *Canadian Journal of Zoology,* **60**, 3400-3408.

Falls, J. B., Dickinson, T. E. & Krebs, J. R. 1990. Contrast between successive songs affects the response of eastern meadowlarks to playback. *Animal Behaviour,* **39**, 717-728.

Gaunt, A. S., Bucher, T. L., Gaunt, S. L. L. & Baptista, L. F. 1996. Is singing costly? *Auk,* **113,** 718-721.

Gibson, R. M. 1996. Female choice in sage grouse: the roles of attraction and active comparison. *Behavioral Ecology and Sociobiology,* **39**, 55-60.

Godfray, H. C. J. 1991. Signalling of need by offspring to their parents. *Nature,* **352**, 328-330.

Gottlander, K. 1987. Variation in the song rate of the male pied flycatcher (*Ficedula hypoleuca*): causes and consequences. *Animal Behaviour,* **35**, 1037-1043.

Grafen, A. 1990. Biological signals as handicaps. *Journal of Theoretical Biology,* **144**, 517-546.

Greig-Smith, P. W. 1982. Song-rates and parental care by individual male stonechats (*Saxicola torquata*). *Animal Behaviour,* **30**, 245-52.

Guilford, T. & Dawkins, M. S. 1995. What are conventional signals? *Animal Behaviour,* **49**, 1689-1695.

Hack, M. 1997. Assessment strategies in the contests of male crickets, *Acheta domesticus* (L.). *Animal Behaviour,* **53**, 733-474.

Hasselquist, D., Bensch, S. & von Schantz, T. 1996. Correlation between male song repertoire, extra-pair paternity and offspring survival in the great reed warbler. *Nature,* **381**, 229-232.

Hasson, O. 1990. Amplifiers and their role in the evolution of male display. *Evolutionary Ecology*, **4**, 277-289.

Hasson, O. 1994. Cheating signals. *Journal of Theoretical Biology,* **167**, 223-238.

Hasson, O. 1997. Towards a general theory of biological signaling. *Journal of Theoretical Biology,* **185**, 139-156.

Hiebert, S. M., Stoddard, P. K. & Arcese, P. 1989. Repertoire size, territory acquisition, and reproductive success int the song sparrow. *Animal Behaviour,* **37**, 266-273.

Hill, G. E. 1990. Female house finches prefer colourful mates: sexual selection for a condition-dependent trait. *Animal Behaviour,* **40**, 563-572.

Hill, G. E. 1991. Plumage color is a sexually selected indicator of male quality. *Nature,* **350**, 337-339.

Hill, G. E. 1996. Redness as a measure of the production cost of ornamental coloration. *Ethology Ecology and Evolution,* **8**, 157-175.

Hinde, R. A. 1958. Alternative motor patterns in chaffinch song. *Animal Behaviour,* **6**, 211-218.

Horn, A. G. & Falls, J. B. 1991. Song switching in mate attraction and territory defense by western meadowlarks (*Sturnella neglecta*). *Ethology*, **87**, 262-268.

Horn, A. G., Leonard, M. L., Ratcliffe, L., Shackleton, S. A. & Weisman, R. G. 1992. Frequency variation in songs of black-capped chickadees *(Parus atricapillus*). *Auk,* **109**, 847-852.

Horn, A. G., Dickinson, T. E. & Falls, B. 1993. Male quality and song repertoires in western meadowlarks (*Sturnella neglecta*). *Canadian Journal of Zoology,* **71**, 1059-1061.

Howard, R. D. 1974. The influence of sexual selection and interspecific competition on mockingbird song (*Mimus polyglottos*). *Evolution,* **28**, 428-438.

Hughes, M., Nowicki, S., Searcy, W. A. & Peters, S. 1998. Song-type sharing in song sparrows: implications for repertoire function and song learning. *Behavioral Ecology and Sociobiology,* **42**, 437-446.

Hultsch, H. & Todt, D. 1981. Repertoire sharing and song-post distance in nightingales (*Luscinia megarhynchos* B.). *Behavioral Ecology and Sociobiology,* **8**, 183-188.

Hurd, P. L. 1997a. Game theoretical perspectives on conflict and biological communication. Ph.D. thesis, Stockholm University, Stockholm, Sweden.

Hurd, P. L. 1997b. Is signalling of fighting ability costlier for weaker individuals? *Journal of Theoretical Biology,* **184**, 83-88.

Hurd, P. L.& Enquist, M. 1998. Conventional signalling in aggressive interactions: the importance of temporal structure. *Journal of Theoretical Biology,* **192**, 197-211.

Hurd, P. L., Wachtmeister, C. A. & Enquist, M. 1995. Darwin's principle of antithesis revisited: a role for perceptual biases in the evolution of intraspecific signals. *Proceedings of the Royal Society of London, Series B,* **259**, 201-205.

Jakobsson, S., Radesäter, T. & Järvi, T. 1979. On the fighting behaviour of *Nannacara anomala* (Pisces, Cichlidae) males. *Zeitschrift für Tierpsychologie,* **25**, 257-314.

Jenkins, P. F. 1977. Cultural transmission of song patterns and dialect development in a free-living bird population. *Animal Behaviour,* **25**, 57-63.

Johnstone, R. 1997. The evolution of animal signals. In: *Behavioral Ecology: an Evolutionary Approach,* 4th edn (Ed. by J. R. Krebs & N. B. Davies), pp. 155-178. London: Blackwell.

Johnstone, R. A. & Grafen, A. 1992. The continuous Sir Philip Sidney game: a simple model of biological signalling. *Journal of Theoretical Biology,* **156**, 215-234.

Johnstone, R. A. & Norris, K. 1993. Badges of status and the cost of aggression. *Behavioral Ecology and Sociobiology,* **32**, 127-134.

King, A. P., West, M. J. & Eastzer, D. H. 1980. Song structure and song development as potential contributions to reproductive isolation in cowbirds (*Molothrus ater*). *Journal of Comparative and Physiological Psychology,* **94**, 1028-1039.

Kramer, H. G. & Lemon, R. E. 1983. Dynamics of territorial singing between neighboring song sparrows (*Melospiza melodia*). *Behaviour,* **85**, 198-223.

Kramer, H. G., Lemon, R. E. & Morris, M. J. 1985. Song switching and agonistic stimulation in the song sparrow (*Melospiza melodia*): five tests. *Animal Behaviour,* **33**, 135-149.

Krebs, J. R., Ashcroft, R. & van Orsdol, K. 1981. Song matching in the great tit *Parus major*. *Animal Behaviour,* **29**, 918-923.

Krebs, J. R. & Kroodsma, D. E. 1980. Repertoires and geographical variation in bird song. *Advances in the Study of Behavior,* **11**, 143-177.

Kroodsma, D. E. 1979. Vocal dueling among marsh wrens. *Auk,* **96**, 506-515.

Kroodsma, D. E. & Byers, B. E. 1991. The function(s) of bird song. *American Zoologist,* **31**, 318-328.

Kroodsma, D. E. & Verner, J. 1978. Complex singing behaviors among *Cistothorus* wrens. *Auk,* **95**, 703-716.

Lambrechts, M. L. 1996. Organization of birdsong and constraints on performance. In: *Ecology and Evolution of Acoustic Communication in Birds* (Ed. by D. E. Kroodsma & E. H.Miller), pp. 305-320. Ithaca, NY: Cornell University Press.

Lambrechts, M. & Dhondt, A. A. 1987. Male quality, reproduction, and survival in the great tit (*Parus major*). *Behavioral Ecology and Sociobiology,* **19**, 57-63.

Lambrechts, M. M. & Dhondt, A. A. 1988. The anti-exhaustion hypothesis: a new hypothesis to explain song performance and song switching in the great tit. *Animal Behaviour,* **36**, 327-3334.

Lampe, H. M. & Espmark, Y. O. 1994. Song structure reflects male quality in pied flycatchers, *Ficedula hypoleuca*. *Animal Behaviour,* **47**, 869-876.

Langmore, N. E. 1997. Song switching in monandrous and polyandrous dunnocks, *Prunella modularis*. *Animal Behaviour,* **53**, 757-766.

Lein, M. R. 1978. Song variation of chestnut-sided warblers: its nature and suggested significance. *Canadian Journal of Zoology,* **56**, 1226-1283.

Lemel, J. & Wallin, K. 1993. Status signalling, motivational condition and dominance: an experimental study in the great tit, *Parus major*. *Animal Behaviour,* **45**, 549-558.

Lemon, R. E. 1968. The relation between organization and function of song in cardinals. *Behaviour,* **32**, 158-177.

Lemon, R. E. 1974. Song dialects, song matching and species recognition by cardinals *Richmondena cardinalis*. *Ibis,* **116**, 545-548.

Lemon, R. E., Perrault, S. & Weary, D. 1994. Dual strategies of song development in American redstarts, *Setophaga ruticilla*. *Animal Behaviour,* **47**, 317-329.

Marden, J. H. & Rollins, R. A. 1994. Assessment of energy reserves by damselflies engaged in aerial contests for mating territories. *Animal Behaviour,* **48**, 1023-1030.

Marden, J. H. & Waage, J. K. 1990. Escalated damselfly territorial contests are energetic wars of attrition. *Animal Behaviour,* **39**, 954-959.

Maynard Smith, J. 1991. Honest signalling: the Philip Sidney game. *Animal Behaviour,* **42**, 1034-1035.

Maynard Smith, J. 1994. Must reliable signals be costly? *Animal Behaviour,* **47**, 1115-1120.

Maynard Smith, J. & Harper, D. G. C. 1988. The evolution of aggression: can selection generate variability? *Philosophical Transactions of the Royal Society of London, Series B,* **319**, 557-570.

Maynard Smith, J. & Harper, D. 1995. Animal signals: models and terminology. *Journal of Theoretical Biology,* **177**, 305-311.

McGregor, P. K., Dabelsteen, T., Shepherd, M. & Pedersen, S. B. 1992. The signal value of matched singing in Great Tits: evidence from interactive playback experiments. *Animal Behaviour,* **43**, 987-998.

Møller, A. P. 1987. Social control of deception among status signalling house sparrows *Passer domesticus*. *Behavioral Ecology and Sociobiology,* **20**, 307-311.

Møller, A. P., Barbosa, A., Cuervo, J. J., de Lope, F., Merino, S. & Saino, N. 1998. Sexual selection and tail streamers in the barn swallow. *Proceedings of the Royal Society of London, Series B*, **165,** 409-414.

Molles, L. E. & Vehrencamp, S. L. 1999. Repertoire size, repertoire overlap, and singing modes in the banded wren, *Thryothorus pleurostictus*. *Auk*, **116**, 677-689.

Mondloch, C. J. 1995. Chick hunger and begging affect parental allocation of feeding in pigeons. *Animal Behaviour,* **49**, 601-613.

Morton, E. S. 1987. The effects of distance and isolation on song-type sharing in the Carolina wren. *Wilson Bulletin,* **99**, 601-610.

Nielsen, B. M. B. & Vehrencamp, S. L. 1995. Responses of song sparrows to song-type matching via interactive playback. *Behavioral Ecology and Sociobiology,* **37**, 109-117.

Nordby, J. C., Campbell, S. E. & Beecher, M. D. 1999. Ecological correlates of song learning in song sparrows. *Behavioral Ecology,* **10**, 287-297.

Nowicki, S., Peters, S. & Podos, J. 1998. Song learning, early nutrition and sexual selection in songbirds. *American Zoologist,* **38,** 179-190.

Nur, N. & Hasson, O. 1984. Phenotypic plasticity and the handicap principle. *Journal of Theoretical Biology,* **110**, 275-297.

Nyström, K. G. K. 1997. Food density, song rate, and body condition in territory-establishing willow warblers (*Phylloscopus trochilus*). *Canadian Journal of Zoology,* **75**, 47-58.

Otter, K., Chruazcz, B. & Ratcliff, L. 1997. Honest advertisement and song output during the dawn chorus of black-capped chickadees. *Behavioral Ecology,* **8,** 167-173.

Parker, G. A. 1979. Sexual selection and sexual conflict. In: *Sexual Selection and Reproductive Competition in Insects* (Ed. by M. S. Blum & N. A. Blum), pp. 173-201. New York: Academic Press.

Payne, R. B. 1981. Song learning and social interactions in indigo buntings. *Animal Behaviour,* **29**, 688-697.

Payne, R. B. 1985. Behavioral continuity and change in local song populations of village indigobirds *Vidua chalybeata. Zeitschrift für Tierpsychologie,* **70**, 1-44.

Poole, J. 1989. Announcing intent: the aggressive state of musth in African elephants. *Animal Behaviour,* **37**, 140-152.

Popp, J. W. 1987. Risk and effectiveness in the use of agonistic displays by American goldfinches. *Behaviour,* **103**, 141-156.

Qvarnström, A. 1997. Experimentally increased badge size increases male competition and reduces male paternal care in the collared flycatcher. *Proceedings of the Royal Society of London, Series B*, **264**, 1225-1231.

Qvarnström, A. & Forsgren, E. 1998. Should females prefer dominant males? *Trends in Ecology and Evolution*, **13**, 498-501.

Radesäter, T., Jakobsson, S., Andbjer, N., Bylin, A. & Nyström, K. 1987. Song rate and pair formation in the willow warbler, *Phylloscopus trochilus. Animal Behaviour,* **35**, 1645-1651.

Ramer, J. D., Jenssen, T. A. & Hurst, C. 1983. Size-related variation in the advertisement call of *Rana clamitans* (Anura: Ranidae) and its effect on conspecific males. *Copeia*, **1983**, 141-155.

Reid, M. L. 1987. Costliness and reliability in the singing vigour of Ipswich sparrows. *Animal Behaviour,* **35**, 1735-1743.

Ritchison, G. 1988. Song repertoires and the singing behavior of male northern cardinals. *Wilson Bulletin,* **100**, 583-603.

Robertson, J. G. M. 1986. Male territoriality, fighting and assessment of fighting ability in the australian frog *Uperoleria rugosa. Animal Behaviour,* **34**, 763-772.

Rohwer, S. A. 1982. The evolution of reliable and unreliable badges of fighting ability. *American Zoologist,* **33**, 1325-1331.

Saino, N., Galeotti, P., Sacchi, R. & Møller, A. P. 1997. Song and immunological condition in male barn swallows (*Hirundo rustica*). *Behavioral Ecology,* **8**, 364-371.

Schroeder, D. J. & Wiley, R. H. 1983. Communication with shared song themes in tufted titmice. *Auk,* **100**, 414-424.

Searcy, W. A. 1992. Song repertoire and mate choice in birds. *American Zoologist,* **32**, 71-80.

Searcy, W. A. & Andersson, M. 1986. Sexual selection and the evolution of song. *Annual Reviews of Ecology and Systematics*, **17**, 507-534.

Searcy, W. A. & Yasukawa, K. 1990. Use of the song repertoire in intersexual and intrasexual contexts by male red-winged blackbirds. *Behavioral Ecology and Sociobiology,* **27**, 123-128.

Searcy, W. A. & Yasukawa, K. 1996. Song and female choice. In: *Ecology and Evolution of Acoustic Communication in Birds* (Ed. by D. E. Kroodsma & E. H. Miller), pp. 454-473. Ithaca, NY: Cornell University Press.

Searcy, W. A., Podos, J., Peters, S. & Nowicki, S. 1995. Discrimination of song types and variants in song sparrows. *Animal Behaviour,* **49**, 1212-1226.

Searcy, W. A., Nowicki, S. & Hughes, M. 1997. The response of male and female song sparrows to geographic variation in song. *Condor,* **99**, 651-657.

Simpson, B. S. 1985. Effects of location in territory and distance from neighbours on the use of song repertoires by Carolina wrens. *Animal Behaviour,* **33**, 793-804.

Slater, P. B. J. 1981. Chaffinch song repertoires: observations, experiments and a discussion of their significance. *Zeitschrift für Tierpsychologie*, **56**, 1-24.

Slater, P. B. J. 1989. Bird song learning: causes and consequences. *Ethology, Ecology and Evolution*, **1,** 19-46.

Smith, W. J. 1991. Singing is based on two markedly different kinds of signaling. *Journal of Theoretical Biology,* **152**, 241-253.

Spector, D. A. 1992. Wood-warbler song systems: a review of Paruline singing behaviors. *Current Ornithology*, **9**, 199-238.

Studd, M. V. & Robertson, R. J. 1985. Evidence for reliable badges of status in territorial yellow warblers (*Dendroica petechia*). *Animal Behaviour,* **33**, 1102-1113.

Taigen, T. L. & Wells, K. D. 1985. Energetics of vocalization by an anuran amphibian (*Hyla versicolor*). *Journal of Comparative Physiology, Series B*, **155**, 163-170.

Thapar, V. 1986. *Tigers: Portrait of a Predator*. London: Collins.

Todt, D. 1981. On functions of vocal matching: effect of counter-replies on song post choice and singing. *Zeitschrift für Tierpsychologie,* **57**, 73-93.

Vehrencamp, S. L., Bradbury, J. B. & Gibson, R. M. 1989. The energetic cost of display in male sage grouse. *Animal Behaviour,* **38**, 885-896.

Verner, J. 1975. Complex song repertoire of male long-billed marsh wrens in eastern Washington. *Living Bird*, **14**, 263-300.

Viljugrein, H. 1997. The cost of dishonesty. *Proceedings of the Royal Society of London, Series B*, **264**, 815-821.

Waas, J. R. 1990. An analysis of communication during aggressive interactions of little blue penguins (*Eudyptula minor*). In: *Penguin Biology* (Ed. by L. S. Davis & J. T. Darby), pp. 366-371. San Diego: Academic Press.

Waas, J. R. 1991. The risks and benefits of signalling aggressive motivation: a study of cave-dwelling little blue penguins. *Behavioral Ecology and Sociobiology,* **29**, 139-146.

Weary, D. M., Krebs, J. R., Eddyshaw, R., McGregor, P. K. & Horn, A. 1988. Decline in song output by great tits: exhaustion or motivation? *Animal Behaviour,* **36**, 1242-1244.

Welling, P. P., Rytkönen, S. D. & Orell, M. I. 1997. Song rate correlates with paternal care and survival in willow tits: advertisement of male quality? *Behaviour,* **134**, 891-904.

Whitney, C. L. 1991. Avoidance of song matching in the wood thrush: a field experiment. *Wilson Bulletin,* **103**, 96-101.

Wiley, R. H., Godard, R. & Thompson, A. D. Jr. 1994. Use of two singing modes by hooded warblers as adaptations for signalling. *Behaviour,* **129**, 243-278.

Wolffgramm, J. 1979. Vergleich von hoch-variablen Gesangsvorträgen benachbarter Amseln. *Verhandlungen der Deutschen Zoologischen Gesellschaft*, **72**, 237.

Yasukawa, K., Blank, J. L. & Patterson, C. B. 1980. Song repertoires and sexual selection in the red-winged blackbird. *Behavioral Ecology and Sociobiology,* **7**, 233-238.

Zahavi, A. 1975. Mate selection - a selection for a handicap. *Journal of Theoretical Biology,* **53**, 205-214.

Zahavi, A. 1977. Reliability in communication systems and the evolution of altruism. In: *Evolutionary Ecology* (Ed. by B. Stonehouse & C. Perrins), pp. 253-259. London: Macmillan.

Zahavi, A. & Zahavi, A. 1997. *The Handicap Principle*. Oxford: Oxford University Press.

IN: ESPMARK,Y., AMUNDSEN,T. & ROSENQVIST,G. (EDS.) 2000
ANIMAL SIGNALS:
SIGNALLING AND SIGNAL DESIGN IN ANIMAL COMMUNICATION, PP. 301-315
TAPIR ACADEMIC PRESS, TRONDHEIM, NORWAY

14

Male-Male Competition and Female Choice in the Evolution of Vocal Signaling

WILLIAM A. SEARCY[1] & STEPHEN NOWICKI[2]

[1]Department of Biology, University of Miami, Coral Gables, FL 33124, USA

E-mail: wsearcy@umiami.ir.miami.edu

[2]Evolution, Ecology, and Organismal Biology Group, Department of Zoology, Duke University, Durham, NC 27708, USA

Abstract

Song in male songbirds has both a male-female function, in attracting females and stimulating them to reproduce, and a male-male function, most often in territory defense. Much is known about the attributes that adapt song to the male-female function; complexity, local song structure, and high song output in particular have been shown to be important. In contrast, relatively little is known about the features adapting song to its male-male function. Various aspects of song and singing behavior have been hypothesized to function in territory defense, notably song type matching and song type switching, but these hypotheses remain little tested. The major problem is that the experimental design typically used to investigate male-male communication, territorial playback, does not test function. We discuss two experimental designs that can be used to investigate adaptation to the male-male function of song in songbirds. In one (speaker occupation), the researcher removes owners from their territories and measures the effect on intruder pressure of playback with and without the feature of interest. In the second, the researcher elicits singing behavior from the subject and tests whether features of the song predict subsequent attacks on a simulated intruder. Further work is needed with both designs.

Males produce elaborate vocalizations termed song in many species of acoustic insects and anurans as well as in the songbirds. In all these groups, the principal functions of song are thought to lie, on the one hand, in attracting females and stimulating them to mate, and on the other in getting the better of rival males in male-male aggressive competition (Searcy & Andersson 1986). Presumably, then, song is adapted to fulfill these two functions – that is, the features of song and singing behavior have been shaped by natural selection to fulfill both a male-female courtship function and a male-male aggressive function. For songbirds, we will argue that we know a substantial amount about how song is adapted for the male-female function: we have experimental evidence from a range of songbird species showing that specific features of singing behavior affect the function of song in attracting and stimulating females, and we are beginning to understand for some of these features why females have evolved to respond to them. By contrast, we know relatively little about parallel adaptations for the male-male function of song. Certain features of song and singing behavior are consistently associated with male-male interactions, but little direct evidence exists to support the hypothesis that these features actually influence the behavior of male receivers in ways that are adaptive to signalers.

Male-female function

In the songbirds, evidence for the male-female function of song is extensive and compelling. The evidence includes both correlations between measures of male mating success and particular aspects of song (e.g. Howard 1974; Catchpole 1980; Hasselquist et al. 1996; Buchanan & Catchpole 1997) and experimental evidence that females are attracted to male song in the field (Eriksson & Wallin 1986; Mountjoy & Lemon 1991; Johnson & Searcy 1996). In addition, extensive experimental evidence exists demonstrating how song features are adapted for the male-female function. Much of this evidence comes from the copulation solicitation display assay, in which captive females are treated with estradiol and then tested for response to song playback (King & West 1977; Searcy & Marler 1981; Searcy 1992). Females respond to appropriate song with copulation solicitation, a display which is given by female birds in nature immediately prior to and during copulation, and which thus provides a convincing measure of the effectiveness of song in stimulating female mating behavior. Additional evidence on the male-female function comes from phonotaxis experiments, in which females are tested for tendency to approach song playback (Miller 1979; Clayton 1988).

Among the features of song that have been demonstrated to affect its male-female function, we will mention three that seem to be of widespread importance.

Complexity

Female songbirds respond preferentially to more complex song relative to less complex song in a number of species. For example, in song sparrows *Melospiza melodia*, females give more solicitation display in response to four song types than to single song types, more to eight song types than to four, and more to 16 than to eight (Searcy & Marler 1981;

Searcy 1984). Preferential response to larger repertoires has also been demonstrated in sedge warblers *Acrocephalus schoenobaenus* (Catchpole et al. 1984), great reed warblers *Acrocephalus arundinaceus* (Catchpole et al. 1986), and great tits *Parus major* (Baker et al. 1986). Further examples are given in Searcy & Yasukawa (1996).

Local song structure

Female songbirds in general respond preferentially to songs recorded near their breeding sites relative to conspecific songs recorded at more distant sites. Again, we can use song sparrows as an example: female song sparrows captured at a site in Pennsylvania gave approximately five times as much solicitation display for local songs than for songs recorded 500 km away (Searcy et al. 1997). More surprisingly, females also gave significantly more display for local songs than for songs recorded just 35 km away (W. A. Searcy, M. Hughes & S. Nowicki, unpublished). Preferences for local songs over foreign have also been shown in brown-headed cowbirds *Molothrus ater* (King et al. 1980), yellowhammers *Emberiza citrinella* (Baker et al. 1987), and red-winged blackbirds *Agelaius phoeniceus* (Searcy 1990). Searcy (1992) gives additional examples.

Song output

Female songbirds often show preferences for high song output, in terms of song rates, song length, or proportion of the day spent singing. In pied flycatchers *Ficedula hypoleuca*, males with song rates artificially increased by food provisioning are more successful in attracting females than are unfed controls (Alatalo et al. 1990). In zebra finches *Taenopygia guttata*, females prefer to approach and associate with males singing at a rapid rate compared to males singing at a slower rate (Collins et al. 1994). Female white-throated sparrows *Zonotrichia albicollis* court more in response to playback with more notes per time than to playback with fewer notes (Wasserman & Cigliano 1991).

The evolution of this last type of preference seems relatively easy to explain. Current theory holds that signals of quality can be honest if they are costly along a relevant dimension (Grafen 1990). High song output presumably is costly, perhaps in energy, but certainly in time taken from foraging and other activities. Song output should therefore be an honest signal of a male's energy balance, and could therefore be used by a female to assess either direct benefits to be obtained from food supplies on a male's territory and from paternal care for her offspring, or indirect benefits in good genes for foraging ability and general viability.

Why song complexity might be reliably linked to male quality is not as readily apparent, but Nowicki et al. (1998) have recently proposed an explanation for how such a link might come about. According to their 'nutritional stress hypothesis', song may be a reliable indicator of male quality because the brain nuclei underlying song learning and production develop during the period just before and after independence, when young birds are especially likely to experience nutritional stress. Individuals will differ in the amount of stress they experience and in their developmental response to a given level of stress, and these individual differences will affect both brain development and the development of the phenotype in general. Variation in brain development will in turn affect song learning. By choosing males based on the quantity and quality of learned song, females can thus choose males of greater developmental stability and higher phenotypic quality. Aspects of song complexity, such as song repertoire size, reflect the amount of learning, while a male's ability to reproduce local song reflects the quality of learning. Thus the nutritional stress hypothesis potentially can explain both the remaining categories of female preferences based on song.

Male-male function

The idea that bird song functions in male-male competition, and in particular in male-male competition for territory, is an old one. Howard (1920) discussed this hypothesis in his classic book on territory in birds, and supported the hypothesis with observational evidence, for example showing that male birds tend to sing when on territory but not when off, and that the seasonal onset of song tends to coincide with the acquisition of territory. Such coincidences have other explanations, of course; for example, both these patterns would be predicted by the female attraction function of song, given that a male needs a territory as well as song to attract a female. We need to go beyond such correlations and seek direct experimental evidence that singing males fare better in territory defense than do non-singing males.

Specific aspects of singing behavior have been hypothesized to function in male-male competition for territory, such as song type matching, repertoire matching, and song type switching (see also Vehrencamp, this volume). Song type matching means that a singing male chooses among multiple song types in his repertoire in order to reply to a rival male with the song type that the rival has just sung. Song type matching has been demonstrated in a variety of species of songbirds (e.g. Lemon 1968; Krebs et al. 1981; Falls 1985; Stoddard et al. 1992), and has been widely interpreted as an aggressive signal (Brémond 1968; Armstrong 1973; Krebs et al. 1981). Repertoire matching means that a male replies to a rival, not with the song type that the rival has just sung, but with any shared type. Beecher et al. (1996, 1997) have demonstrated repertoire matching in song sparrows, and have suggested that this signal is more aggressive than non-matching but less aggressive than song type matching. Switching refers simply to changing song types in the context of repeated singing. Males are known to increase switching rates in aggressive contexts in a number of species (Falls & D'Agincourt 1982; Kramer & Lemon 1983; Simpson 1985). Song type matching, repertoire matching, and song type switching have all been hypothesized to function in male-male competition for territories, but direct experimental tests of these functions have not been made.

Male-male communication via song in birds has been subject to extensive experimentation, but the great majority of experiments in this area do not bear on adaptive function. The predominant experimental design used in investigating male-male communication is territorial playback, in which song is played from a speaker placed on a male's territory, and the owner's response is monitored. This design works well for measuring discrimination and inferring recognition, and forms the basis for most studies of species recognition, dialect discrimination, neighbor-stranger discrimination, and individual recognition in birds (Becker 1982; Falls 1982; Godard 1991; Stoddard 1996). The design has also been used to compare response to matching and non-matching song types (e.g. McGregor et al. 1992) and to different switching patterns (Nielsen & Vehrencamp 1995). Territorial playback does not, however, address function in any direct way.

The function of song in male-male interactions is to aid in territory defense, which presumably means repelling or intimidating other males, both neighboring territory owners and non-territorial floaters. Territorial playback, however, measures the response of owners on their own territories rather than the response of outside males, and what is measured is tendency to approach the song and give aggressive displays, not repulsion or submissiveness. In the solicitation display assay, the behavioral response that is measured, female copulation solicitation, is of obvious benefit to a singing male, and we can reasonably assume that the greater the response elicited the better for the singer. In territorial playback, by contrast, the behavioral response that is measured, aggressive behavior from a territory owner, is not of any direct benefit to a singing male, and we cannot ass-ume that the greater the response elicited the better. In fact, we are not safe in inferring from response to territorial playback that song has any male-male function at all. Suppose instead that song in a particular species functions purely in attracting females for mating; if so, it would still be adaptive for a territory owner to respond aggressively to any conspecific song heard on his territory, as hearing such song would signal to the owner the presence of an intruder attempting to mate with resident females.

Furthermore, we know that the strongest aggressive signals do not always elicit the greatest response in territorial playback. In great reed warblers, for example, males produce short songs in natural and simulated encounters with other males (Catchpole 1983). Males singing undisturbed on their territories produce longer, more complex songs, but only up to the time of pairing (Catchpole 1983). In territorial playbacks, the apparently female-directed long songs produce a greater response in territory owners, i.e. closer approach, than do the seemingly male-directed short songs (Catchpole 1983). Similarly, in willow warblers *Phylloscopus trochilus*, songs with 'Type A' syllables are produced more frequently in agonistic encounters than in undisturbed singing, and are especially associated with imminent attack (Järvi et al. 1980). In territorial playback, Type A songs are more likely than normal song to elicit withdrawal, rather than approach. In chestnut-sided warblers *Dendroica pensylvanica* a male-directed song type again produces lower response in territorial playbacks than does a female-directed type (Ficken & Ficken 1970). The opposite pattern, of male-directed song producing a stronger response, does occur (e.g. Nelson & Croner 1991), but cannot be said to predominate. These examples illustrate the difficulty of drawing inferences about the male-male function of song from responses to territorial playback.

Two experimental designs have been used that do test the male-male function of song more directly: muting and speaker occupation. In muting experiments, territory owners are deprived of the ability to vocalize, either by sectioning the hypoglossal nerve (Peek 1972;

Smith 1976), which innervates the syrinx, or by puncturing the interclavicular air sac (Smith 1979), which when under pressure pushes the syringeal membranes into the bronchial air flow. Muting provides very convincing evidence of the territory defense function of song, as in McDonald's (1989) work with Scott's seaside sparrow *Ammodramus maritimus*.

McDonald (1989) muted a total of 21 territorial males over two years; 13 sham-operated and 30 undisturbed males served as controls. Rates of intrusions by other males were about 25-30 times higher on the territories of muted males compared to controls. Muted males on average lost 79% of their territories while devocalized, whereas control males tended to increase their territory size by encroaching on territories of muted males. Muted males gradually regained the ability to sing over about a 12 day period, and most were able to regain their lost territory once able to sing again. Fortuitously, 'muted' males were able to produce vocalizations other than song, thus eliminating a possible criticism of the experiment as a test specifically of song function.

A limitation of muting experiments is that the manipulation is all or nothing; singing is either eliminated or not, and there is no way to manipulate particular aspects of singing behavior, such as matching, switching, or repertoire size. The second experimental method for investigating male-male function, speaker occupation, does not suffer this limitation. In speaker occupation, a territorial male is removed from his territory, and loudspeakers are set out in his place. Songs can then be broadcast from the speakers, and their effect on intruder pressure compared to some control treatment, such as silence or a presumably neutral sound stimulus. Alternatively, more complex treatments can be used to test for more subtle effects; for example, one could compare effects on intrusion of broadcasting large vs. small repertoires, local vs. foreign songs, and so forth.

The speaker occupation design can be illustrated using a study we recently completed with song sparrows (Nowicki et al. 1998). Male song sparrows are territorial and monogamous (Nice 1937, 1943). Our experiments were performed in northwestern Pennsylvania, where male song sparrows hold territories primarily in old fields. Our procedure was first to choose a pair of nearby territories, and randomly designate one as an experimental territory and one as a control. Pairs of territories were chosen in the same old field, so that they would be matched as far as possible for habitat and local song sparrow density. Next, we removed the owners of both territories as simultaneously and as near dawn as possible, using mist nets and playback of the owners' own songs. We then placed two speakers on the experimental territory, and from them played songs we had recorded from the original owner, with playback alternating between the two speakers. We left the second territory as a silent control, unoccupied by speakers. Finally, we observed each of the two territories for intrusion by other male song sparrows, either neighbors or unmarked males that we presumed were floaters. In the first year of the study, we continued playback and observations until 20:00 on the day of removal. During the second and third year, we continued playback until dusk on the day of removal, resumed playback at dawn the following day, and continued playback and observations until 10:00.

We performed the experiment on a total of 11 pairs of territories over the three years of the study. Partial or full takeovers occurred first on the control (silent) territory in eight pairs, first on the experimental territory in no pairs, and on neither territory in three pairs. The null hypothesis that takeovers are equally likely to occur first on control and experimental territories can be rejected ($P < 0.005$ by a one-tailed sign test).

Speaker occupation experiments such as this have been carried out in only a handful of species (Table 1). Most of the experiments have been tests of whether song *per se* is effective in territory defense, and thus have used treatments such as song vs. no song, or song vs. noise, as in our experiment (Table 1, section A). These experiments in general have supported the effectiveness of song in territory defense, though not all have used sample sizes large enough to allow statistical tests. Only three experiments have examined the importance of specific features of song or singing behavior in competition for territory (Table 1, section B). In great tits, Krebs et al. (1978) found that reoccupation was slower on territories defended by repertoires than on territories defended by single song types, but no statistical test was presented. The remaining two experiments, using red-winged blackbirds, did find statistically significant effects: trespass rates were lower on territories defended by repertoires compared to ones defended by single song types (Yasukawa 1981), and songs of the original owners were better at deterring neighbors from trespassing than were songs of strangers (Yasukawa et al. 1982). No results are available on the patterns of singing most often hypothesized to function in male-male competition, such as matching and switching.

It is understandable that no speaker occupation experiments have been done to test the function of song matching or switching, because such experiments pose considerable technical difficulties. With matching, one problem is whom one would match with playback, and when. Neighboring territory owners could be matched via interactive playback (McGregor et al. 1992) as soon as the territory owner was removed, but non-neighbors such as floaters could be matched only when they arrived on the focal territory and started to sing. Anticipating which non-neighbors would intrude, so that recordings of their songs could be readied for playback, might often be impossible.

A second problem is more general, and stems from the low rate of measurable response that occurs in many speaker occupation experiments. Our song sparrow experiment can be used to illustrate this problem. We succeeded in demonstrating a significant effect in this experiment with only 11 pairs of trials, but the two treatments that we used, song vs. no song, are just those two we would expect to produce the largest treatment effect. Effects with other pairs of treatments, such as matching vs. non-matching songs, would probably be smaller, perhaps considerably smaller. Suppose, for example, that matching turns out to be twice as effective as non-matching, in the sense that a territory defended by a non-matching playback is twice as likely to be occupied first by an intruder compared to a territory defended by a matching playback. Then a simple power analysis tells us that to get a significant result at a minimal alpha level (0.05 one-tailed), we would need a sample of 30 pairs of territories. This is 30 pairs of territories in which some intrusion or takeover occurs; in our experiments we had no intrusions in 27% of our pairs, so we would need

Table 1. *The results of speaker occupation experiments in passerine birds. Section A gives studies testing whether song functions in territory defense; section B gives studies testing whether particular song features affect defense.*

Species	Effects found	Statistically Significant?	Reference
A. Effects of song vs. no song			
Thrush nightingale *Luscinia luscinia*	song delays reoccupation relative to no song	?	Göransson et al. 1974
Great tit *Parus major*	song delays reoccupation relative to no song or noise	?	Krebs 1977
Red-winged blackbird *Agelaius phoeniceus*	fly throughs lowered by single song type relative to no song	yes	Yasukawa 1981
White-throated sparrow *Zonotrichia albicollis*	song delays intrusions and limits number of intrusions relative to no song	yes	Falls 1988
Song sparrow *Melospiza melodia*	song delays intrusions relative to no song	yes	Nowicki et al. 1998
B. Effects of song features			
Great tit *Parus major*	multiple song types delay reoccupation relative to single song types	?	Krebs et al. 1978
Red-winged blackbird *Agelaius phoeniceus*	multiple song types lower non-neighbor trespass rates relative to single type	yes	Yasukawa 1981
	playback of owner's songs lowers neighbor trespass relative to playback of stranger songs	yes	Yasukawa et al. 1982

about 41 pairs to get a sufficient number of results. These 41 pairs mean 82 removals. Given that each experiment takes 18 hours of observation plus many days of preparatory work in banding and recording birds, mapping territories, etc., this is a formidable amount of work that is projected.

Nevertheless, additional experimentation is needed in order to make progress on studying adaptation for the male-male function of song in songbirds. We suggest two approaches. The first is to bite the bullet and proceed with the relevant speaker occupation experiments. If this course is taken, we recommend using species and populations in which intrusion pressure is known beforehand to be high. Some aspect of intrusion by rival males is the responding variable measured in all speaker occupation experiments, and if intrusion rates are low enough that no intrusion is observed in a given trial, even on silent control territories, then nothing is learned from that trial. High intrusion pressure is one key to the success of the red-winged blackbird experiments (Table 1). We know that in red-winged blackbirds more than half the males in most populations lack territories, that rates of trespassing on occupied territories are high, and that takeovers of empty territories can occur extremely rapidly (Searcy & Yasukawa 1995). Because intrusion pressure is high in redwings, Yasukawa (1981a, b; Yasukawa et al. 1982) was able in his speaker occupation experiments to use relatively short observation periods and still observe enough response that he could employ quantitative measures of intrusion, thereby increasing the power of his statistical tests.

Thus to perform a test of whether switching, for example, functions in territory defense, one should choose a species in which switching seems to be used as a signal, and a population in which intrusion pressure is thought to be high. One could then use relatively short observation periods and so perform greater numbers of trials, while getting fewer null results, and thus arrange a statistically powerful test of the effect of switching. Unfortunately, for song type matching and repertoire matching the problem of whom one would match and when would still remain.

The second approach, which we recommend when speaker occupation seems impractical, is to test whether particular song features predict aggression from the singer. If one could demonstrate that matching, for example, was a signal of aggressive intent towards other males, this would be strong circumstantial evidence that matching functions in male-male aggressive competition, though not as direct a demonstration as in speaker occupation. Some earlier work with song features approximates such a test. For example, Capp & Searcy (1991) gave 23 male bobolinks *Dolichonyx oryzivorus* an opportunity to attack a caged male on their territories, and compared the song (prior to attack) of the 12 males that did attack to the 11 that did not. Attackers sang shorter songs than non-attackers, in particular immediately prior to attacks. We know of no similar experiments that have yet been done with matching or switching.

Conclusion

Much is known already about the features that adapt song and singing behavior to the male-female function of song in songbirds, and we are beginning to learn why females have evolved to respond to these features. Progress has lagged in understanding adaptations of song to its male-male function, both in terms of how particular features enhance this function and also in terms of why males have evolved to respond to these features. This lack of progress is not due to lack of interest in male-male function, but rather to the lack of simple methods for studying the effect of song on male response in an evolutionarily relevant context. We have suggested some partial solutions to this problem, but these will require much time-consuming work. Perhaps by better defining the questions we are asking about male-male function, we eventually will develop both better methods and better theory.

An additional point we have not discussed, but which bears on our understanding of male-male competition and female choice in the evolution of vocal signaling, is to consider whether male function or female function might have evolved first in a 'dual function' signal such as bird song (Berglund et al. 1996). Although it is generally agreed that song currently serves a dual function – as a signal to females on the one hand and to males on the other – the signal could have evolved to serve this dual function in different ways (Berglund et al. 1996; Pope 1998). Song could have evolved first as a signal to females, with males responding as eavesdroppers (McGregor et al., this volume). Once males start responding as eavesdroppers, then new selective forces on signal evolution might arise due to this male response, but in the context of adaptations for male-female function that already have arisen. Alternatively, song may have evolved first as a signal of competitive ability in male-male interactions. In this case, females may respond to song as eavesdroppers if information about competitive ability also is useful in assessing potential mates, and then female response could begin to further shape song evolution. A third possibility, of course, is that song arose initially as a dual function signal, and that neither female or male function evolutionarily preceded the other. It may be impossible to determine which of these three evolutionary scenarios applies to bird song, although comparative methods may provide some insight into this question (see Pope 1998 for an example in a different signaling system). Berglund et al. (1996) suggest that male function may generally precede female function in the evolution of dual signals, but we are tempted to argue that the opposite may be true for bird song. We must admit, however, that our position on this issue is not yet substantiated by data, and perhaps stems simply from the fact that more is known about the female side of the story. In any case, further investigation of which function is primary will be an important step in understanding the evolution of song.

Acknowledgements

We thank the organizers, Yngve Espmark, Gunilla Rosenqvist, and Trond Amundsen, for inviting us to participate, and for all their assistance. We thank Sandy Vehrencamp and Brian Wisenden for comments on the manuscript. Financial support for our work was provided by National Science Foundation Grants IBN-9523635 to WAS and IBN-9408360 to SN.

References

Alatalo, R. V., Glynn, C. & Lundberg, A. 1990. Singing rate and female attraction in the pied flycatcher: an experiment. *Animal Behaviour*, **39**, 601-603.

Armstrong, E. A. 1973. *A Study of Bird Song*. New York: Dover.

Baker, M. C., Bjerke, T. K., Lampe, H. & Espmark, Y. 1986. Sexual response of female great tits to variation in size of males' song repertoires. *American Naturalist*, **128**, 491-498.

Baker, M. C., Bjerke, T. K., Lampe, H. U. & Espmark, Y. O. 1987. Sexual response of female yellowhammers to differences in regional song dialects and repertoire sizes. *Animal Behaviour*, **35**, 395-401.

Becker, P. H. 1982. The coding of species-specific characteristics in bird sounds. In: *Acoustic Communication in Birds*, Vol. 1 (Ed. by D. E. Kroodsma & E. H. Miller), pp. 213-252. New York: Academic Press.

Beecher, M. D., Stoddard, P. K., Campbell, S. E. & Horning, C. L. 1996. Repertoire matching between neighbouring song sparrows. *Animal Behaviour*, **51**, 917-923.

Beecher, M. D., Nordby, J. C., Campbell, S. E., Burt, J. M., Hill, C. E. & O'Loghlen, A. L. 1997. What is the function of song learning in songbirds? *Perspectives in Ethology*, **12**, 77-97.

Berglund, A., Bisazza, A. & Pilastro, A. 1996. Armaments and ornaments: an evolutionary explanation of traits of dual utility. *Biological Journal of the Linnean Society*, **58**, 385-399

Brémond, J. C. 1968. Recherches sur la sémantique et les éléments vecteurs d'information dans les signaux acoustiques du rouge-gorge (*Erithacus rubecula* L.) *La Terre et la Vie*, **2**, 109-220.

Buchanan, K. L. & Catchpole, C. K. 1997. Female choice in the sedge warbler, *Acrocephalus shoenobaneus*: multiple cues from song and territory quality. *Proceedings of the Royal Society of London, Series B*, **264**, 521-526.

Capp, M. S. & Searcy, W. A. 1991. Acoustical communication of aggressive intentions by territorial male bobolinks. *Behavioral Ecology*, **2**, 319-326.

Catchpole, C. K. 1980. Sexual selection and the evolution of complex songs among European warblers of the genus *Acrocephalus*. *Behaviour*, **74**, 149-166.

Catchpole, C. K. 1983. Variation in the song of the great reed warbler *Acrocephalus arundinaceus* in relation to mate attraction and territorial defence. *Animal Behaviour*, **31**, 1217-1225.

Catchpole, C. K., Leisler, B. & Dittami. J. 1986. Sexual differences in the responses of captive great reed warblers (*Acrocephalus arundinaceus*) to variation in song structure and repertoire size. *Ethology,* **73**, 69-77.

Catchpole, C. K., Dittami, J. & Leisler, B. 1984. Differential responses to male song repertoires in female songbirds implanted with oestradiol. *Nature*, **312**, 563-564.

Clayton, N. S. 1988. Song discrimination learning in zebra finches. *Animal Behaviour*, **36**, 1016-1024.

Collins, S. A., Hubbard, C. & Houtman, A. M. 1994. Female mate choice in the zebra finch: the effect of male beak colour and male song. *Behavioral Ecology and Sociobiology*, **35**, 21-25.

Eriksson, D. & Wallin, L. 1986. Male bird song attracts females: a field experiment. *Behavioral Ecology and Sociobiology*, **19**, 297-299.

Falls, J. B. 1982. Individual recognition by sound in birds. In: *Acoustic Communication in Birds*, Vol. 2 (Ed. by D. E. Kroodsma & E. H. Miller), pp. 237-278. New York: Academic Press.

Falls, J. B. 1985. Song matching in western meadowlarks. *Canadian Journal of Zoology*, **63**, 2520-2524.

Falls, J. B. 1988. Does song deter intrusion in white-throated sparrows (*Zonotrichia albicollis*)? *Canadian Journal of Zoology*, **66**, 206-211.

Falls, J. B. & D'Agincourt, L. G. 1982. Why do meadowlarks switch song types? *Canadian Journal of Zoology*, **60**, 3400-3408.

Ficken, M. S. & Ficken, R. W. 1970. Responses of four warbler species to playback of their two song types. *Auk,* **87**, 296-304.

Godard, R. 1991. Long-term memory of individual neighbours in a migratory songbird. *Nature*, **350**, 228-229.

Göransson, G., Högstedt, G., Karlsson, J., Källander, H. & Ulfstrand, S. 1974. Sångens roll för revirhållandet hos näktergal *Luscinia luscinia* — några experiment med play-back-teknik. *Vår Fågelvärld*, **33**, 201-209.

Grafen, A. 1990. Sexual selection unhandicapped by the Fisher process. *Journal of Theoretical Biology*, **144**, 473-516.

Hasselquist, D., Bensch, S. & von Schantz, T. 1996. Correlation between male song repertoire, extra-pair paternity and offspring survival in the great reed warbler. *Nature*, **381**, 229-232.

Howard, E. 1920. *Territory in Bird Life*. London: Collins.

Howard, R. D. 1974. The influence of sexual selection and interspecific competition on mockingbird song (*Mimus polyglottos*). *Evolution*, **28**, 428-438.

Järvi, T., Radesäter, T. & Jakobsson, S. 1980. The song of the willow warbler *Phylloscopus trochilus* with special reference to singing behaviour in agonistic situations. *Ornis Scandinavica*, **11**, 236-242.

Johnson, L. S. and Searcy, W. A. 1996. Female attraction to male song in house wrens (*Troglodytes aedon*). *Behaviour*, **133**, 357-366.

King, A. P. and West, M. J. 1977. Species identification in the North American cowbird: appropriate responses to abnormal song. *Science*, **195**, 1002-1004.

King, A. P., West, M. J. & Eastzer, D. H. 1980. Song structure and song development as potential contributors to reproductive isolation in cowbirds (*Molothrus ater*). *Journal of Comparative and Physiological Psychology*, **94**, 1028-1039.

Kramer, H. G. & Lemon, R. E. 1983. Dynamics of territorial singing between neighboring song sparrows (*Melospiza melodia*). *Behaviour*, **85**, 198-223.

Krebs, J. R. 1977. Song and territory in the great tit *Parus major*. In: *Evolutionary Ecology* (Ed. by B. Stonehouse & C. Perrins), pp. 47-62. Baltimore: University Park Press.

Krebs, J. R., Ashcroft, R. & van Orsdol, K. 1981. Song matching in the great tit *Parus major* L. *Animal Behaviour*, **29**, 918-923.

Krebs, J. R., Ashcroft, R. & Webber, M. 1978. Song repertoires and territory defence in the great tit. *Nature*, **271**, 539-542.

Lemon, R. E. 1968. The relation between organisation and function of song in cardinals. *Behaviour*, **32**, 158-178.

McDonald, M. V. 1989. Function of song in Scott's seaside sparrow, *Ammodramus maritimus peninsulae*. *Animal Behaviour*, **38**, 468-485.

McGregor, P. K., Dabelsteen, T., Shepherd, M. & Pedersen, S. B. 1992. The signal value of matched singing in great tits: evidence from interactive playback experiments. *Animal Behaviour*, **43**, 987-998.

Miller, D. B. 1979. Long-term recognition of father's song by female zebra finches. *Nature*, **280**, 389-391.

Mountjoy, D. J. and Lemon, R. E. 1991. Song as an attractant for male and female European starlings and the influence of song complexity on their response. *Behavioral Ecology and Sociobiology*, **28**, 97-100.

Nelson, D. A. & Croner, L. J. 1991. Song categories and their functions in the field sparrow (*Spizella pusilla*). *Auk*, **108**, 42-52.

Nice, M. M. 1937. Studies in the life history of the song sparrow. I. A population study of the song sparrow. *Transactions of the Linnean Society of New York*, **4**, 1-247.

Nice, M. M. 1943. Studies in the life history of the song sparrow. II. The behavior of the song sparrow and other passerines. *Transactions of the Linnean Society of New York*, **6**, 1-328.

Nielsen, B. M. B. & Vehrencamp, S. L. 1995. Responses of song sparrows to song-type matching via interactive playback. *Behavioral Ecology and Sociobiology*, **37**, 109-117.

Nowicki. S., Peters, S. & Podos, J. 1998. Song lcarning, carly nutrition and sexual selection in songbirds. *American Zoologist*, **38**, 179-190.

Nowicki, S., Searcy, W. A. & Hughes, M. 1998. The territory defense function of song in song sparrows: a test with the speaker occupation design. *Behaviour*, **135**, 615-628.

Peek, F. W. 1972. An experimental study of the territorial function of vocal and visual display in the male red-winged blackbird (*Agelaius phoeniceus*). *Animal Behaviour*, **20**, 112-118.

Pope, D. S. 1998. The fiddler crab claw waving display: function and evolution of a sexually selected signal. Ph.D. Thesis, Department of Zoology, Duke University, Durham, NC, USA.

Searcy, W. A. 1984. Song repertoire size and female preferences in song sparrows. *Behavioral Ecology and Sociobiology*, **14**, 281-286.

Searcy, W. A. 1990. Species recognition of song by female red-winged blackbirds. *Animal Behaviour*, **40**, 1119-1127.

Searcy, W. A. 1992. Measuring responses of female birds to male song. In: *Playback Studies of Animal Communication* (Ed. by P. K. McGregor), pp. 175-189. New York: Plenum.

Searcy, W. A. & Andersson, M. 1986. Sexual selection and the evolution of song. *Annual Review of Ecology and Systematics*, **17**, 507-533.

Searcy, W. A. & Marler, P. 1981. A test for responsiveness to song structure and programming in female sparrows. *Science*, **213**, 926-928.

Searcy, W. A., Nowicki, S. & Hughes, M. 1997. The response of male and female song sparrows to geographic variation in song. *Condor*, **99**, 651-657.

Searcy, W. A. & Yasukawa, K. 1995. *Polygyny and Sexual Selection in Red-winged Blackbirds.* Princeton, NJ: Princeton University Press.

Searcy, W. A. & Yasukawa, K. 1996. Song and female choice. In: *Ecology and Evolution of Acoustic Communication in Birds* (Ed. by D. H. Kroodsma & E. H. Miller), pp. 454-473. Ithaca, NY.: Cornell University Press.

Simpson, B. S. 1985. Effects of location in territory and distance from neighbors on the use of song repertoires by Carolina wrens. *Animal Behaviour*, **33**, 793-804.

Smith, D. G. 1976. An experimental analysis of the function of red-winged blackbird song. *Behaviour*, **56**, 136-156.

Smith, D. G. 1979. Male singing ability and territory integrity in red-winged blackbirds (*Agelaius phoeniceus*). *Behaviour*, **68**, 193-206.

Stoddard, P. K. 1996. Vocal recognition of neighbors by territorial passerines. In: *Ecology and Evolution of Acoustic Communication in Birds* (Ed. by D. H. Kroodsma & E. H. Miller), pp. 356-376. Ithaca, NY.: Cornell University Press.

Stoddard, P. K., Beecher, M. D., Campbell, S. E. & Horning, C. L. 1992. Song-type matching in the song sparrow. *Canadian Journal of Zoology*, **70**, 1440-1444.

Wasserman, F. E. & Cigliano, J. A. 1991. Song output and stimulation of the female in white-throated sparrows. *Behavioral Ecology and Sociobiology*, **29**, 55-59.

Yasukawa, K. 1981a. Song and territory defense in the red-winged blackbird. *Auk*, **98**, 185-187.

Yasukawa, K. 1981b. Song repertoires in the red-winged blackbird (*Agelaius phoeniceus*): a test of the Beau Geste hypothesis. *Animal Behaviour*, **29**, 114-125.

Yasukawa, K., Bick, E. I., Wagman, D.W. & Marler, P. 1982. Playback and speaker-replacement experiments on song-based neighbor, stranger and self discrimination in male red-winged blackbirds. *Behavioral Ecology and Sociobiology*, **10**, 211-215.

IN: ESPMARK,Y., AMUNDSEN,T. & ROSENQVIST,G. (EDS.) 2000
ANIMAL SIGNALS:
SIGNALLING AND SIGNAL DESIGN IN ANIMAL COMMUNICATION, PP. 317-327
TAPIR ACADEMIC PRESS, TRONDHEIM, NORWAY

15

Why Female Birds Sing

NAOMI E. LANGMORE

Department of Zoology,University of Cambridge,
Downing Street, Cambridge, CB2 3EJ, UK

Present address: Division of Botany and Zoology, Australian National University,
Camberra, ACT 0200, Australia

E-mail: naomi.langmore@anu.edu.au

Abstract

In birds, males are usually the more competitive sex so it is usually the male that sings to attract a mate and defend a territory. However, when competition between females arises, female birds may also sing for broadly the same functions as males. Recent studies show that female birds sing to defend territories, attract mates and perhaps also to defend their mate against other females. Although these functions are broadly the same as the functions of male song, the causes of intrasexual competition differ between the sexes. It was shown that females sang to defend resources when male defence was lacking or when breeding vacancies were unusually scarce. Females sang to attract mates and possibly to defend mates in polygynous and polygynandrous mating systems, when there was competition between females for male parental care.

Bird song is generally considered to be the preserve of male birds and song is often defined and distinguished from calls by the fact that it is produced only by males. For example, Catchpole & Slater (1995) defined bird song as 'long, complex vocalizations produced by males during the breeding'.

The reason that males rather than females sing is related to the operational sex ratio; there are usually more males than females available to mate at any one time, so males have to compete with one another for matings (Clutton-Brock & Vincent 1991). Since males are the more competitive sex, it makes sense that males rather than females use song to defend territories and attract mates.

The key point here is competition; males use song to compete with one another. The argument that I will make in this chapter is that although males are usually the more competitive sex, competition between females is also common, and when competition arises between female birds they often sing (see also Langmore 1998). As the following examples illustrate, females may have to compete in a wide range of circumstances (Slagsvold & Lifjeld 1994), for example when nest sites or breeding vacancies are scarce, when male territorial defence is periodically lacking, in species that defend year-round territories, in polygynous or polygynandrous species, and in species that experience intraspecific brood parasitism. Under at least some of these circumstances, females sing in the same contexts as males, that is for mate attraction and territorial defence. Here I review some of the evidence from recent experimental studies that lead to this conclusion.

Territorial defence

This is a broad term which incorporates both resource defence and mate defence. Male song may serve either or both of these functions. In most songbird species the mate and resources are contained in the same area, and males probably sing to defend both simultaneously (e.g. great tits *Parus major*; Slagsvold et al. 1994). By contrast, in the dunnock *Prunella modularis* male song appears to function primarily for mate defence. Males compete to defend female territories (Davies 1992) and song rate is positively correlated with the number of females a male is defending (Langmore 1996). Males cease singing if the females are removed from the territory during the breeding season, and also after cessation of breeding even though they remain on the territory (Langmore 1996). In other species males occupy solitary winter territories, and sing to defend resources in the absence of a mate (e.g. European robins *Erithacus rubecula*; Lack 1965). Resource defence and mate defence are discussed separately here because the circumstances that favour female song for resource defence appear to differ from those that favour female song for mate defence.

Resource defence

Resource defence appears to be a common function of female song. Females may use song to defend resources throughout the year, seasonally, or only in exceptional circumstances when female-female competition is unusually intense.

For example, in the superb fairy-wren *Malurus cyaneus* females sing at high rates throughout the year (Cooney & Cockburn 1995). Females sang at the highest rates when establishing a new territory, and they sang in response to playbacks of both male and female song, so Cooney and Cockburn suggested that females used their songs for territorial defence. Why do female fairy-wrens need to defend their territories, rather than leaving it to their mate as females of other species do? The answer probably lies in the unusual mating system of this species. Superb fairy-wrens have one of the highest known rates of cuckoldry in birds; 76% of offspring are fathered by males outside the social group (Mulder et al. 1994). In order to obtain extra-pair matings, males spend large amounts of time absent from the territory engaging in extra-pair courtship displays, leaving the territory undefended. Therefore, in this species females can not always rely on males to defend the territory (Cooney & Cockburn 1995).

Another species in which females are accomplished songsters is the European robin (Lack 1965). During the breeding season, male robins defend the territory and females do not sing. However, in the autumn the pair separates and the male and female set up individual winter territories. Female testosterone levels rise during this period (Kriner & Schwabl 1991) and they use song to defend their winter territories. Females will also sing in response to playback of male or female song during this period (Hoelzel 1986). The defence of individual winter territories by robins is thought to be necessary because of the scarcity of food during the winter months. Females of several other species also appear to use song for this function, including mockingbirds *Mimus polyglottos* (Lack 1965), loggerhead shrikes *Lanius ludovicianus* (Lack 1965) and wren tits *Chamaea fasciata* (Inman 1986).

Females may also use song to defend resources in exceptional circumstances when competition is unusually intense. For example, female song sparrows *Melospiza melodia* do not normally sing, but Arcese et al. (1988) observed that in some years females sang prior to nest-building, the period when territory turnover and settlement by yearlings was most common. Most females sang during territorial conflicts between female intruders and female territory owners, and females never sang when males intruded on the territory. Moreover, the years in which female song occurred were characterised by high population densities, depressed average breeding success and increased variation in breeding success. In this species, it seems that female song is related to competition with other females for breeding space and occurs only at times when breeding vacancies are unusually scarce.

Finally, resource defence also appears to be a function of the female component of the duet of bay wrens *Thryothorus nigricapillus*. In this species the female always initiates the duet and the male replies. Levin (1996a, b) found that both paired and unpaired females responded strongly to playback of female song and weakly to playback of male song. Female song rates were unaffected by the removal of their mates and females successfully defended their territories in the absence of their mates. She suggested that in bay wrens the songs of females function in resource defence against other females, and a duet is formed because the male answers female songs as a means of advertising her mated status and thereby guarding her from other males. Why do female bay wrens need to defend their territories? One explanation has been suggested by Morton (1996), who argues that when the territory is maintained year-round over many years, as is often the case in tropical birds, females would benefit by being able to retain the territory in the event of the death or divorce of their mate. The ability to retain the territory would be an advantage to female bay wrens, because Levin (1996a) observed a very high divorce or death rate, with only 62% of all pairs remaining together for even the first two months of the breeding season. This argument might also explain the female component of duets in other tropical species.

These four examples illustrate the different forms that female song for resource defence can take; year-round solo song in fairy-wrens, seasonal solo song in robins, facultative solo song in song sparrows, and year-round duetting in bay wrens. Despite these different forms, it seems that female song for resource defence mainly occurs in species where male defence is periodically lacking or when intrasexual competition for breeding sites is unusually intense.

Mate defence

The evidence that females use song for mate defence is suggestive, but not conclusive so far. Females may need to defend their mate from other females if there is a risk of polygyny, which can entail reduced parental care from the male (Slagsvold & Lifjeld 1994). Mate defence has been proposed to explain both duetting and solo song of female birds.

In duetters, the argument is that members of a mated pair answer each others' songs to advertise their mated status and thus deter same-sex rivals (Brown & Lemon 1979). As discussed previously, this may be the function of the male replies to female songs in bay wrens. However, if females are defending their mates we would expect the reverse pattern; duets comprising female replies to male songs. This is not always easy to detect because many duetters are sexually monomorphic, but studies of several duetters revealed that females often do answer their mate's songs. P.J.B. Slater et al. (unpublished) used molecular techniques to sex yellow-crowned gonoleks *Laniarius barbarus* and mohos *Hypergerus atriceps* and found that all duets comprised female replies to male songs. Similarly, Sonnenschein & Reyer (1983) found that many duets of the slate-coloured boubou *Laniarius funebris* comprised female replies to male songs. There was also some circumstantial evidence that females sang to guard their mate in this species, because females answered the song of their mate more often at times when the risk of losing him to other females was highest, that is, before commencement of breeding and after nest failure.

Mate defence might also be the function of female solo songs in polygynous species. A literature survey shows a marked association between opportunistic polygyny and female song, with almost all species for which female solo song is noted also having opportunistically polygynous mating systems (Table 1). In all species female song occurred in the context of female-female aggression and in six species female song was also elicited experimentally using simulated or actual female intrusions (Table 1).

Table 1. *Polygynous species in which female song has been observed, showing the percentage of females that bred in polygynous groups. In all species female song was associated with female-female aggression and in six species female song was elicited by simulating female intrusions using either playbacks or model, caged or released females.*

Species with female song	% female in polygynous groups	Female song associated with females aggession?	Female song in response to simulated female intrusions?	Female song reference	Polygyny reference
Red-winged blackbird *Agelaius phoeniceus*	69-99	Yes	Yes	Beletsky 1983	Searcy & Yasukawa 1995
Starling *Sturnus vulgaris*	33-75	Yes	Yes	Sandell & Smith 1997	Pinxten & Eens 1990
Great reed warbler *Acrocephalus arundinaceus*	40-50	Yes	Yes	Bensch & Hasselquist 1992	Bensch & Hasselquist 1992
Yellow warbler *Dendroica petechia*	23	Yes	Yes	Hobson & Sealy 1990	DellaSala 1986
Song sparrow *Melospiza melodia*	21	Yes	-	Arcese et al. 1988	Arcese 1989
Bluethroat *Luscinia svecica*	10-40	Yes	Yes	S. A. Sæther & T. Amundsen,unpublished	T. Amundsen & J. T. Lifjeld, unpublished
White-crowned sparrow *Zonotrichia leucophrys*	several cases	Yes (& male - female)	Yes (& male)	Baptista et al. 1993	Baptista et al. 1993
House wren *Troglodytes aedon*	15-25	Yes	-	Johnson & Kermott 1990	Soukup & Thompson 1998

Although female song is clearly associated with competition between females in these species, and the incidence of polygyny in all species suggests that female song functions for mate defence, it is also possible that females are singing to defend resources or to prevent intraspecific brood parasitism. However, two experimental studies on polygynous species suggest that polygynous females sing to defend mates rather than resources.

In one of the most highly polygynous species of birds, the red-winged blackbird *Agelaius phoeniceus*, polygynous females were aggressive towards one another, and females regularly produced a 'teer' song in the context of female-female aggression and in response to playback of female songs (Beletsky 1983). Two experiments suggest that females were defending their mate rather than resources (Yasukawa & Searcy 1982). First, nesting females attacked a model female mounted in a copulation solicitation posture significantly more often than a model mounted in a normal, perched posture. Second, primary females receive the bulk of the male's parental investment, so a primary female risks the loss of her mate's parental care if she allows another female to nest within her mate's territory. In accordance with this, primary females were more aggressive to a model female than secondary females.

Further evidence that female vocalizations function for mate defence rather than resource defence comes from a study on dunnocks (Langmore & Davies 1997). In this species, groups can be monogamous, polyandrous, polygynous or polygynandrous, and of these mating systems polygyny is the rarest. Females defended separate territories and groups were formed according to the number of female territories a male could monopolize (Davies 1992). Therefore, although polygynous females shared the same male, they did not compete for resources. In order to increase levels of female-female competition, we removed males to increase the number of polygynous groups. We predicted that competing, polygynous females would call at higher rates than non-competing, monogamous females, and that female-female competition might also elicit song.

Female dunnocks produced three simple calls, a tseep call, a trill call and an alarm call, but they rarely sang. Tseep calls were used in aggression with other females and bouts of countertseeping between two females sometimes escalated into fights. Females used trill calls to attract males and they only trilled during the pre-laying and laying periods.

Following the male removals, competing, polygynous females produced more simple trills to attract males and more 'tseep' calls during female-female conflicts than non-competing, monogamous females. Although female song is rare in this species, 23% of polygynous females also produced complex songs when their mate left them to join another female, whereas no monogamous females sang, but this was not a statistically significant difference. We suggest that trills, tseeps and songs are forms of mate defence (Langmore & Davies 1997). Females deter rivals directly, using tseeps and overt aggression. Females also constantly trilled to attract their mate during the pre-breeding and fertile periods, probably reducing the mate's opportunity to encompass and defend a second female's territory.

Mate attraction

Another main function of male bird song is mate attraction (Catchpole & Slater 1995). In theory, it seems unlikely that female birds should have to sing to attract a mate because there is usually more competition between males than between females for mates (Clutton-Brock & Vincent 1991), so females need not invest in advertising. As expected, recent male removal experiments in northern cardinals *Cardinalis cardinalis* (McElroy & Ritchison 1996), bay wrens (Levin 1996a) and dunnocks (Langmore & Davies 1997) have demonstrated that female song does not function for mate attraction in these species. However, in some circumstances female competition does result in female song for mate attraction.

A study of alpine accentors *Prunella collaris* in the French Pyrénées revealed that females sing to attract males (Langmore et al. 1996). Alpine accentors bred in non-territorial, polygynandrous groups of up to four males and four females. All the males in the group attempted to copulate with every female and there was a dominance hierarchy among the males. Males mate guarded females during their fertile periods to prevent other males from gaining paternity of offspring. Males helped provision offspring of any females with whom they had obtained a threshold level of copulations (Hartley et al. 1995), so females benefited by mating with as many males in the group as possible. Since females all started breeding as soon as the snow melted, breeding was highly synchronous. This resulted in competition between females for males during the fertile period and females used two displays to attract males. First, they solicited copulations from mate guarding males at an extraordinarily high rate compared to most other species (see Birkhead et al. 1987), on average once every eight and a half minutes throughout the fertile period (Davies et al. 1996). Second, they produced simple trills and complex songs to attract males (Langmore et al. 1996). Females sang only during their fertile period, and they sang at a greater rate when alone than when they were accompanied by one or more mate guarding males. Female songs and trills elicited approaches from males, but never from females, and females often solicited copulations immediately after the male arrived. Males were also attracted to playbacks of female song.

Our data suggest that females use song and solicitation in quite sophisticated ways to regulate the mating access of each male in the group. Females increased their chances of mating with all the males, despite the dominant male's mate guarding, by soliciting at a greater rate to males who had obtained the least mating access (Davies et al. 1996). Furthermore, solicitation rate was inversely related to song rate. In other words, once a particular male had accompanied the female for a sufficient length of time to ensure male parental care, she decreased her solicitation rate to him, whilst simultaneously increasing her song rate to attract a different male (Langmore et al. 1996).

As in the polygynous species discussed above, female song in alpine accentors appears to be a facultative trait that is associated with competition between females. Singing to attract males, rather than to repel rival females, makes sense in alpine accentors because home range sizes were extremely large (ca. 30 ha). With such a large home range territorial defence against rival females may not be feasible, and it may also be difficult for males to locate receptive females without some form of long range advertisement.

Mate attraction may also be the function of a rare 'courtship song' of male and female dusky antbirds *Cercomacra tyrannina (*Morton 1996). This song is different from the usual territorial duet of the species, and is only produced following the divorce or death of a mate. When mates were removed experimentally, both male and female courtship songs appeared to attract new mates. Morton suggested that a mate attraction song is necessary for both males and females in species that defend year-round territories, because it facilitates acquisition of a new mate whilst minimizing the risk of losing the territory. If so, we would predict that female mate attraction songs should be particularly widespread in tropical, territorial species.

Coordination of breeding activities

Another function of both male and female song is the coordination of breeding activities. Males and females use song to coordinate nest relief in slate-coloured boubous (Sonnenschein & Reyer 1983) and black-headed grosbeaks *Pheucticus melanocephalus*

(Ritchison 1983) and provisioning of young in northern cardinals (Halkin 1997). Female song was also used to maintain contact with fledglings in black-headed grosbeaks (Ritchison 1983) and to encourage male nest defence in red-winged blackbirds (Yasukawa 1989).

Conclusions

In all these examples females used their song for the same functions as male song. Like males, females competed for resources and mates. However unlike males, female competition was not the outcome of a bias in the operational sex ratio, but occurred under a completely different set of conditions. First, female resource defence was largely restricted to species where male defence was periodically lacking, or when female breeding vacancies were unusually scarce. Second, whereas males compete for mates to increase their chances of producing offspring, females in these studies competed for mates to ensure male parental care. This competition took the form of direct aggression towards rival females, and indirect competition by singing to attract males.

We can predict that female song will occur in other bird species with high levels of female-female competition for resources or mates, or where the ecology or mating system result in periodic male absence from the territory. In particular, female song is likely to be common in many species which defend territories year-round and in polygynous species where polygyny carries a cost to the females in terms of reduced paternal care.

Acknowledgements

I am most grateful to the Royal Norwegian Society of Sciences and Letters, The Foundation, for inviting me to participate in this symposium and to the organizers and participants for such an interesting and enjoyable week. I thank Yngve Espmark, Becky Kilner, and Peter McGregor for comments on this chapter.

References

Arcese, P. 1989. Intrasexual competition and the mating system in primarily monogamous birds: the case of the song sparrow. *Animal Behaviour,* **38**, 96-111.

Arcese, P., Stoddard, P. K. & Hierbert, S. M. 1988. The form and function of song in female song sparrows. *Condor,* **90**, 44-50.

Baptista, L. F., Trail, P. W., Dewolfe, B. B. & Morton, M.L. 1993. Singing and its functions in female white-crowned sparrows. *Animal Behaviour,* **46**, 511-524.

Beletsky, L. D. 1983. Aggressive and pair-bond maintenance songs of female red-winged blackbirds (*Agelaius phoeniceus*). *Zeitschrift für Tierpsychologie,* **62**, 47-54.

Bensch, S. & Hasselquist, D. 1992. Evidence for active female choice in a polygynous warbler. *Animal Behaviour,* **44**, 301-311.

Birkhead, T. R., Atkin, L. & Møller, A. P. 1987. Copulation behaviour of birds. *Behaviour,* **101**, 101-138.

Brown, R. N. & Lemon, R. E. 1979. Structure and evolution of song form in the wrens *Thryothorus sinaloa* and *T. felix*. *Behavioral Ecology and Sociobiology,* **5**, 111-131.

Catchpole, C. K. & Slater, P. J. B. 1995. *Bird Song: Biological Themes and Variations*. Cambridge: Cambridge University Press.

Clutton-Brock, T. H. & Vincent, A. C. J. 1991. Sexual selection and the potential reproductive rates of males and females. *Nature,* **351**, 58-60.

Cooney, R. & Cockburn, A. 1995. Territorial defence is the major function of female song in the superb fairy-wren, *Malurus cyaneus*. *Animal Behaviour,* **49**, 1635-1647.

Davies, N. B. 1992. *Dunnock Behaviour and Social Evolution.* Oxford: Oxford University Press.

Davies, N. B., Hartley, I. R., Hatchwell, B. J. & Langmore, N. E. 1996. Female control of copulations to maximize male help: a comparison of polygynandrous alpine accentors *Prunella collaris* and dunnocks *Prunella modularis*. *Animal Behaviour,* **51**, 27-47.

DellaSala, D. A. 1986. Polygyny in the yellow warbler. *Wilson Bulletin,* **98**, 152-154.

Halkin, S. L. 1997. Nest-vicinity song exchanges may coordinate biparental care of northern cardinals. *Animal Behaviour,* **54**, 189-198.

Hartley, I. R., Davies, N. B., Hatchwell, B. J., Desrochers, A., Nebel, D. & Burke, T. 1995. The polygynandrous mating system of the alpine accentor *Prunella collaris.* II. Multiple paternity and parental effort. *Animal Behaviour,* **49**, 769-788.

Hobson, K. A. & Sealy, S. G. 1990. Female song in the yellow warbler. *Condor,* **92**, 259-261.

Hoelzel, A. R. 1986. Song characteristics and response to playback of male and female robins *Erithacus rubecula*. *Ibis,* **128**, 115-127.

Inman, B. L. 1986. Female vocalizations and their role in the avian breeding cycle. *Annals of the New York Academy of Sciences,* **474,** 44-52.

Johnson, L. S. & Kermott, L. H. 1990. Structure and context of female song in a north-temperate population of house wrens. *Journal of Field Ornithology,* **61**, 273-284.

Kriner, E. & Schwabl, H. 1991. Control of winter song and territorial aggression of female robins *(Erithacus rubecula)* by testosterone. *Ethology,* **87**, 37-44.

Lack, D. 1965. *The Life of the Robin.* London: Witherby.

Langmore, N. E. 1996. Territoriality and song as flexible paternity guards in dunnocks and alpine accentors. *Behavioral Ecology,* **7,** 183-188.

Langmore, N. E. 1998. Functions of duet and solo songs of female birds. *Trends in Ecology and Evolution,* **13**, 136-140.

Langmore, N. E. & Davies, N. B. 1997. Female dunnocks use vocalizations to compete for males. *Animal Behaviour,* **53**, 881-890.

Langmore, N. E., Davies, N. B., Hatchwell, B. J. & Hartley, I. R. 1996. Female song attracts males in the alpine accentor *Prunella collaris. Proceedings of the Royal Society of London, Series B,* **263**, 141-146.

Levin, R. N. 1996a. Song behaviour and reproductive strategies in a duetting wren, *Thryothorus nigricapillus*: I. Removal experiments. *Animal Behaviour,* **52**, 1093-1106.

Levin, R. N. 1996b. Song behaviour and reproductive strategies in a duetting wren, *Thryothorus nigricapillus*: II Playback experiments. *Animal Behaviour,* **52**, 1107-1117.

McElroy, D. B. & Ritchison, G. 1996. Effect of mate removal on singing behavior and movement patterns of female northern cardinals. *Wilson Bulletin,* **108**, 550-555.

Morton, E. S. 1996. A comparison of vocal behavior among tropical and temperate passerine birds. In: *Ecology and Evolution of Acoustic Communication in Birds* (Ed. by D.E. Kroodsma & E.H. Miller), pp. 258-268. Ithaca, NY: Cornell University Press.

Mulder, R. A., Dunn, P. O., Cockburn, A., Lazenby-Cohen, K. A. & Howell, M. J. 1994. Helpers liberate female fairy-wrens from constraints on extra-pair mate choice. *Proceedings of the Royal Society of London, Series B,* **255**, 223-229.

Pinxten, R. & Eens, M. 1990. Polygyny in the European starling: effect on female reproductive success. *Animal Behaviour,* **40**, 1035-1047.

Ritchison, G. 1983. The function of singing in female black-headed grosbeaks *(Pheucticus melanocephalus):* family-group maintenance. *Auk,* **100**: 105-116.

Sandell, M. I. & Smith, H. G. 1997. Female aggression in the European starling during the breeding season. *Animal Behaviour,* **53**, 13-23.

Searcy, W. A. & Yasukawa, K. 1995. *Polygyny and Sexual Selection in Red-winged Blackbirds.* Princeton, NJ: Princeton University Press.

Slagsvold, T., Dale, S. & Sætre, G.-P. 1994. Dawn singing in the great tit *(Parus major)*: mate attraction, mate guarding, or territorial defence? *Behaviour,* **131**, 115-138.

Slagsvold, T. & Lifjeld, J. T. 1994. Polygyny in birds: the role of competition between females for male parental care. *American Naturalist,* **143**, 59-80.

Sonnenschein, E. & Reyer, H.-U. 1983. Mate-guarding and other functions of antiphonal duets in the slate-coloured boubou (*Laniarius funebris*). *Zeitschrift für Tierpsychologie,* **63**, 112-140.

Soukup, S. S. & Thompson, C. F. 1998. Social mating system and reproductive success in house wrens. *Behavioral Ecology,* **9**, 43-48.

Yasukawa, K. 1989. The costs and benefits of a vocal signal: the nest-associated 'Chit' of the red-winged blackbird, *Agelaius phoeniceus. Animal Behaviour,* **38**, 866-874.

Yasukawa, K. & Searcy, W. A. 1982. Aggression in female red-winged blackbirds: a strategy to ensure male parental investment. *Behavioral Ecology and Sociobiology,* **11**, 13-17.

IN: ESPMARK,Y., AMUNDSEN,T. & ROSENQVIST,G. (EDS.) 2000
ANIMAL SIGNALS:
SIGNALLING AND SIGNAL DESIGN IN ANIMAL COMMUNICATION, PP. 329-340
TAPIR ACADEMIC PRESS, TRONDHEIM, NORWAY

16

Communication Networks: Receiver and Signaller Perspectives

PETER K. McGREGOR*, KEN OTTER & TOM M. PEAKE

Department of Animal Behaviour, Zoological Institute,
University of Copenhagen, Tagensvej 16, DK-2200 Copenhagen N, Denmark

**E-mail: pkmcgregor@zi.ku.dk*

Abstract

This chapter explores the consequences of the idea of communication networks (several individuals in communication range) for receivers and signallers. The emphasis is on receiver behaviour, particularly eavesdropping (defined as extracting information from signalling interactions). The contention that eavesdropping is an important receiver behaviour is supported by the results of experiments on male siamese fighting fish Betta splendens *and female great tits* Parus major, *for visual and acoustic signals respectively. Two aspects of signalling in communication networks are discussed briefly; how to direct signals to a particular individual and whether it is possible to restrict the spread of long-range signals in networks. Relative timing of signals is proposed as a way signallers of any taxon employing signals of most modalities can direct their signals. Restricting the spread of signals through a network requires either a switch in signal modality or signal structure, except in the case of highly directional signals. Finally, receiver psychology is discussed with respect to effects such as precedence and visual object grouping, and whether communication networks impose trade-offs for receiver psychology.*

The idea of communication networks follows logically from the observation that the long-range signals of animals travel considerable distances in relation to the average spacing between individuals. Therefore, several individuals are able to communicate with one another, forming a communication network. Such networks are possible in any signalling modality and in any taxon. It has been argued that communication networks are the context in which long-range communication behaviour has evolved (McGregor 1993; McGregor & Dabelsteen 1996). The major aim of this paper is to present our experimental evidence of communication behaviour adapted to a network context. Such studies provide considerable logistic and analysis challenges and therefore, as a first step, we report examples from subsets of networks, often featuring just three individuals.

In this chapter we look first at receiver behaviour in the context of a communication network, then briefly discuss two aspects of signalling behaviour and finish with aspects of receiver psychology. Where possible we try to present a range of both modalities and taxa, although we acknowledge a bias towards acoustic and visual signals of vertebrates. The bias towards receiving behaviour in communication networks is intended, partly to balance the attention that signalling behaviour in choruses has attracted, but mainly because we think that this is the area that poses the greatest experimental challenge, particularly in a field setting.

Receiver behaviour in communication networks

There are several aspects of receiver behaviour in communication networks that are of interest. For example, how receivers discriminate information from individual signallers when signals from several individuals occur simultaneously (e.g. Robisson et al. 1993; Aubin & Mathevon 1995; Klump 1996). We consider that one of the most interesting receiver behaviours, and one that is open to experimental investigation, is how information is gathered from interactions between signallers. We have used the term *eavesdropping* for this receiving behaviour (McGregor 1993) and define it as extracting information from signalling interactions between others (McGregor & Dabelsteen 1996). The advantages to a receiver from eavesdropping are that information on relative features of the signallers (such as fighting ability) can be gathered at little cost, with no risk and before the information is needed (for example, in a subsequent interaction with one of the interacting signallers). Interestingly, it has been suggested that several features of signals used in aggressive interactions have characteristics that seem designed for reception by eavesdroppers rather than by the obvious participants (Zahavi 1979). Previous discussions of the advantages of eavesdropping have been mainly confined to an aggressive context (e.g. McGregor 1993) although eavesdropping by females during mate choice has also been suggested (e.g. McGregor & Dabelsteen 1996). We present experimental evidence for eavesdropping in both contexts.

Eavesdropping by Siamese fighting fish

Siamese fighting fish *Betta splendens* are anabantids from South-East Asia. Males defend territories in the wild and in the lab (Forselius 1957) using striking visual displays

Simpson 1968) that include gill cover erection and fin spread. These displays precede fighting and such fights often involve physical damage to the fish, sometimes resulting in death. Therefore, we would expect strong selection pressure for eavesdropping, i.e. gathering information on the fighting ability (relative dominance) of potential opponents before a fight. The close proximity of territorial males and the clarity of the water they inhabit (Forselius 1957) would allow eavesdropping in the wild.

In our laboratory experiments (Oliveira et al. 1998), the male subject was placed in a central compartment flanked on two sides by a pair of smaller compartments, each of which contained a single male, therefore there were four other males in addition to the subject. One-way glass ensured that the male subject could observe, but not interact with, the four other males (see Fig. 1 in Oliveira et al. 1998 for further details). We first investigated whether subjects paid more attention to two males interacting with (i.e. visually displaying to) one another than to two males not displaying to one another. The displaying males were in one of the pair of flanking compartments, separated from each other by a clear divider. The other two males were in the other pair of flanking compartments, separated from each other by an opaque divider and unaware of the presence of each other. Male subjects spent significantly more time within 5 cm and gazing at the two interacting males (% of time gazing at interacting neighbours = 13.1 ± 3.4 *vs.* 5.2 ± 0.9 for non-interacting neighbours; values are mean ± se; $N = 17$, $P = 0.01$, Wilcoxon matched pairs test). Thus subjects pay more attention to interacting neighbouring males, fulfilling a prerequisite for eavesdropping.

To test if the subjects' behaviour was influenced by information on relative fighting ability obtained by eavesdropping on the interactions they observed, subjects were allowed to observe interactions with a clear winner and a clear loser. (The eventual winners and losers of fights are reliably indicated by their visual display behaviour (Simpson 1968)). Once again one-way glass allowed subjects to see displaying males while preventing displaying males from seeing the subjects. A comparable interaction took place simultaneously which subjects could not see. In this way, we were able to exclude the possibility that it was the difference in behaviour of fish that had recently won or lost that subsequently elicited different behaviour from the subject. After the 15 mins taken for clear winners and losers to be apparent, all males were visually isolated in their respective compartments. Then each of the four other males was introduced sequentially into a transparent container placed within the subject's compartment. The subject's response was noted for 10 mins after introduction for each of the four other males. (Full details of the experiment are given in Oliveira et al., 1998).

If subjects have eavesdropped and use the information obtained in a subsequent interaction with one of the observed individuals, we would predict an initial difference in response towards the seen winner (W) when compared to the seen loser (L). However, we would predict no such difference towards the unseen winner (w) when compared to the unseen loser (l). The two measures of initial response differed as predicted. The subjects took significantly longer to approach and longer to display to seen winners than losers, but there was no such difference for unseen winners and losers (see Fig. 2a in Oliveira et al. 1998). Latency (approach and display) could be taken as an indication of willingness to engage in a fight, therefore subjects are more willing to fight individuals which they have seen lose than to fight individuals they have seen win, relative to their intermediate response to unseen individuals. Information gathered by eavesdropping would not be expected to influence measures of response over the entire interaction, because such measures are strongly influenced by the dynamics of the interaction. In support of this expectation, no significant differences were found in the two overall measures taken (display duration and display frequency, see Fig. 2b in Oliveira et al. 1998). Moreover, although subjects took longer to respond to seen winners (W) they tended to escalate to a higher level (as measured by the proportion of subjects that attempted to bite the intruder; seen interactions: W = 29.1% *vs.* L = 8.3%, $N = 24$, two-tailed difference between the two proportions $P = 0.06$; unseen fights: w = 29.1% *vs.* l = 29.1%, $N = 24$, two-tailed difference between the two proportions $P = 1.0$). These results are consistent with subjects having identified the seen winner as a stronger opponent than the seen loser.

We conclude therefore that male fighting fish eavesdrop on the visual displays of potential future opponents and use the information gathered on relative fighting ability to alter their behaviour in the initial stages of contests occurring up to 30 minutes later.

Eavesdropping by female great tits

In the same way that males can gain information on the relative ability of future aggressive opponents by eavesdropping, females could gain information on the relative quality of mates by eavesdropping on male-male aggressive interactions (McGregor & Dabelsteen 1996). We explored this possibility by experimentally manipulating a female great tit's *Parus major* perception of the relative quality of her mate. The experiment took place during the fertile period of the female, when female song birds are known to engage in varying levels of extra-pair copulations resulting in extra-pair paternity (e.g. Birkhead & Møller 1992). It has been argued that females should use singing interactions in such decisions for three main reasons (K.A. Otter et al. unbublished). First, because singing interactions allow long-range assessment of males, thereby avoiding the possibility of sexual coercion (Clutton-Brock & Parker 1995). Second, current relative quality can be assessed. Finally, assessment is particularly easy to accomplish among neighbouring males, and these males are the most likely extra-pair sires (e.g. Strohbach et al. 1998).

We used interactive playback (Dabelsteen & McGregor 1996) to simulate intrusion by an unknown male into the territory of the female subject's mate. This intruder (stranger) either used song in a more aggressive fashion (by overlapping the territory holder and increasing relative song length) or in a less aggressive fashion (by alternating and decreasing relative song length) (for details of why we think such patterns of song use encode this information see McGregor et al. 1992; Dabelsteen et al. 1996, 1997). For brevity, and because we think this is how females probably perceive such interactions, we will refer to territory holding males 'winning' interactions with playback (i.e. subjected to less aggressive playback) or 'losing' (i.e. subjected to more aggressive playback).

Playback subjects were grouped into experimental dyads, with males in a dyad holding neighbouring territories, one of the dyad would win and then, a few minutes later, the other would lose to playback (balanced for order over the course of the experiment). The aim was to change the subject female's perception of her mate relative to a potential extra-pair partner (the neighbour). On two occasions (one in the morning and one in the afternoon) after the two playbacks (usually the afternoon of the playback day and the next morning, but sometimes solely on the day after playback), females were followed for two 15min periods and intrusions into neighbouring territories noted. Such intrusions are a precursor to extra-pair copulation.

The results were striking (Otter et al. 1999). Females with mates that had lost to playback visited neighbours' territories significantly more commonly than females with mates that won. Furthermore, females of losing males tended to visit the winning neighbour's territory. These results clearly suggest that females eavesdrop on singing interactions between their mate, neighbouring males and an intruder, and information gained from such eavesdropping influences their subsequent extra-pair behaviour during the next two days.

Conclusion

Overall, we think these two experiments demonstrate the validity of the idea of eavesdropping as an important component of receiver behaviour. The fact that eavesdropping is only possible in the context of a communication network is an instance of social organization facilitating communication behaviour, an idea discussed in more detail by Oliveira & McGregor (in press). Although our experiments were not designed to investigate how long eavesdropping information can be retained, they showed retention for up to 30 minutes by *Betta splendens* and up to 48 hours by *Parus major*. We suggest that information gathered by eavesdropping is likely to be remembered for much longer than these periods. In territorial songbirds two species have been shown to retain an ability to identify neighbours by song for several months (McGregor & Avery 1986; McGregor 1989; Godard 1991). The information derived from eavesdropping is likely to have comparable value to that derived from direct interactions with neighbours and is therefore likely to be remembered for comparable periods of time (see von Fersen et al. 1991 and McGregor 1993 for more detailed discussion and the role of Fisher's (1954) *Dear Enemy* effect in identifying territorial neighbours).

Signaller behaviour in communication networks

The first studies of communication to consider groups of signallers and the consequences for receivers were investigations of insect and amphibian choruses (e.g. Otte 1974; Alexander 1975; Ryan et al. 1981). Such choruses function in mate attraction and therefore much interest has focussed on the signal features that characterize a successful signaller, given that a female has so many signallers from which to choose. We will return to aspects of this question later under the heading of receiver psychology. In this section we will consider two aspects of signalling in communication networks that are amenable to experimental investigation, but which have received little attention (with the possible exception of some bird song studies).

Directing signals

The far-travelling nature and omni-directional characteristics of almost all long-range signals render them inherently unsuitable for directing a signal towards a single receiver. Such directing is likely to play an important role in male-male contexts such as territory defence, where directing signals at a male that is perceived as a particular threat at the moment (e.g. one that has crossed a mutual territory boundary) may prevent escalation to physical aggression. The same could be true also in male-female contexts, where directing signals to a female that is perceived as a mating partner at the moment (e.g. encouraging approach to elicit extra-pair behaviour) may increase reproductive success.

In the field of bird song, the phenomenon of song type matching (also referred to as matched counter-singing) has long been considered to play the role of directing song (e.g. Brémond 1968; Horn & Falls 1988; Stoddard et al. 1992; Beecher et al. 1996). However, matched counter-signalling is unlikely to be a common method of directing signals. Songbirds are unusual in having repertoires of song types with which to match; animals in most other taxa have only one long-range signal. A more likely general candidate for directing signals is relative timing, with consistently close association of signals indicating the individuals involved in an interaction. Two or more individuals are easily incorporated into this scheme and the idea can be applied to any signal modality (chemical signals are a possible exception) and any taxon. Studies of relative timing in song birds have tended to view close temporal associations of signals as aggressive, with signal overlap as the most aggressive signal (e.g. McGregor et al. 1992; Dabelsteen et al. 1996, 1997; Naguib & Todt 1997; Todt & Naguib, in press). Studies of leader-follower interactions have also tended to interpret such signalling as encoding relative dominance (Todt & Naguib, in press). However, it is possible that the responses which experimenters interpret as showing aggressive intent are simply the responses elicited when signallers are aware that they are engaged in an interaction. In principle it should be possible to design playback experiments to study whether directing signals and willingness to escalate are encoded differently in signal timing behaviour or whether these two features are inextricably linked.

If such attempts at directing signals are successful, then this information is also available to eavesdroppers and may be used to enhance the success of the eavesdropper. For example, if it is clear from the extent of overlapping that two males are engaged in a territorial singing interaction that may escalate into physical aggression, a male eavesdropper (e.g. a neighbour) should choose such a time to intrude into their territories, perhaps to solicit extra-pair copulations from their mates. For this reason we might expect the individuals engaged in such signalling to attempt to keep this interaction hidden from potential eavesdroppers. Such 'privatising' is discussed in the next section.

Private signals

The main question we wish to raise here is whether it is ever possible to restrict transmission of long-range signals to the intended receiver. Our initial view is that it is not possible, with the exception of long-range signals that are inherently strongly directional - for example, the beamed sound signals of cetaceans such as dolphins and sperm whales (*Physeter* spp.). The more common solution to the problem of preventing unwanted eavesdropping is to change the nature of the signal to restrict its transmission range. An example is the switch from advertising song in song birds to low amplitude, higher frequency almost continuous vocalizing in contexts when preventing eavesdropping would be advantageous (e.g. the robin *Erithacus rubecula*, Dabelsteen et al. 1997; see also figures in Dabelsteen et al. 1998).

Conclusion

Although much signalling behaviour in communication networks is inextricably linked with receiver behaviour (and therefore best dealt with under the heading of receiver psychology effects), there are interesting questions that are raised by considering signal evolution in the context of communication networks (for a theoretical approach to this question see Johnstone, this volume). Many of the technical approaches needed to investigate signalling in networks already exist (e.g. passive location systems (McGregor et al. 1997) and interactive playback), what is now required is the sophisticated experimental approach that has characterized much work in communication.

Receiver psychology and communication networks

The acoustic choruses of insects and frogs, together with the visual choruses of some fireflies, are notable for the high degree of synchrony of the signals in the chorus. Recent work has suggested that such synchrony is a byproduct of receiver psychology (Guilford & Dawkins 1991); specifically, the precedence effect means that the signal most likely to attract an approaching mate is the first of several to be detected. The outcome of many individuals trying to ensure that their signal is the first detected can produce the degree of synchrony for which choruses are noted (Greenfield 1994; Greenfield et al. 1997).

We recently studied the visual signals (claw waving) of male fiddler crabs *Uca tangeri* in Portugal, fully expecting to find a precedence effect. In fact, we found that males of different sizes (and therefore different attractiveness to females) differed in the timing of their waves when waving alone, as compared with waving with other males close by. Specifically, the largest male in a group of three increased inter-wave interval when waving in a group so that its wave occurred after the other two males (Burford et al. 1998). This can be explained by another receiver psychology effect known as visual object grouping (Koffka 1935). The point of this example is that although the psychological effect is different from the precedence effect, the selection pressure for its implementation is the same - enhanced signal effectiveness. We suggest that several aspects of receiver psychology will be best understood in terms of receiving in a communication network. Once again the effects must be investigated experimentally before an understanding of the system can be claimed (see Milinski 1997 for a general discussion of correlational and experimental approaches).

Receiver psychology and trade-offs

Mate choice decisions, at least in most vertebrates, are rarely as simple as detecting a mate; with several aspects of mate quality being important (e.g. Andersson 1994). Given that different aspects of signalling may be perceived by receivers using several rules to decide relative mate quality, the possibility exists that signallers have to trade-off these effects. An example that may represent such a trade-off became apparent during recent work on a non-songbird, the corncrake *Crex crex*. Males of this species use a long-range signal (the *crex-crex* of the Latin binomial) at night to attract mates and to interact with other males (Peake 1997). In a local cluster of several males, the male with the highest calling rate attracts a female first (T.M. Peake & P.K. McGregor unpublished) and such males also show least variation in inter-call interval (which determines calling rate) (Peake et al. 1998, T.M. Peake et al. unpublished). Another interesting feature is that although the calling rate of males is fixed over several nights (Peake 1997), males attempt to overlap higher rate competitors and also playback, by stopping and starting so as to remain in phase, thus varying the inter-call interval (T.M. Peake et al., unpublished). However, this signalling strategy runs the risk of giving the wrong signal to a potential mate; the variation created by stopping and starting in order to overlap is very similar to the variable inter-call interval characteristic of the males with the lowest calling rate which attract a mate last. It would be interesting to know whether females apply different rules when assessing calling rate variation heard in the context of overlapping versus the context of solo spontaneous calling. If they do not, then males have a conflict between the advantages of overlapping other males (interpreted as an aggressive signal that may deter them from settling in the area) and of producing calls with little variation in inter-call interval (which are most likely to attract females).

Overall conclusion

We would like to finish by returning to the receiver perspective. The perceptual abilities of receivers are the key to understanding the importance of communication networks as the context in which long-range signals have evolved. (For a similar argument, see Hasson, this volume). We say this because the approach of indirectly deducing what receivers can do from considerations of, say, the physical properties of signals can be misleading. For example, it was assumed that if a signal was completely embedded in noise then it could not be discriminated. However, psycho-acoustic experiments on birds have shown that if the sources of signal and noise are separated in space then birds can perform such discriminations (Larsen et al. 1994). Similarly, aerial predator alarm calls of birds (*seet* calls) are assumed to be difficult for avian predators (and indeed conspecifics) to detect. However, determinations of the spectral sensitivity of a common aerial predator (sparrowhawk *Accipter nisus*) and one species of their *seet*-calling prey (great tit) show increased sensitivity in the frequency range of this call, with the prey currently ahead in this perceptual arms race (Klump et al. 1986). These elegant studies demonstrate that while a knowledge of signal (and indeed signaller) characteristics is a useful first step, it is no substitute for directly investigating the perceptual capabilities of receivers.

Acknowledgements

It is a privilege to acknowledge that many of the ideas in this paper have resulted from discussions with others working on communication, in particular Torben Dabelsteen and, more recently, Rui Oliveira and the 'Communication Crew'. We are very grateful to all the co-authors of published work reported in this paper for allowing us to summarize our results: Fiona Burford, Torben Dabelsteen, Claire Latruffe, Rui Oliveira and Andrew Terry. Most of the research was carried out while at the University of Nottingham and PKM thanks his colleagues Chris Barnard, Francis Gilbert and Jane Hurst for making the Behaviour and Ecology Research Group such a stimulating environment. PKM also thanks the Centre for Sound Communication, Danish National Research Foundation, for sabbatical funding which allowed the ideas to be developed and investigations to be initiated. Our research has been funded by ASAB (UK), JNICT (Portugal), NERC (UK), NSERC (Canada), The Royal Society (UK) and most recently by SNF (Denmark) and the University of Copenhagen. We thank an anonymous referee for comments on a previous draft. Finally, but most importantly, PKM thanks all those who have made up the 'Communication Crew' over the years, it has been a privilege to work with them.

References

Alexander, R. D. 1975. Natural selection and chorusing behavior in acoustical insects. In: *Insects, Science and Society* (Ed. by D. Pimentel), pp. 35-77. New York: Academic Press.

Andersson, M. 1994. *Sexual Selection.* Princeton: Princeton University Press.

Aubin, T. & Mathevon, N. 1995. Adaptation to severe conditions of propagation: long-distance distress calls and courtship calls of a colonial seabird. *Bioacoustics,* **6**, 153-161.

Beecher, M. D., Stoddart, P. K., Campbell, S. E. & Horning, E. L. 1996. Repertoire matching between neighboring song sparrows. *Animal Behaviour,* **51**, 917-923.

Birkhead, T. R. & Møller, A. P. 1992. *Sperm Competition in Birds: Evolutionary Causes and Consequences.* London: Academic Press.

Brémond, J.-C. 1968. Recherches sur la semantique et les elements vecteurs d'information dans les signaux acoustiques du rouge-gorge (*Erithacus rubecula* L.). *La Terre et la Vie,* **2**, 109-220.

Burford, F. R. L., McGregor, P. K. & Oliveira, R. F. 1998. Chorusing by male European fiddler crabs, *Uca tangeri*: a study of visual communication networks. *Acta Ethologica,* **1**, 33-41.

Clutton-Brock, T. H. & Parker, G. A. 1995. Sexual coercion in animal societies. *Animal Behaviour,* **49**, 1345-1365.

Dabelsteen, T. & McGregor, P. K. 1996. Dynamic acoustic communication and interactive playback. In: *Ecology and Evolution of Acoustic Communication in Birds* (Ed. by D.E. Kroodsma & E.H. Miller), pp. 398-408. Ithaca, NY: Cornell University Press.

Dabelsteen, T., McGregor, P. K., Shepherd, M., Whittaker, X. & Pedersen, S. B. 1996. Is the signal value of overlapping different from that of alternating during matched singing in great tits? *Journal of Avian Biology,* **27**, 189-194.

Dabelsteen, T., McGregor, P. K., Holland, J., Tobias, J. A. & Pedersen, S. B. 1997. The signal function of overlapping singing in male robins (*Erithacus rubecula*). *Animal Behaviour,* **53**, 249-256.

Dabelsteen, T., McGregor, P. K., Lampe, H. M., Langmore, N. E. & Holland, J. 1998. Quiet song in song birds: an overlooked phenomenon. *Bioacoustics,* **9**, 89-106.

Fersen, L. von, Wynne, C. D. L. & Delius, J. D. 1991. Transitive inference formation in pigeons. *Journal of Experimental Psychology: Animal Behavior Processes,* **17**, 334-341.

Fisher, R. A. 1954. Evolution and bird sociality. In: *Evolution as a Process* (Ed. by J. Huxley, A.C. Hardy & E.B. Ford), pp. 71-83. London: Allen & Unwin.

Forselius, S. 1957. Studies of Anabantid fishes. *Zoologiska Bidrag från Uppsala,* **32**, 95-597.

Godard, R. 1991. Long-term memory of individual neighbours in a migratory songbird. *Nature,* **350**, 228-229.

Greenfield, M. D. 1994. Synchronous and alternating choruses in insects and anurans: common mechanisms and diverse functions. *American Zoologist,* **34**, 605-615.

Greenfield, M. D., Tourtellot, M. K. & Sneddon, W. A. 1997. Precedence effects and the evolution of chorusing. *Proceedings of the Royal Society of London, Series B,* **264**, 355-361.

Guilford, T. & Dawkins, M. S. 1991. Receiver psychology and the evolution of animal signals. *Animal Behaviour,* **42**, 1-14.

Horn, A. G. & Falls, J. B. 1988. Repertoires and countersinging in western meadowlarks *Sturnella neglecta. Ethology,* **77**, 337-343.

Klump, G. M, Kretzschmar, E. & Curio, E. 1986. The hearing of an avian predator and its avian prey. *Behavioral Ecology and Sociobiology,* **18**, 317-323.

Klump, G. M. 1996. Communication in the noisy world. In: *Ecology and Evolution of Acoustic Communication in Birds* (Ed. by D.E. Kroodsma & E.H. Miller), pp. 321-338. Ithaca, NY: Cornell University Press.

Koffka, K. 1935. *Principles of Gestalt Psychology*. New York: Harcourt, Brace & World.

Larsen, O. N., Dent, M. L. & Dooling, R. J. 1994. Free-field release from masking in the budgerigar (*Melopsittacus undulatus*). In: *Sensory Transduction.* Vol. 2 (Ed. by H. Breer & N. Elsner), p. 370. Stuttgart: Thieme Verlag.

McGregor, P. K. 1989. Pro-active memory interference in neighbour recognition by a songbird. *Proceedings of the 19th International Ornithological Congress,* 1391-1397.

McGregor, P. K. 1993. Signalling in territorial systems: a context for individual identification, ranging and eavesdropping. *Philosophical Transactions of the Royal Society of London, Series B,* **340**, 237-244.

McGregor, P. K. & Avery, M. I. 1986. The unsung songs of great tits (*Parus major*): learning neighbours' songs for discrimination. *Behavioral Ecology and Sociobiology,* **18**, 311-316.

McGregor, P. K. & Dabelsteen, T. 1996. Communication networks. In: *Ecology and Evolution of Acoustic Communication in Birds* (Ed. by D.E. Kroodsma & E.H. Miller), pp. 409-425. Ithaca, NY: Cornell University Press.

McGregor, P. K., Dabelsteen, T., Clark, C. W., Bower, J. L., Tavares, J.-P. & Holland, J. 1997. Accuracy of a passive acoustic location system: empirical studies in terrestrial habitats. *Ethology, Ecology and Evolution,* **9**, 269-286.

McGregor, P. K., Dabelsteen, T., Shepherd, M. & Pedersen, S. B. 1992. The signal value of matched singing in great tits: evidence from interactive playback experiments. *Animal Behaviour,* **43**, 987-998.

Milinski, M. 1997. How to avoid the seven deadly sins in the study of behavior. *Advances in the Study of Behavior,* **26**, 159-180.

Naguib, M. & Todt, D. 1997. Effects of dyadic interactions on other conspecific receivers in nightingales. *Animal Behaviour,* **54**, 1535-1543.

Oliveira, R. F. & McGregor, P. K. In press. Redes de comunicação e estruturas sociais. In: *Comunicação e etologia social - Actas do seminário sobre Comunicação e etologia social* (Lisboa, 7 Maio, 1996) (Ed. by R.F. Oliveira & A.J. Santos). Lisboa: ISPA.

Oliveira, R. F., McGregor, P. K., & Latruffe, C. 1998. Know thine enemy: fighting fish gather information from observing conspecific interactions. *Proceedings of the Royal Society of London, Series B,* **265**, 1045-1049.

Otte, D. 1974. Effects and functions in the evolution of signalling systems. *Annual Review of Ecology and Systematics,* **5**, 385-417.

Otter, K. A., McGregor, P. K., Terry, A. M. R., Burfjord, F. R. L., Peake,T. M. & Dabelsteen, T. 1999. Do female great tits (*Parus major)* assess extra-pair males by eavesdropping? A field study using interactive song playback. *Proceedings of the Royal Society of London, Series B*, **266**, 1305-1309.

Peake, T. M., McGregor, P. K., Smith, K. W., Tyler, G., Gilbert, G. & Green, R. E. 1998. Individuality in corncrake *Crex crex* vocalizations. *Ibis,* **140**, 121-127.

Robisson, P., Aubin, T. & Brémond, J. C. 1993. Individuality in the voice of the emperor penguin *Aptenodytes forsteri*: adaptation to a noisy environment. *Ethology,* **94**, 279-290.

Ryan, M. J., Tuttle, M. D. & Taft, L. K. 1981. The costs and benefits of frog chorusing behaviour. *Behavioral Ecology and Sociobiology,* **8**, 273-278.

Simpson, M. J. A. 1968. The display of the Siamese fighting fish, *Betta splendens. Animal Behaviour Monographs*, **1**, 1-73.

Stoddard, P. K., Beecher, M. D., Campbell, S. E. & Horning, C. L. 1992. Song-type matching in the song sparrow. *Canadian Journal of Zoology,* **70**, 1440-1444.

Strohbach, S., Curio, E., Bathen, A., Epplen, J. T. & Lubjuhn, T. 1998. Extrapair paternity in the great tit (*Parus major*): a test of the 'good genes' hypothesis. *Behavioral Ecology,* **9**, 388-396.

Todt, D. & Naguib, M. In press. Vocal interactions in birds: the use of song as a model in communication. *Advances in the Study of Behavior,* **28**.

Zahavi, A. 1979 Why shouting. *American Naturalist,* **113**, 155-156.

IN: ESPMARK,Y., AMUNDSEN,T. & ROSENQVIST,G. (EDS.) 2000
ANIMAL SIGNALS:
SIGNALLING AND SIGNAL DESIGN IN ANIMAL COMMUNICATION, PP. 341-352
TAPIR ACADEMIC PRESS, TRONDHEIM, NORWAY

17

The Significance of Learning in Signal Development: The Curious Case of the Chaffinch

PETER J. B. SLATER*, ROBERT F. LACHLAN & KATHARINA RIEBEL

School of Environmental & Evolutionary Biology,
University of St Andrews, St Andrews, Fife, Scotland KY16 9TS

* *E-mail: pjbs@st-andrews.ac.uk*

Abstract

*Vocal learning, in the strict sense that individuals have been shown to copy the detailed structure of the sounds they produce from other individuals, has evolved in at least three groups of birds and four of mammals. While its origins are hard to examine, a number of suggestions have been made about why it is maintained in these groups. These ideas are reviewed and the evidence for them examined. While some appear to apply in certain instances, the evidence is against any of them providing a widespread explanation for vocal learning. In particular, we have been unable to confirm the importance of any hypothesis for the functional significance of vocal learning in our chosen study species, the chaffinch (*Fringilla coelebs*). Instead we outline evidence for a new suggestion, the 'cultural trap hypothesis', which proposes that vocal learning once it has evolved may persist through an interaction between genes and culture without necessarily having any direct selective advantage of its own.*

The reasons why some animals copy the sounds that they produce from other individuals is one of the more perplexing aspects of animal communication. There have been numerous studies on vocal learning and, although many ideas have been put forward, no widely accepted theory of its functional significance has emerged (Slater 1989). In this article we will start by reviewing the nature and occurrence of vocal learning, we will then go on to outline the various reasons that have been advanced as to why selection may have favoured it. In each case, supporting evidence will be reviewed and we will also consider how well the suggested function matches our knowledge of song in our chosen study species, the chaffinch *Fringilla coelebs*. Finally we will briefly outline a new theory suggesting that vocal learning, once it has evolved, may persist even if it conveys no direct benefits, the so-called 'cultural trap hypothesis'.

The occurrence of vocal learning

There is, as yet, no evidence that vocal learning occurs except in birds and mammals. Among birds it has evolved at least three times (Slater 1989). The most striking case is that of the oscine passerines, representing nearly half of all extant bird species, among which vocal learning has been found in every species examined so far (Kroodsma & Baylis 1982). This is in stark contrast to the sub-oscine passerines, in which several species have now been examined but no evidence found for vocal learning (e.g. Kroodsma & Konishi 1991). It seems likely therefore that it evolved among the oscines shortly after these two groups separated. As is well known, vocal learning is also found among the parrots (Psittaciformes: Todt 1975; Pepperberg 1994), and recent evidence points to its existence in hummingbirds (Apodiiformes: Baptista & Schuchmann 1990; Gaunt et al. 1994).

All these examples concern cases where birds have been shown experimentally to learn the detailed structure of their sounds from other individuals. In some other groups, such as owls and petrels, there is evidence for geographical variation in calls (James 1985; Galeotti et al. 1996). While such variation is most likely to arise through learning, especially where short distances are involved, differences over long distances, particularly between isolated populations, could equally arise through genetic differentiation. Thus, while the learning of sounds in other groups of birds remains a possibility, clinching evidence is as yet lacking.

Information on mammals is more scanty (Janik & Slater 1997). While there is no doubt that many species do modify their use of sounds depending on the context, and learning can play a substantial rôle here, the copying of sound structure from other individuals is rarer. Humans are obviously very good at it, but the evidence for other primates is slight (Mitani et al. 1992). However, there is now good evidence in three other groups: bats (Chiroptera: Jones & Ransome 1993), seals (Pinnipedia: Ralls et al. 1985) and whales and dolphins (Cetacea: Richards et al. 1984). In these cases the changes observed go beyond alterations in speed or repetition, which might arise just through changing excitement, or adjustments in the ordering of elements or the matching of frequency between individuals, all of which may be relatively simple processes. Instead these groups show the copying of detailed and complex structures in a similar way to that which has been so extensively described in birds.

For several bird species there is evidence that a much greater variety of sounds is produced early in song development than in the finished song (Marler & Pickert 1984). This led Marler & Peters (1982) to describe the crystallisation of the final song as a 'culling' process during which unwanted sounds are rejected. Marler & Nelson (1993) went on to put forward the hypothesis that 'action based learning' is involved, whereby the repertoire of a bird becomes selected from a larger set of pre-existing possibilities. In some cases the songs selected for retention are clearly those shared with neighbours (Nelson 1992), enabling the young bird to countersing with them using matched song types. While the songs from which those to be used are selected may be learnt earlier, it remains possible that some species may hatch with a large pre-existing repertoire of sounds, and the important developmental process may be selection from among these rather than copying sounds from other individuals (see Table 1). This is a challenging idea which certainly requires more investigation, especially at the level of individual sound elements and in animals with relatively simple songs. But copying of sound structures and sequences must certainly have a rôle in species where these are highly complex and varied, and where they match tutors with an impressive degree of precision. That birds will copy sounds quite distinct from those in the normal species repertoire also argues for a process of imitation rather than one of selection.

Table 1. *Three models of song learning in birds. The Memory Based Learning model is that most usually put forward, while the Action Based Learning one might, at an extreme, involve entirely selection rather than the learning of sounds. The Mixed Model is one for which there is now evidence in a number of species.*

Models	Sequence of Events		
Memory-based Learning	Learning of selected sounds	Production of learned sounds	Finished song repertoire
Mixed Model	Learning of many sounds	Production and selective attrition	Finished song repertoire
Action-based Learning	Predetermined sound repertoire	Production and selective attrition	Finished song repertoire

Possible advantages of vocal learning

In discussing the functional significance of vocal learning it must be borne in mind that the current reasons why sounds are learnt may be quite different from those that led to the evolution of vocal learning in the first place. Indeed, to take the example of passerine birds, the origin of vocal learning shortly after the oscines and sub-oscines split suggests that the circumstances of its evolution may have been very different from those that exist today. Arguments about origin are, however, speculative and we will limit ourselves here to discussing reasons that have been suggested for why it persists. We will discuss the main hypotheses that have been put forward, in each case citing what seem to us the best examples and how the chaffinch data may match up to that particular idea.

Complex information transmission

Some bird species have very large song repertoires (e.g. Kroodsma & Parker 1977), and similar long and complex songs have also been found in whales (Payne & McVay 1971). Most such repertoires are achieved by copying from other individuals, frequently including mimicry of the sounds of other species (Baylis 1982). The most likely reason for such large and varied repertoires is sexual selection, and there is evidence in some species that females do indeed prefer males with more varied songs (e.g. Lampe & Sætre, 1995; Buchanan & Catchpole 1997). Learning from others, including interspecific mimicry, may be a simple and economical way to achieve this variety. While Kroodsma et al. (1997) argue that grey catbirds *Dumetella carolinensis* build up large repertoires by improvisation, this is certainly unusual, and learning in the sense of copying from others appears to be essential for the development of large repertoires in other species.

The complexity argument cannot however apply to all vocal learning. Many learnt songs, like that of the chaffinch, are relatively simple, with a repertoire that only ranges from one to a few types. They are not markedly more complex than the unlearnt songs of the suboscines. Less complex vocalisations, such as the signature whistles of dolphins (Tyack 1986) and the rain calls of chaffinches (Riebel & Slater 1998a), are also often copied. The correlation of learning with complexity is thus not a very good one: even though complexity may require learning, sounds that are not complex are often learnt as well.

Individual recognition

Learning may lead to variety, especially as copying errors are more frequent than with genetic transmission. In the case of the chaffinch, Slater et al. (1980) estimated that 15% of the songs in their population had either arisen through cultural mutations or been introduced by immigrants. There is clearly a much higher rate of song turnover than would be true with genetic transmission (Ince et al. 1980). This variety leads to differences between individuals which might in turn permit individual recognition, at least where repertoire sizes are small enough for within-individual variation not to lead to confusion.

Nottebohm (1972) argued against this idea on the grounds that copying is often precise and would thus lead to similarities not differences. Furthermore, slight differences in morphology between animals could lead to recognisable differences in sound production without learning being involved (see Beer 1970). Despite these arguments, Janik & Slater (1997) suggest that learning may be advantageous in enhancing differences between individuals especially where animals combine features from a number of others, and this may be beneficial particularly in the high background noise environments typically occupied by mammals that learn their vocal signals. Learning is certainly involved in the development of signature whistles in bottlenose dolphins *Tursiops truncatus*, and the fact that males, which disperse, have calls more similar to those of their mothers than females, which do not, argues for individual distinctiveness being an important factor (Sayigh et al. 1995). But sea mammals and bats may be a special case here: a remarkable amount of detailed specific information, including individual identity, can be encoded in a simple unlearnt call (see, for example, Nuechterlein 1981).

If learning is indeed involved in generating the differences on which individual recognition is based, this will only be a valid functional explanation if it benefits individuals to reveal their identities to others. In closely-knit groups of dolphins which stay together for long periods of time, and must maintain contact despite often being out of sight, the importance of individual recognition to all parties seems clear (see Janik & Slater 1998). This is less obvious in the more anonymous flocks of birds or ungulates.

Group recognition/Matching neighbours

Learning leads to similarities between the calls of animals in a particular area and, because of accumulating errors, to differences between areas. Where animals learn from neighbours, individuals in a group (Feekes 1977, 1982), or on adjacent territories (Bertram 1970; Payne 1985), often share signals. However, it is very unlikely that such sharing is a benefit of learning, as the less error-prone transmission of unlearnt sounds would lead to a greater degree of sharing not a lesser one. Nevertheless learnt sounds may act as 'passwords' within a group, as Feekes (1977) proposed, so that intruders unable to match the local signal can be excluded. Such a function for vocal learning is, however, most likely to be restricted to tightly integrated social groups. There is no doubt that it does not apply to the chaffinch, as birds of this species do not have songs that match those of their neighbours to more than a chance extent (Slater & Ince 1982).

Kin recognition

The idea that young birds might learn their songs from their fathers and that they might subsequently be used in kin recognition is an attractive one. Because of cultural mutations, such transmission would lead to stronger differences between lineages than would be possible without learning. Birds able to identify kin in this way might thus be able to mate assortatively or apportion helping behaviour preferentially towards relatives. However, clear evidence of young birds learning song from their fathers has been obtained in very few species, and it may be a laboratory artefact in some such cases (Slater & Mann 1990). The major exception here is in Darwin's finches *Geospiza fortis* and *G. scandens*, where the great majority of males sing the same song type as their father, and as their paternal grandfather, thus showing inheritance to be cultural (Grant & Grant 1996). Earlier work on the same two species by Millington & Price (1985) found no evidence for assortative mating by song, but Grant & Grant (1996) found females of *G. fortis* in their second breeding year to avoid mating with males having similar songs to their fathers. They argue that it is necessary to exclude first year females from the analysis as these pair later in the season and have a limited choice of mates. Their data do thus suggest that culturally inherited song in this species may function in mate choice to ensure outbreeding. Lack of other examples of song being passed from father to son make it unlikely that kin recognition is a widespread function of song learning. In the chaffinch, song repertoires are often built up by copying from more than one other individual (Slater et al. 1980) and, while some learning does occur in fledglings and may be from the father, much of it takes place the following spring when birds are setting up their territories (Slater & Ince 1982).

Within group mating

This suggestion is based on a similar idea to the one mentioned above, which proposed that animals might use song to choose a mate and thus achieve outbreeding. However, here the alternative suggestion, originally put forward by Nottebohm (1972), is that birds might choose mates from the same dialect area as themselves and thus preserve local adaptations. A suggested example here was the white-crowned sparrow *Zonotrichia leucophrys*, in which males have only a single song type, and this is shared widely within an area. Baker & Cunningham (1985) argue in support of the idea, but it has proved controversial, with lack of evidence that dialect boundaries act as barriers to dispersal (Zink & Barrowclough 1984) and evidence that females injected with testosterone often sing different dialects to those of their mates (Baptista & Morton 1982). It would in any case only be applicable to species where there are clear regional dialects. In the chaffinch, while song does vary from place to place it does not generally do so in a systematic manner, so that all songs from a particular area do not share a feature that songs from other areas lack (Slater et al. 1984). The closest it comes to this is in the case of the 'kit', an extra note like the call of a great-spotted woodpecker *Dendrocopos major*, that sometimes occurs at the end of the song. This is common on continental Europe (e.g. Mundry et al. 1994) but, until recently, unrecorded in Britain. However, even where it does occur, not all birds possess it, so its presence or absence is still not a clear cue as to the location.

Rather little is known about why female songbirds choose one male rather than another. If they are played song, especially when injected with oestrogen, they will often show courtship solicitation displays. Using this technique, preferences for their own species or for local song, and for a song repertoire rather than a single song type, have sometimes been shown (e.g. Baker et al. 1987). Recent work on canaries *Serinus canaria* also suggests that certain types of song phrase are more stimulating than others (Vallet & Kreutzer 1995, Vallet et al. 1998). An alternative technique we have used on chaffinches is operant conditioning, originally pioneered with males by Stevenson (1969) who found that song itself was a reinforcer in this species. Our recent experiments have shown that this is also true of young females (Riebel & Slater 1998b), which will work to hear song and were found to prefer songs which included the terminal flourish to ones lacking this distinctive feature. However, we were unable to find any difference in preference for songs that young females had experienced early in life compared with those unfamiliar to them. This also argues against song being used as an assortative mating cue in this species, a finding that is not surprising given that males often learn songs after dispersal.

Habitat matching

Hansen (1979) suggested that vocal learning in birds may match their song to the habitat in which it is sung. If birds learn songs at a distance from their tutor, those sounds that carry best through the habitat will be preferentially learnt, and song should become progressively moulded to match the habitat. In line with this, Gish & Morton (1981) showed that Carolina wren *Thryothorus ludovicianus* songs degraded least in the habitat in which they were recorded. In both the great tit *Parus major* (Hunter & Krebs 1979) and the rufous-collared sparrow *Zonotrichia capensis* (Nottebohm 1975; Handford 1981) songs vary between habitats. The most striking difference, easily understood in terms of Hansen's hypothesis, is that rapid trills are less common where vegetation is dense, and would thus be likely to distort them. Despite these examples, evidence for habitat matching in song is sparse. A species such as the chaffinch would be a good candidate for such an effect, with birds holding territories in areas varying from open scrub, to town gardens, to dense coniferous forest. However, while song varies considerably between individuals, the general structure of chaffinch song is remarkably fixed, with little overall geographical variation (Slater et al. 1984) and no obvious link of form with habitat (Williams & Slater 1993).

Conclusion

We have reviewed the main ideas that have been put forward to account for vocal learning in animals. Several of them have evidence in their favour for one species or another, but none appears to have widespread applicability. Must one therefore resort to the idea that song learning fulfils one function in some species and another in others? This would seem unlikely to be the case. In songbirds, for example, there are well over 4000 species, and every one that has been examined, an increasingly long list, has been found to learn its song. Yet many features of song are extraordinarily different between them, for example whether one sex sings or both, repertoire size, and the timing and length of song season. Might there be something that they all have in common that leads them to share song learning?

We have recently carried out computer simulations that suggest this may indeed be the case (Lachlan & Slater 1999). The suggestion is that once a group has evolved song learning, for whatever reason, it may persist without any immediate function. The basis for this hypothesis lies in the evolutionary interaction of culturally transmitted song and a genetically based trait. The latter is a filter that creates a preference for conspecific song. Song learning experiments have revealed that male chaffinches possess such filters which act during song learning (e.g. Thorpe 1958). Mate choice and song tutoring experiments suggest that females also possess the same filter, which they use to recognise potential mates. The most likely reason for song learning to disappear would be if selection favoured a progressive reduction in the breadth of the filter (i.e. the range over which songs are recognised as species typical), since very narrow genetic filters would allow very little potential for variation through learning. Therefore, we used a simulation model to explore whether individuals with narrower filters might be able to invade populations consisting of those with broader filters and vice versa. Over a wide range of conditions, broader filters were more successful. The reason for this is as follows. Males with wider filters have the potential to learn songs that females with narrower filters cannot recognize. Therefore, in a mixed population of individuals with wide and narrow filters, these two classes of individual will be less successful than males with narrow filters, or females with wide filters. However, when one or other predisposition is uncommon, these selection pressures against narrow and wide filters will not be equal. When the narrow filter is rare, many males may be singing unusual songs and females with the narrow filter will be heavily disadvantaged (Fig. 1a). When the wide filter is rare, though, males with this filter will tend to copy the songs of narrow filter males around them - avoiding discrimination by females (Fig. 1b). Overall, this should lead to a selection pressure for wider filters, which will probably eventually be limited by increasing confusion with other species.

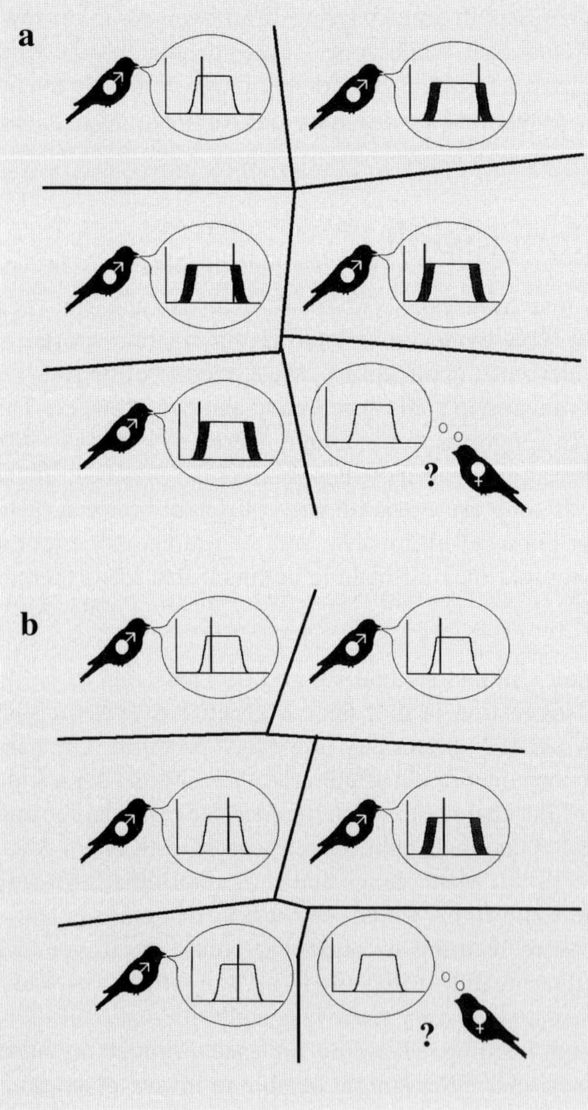

***Figure 1. (a)** A female with narrow filter is shown trying to decide among five territorial males, four of them with wide filters and one with a narrow filter. The songs of the males, symbolised here by the vertical line, will often be outside the acceptable range for a female whose filter is narrow. In the case shown here only the single male in the five territories with a narrow filter has a song she would find acceptable. As a result narrow filter females will have fewer mating opportunities and become rarer in the population.*

***Figure 1. (b)** In a population consisting largely of males with narrow filters, broad filter males will fare equally well because they are most likely to learn from a narrow filter tutor and so have a song that conforms to the population norm.*

These simulations suggest that the learning of signals such as songs used in mate choice may persist in a population even if it does not have any direct functional significance, simply through an interaction between genes and culture. The animals are caught in a 'cultural trap'. While the hypothesis needs to be tested, it may provide a novel and rather unexpected answer to the old question of why birds learn their songs. Given the failure of earlier hypotheses to come up with plausible theories for species such as the chaffinch, it does seem to be the time for some more radical suggestions.

Acknowledgements

PJBS is most grateful to the organisers for inviting him to the excellent meeting at Kongsvoll, and to the Royal Society of Edinburgh for a travel grant. Funding for their research has been provided through a BBSRC studentship to RFL and a Marie-Curie Fellowship from the EU to KR. Bill Searcy's perceptive comments on the manuscript were much appreciated.

References

Baker, M. C., Bjerke, T. K., Lampe, H. M. & Espmark, Y. O. 1987. Sexual response of female yellowhammers to differences in regional song dialects and repertoire sizes. *Animal Behaviour,* **35**, 395-401.

Baker, M. C. & Cunningham, M. A. 1985. The biology of bird-song dialects. *Behavioral and Brain Sciences,* **8**, 85-133.

Baptista, L. F. & Morton, M. L. 1982. Song dialects and mate selection in montane white-crowned sparrows. *Auk,* **99**, 537-547.

Baptista, L. F. & Schuchmann, K. 1990. Song learning in the Anna hummingbird *(Calypte anna). Ethology,* **84**, 15-26.

Baylis, J. R. 1982. Avian vocal mimicry: its function and evolution. In: *Acoustic Communication in Birds.* (Ed. by D. E. Kroodsma & E. H. Miller), pp. 51-83. New York: Academic Press.

Beer, C. G. 1970. Individual recognition of voice in the social behavior of birds. *Advances in the Study of Behavior,* **3**, 27-74.

Bertram, B. 1970. The vocal behaviour of the Indian hill mynah. *Animal Behaviour Monographs,* **3**, 81-192.

Buchanan, K. L. & Catchpole, C. K. 1997. Female choice in the sedge warbler, *Acrocephalus schoenobaenus*: multiple cues from song and territory quality. *Proceedings of the Royal Society of London, Series B,* **264**, 521-526.

Feekes, F. 1977. Colony specific song in *Cacicus cela* (Icteridae, Aves): the password hypothesis. *Ardea,* **65**, 197-202.

Feekes, F. 1982. Song mimesis within colonies of *Cacicus c. cela* (Icteridae, Aves). A colonial password? *Zeitschrift für Tierpsychologie,* **58**, 119-152.

Galeotti, P. R., Appleby, B. M. & Redpath, S. M. 1996. Macro and microgeographical variations in the 'hoot' of Italian and English tawny owls (*Strix aluco*). *Italian Journal of Zoology,* **63**, 57-64.

Gaunt, S. L. L., Baptista, L. F., Sanchez, J. E. & Hernandez, D. 1994. Song learning as evidenced from song sharing in two hummingbird species (*Colibri coruscans* and *C. thalassinus*). *Auk,* **111**, 87-103.

Gish, S. L. & Morton, E. S. 1981. Structural adaptations to local habitat acoustics in Carolina wren songs. *Zeitschrift für Tierpsychologie,* **56**, 74-84.

Grant, B. R. & Grant, P. R. 1996. Cultural inheritance of song and its role in the evolution of Darwin's finches. *Evolution,* **50**, 2471-2487.

Handford, P. 1981. Vegetational correlates of variation in the song of *Zonotrichia capensis*. *Behavioral Ecology and Sociobiology,* **8**, 203-206.

Hansen, P. 1979. Vocal learning: its role in adapting sound structures to long-distance propagation and a hypothesis on its evolution. *Animal Behaviour,* **27**, 1270-1271.

Hunter, M. L. & Krebs, J. R. 1979. Geographical variation in the song of the great tit (*Parus major*) in relation to ecological factors. *Journal of Animal Ecology,* **48**, 759-785.

Ince, S. A., Slater, P. J. B. & Weismann, C. 1980. Changes with time in the songs of a population of chaffinches. *Condor,* **82**, 285-290.

James, P. C. 1985. Geographical and temporal variation in the calls of the manx shearwater *Puffinus puffinus* and the British storm petrel *Hydrobates pelagicus*. *Journal of Zoology,* **207**, 331-344.

Janik, V. M. & Slater, P. J. B. 1997. Vocal learning in mammals. *Advances in the Study of Behavior,* **26**, 59-99.

Janik, V. M. & Slater, P. J. B. 1998. Context specific use suggests that bottlenose dolphin signature whistles are cohesion calls. *Animal Behaviour,* **56**, 829-838.

Jones, G. & Ransome, R. D. 1993. Echolocation calls of bats are influenced by maternal effects and change over a lifetime. *Proceedings of the Royal Society of London, Series B,* **252**, 125-128.

Kroodsma, D. E. & Baylis, J. R. 1982. A world survey of evidence for vocal learning in birds. In: *Acoustic Communication in Birds*. (Ed. by D. E. Kroodsma & E. H. Miller), pp. 311-337. New York: Academic Press.

Kroodsma, D. E., Houlihan, P. W., Fallon, P. A. & Wells, J. A. 1997. Song development in grey catbirds. *Animal Behaviour,* **54**, 457-464.

Kroodsma, D. E. & Konishi, M. 1991. A suboscine bird (eastern phoebe, *Sayornis phoebe*) develops normal song without auditory feedback. *Animal Behaviour,* **42**, 477-487.

Kroodsma, D. E. & Parker, L. D. 1977. Vocal virtuosity in the brown thrasher. *Auk,* **94**, 783-785.

Lachlan, R. F. & Slater, P. J. B. 1999. The maintenance of vocal learning by gene-culture interaction: the cultural trap hypothesis. *Proceedings of the Royal Society of London, Series B*, **266**, 701-706.

Lampe, H. M. & Sætre, G.-P. 1995. Female pied flycatchers prefer males with larger song repertoires. *Proceedings of the Royal Society of London, Series B,* **262**, 163-167.

Marler, P. & Nelson, D. A. 1993. Action-based learning: a new form of developmental plasticity in bird song. *Netherlands Journal of Zoology,* **43**, 91-103.

Marler, P. & Peters, S. 1982. Developmental overproduction and selective attrition: new processes in the epigenesis of bird song. *Developmental Psychobiology,* **15**, 369-378.

Marler, P. & Pickert, R. 1984. Species-universal microstructure in a learned birdsong: the swamp sparrow (*Melospiza georgiana*). *Animal Behaviour,* **32**, 673-689.

Millington, S. J. & Price, T. D. 1985. Song inheritance and mating patterns in Darwin's finches. *Auk,* **102**, 342-346.

Mitani, J. C., Hasegawa, T., Gros-Louis, J., Marler, P. & Byrne, R. 1992. Dialects in wild chimpanzees? *American Journal of Primatology,* **27**, 233-243.

Mundry, R., Hau, B. & Böhner, J. 1994. Individual- und strophenspezifischer Einsatz des 'kit' im Gesang des Buchfinken (*Fringilla coelebs*). *Journal für Ornithologie,* **135**, 223-231.

Nelson, D. A. 1992. Song overproduction and selective attrition lead to song sharing in the field sparrow (*Spizella pusilla*). *Behavioral Ecology and Sociobiology,* **30**, 415-424.

Nottebohm, F. 1972. The origins of vocal learning. *American Naturalist,* **106**, 116-140.

Nottebohm, F. 1975. Continental patterns of song variability in *Zonotrichia capensis*: some possible ecological correlates. *American Naturalist,* **109**, 605-624.

Nuechterlein, G. L. 1981. Variations and multiple functions of the advertising display of western grebes. *Behaviour,* **76**, 289-317.

Payne, R. B. 1985. Behavioral continuity and change in local song populations of village indigobirds *Vidua chalybeata. Zeitschrift für Tierpsychologie,* **70**, 1-44.

Payne, R. S. & McVay, S. 1971. Songs of humpback whales. *Science,* **173**, 585-597.

Pepperberg, I. M. 1994. Vocal learning in gray parrots (*Psittacus erithacus*): effects of social interaction, reference, and context. *Auk*, **111**, 300-313.

Ralls, K., Fiorelli, P. & Gish, S. 1985. Vocalizations and vocal mimicry in captive harbor seals, *Phoca vitulina. Canadian Journal of Zoology,* **63**, 1050-1056.

Richards, D. G., Wolz, J. P. & Herman, L. M. 1984. Vocal mimicry of computer-generated sounds and vocal labelling of objects by a bottlenosed dolphin, *Tursiops truncatus. Journal of Comparative Psychology,* **98**, 10-28.

Riebel, K. & Slater, P. J. B. 1998a. Male chaffinches (*Fringilla coelebs*) can copy calls from a tape tutor. *Journal für Ornithologie,* **139**, 353-355.

Riebel, K. & Slater, P. J. B. 1998b. Testing female chaffinch song preferences by operant conditioning. *Animal Behaviour,* **56**, 1443-1453.

Sayigh, L. S., Tyack, P. L., Wells, R. S., Scott, M. D. & Irvine, A. B. 1995. Sex difference in signature whistle production of free-ranging bottlenose dolphins, *Tursiops truncatus. Behavioral Ecology and Sociobiology,* **36**, 171-178.

Slater, P. J. B. 1989. Bird song learning: causes and consequences. *Ethology, Ecology and Evolution,* **1**, 19-46.

Slater, P. J. B., Clements, F. A. & Goodfellow, D. J. 1984. Local and regional variations in chaffinch song and the question of dialects. *Behaviour,* **88**, 76-97.

Slater, P. J. B. & Ince, S. A. 1982. Song development in chaffinches: what is learnt and when? *Ibis,* **124**, 21-26.

Slater, P. J. B., Ince, S. A. & Colgan, P. W. 1980. Chaffinch song types: their frequencies in the population and distribution between the repertoires of different individuals. *Behaviour,* **75**, 207-218.

Slater, P. J. B. & Mann, N. I. 1990. Do male zebra finches learn their fathers' songs? *Trends in Ecology and Evolution,* **5**, 415-417.

Stevenson, J. G. 1969. Song as a reinforcer. In: *Bird Vocalizations.* (Ed. by R. A. Hinde), pp. 49-60. Cambridge: Cambridge University Press.

Thorpe, W. H. 1958. The learning of song patterns by birds, with especial reference to the song of the chaffinch *Fringilla coelebs. Ibis,* **100**, 535-570.

Todt, D. 1975. Social learning of vocal patterns and modes of their application in grey parrots (*Psittacus erithacus*). *Zeitschrift für Tierpsychologie,* **39**, 178-188.

Tyack, P. 1986. Whistle repertoires of two bottlenosed dolphins, *Tursiops truncatus*: mimicry of signature whistles? *Behavioral Ecology and Sociobiology,* **18**, 251-257.

Vallet, E., Beme, I. & Kreutzer, M. 1998. Two-note syllables in canary songs elicit high levels of sexual display. *Animal Behaviour,* **55**, 291-297.

Vallet, E. & Kreutzer, M. 1995. Female canaries are sexually responsive to special song phrases. *Animal Behaviour,* **49**, 1603-1619.

Williams, J. M. & Slater, P. J. B. 1993. Does chaffinch *Fringilla coelebs* song vary with the habitat in which it is sung? *Ibis,* **135**, 202-208.

Zink, R. M. & Barrowclough, G. F. 1984. Allozymes and song dialects: a reassessment. *Evolution,* **38**, 444-448.

IN: ESPMARK,Y., AMUNDSEN,T. & ROSENQVIST,G. (EDS.) 2000
ANIMAL SIGNALS:
SIGNALLING AND SIGNAL DESIGN IN ANIMAL COMMUNICATION, PP. 353-364
TAPIR ACADEMIC PRESS, TRONDHEIM, NORWAY

18

Multiple Nestling Begging Signals: Reed Warbler Chicks and Cuckoo Tricks

REBECCA M. KILNER

Department of Zoology, University of Cambridge,
Downing Street, Cambridge, CB2 3EJ, UK

E-mail: rmk1002@hermes.cam.ac.uk

Abstract

Nestling birds solicit food using complex begging displays in which the presentation of a brightly coloured gape is accompanied by repetitive calling. The solicitation display of the common cuckoo nestling, a brood parasite, is particularly striking involving the use of a rapid begging call rate, far more rapid than that produced by chicks reared by their own parents. Here I describe experiments which reveal the function of the cuckoo's excessive begging call when it parasitizes the parental care of the reed warblers. The experiments show that reed warbler parents integrate visual and vocal cues from their young, varying the provisioning rate in response to the gape area they display and the rate at which they call. Together these two signals provide parents with a more accurate picture of the state of their young than either would alone. Moreover, parents use exactly the same integration rules whether provisioning a cuckoo chick or their own young. Reared alone in the nest, the cuckoo presents a deficient visual display. Only by calling at an excessively fast rate can the cuckoo compensate and so elicit sufficient care to grow and develop successfully.

Many animal displays are strikingly complex, often involving a variety of signals (e.g. Zuk et al. 1992; Rowe & Guilford 1996; Scheffer et al. 1996). The solicitation displays of young passerines in the nest provide a good example. Superficially, these displays are very alike, varying little between species. Nestlings typically stretch and posture (Redondo & Castro 1992), open their mouths to display a brightly coloured gape and accompany this visual display with incessant calling. Between species, however, the details of the display can vary considerably. Begging calls may be short and high-pitched as in chaffinches *Fringilla coelebs,* or lower and longer as in the European starlings *Sturnus vulgaris* (Redondo & Arias de Reyna 1988). Nestling mouth colour is equally variable, ranging from yellow in robins *Erithacus rubecula* to orange in pied wagtails *Motacilla alba* and red in reed buntings *Emberiza schoeniclus* (Kilner & Davies 1998).

A particularly striking and much discussed nestling solicitation display is that of the common cuckoo *Cuculus canorus*. This brood parasite lays its eggs in the nests of various host species, relying on them entirely to incubate its eggs and rear its young to independence. The bizarre sight of a small warbler feeding a cuckoo nestling, up to seven times its size and utterly different in appearance to its own young, has led to all kinds of speculation about the narcotic lure and slave-driving propensity of the cuckoo chick's begging display (see, for example, Dawkins 1976). In this paper, I discuss the results of some field experiments that tackle the issue of how the cuckoo's begging display contributes to its success at eliciting care from its host species, in this case the reed warbler *Acrocephalus scirpaceus*. Previous studies of brood parasite begging have considered cases where the parasite is raised alongside the host young (Redondo 1993; Soler et al. 1995; Lichtenstein & Sealy 1998). Here, a major selective pressure on begging is competition within the brood for food. Since newly hatched common cuckoo nestlings evict their reed warbler nestmates, and so end up being reared alone, this system is ideal for isolating the signals that affect provisioning rates independent of competition in the nest.

Multiple signals: the chicks' perspective

A common assumption is that the cuckoo nestling is something of a slave driver, working parents, and indeed any other bird that is passing by (see, for example, Dawkins 1976), into a frenzy of feeding activity. Contrary to popular belief, however, observations of nest provisioning rates show that reed warbler parents feed a cuckoo chick similar food and at roughly the same rate as they would a brood of four of their own young, the modal brood size in this species (Brooke & Davies 1989). Although the cuckoo spends longer in the nest, 17 days compared with 11 for a reed warbler brood, it is only at the very end of its nestling period that it commands a significantly higher provisioning rate than a brood of four reed warblers. The overall similarity between the provisioning rate of a brood and a cuckoo nestling can be seen when provisioning rate is compared with nestling mass (Fig. 1; throughout the paper, unless stated otherwise, provisioning rate means the number of nest visits per hour at which a nestling was fed).

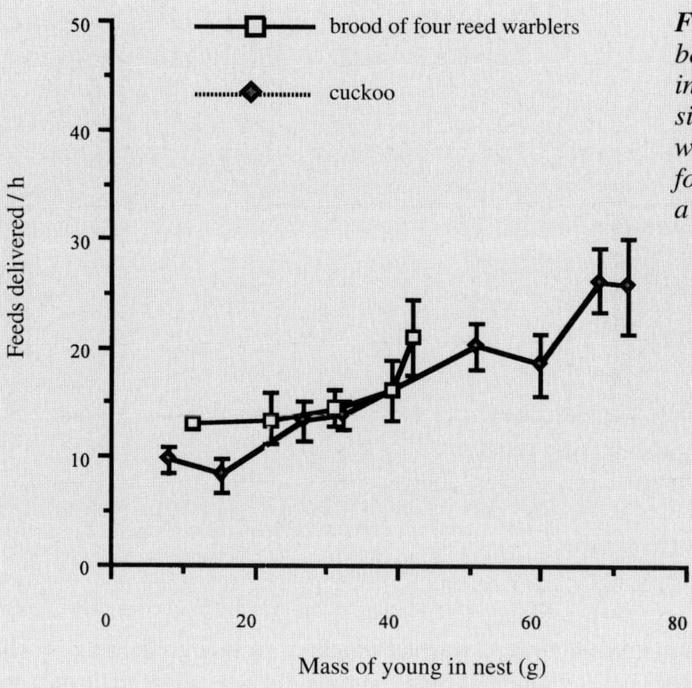

Figure 1. The relationship between the mass of young in the nest and provisioning rate by adult reed warblers to a brood of four reed warblers and to a single cuckoo nestling.

Chick mass

An obvious interpretation of these data is that parents use a rough rule of thumb to set their provisioning rate, simply assessing the mass of the young in their nest. So perhaps the reason why a cuckoo chick commands the same visit rate as a brood simply lies in its relatively large size, because the total mass of four reed warblers is roughly the same as the mass of a cuckoo (Kilner & Davies, 1999). We tested this idea by substituting cuckoo chicks with blackbird *Turdus merula* or song thrush *T. philomelos* nestlings. *Turdus* young (mean ± 1 s.d. = 25.5 ± 6.8g) are roughly the same mass as a cuckoo chick (mean ± 1 s.d. = 25.5 ± 6.0g), and are readily accepted and fed by reed warbler parents (Davies et al. 1998). In our first experiment, we tested the hypothesis that nestling mass influences provisioning rate by comparing the loads of food brought to the nest per hour by a single reed warbler chick, a single *Turdus* chick and a single cuckoo chick. In the *Turdus* experiments, not all the food brought to the nest was given to chicks, but since the experiments were designed to test parental behaviour we decided to measure the loads brought as their response to each treatment. Scoring feeds delivered per hour to nestlings yields the same result.

If the parents visited the nest in relation to chick mass alone, then we would expect them to visit the larger *Turdus* and cuckoo chicks equally frequently, and much more often than the reed warbler chick. We found a significant difference in number of loads brought to the three treatments ($F_{2,41} = 5.36$, $P < 0.01$) but *post hoc* analyses showed that, in fact, the cuckoo chick was fed significantly more than the *Turdus* chick, and a *Turdus* chick was

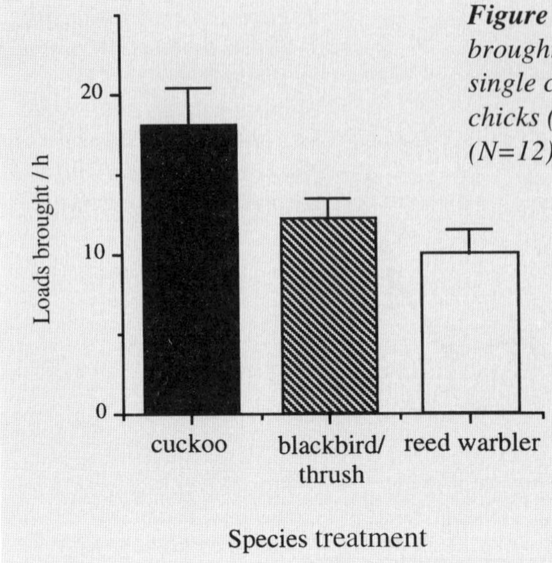

Figure 2. *Number of food loads (mean+1 s.e.) brought per hour by pairs of reed warblers to single cuckoo chicks (N=13), single* Turdus *chicks (N=19) and single reed warbler chicks (N=12). (Modified from Davis et al. 1998).*

no better at attracting feeds than a single reed warbler chick, even though the reed warbler was less than a third its mass (Fig. 2). Quite clearly, parents did not adjust their nest visit rate simply in relation to the mass of young in their nest.

Vocal begging displays

Since parents appear not to respond directly to the size of the cuckoo chick, perhaps they adjust their provisioning rate in relation to its begging display. Cuckoos start to call when they are 6 days old (Kilner & Davies, 1999), and they call at an unusually fast rate. Whereas a single reed warbler chick has a relatively slow begging call, the cuckoo begging call is much faster and is similar to that of a brood when both are about a week old (Davies et al. 1998). In fact, the begging call of young cuckoo chicks sounded to our ears quite like the calls of a brood of four reed warblers. So perhaps this part of the begging display is particularly important in eliciting care from the reed warbler parents.

We examined the significance of the cuckoo's begging call by testing whether parents set their provisioning rate by listening to the rate of calling from the nest. In this experiment we gave parents a single *Turdus* chick again, to simulate a cuckoo-sized nestling, but this time we also played back additional begging calls through a speaker fixed at the side of the nest. We broadcast either the calls of a nestling cuckoo, or the calls of a brood of four reed warblers. For a control, we broadcast nothing. We repeated these three treatments at 10 different nests and scored the rate at which food loads were brought to the nest in each treatment. Playback tapes were made by recording a sequence lasting 20 sec from broods of four reed warblers or cuckoo nestlings when both were roughly a week old. The sequences were edited with the SoundEdit application on a Macintosh computer, with a high pass filter to remove background noise. The resulting sequence (ca. 15 sec) was then re-recorded onto metal audiotapes as a continuous 10 min loop and broadcast at natural amplitude. A different brood and cuckoo recording was used for each of the ten experiments (Davies et al. 1998).

The experiment showed that parents clearly did respond to the playbacks when deciding how often to visit the nest, increasing their visits in response to a high call rate (Fig. 3). Interestingly, parents responded similarly to playback of the brood and playback of the cuckoo calls. We know from other experiments that any playback is not sufficient to make parents visit the nest more frequently. For example, if we broadcast the begging calls of a single reed warbler chick calling while reed warbler young in the nest solicit food, then parents visit the nest less frequently in response than if we broadcast the calls of a four reed warbler brood. So the results of our experiment suggest that perhaps the cuckoo commands the same visit rate as a brood because it mimics a brood's vocalizations (Davies et al. 1998).

We tested this hypothesis by comparing the visit rate to a *Turdus* nestling accompanied by playback of cuckoo vocalizations with that of a cuckoo. We predicted that there should be no significant difference between the two, because in both tests, parents were faced with a similar sized nestling making or appearing to make exactly the same vocalizations. In contrast to our prediction, however, we found that the *Turdus* nestling was fed significantly more frequently than the cuckoo (Kilner et al. 1999). Now we know from other experiments that the *Turdus* nestling's begging call cannot explain the difference in provisioning rate, because when we compare the nest visit rate to a *Turdus* chick accompanied by brood vocalizations with that to a brood, the brood is provisioned significantly more often (Kilner et al. 1999). This suggests that parents must use some other component of the begging display, in addition to calls, when deciding how often to visit the nest with food.

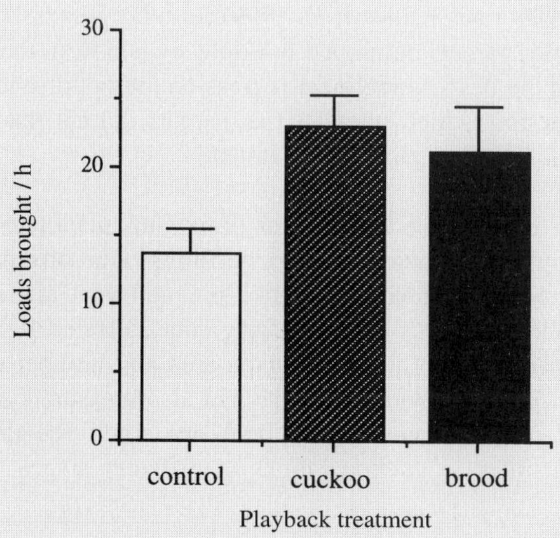

Figure 3. *Loads delivered per hour (mean +1 s.e.) to single blackbird or song thrush chicks, each tested under three conditions: control (no playback), playback of cuckoo begging calls or calls of a brood of four reed warblers. (Modified from Davis et al. 1998).*

Visual and vocal begging displays

Although the *Turdus* chick and cuckoo were matched for mass in our experiments, they differed significantly in their gape area. In fact, for a given mass, a *Turdus* chick nestling had a larger gape area than a cuckoo nestling (Kilner et al. 1999). So perhaps this difference in gape area can account for the difference in feeding success between *Turdus* nestlings with cuckoo playback and cuckoo nestlings. Perhaps parents use the total gape area on display when deciding how often to visit the nest, as well as the calls that they hear. To test whether this visual signal was also used by reed warbler parents when deciding how often to feed their own young we had to do a second experiment.

The principle aim of this experiment was to tease apart the relative significance of visual and vocal signals on provisioning rate. We did this using a combination of brood size manipulations (one to eight young, $N = 59$), which affected both signals equally, and playbacks of different numbers of reed warbler chicks calling (one or four chicks) at broods of two to five young ($N = 52$), which simply increased vocal signals. The manipulation is perhaps most easily understood by means of an illustration. For example, if we played back the calls of four reed warblers at a brood of three chicks, parents would effectively be hearing seven chicks but seeing only three. But if we increased the brood size to seven, then parents hearing seven chicks would also see seven chicks. In this way we could tease apart parental response to visual and vocal signals. For each manipulation, we scored nest visit rate per hour, and we were careful to do all manipulations when chicks were the same age. To explain the variance in nest visit rate, we used a multiple regression with maximum number of chicks calling per nest and maximum number of gapes on display per nest as the two independent factors. (Our limited data for broods of two to four chicks show that the mean proportion of the brood which gaped per visit was 0.8 ($N = 13$)).

The multiple regression showed that both the number of gapes and the number of chicks calling had independent effects on nest visit rate (Kilner et al. 1999). In other words, the experiment demonstrated that reed warbler parents combined multiple signals from their young to adjust their provisioning rate at the nest. A nonlinear regression did not provide a better fit than the linear multiple regression, which suggests that parents did not reach their provisioning limit within the range of our experimental treatments.

An exciting corollary of this finding was that it provided a means of making quantitative predictions about how often parents should provision cuckoo nestlings. Our original regression equation from the multiple regression analysis defined parental visit rate in terms of number of gapes and the number of chicks calling. Using data collected from broods in the field, we translated number of gapes into total gape area and number of chicks calling into call rate (calls produced per 6 sec). Since we had also measured the gape area and call rate of cuckoos at different ages in the field, we could use this regression equation to predict the visit rates to cuckoo chicks at different stages of development.

Exploitation by cuckoo chicks

To test the accuracy of these predicted values, we plotted observed values of cuckoo provisioning rate against the predicted values we had derived using the regression equation. We found that the regression line did not differ significantly from the line y = x (Kilner & Davies, 1999), supporting the hypothesis that host reed warblers use exactly the same cues to decide how often to feed a nestling cuckoo, as they would when rearing their own brood. The cuckoo's success at attracting sufficient food comes through exploiting the reed warbler's usual rules for feeding its own young.

Precisely how well the cuckoo has adapted its begging display to the receiver psychology of the parent became clearer when we compared the two components of the display, gape area and call rate, in broods of four reed warblers and cuckoos. As cuckoo nestlings grow, the total gape area they are able to display is much less than that displayed by a brood of four reed warblers of equivalent mass, and the difference becomes progressively more pronounced with increasing chick age. The difference is not surprising, since the brood may display four gapes at once while a cuckoo can only display one. However, to compensate for the shortfall in gape area, the cuckoo calls at a faster rate than a brood of four reed warblers, the positive difference in call rate between a cuckoo and reed warbler brood offsetting the negative difference in gape area. In other words, it is to compensate for its subnormal visual stimulus, that the cuckoo produces a supernormal vocal stimulus. We tested this idea by calculating how frequently a cuckoo chick should call given our observations of its provisioning rate and measurements of the gape area (Kilner et al. 1999). For each day of the nestling period (day 4 to day 11, day 13 and day 15) we calculated the mean gape area displayed by a cuckoo nestling and the mean rate at which it was provisioned per hour. We then substituted these means into the general regression equation described above to calculate predicted calling rates for cuckoos on each day of the nestling period. When we compared the regression line through these predicted values with the regression line through our observations of cuckoo calling rate, at equivalent ages, we found no significant difference in either the slope or the elevation (Kilner et al. 1999). Our calculations therefore support the idea that the cuckoo nestling's excessive begging call during the latter half of its nestling life serves to compensate for its deficient visual stimulus.

Our data also lay to rest the popular idea that the cuckoo chick entices parents into feeding it by displaying its bright red supernormal gape (Dawkins 1976). Since a cuckoo nestling displays a smaller gape than a blackbird nestling of the same mass, if anything the cuckoo has a subnormal gape. We also know from gape painting experiments that the red gape of the cuckoo is nothing more than a red herring; altering a reed warbler nestling's mouth from its natural yellow colour to red does nothing for its prospects of being fed (D. Noble et al., unpublished).

Multiple signals: the parents' perspective

By trying to understand how a cuckoo elicits sufficient care from its hosts, we have stumbled upon the fact that adult reed warblers use multiple begging signals to regulate their rate of provisioning at the nest. From the chick's perspective, gape area and call rate have separate important roles in regulating how often parents come to the nest with food. But from the parent's perspective, what information do these signals carry? In other words, what can adults find out about their chicks from these signals?

We know from separate experiments on canaries (R. M. Kilner, unpublished) and blackbirds (R. M. Kilner & N. B. Davies, unpublished) that the best single correlate of any measure of begging intensity is the amount of food a chick will eat before it stops begging. The reason that it is such a good predictor of begging intensity is that it varies both with how food deprived a chick is and how big it is, simultaneously combining 'short term' need and 'long term' need (Price et al. 1996). In our next experiment, we tested whether gape area and begging call rate were correlated with the amount of food required by reed warblers to stop begging. The aim of this experiment was to test whether gape area and begging call rate functioned as 'back-up signals' (Møller & Pomiankowski 1993; Johnstone 1996); that is, whether the two signals together could provide parents with a more accurate assessment of the needs of their young than either could alone.

The protocol was very straightforward. We temporarily removed reed warblers from their nests at a range of different ages, and in a range of different brood sizes ($N = 49$). We fed them until they stopped begging with chick rearing mix, and then kept them warm but unfed for 110 min. Then we stimulated them to beg and recorded their begging behaviours on video and audiotape and recorded how much food the chicks ate before they stopped begging again. Finally, to explain the variance in the amount of food consumed, we used a multiple regression with total gape area and begging call rate as the two independent factors.

The result of the experiment was that both gape area and call rate explained a significant amount of variance in chick need (Kilner et al. 1999). Since both these signals could explain different components of variance in the same aspect of chick condition, they were functioning as back-up signals. Put another way, the variance in food consumption explained by both signals was greater than the variance explained by either signal alone (Kilner et al. 1999). It appears that reed warbler parents use both signals to provide information about their young, because it gives them a more accurate assessment of their chicks' nutritional condition than either signal could alone.

Combining the two perspectives

Theoretically, we would expect that for a given level of need, the cuckoo should solicit more feeds than a reed warbler brood (Briskie et al. 1994; Godfray 1995). Reed warbler nestlings must trade-off the benefits they would accrue from a higher provisioning rate with the potential loss of future siblings. But because the cuckoo has no genetic stake in

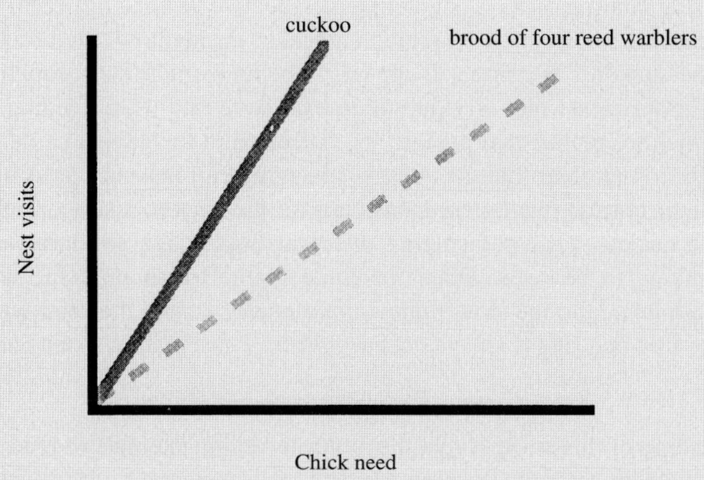

Figure 4. *A simple illustration of the hypothesized relationship between parental provisioning rate and chick need for cuckoo and reed warbler nestlings.*

the future young of its host, it loses nothing by demanding more parental feeds per unit of need than the host young (Fig. 4).

We tested this prediction by using data from the chick feeding experiments described above. These experiments tell us the amount of food required by a chick to stop begging for a given intensity of begging (gape area displayed and call rate). Using the regression equation which relates begging call rate and gape area displayed to adult provisioning rate, we can thus match the amount of food required by chicks to stop begging ('chick need' in Fig. 4) with the rate at which a parent would provision those young ('nest visits' in Fig. 4). Since we had these data for both cuckoo (N = 21) and reed warbler nestlings, we could test the prediction that cuckoos should demand more parental feeds per unit of need than the host young.

The results turned out to be quite surprising and very different from those predicted by theory. Far from behaving selfishly, the cuckoo did not demand significantly more feeds in relation to its need than the hosts' own young (Kilner & Davies, 1999). Indeed, the parents' response to the cuckoos' need was very similar to the way in which they responded to their own chicks' needs.

Conclusions and discussion

The picture which emerges is that the cuckoo chick's lifestyle in a reed warbler's nest is remarkably similar to that of the host's own young. The cuckoo's growth rate (Kilner & Davies, 1999) and the rate at which it is provisioned are similar to those for a brood of four reed warblers. Exactly why this should be remains unclear (see Kilner et al. 1999).

The interesting finding from a signalling perspective is that the reed warbler parents combined information from two begging signals to decide how often to visit the nest. At first sight, the cuckoo's begging call appears to be an extraordinary supernormal stimulus designed to fool parents into working at an excessive rate when feeding the imposter in their nest. Once we realise that parents are interested in the gape area on display as well, however, it becomes clear that the cuckoo's begging call instead serves to induce parents to work at their normal rate by compensating for the cuckoo's subnormal visual display. Whether the cuckoo gens parasitizing different hosts have similarly adapted begging displays remains to be seen. Since the three main British host species, the dunnock *Prunella modularis*, the meadow pipit *Anthus pratensis* and the reed warbler have very different nestling begging displays, this would be a fascinating extension of our current work.

A second finding of this study is that the multiple begging signals of reed warblers appear to function as 'back-up' signals, together providing the parent with accurate information about the nutritional need of its young. It is not yet clear whether this will turn out to be a universal function for multiple begging signals. We do know, however, that the relative significance of the different components of the display is likely to vary between species. For example, although gape area significantly influenced adult behaviour in reed warblers, gape colour did not. This contrasts with the results of other work with canaries (Kilner 1997), and other open-nesting species which feed their young by regurgitation like ravens, linnets and bullfinches. In these species, gape colour is an important component of the display and accurately signals offspring state (Kilner & Davies 1998). In canaries, at least, parents adjust their provisioning behaviour in relation to the colour of the gape (Kilner 1997). So signals of 'need' though mouth colour are correlated with feeding chicks by regurgitation (Kilner & Davies 1998).

It may be that the manner in which parents deliver food at each nest visit is important in affecting the design and function of the different components of the begging display. For example, only when parents feed by regurgitation can they subdivide each load brought to the nest and closely match food allocation to the precise needs of each of their young. Selection for finely-tuned signals of need, such as subtle changes in mouth colour, may only occur when parents bring food items to the nest that are readily divisible among the brood. In other words, the feeding ecology of parents, in addition to more obvious factors like whether the nestling is signalling in an open or hole nest (Redondo & Arias de Reyna 1988; Kilner & Davies 1998), may explain which of the nestling begging signals function to signal need (Kilner & Davies 1998).

A final point to make is that these experiments suggest that the third component of the begging display, begging posture, is irrelevant in influencing parental feeding rate at the nest. Its function may instead be related to influencing the allocation of food among chicks at each nest visit (Kilner 1995). Alternatively, it may serve no direct signalling function at all, acting instead as an amplifier of the other two signals, gape area and call rate (Hasson 1989).

Acknowledgments

First and foremost, I thank Nick Davies and Dave Noble for generously allowing me to work with them on the cuckoos and reed warblers at Wicken Fen, and for lengthy discussions of all the ideas presented in this paper. I also thank Bill Searcy and Trond Amundsen for their comments on an earlier draft of the paper. The cuckoo project was funded by NERC, and I was supported by a Junior Research Fellowship at Magdalene College, Cambridge and a Royal Society Dorothy Hodgkin Research Fellowship.

References

Briskie, J. V., Naugler, C. T. & Leech, S. M. 1994. Begging intensity of nestling birds varies with sibling relatedness. *Proceeding of the Royal Society of London, Series B,* **258**, 73-78.

Brooke, M. de L. & Davies, N. B. 1989. Provisioning of nestling cuckoos *Cuculus canorus* by reed warbler *Acrocephalus scirpaceus* hosts. *Ibis,* **131**, 250-256.

Davies, N. B., Kilner, R. M., & Noble, D. G. 1998. Nestling cuckoos, *Cuculus canorus*, exploit hosts with begging calls that mimic a brood. *Proceedings of the Royal Society of London, Series B,* **265,** 673-678.

Dawkins, R. 1976. *The Selfish Gene.* Oxford: Oxford University Press.

Godfray, H. C. J. 1995. Signaling of need between parents and young: parent-offspring conflict and sibling rivalry. *American Naturalist,* **146**, 1-24.

Hasson, O. 1989. Amplifiers and the handicap principle in sexual selection: a different emphasis. *Proceedings of the Royal Society of London, Series B,* **253**, 205-209.

Johnstone, R. A. 1996. Multiple displays in animal communication: 'backup signals' and 'multiple messages'. *Philosophical Transactions of the Royal Society of London, Series B,* **351**, 329-338.

Kilner, R. 1995. When do canary parents respond to nestling signals of need? *Proceedings of the Royal Society of London, Series B,* **260**, 343-348.

Kilner, R. 1997. Mouth colour is a reliable signal of need in begging canary nestlings. *Proceedings of the Royal Society of London, Series B,* **264**, 963-968.

Kilner, R. & Davies, N. B. 1998. Ecological correlates of a begging signal: nestling mouth colour. *Animal Behaviour,* **56**, 705-712.

Kilner, R. M. & Davies, N. B. 1999. How selfish is a cuckoo chick? *Animal Behaviour,* **58**, 797-808.

Kilner, R. M., Noble, D. G. & Davies, N. B. 1999. Signals of need in parent-offspring communication and their exploitation by the cuckoo. *Nature,* **397**, 667-672 .

Lichtenstein, G. & Sealy, S. G. 1998. Nestling competition, rather than supernormal stimulus, explains the success of parasitic brown-headed cowbird chicks in yellow warbler nests. *Proceedings of the Royal Society of London, Series B,* **265**, 249-254.

Møller, A. P. & Pomiankowski, A. 1993. Why have birds got multiple sexual ornaments? *Behavioral Ecology and Sociobiology,* **32**, 167-176.

Price, K., Harvey. H. & Ydenberg, R. 1996. Begging tactics of nestling yellow-headed blackbirds *Xanthocephalus xanthocephalus* in relation to need. *Animal Behaviour,* **51**, 421-435.

Redondo, T. & Arias de Reyna, L. 1988. Locatability of begging calls in nestling altricial birds. *Animal Behaviour,* **36**, 653-661.

Redondo, T. & Castro, F. 1992. Signalling of nutritional need by magpie nestlings. *Ethology,* **92**, 193-204.

Redondo, T. 1993. Exploitation of host mechanisms for parental care by avian brood parasites. *Etologia,* **3**, 225-297.

Rowe, C. & Guilford, T. 1996. Hidden colour aversion in domestic chicks triggered by pyrazine odours of insect warning displays. *Nature,* **383**, 520-522.

Scheffer, S. J., Uetz, G. W., & Stratton, G. E. 1996. Sexual selection, male morphology, and the efficacy of courtship signalling in two wolf spiders (Araneae: Lycosidae). *Behavioral Ecology and Sociobiology,* **38**, 17-23.

Soler, M., Martinez, J. G., Soler, J. J. & Møller, A. P. 1995. Preferential allocation of food by magpies *Pica pica* to great spotted cuckoo *Clamator glandarius* chicks. *Behavioral Ecology and Sociobiology,* **37**, 7-13.

Zuk, M., Ligon, J. D. & Thornhill, R. 1992. Effects of experimental manipulation of male secondary sex characters on female mate preference in red jungle fowl. *Animal Behaviour,* **44**, 999-1006.

IN: ESPMARK,Y., AMUNDSEN,T. & ROSENQVIST,G. (EDS.) 2000
ANIMAL SIGNALS:
SIGNALLING AND SIGNAL DESIGN IN ANIMAL COMMUNICATION, PP. 365-386
TAPIR ACADEMIC PRESS, TRONDHEIM, NORWAY

19

Scents of Danger: the Evolution of Olfactory Ornamentation in Chemically-mediated Predator-prey Interactions

BRIAN D. WISENDEN

Department of Biology, Moorhead State University,
Moorhead, MN 56563, USA

E-mail: wisenden@mnstate.edu

Abstract

Aquatic environments favour the elaboration of chemical signals and the evolution of receptors to detect them. Encounters between predators and prey escalate from initial detection to attack, capture, and finally ingestion. Chemical cues released at each stage in the predation sequence serve as public information about predation risk and allow prey to assess risk, to reduce the probability of predator attack and to learn the chemical identity of unfamiliar predators. Discussed here are disturbance pheromones, predator odours and chemical cues released upon injury from predator attacks on fellow conspecifics or heterospecific prey with which habitat and predators are shared. Following prey ingestion, chemical cues of the prey leak from the predator's digestive tract. Chemical cues from a predator's diet make injury-released chemical cues available at the detection stage of the predation sequence. Each of these chemical cues represents a reliable form of public information. Publicly available chemical cues are not signals because they do not necessarily benefit the sender. Fishes of the superorder Ostariophysi (minnows, catfish, characins, etc.) possess specialised epidermal cells containing an alarm pheromone. This pheromone amplifies injury-released chemical cues and attracts secondary predators (receivers) that interfere with the primary predator, allowing the prey (signaller) to escape. Alarm substance cells are metabolically expensive and allocation of resources to these cells is facultative. As such, alarm substance cells of the ostariophysan fishes may be considered as olfactory ornaments that have been selected as signals to predators for the benefit of individual prey.

There are three ecological problems faced by all organisms: getting food, avoid being food, and reproduction. The first two are necessary in order to attain the third, ultimate, goal of successfully propagating one's genes. Communication and signals contribute to success in solving all three of these problems. In this discussion, I will focus on evolutionary pressures that have shaped risk assessment (avoiding being food) in aquatic environments.

Antipredator behaviour, like any behaviour, is a result of an interaction between a genetically-based propensity or capacity to perform a behaviour, and environmental cues that signal the appropriate timing or context for the behaviour (Fig. 1). These are the proximate causes of behaviour. The consequences of the behaviour, in this case surviving or not surviving an encounter with a predator, determine the ultimate function of the behaviour. Animals that detect, correctly interpret and appropriately respond to information about predation risk in their environment will on average have greater reproductive success than individuals that do not do so. The nature of the medium in which an animal lives influences to a large extent the types of signals that best serve as reliable indicators of risk and thus, sensory modalities upon which selection should be greatest.

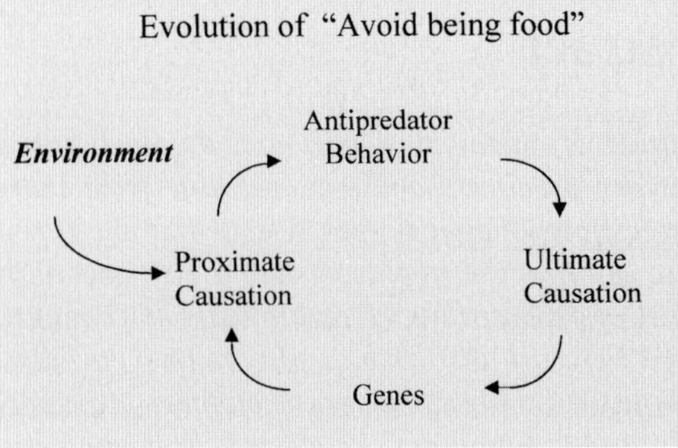

***Figure 1.** Evolution of antipredator behaviour occurs when the fitness consequence (ultimate causation) of that behaviour differentially promotes the frequency of genes that code for the propensity and capacity to perform the behaviour (proximate causation). Environment exerts a constraint on the type of signals that best indicate risk.*

Aquatic environments and chemical cues

The physical and chemical properties of aquatic habitats favour the evolution of chemical cues for the assessment of risk. Chemical cues are better than cues in other sensory modalities when (1) the water contains a high level of suspended solids and is hard to see through, (2) the microhabitat is highly structured, (3) at night, and (4) for animals that have a poorly developed visual sense. In addition, water, the universal solvent, is an excellent medium for the solution and dispersal of chemical cues. Unlike acoustic or visual signals, chemical cues are long-lasting and persist for some time after a predation event, but not necessarily predation risk, has passed.

Several classes of chemical stimuli that are used to assess predation risk will be discussed here (Smith 1992; Chivers & Smith 1998; Kats & Dill 1998). The odour of a predator, or kairomone, may serve to warn prey of the presence of predation risk. Detection of disturbance pheromones, chemicals released by uninjured prey when disturbed by a predator, can also indicate the presence of risk. Another common signal source is chemical cues released upon injury. These alarm pheromones are reliable indicators of risk because they are generally only released in the context of a predation event.

Injury-released chemical cues warn conspecifics of risk, and also any heterospecifics that may share habitat and vulnerability to the same suite of predators (Mathis & Smith 1993a; Wisenden et al. 1995a). These are termed cross-species reactions among members of a prey guild. Finally, injury-released chemical cues may survive passage through the predator's gut and serve to warn prey of a predator's presence when the predator is not actively foraging (Mathis & Smith 1993b).

Aquatic environments are ubiquitous and the evolutionary origin of life. As such, chemical stimuli are the most ancient form of environmental information. Organisms with relatively rudimentary nervous development can accrue great fitness benefits by recognising and responding to predator odours and injury-released cues from members of their prey guild. Chemical cues serve as indicators of predation risk in a broad range of aquatic taxa (Dodson et al. 1994; Chivers & Smith 1998; Kats & Dill 1998).

The *Schreckstoff*-alarm reaction system of ostariophysan fishes (von Frisch 1938, 1941; Smith 1992) acts as amplification of publicly released chemical cues resulting from injury and qualifies as a true signal.

The predation sequence

Predation events typically pass through a series of escalating steps that ultimately result in ingestion of the prey (Lima & Dill 1990; Smith 1992). These steps can be described simply as encounter, attack, capture, and ingestion (Fig. 2). At every step, chemical cues are released by prey that serve to reduce their probability of advancing to the next level in the predation sequence.

Disturbance pheromones and predator odours inform prey of risk early in the predation sequence. This confers greater lead time to prey to adopt behavioural strategies to avoid detection by the predator. Responses generally result in increased vigilance for the approach of a predator and better preparation to initiate an escape should a predator attack. Disturbance pheromones may also be released after a predator has attacked but not yet

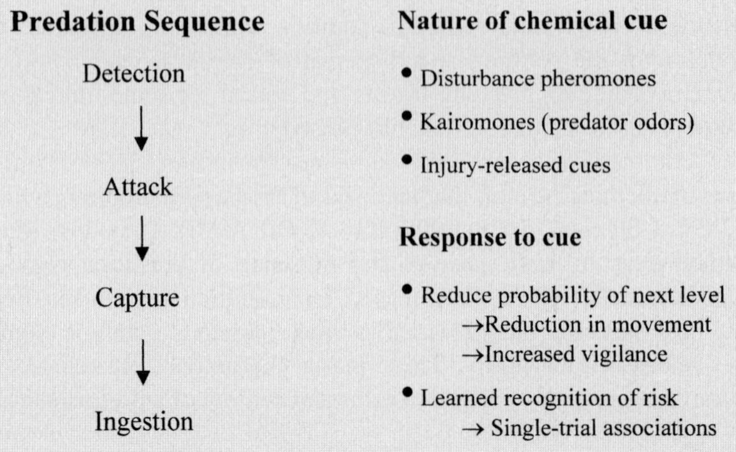

Figure 2. *This simplified flow chart describes the principal stages of a predation event over which natural selection can act. The predation sequence starts with detection and potentially escalates to attack and capture of the prey. Chemical cues released upon disturbance, injury or from a predator's diet induce prey to increase crypsis and vigilance.*

captured prey. Successful attacks produce injury-released chemical cues and indicate a more imminent risk to nearby prey because these cues indicate active foraging by the predator. Injury-released cues generally invoke immediate initiation of antipredator behaviours such as a reduction in activity, movement out of the water column and seeking shelter, behaviours that reduce foraging success of predators. Chemical cues are released from ingested prey from the predator's anus and allow prey to use injury-released cues at the detection level of the predation sequence. Learned recognition of novel indicators of risk occurs when odours that are known to indicate risk, i.e., injury-released cues from conspecifics, are encountered simultaneously with a novel visual or chemical cue (Göz 1941; Magurran 1989; Mathis & Smith 1993b, c; Chivers & Smith 1994a, b, 1995a). Thereafter, previously neutral cues presented alone elicit a full complement of antipredator behaviours. Typically, association between risk and a novel cue occur after a single simultaneous exposure to both cues.

The following experiments illustrate ways in which aquatic prey use chemical cues to reduce the probability of predation at various stages of the predation sequence. This discussion sets the stage for a discussion of what constitutes a true signal and how the specialised epidermal cells of Ostariophysan fishes qualify as an olfactory signal.

Chemical cues and predator-prey interactions

Disturbance pheromones in Iowa darters

Iowa darters *Etheostoma exile* (Pisces: Percidae) are cylindrical fish measuring about seven cm in length. They lack swim bladders and are negatively buoyant. Darters spend most of their time resting on the bottom of lakes and streams, with their anterior ends propped up by their large pectoral fins. They use their pectoral fins to make darting lunges on benthic invertebrate prey. Predator vigilance is constrained by close association with the substratum and manifests itself as a reduction in movement and 'head-up' behaviour. In head-up behaviour, an individual maximally extends its pectoral fins to fully raise its anterior end off of the substratum and arches its back until the tip of its snout is directed vertically upward. This behaviour may enhance both visual assessment of the environment and sampling of chemical cues present in the water column above the boundary layer. This posture allows darters to remain motionless and cryptic while assessing their environment.

We set up 37 litre aquaria in pairs, with tanks within each pair separated by an opaque barrier but connected by a siphon (Wisenden et al. 1995b). A pump drained water from the far end of one tank and returned it to the far end of the other. Water flowed passively between tanks via the siphon. In the downstream tank we placed three 'receiver' darters. In the upstream tank we placed no darters (control) or 12 'sender' darters. In all trials, a model predator was hidden in the upstream tank under a small sheet of plastic before each trial. The model could be made to move around the upstream tank by pulling on a monofilament line that extended about 6 m away from the tank. Each trial lasted for 35 min. For the first 15 min, the number of darts and the number of fish in the head-up posture were recorded. Then, for 5 min, the model predator was made to move around the upstream 'sender' tank. The model did not contact or injure sender fish. For the final 15 min the receiver fish were again observed and the frequency of darts and head-up behaviour recorded.

Receiver darters downstream from 12 sender darters showed a significant decrease in activity ($P < 0.001$, Fig. 3) and a significant increase in head-up behaviour ($P < 0.01$, Fig. 4; Wisenden et al. 1995b). The absence of sender darters in control trials allowed for any disturbing stimuli that may have been produced by the movement of the predator model. The increase in predator vigilance observed in trials with receiver fish strongly suggest the presence of a disturbance pheromone, i.e., a chemical cue released by disturbed fish that alerts others to the presence of risk.

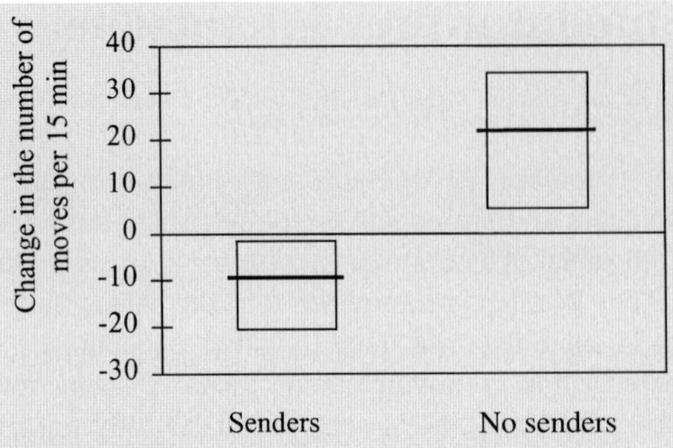

Figure 3. *The change in activity from pre-predator to post-predator periods by receiver darters, expressed as the change in the number of moves. Horizontal line indicates the median, bar represents 25% and 75% quartiles. (Modified from Wisenden et al. 1995b).*

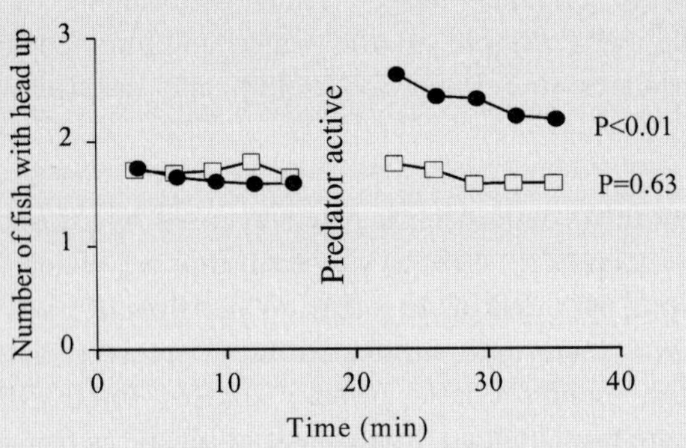

Figure 4. *The number of receiver darters exhibiting head-up posture before and after a model predator was active in the sender tank. Closed circles: sender darters in upstream tank; open squares: no sender darters in upstream tank. (From Wisenden et al. 1995b).*

Detection of predator odours by *Gammarus* amphipods

Gammarus amphipods are small freshwater crustaceans that commonly occur in ponds, lakes and streams. They spend most of their time crawling through organic litter on the substratum where visual assessment of the environment is obstructed. *Gammarus* are prey to many invertebrate and vertebrate predators and therefore should be expected to have undergone selection to use environmental indicators of risk.

We placed individual *Gammarus lacustris* into 37 litre aquaria fitted with twin airlines (Wudkevich et al. 1997). One line powered an airstone, the second line served as a tube through which chemical stimuli could be injected. Injected stimuli were quickly dispersed throughout the tank by water currents created by the airstone.

We performed two experiments. In the first, we compared the behavioural response of individual *Gammarus* to three test stimuli: (i) the odour of predatory northern pike (*Exox lucius*; Pisces: Esocidae), (ii) injury-released chemical cues from conspecific *Gammarus*, or (iii) water (control). The water treatment controlled for any effect of stimulus injection on *Gammarus* behaviour. The injured conspecific treatment was necessary to ascertain if it was truly an alarm pheromone in that it induced the same behavioural response as predator odour (Smith 1992). The experimental protocol was 8 min pre-stimulus observation,

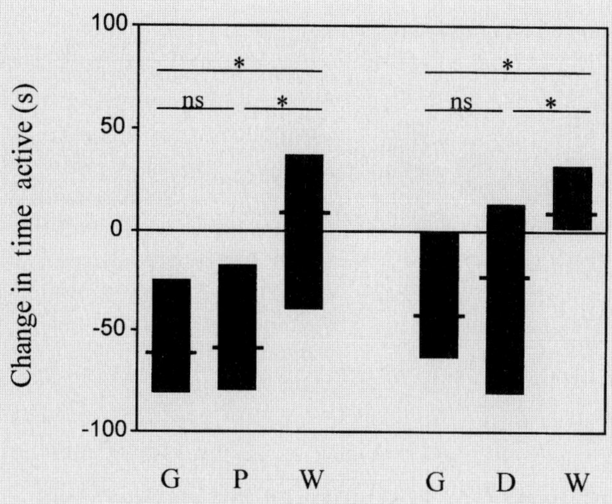

Figure 5. *Median±25% quartiles of change in* Gammarus *activity after introduction of chemical cues from injured* Gammarus *(G), pike odour (P), dragonfly odour (D) or blank water control (W). Lines above bars indicate pairwise comparisons; ns, non significant (P > 0.05), asterisk, P < 0.05. (Modified from Wudkevich et al. 1997).*

1 min over which one stimulus was injected, then 8 min post-stimulus observation. The response variables were the difference between post and pre-stimulus periods in time spent active, and time spent near the bottom (out of the water column).

Relative to the water control, *Gammarus* significantly reduced activity and spent significantly more time near the bottom in response to injured conspecifics and in response to pike odour (Fig. 5, Wudkevich et al. 1997). These data show that *Gammarus* respond to the odour of a predator in the same manner as they do to injury-released cues of conspecifics.

We repeated this experiment, this time using the odour of larval dragonflies, *Aeshna eremita* (Insecta: Odonata), as the source of the predator cue. Experimental protocol was as described for the first experiment. *Gammarus* significantly reduced activity in response to dragonfly odour or injured *Gammarus*, relative to water control (Fig. 6). In contrast to the previous experiment, *Gammarus* did not spend significantly more time near the bottom when the predator cue came from dragonfly larvae. This result suggests that *Gammarus* recognize and distinguish among predator species on the basis of odour, and respond in accordance to the hunting style of each predator. Pike forage in the water column. When pike odour is detected, the best response is to move out of the water column. Dragonfly larvae wait in ambush and stalk prey among debris on the bottom. When dragonfly odour is detected, the best response is to reduce activity but also to frequently swim back into the water column where dragonflies will not pursue prey.

The antipredator response to predator odours and injury-released chemical cues of conspecifics presumably reduces the probability of detection and attack.

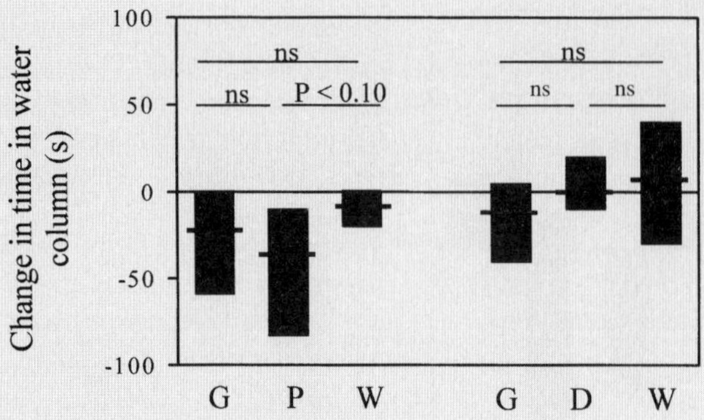

Figure 6. Median±25% quartiles change in time in the water column by Gammarus *after the introduction of chemical cues from injured* Gammarus *(G), pike odour (P), dragonfly odour (D) or blank water control (W). Lines above bars are described in the legend of Fig. 5. (Modified from Wudkevish et al. 1997).*

Effect of *Gammarus* antipredator response on predation risk

The ultimate function of antipredator behaviour is reduction of the probability of predation (e.g. Hews 1988; Mathis & Smith 1993d). We used *Gammarus minus* as the study animal (Wisenden et al. 1999). We divided large aquaria into three sections. The first 37.5 cm was contained by a tight-fitting opaque barrier. Ten *Gammarus* were placed into this compartment. The *Gammarus* section received a twin airline, one to power an airstone and one for stimulus injection. There were five shelter objects placed in this section over a thin layer of silica sand. In the opposite end of the aquarium, a coarse mesh screen created a compartment 30 cm wide that contained an adult green sunfish *Lepomis cyanellus* (Pisces: Centrarchidae) and an overturned clay flowerpot over a thin layer of silica sand. The centre section was 7.5 cm wide and received a constant supply of fresh dechlorinated tap water that passed through the coarse screen to a drain at the distal end of the sunfish section. The flow-through design continuously flushed away sunfish odour. *Gammarus* used in this study were collected from a sunfish-free stream and were not likely to recognize sunfish odour as an indicator of danger (see below). As an additional precaution, sunfish were maintained on a diet of earthworms and mosquitofish, *Gambusia affinis* (Pisces: Poeciliidae) to ensure that *Gammarus* did not respond to chemical cues associated with the sunfish's diet.

The experimental protocol was as follows: we observed *Gammarus* behaviour for 8 min, injected a test stimulus over the ninth min, then observed 2 min of post-stimulus behaviour. Then we removed both barriers separating the *Gammarus* and sunfish sections, which allowed the sunfish to enter the *Gammarus* section to forage. We recorded latency to first attack and overall attack success.

We used three test stimuli in this experiment: (i) injury-released chemical cues from conspecific *Gammarus*, (ii) injury-released chemical cues of sympatric *Lirceus fontinalis* (Isopoda), and (iii) water (control). If injury-released cues are used by conspecifics to reduce the probability of predation, then sunfish should have lower foraging efficiency on *Gammarus* 'warned' by these cues than on *Gammarus* 'warned' with water. The isopod

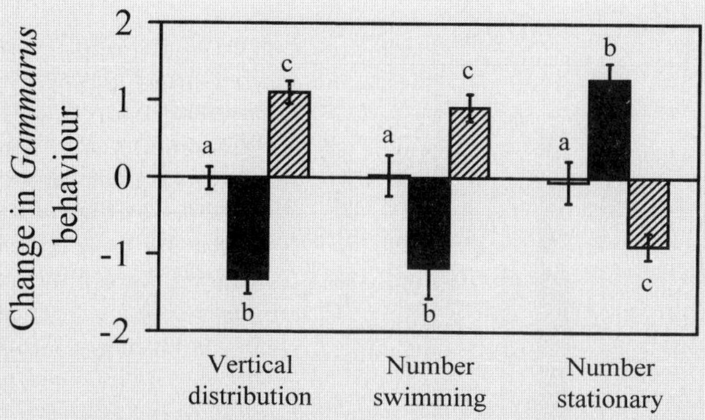

Figure 7. *Change in* Gammarus *behaviour after the introduction of water (open bars), chemical cues from injured* Gammarus *(solid bars) or chemical cues from injured isopods (hatched bars). Different letters near each bar indicate significant differences ($P < 0.05$; separate tests performed for each behaviour). (Modified from Wisenden et al. 1999).*

cue was used to test if *Gammarus* respond to general chemical compounds released from injured crustaceans or a narrow response specifically to cues from injured conspecifics. There was also the possibility of a cross-species reaction between sympatric crustaceans that share habitat and predators.

Relative to the water control, injury-released chemical cues of conspecific *Gammarus* induced a significant reduction in activity, a significant reduction in the number of *Gammarus* swimming and a significant increase in the number of *Gammarus* on the bottom (Fig. 7, Wisenden et al. 1999). In contrast, injury-released chemical cues from isopods caused *Gammarus* to significantly increase activity, significantly increase swimming behaviour and significantly decrease time near the bottom (Fig. 7). Instead of acting as a general crustacean cue or as a cross-species alarm cue, the isopod cue appeared to function as a food stimulus. *Gammarus* eat freshly molted isopods (Bengtsson 1982) and injury-released isopod cues may have triggered a feeding response.

Sunfish were released two min after the introduction of each test stimulus. Latency to first attack was 38% longer than controls ($P < 0.05$) if the *Gammarus* were warned with injury-released cues from conspecifics, relative to the water control (Fig. 8). Latency to first attack was 16% shorter than controls ($P < 0.05$) when *Gammarus* were first presented with chemical cues from injured isopods (Fig. 8). Sunfish attacks were almost always successful regardless of treatment (Fig. 8); therefore a significant difference in latency to first attack indicates an adaptive benefit to antipredator behaviour.

These data show that *Gammarus* engage in antipredator behaviour when they recognize species-specific cues associated with injured conspecifics, and that these antipredator behaviours potentially confer a fitness benefit. In a natural habitat, delayed sunfish attacks would translate into reduced probability of predation.

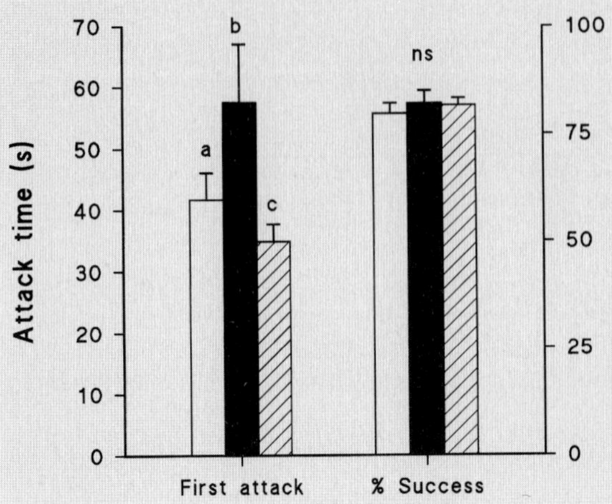

Figure 8. Mean±SE time to first sunfish attack after the removal of barriers separating sunfish and Gammarus*, and percent success of sunfish attacks. Time to first attack differed significantly depending on whether the* Gammarus *were first 'warned' with water (open bar), injury-released cues from Gammarus (solid bar) or isopods (hatched bar). Letters above bars as described for Fig. 7. (From Wisenden et al. 1999).*

Cross-species reactions

Although the previous experiment did not indicate any evidence of a cross-species response to injury-released cues, they occur commonly in aquatic environments. The following is one such example. Fathead minnows *Pimephales promelas* (Pisces: Cyprinidae) and Iowa darters co-occur in Feedlot Pond, Saskatoon, SK, Canada. Minnows and darters are similar in size, occupy similar habitat and are subject to many of the same predators. Under these conditions, injury-released chemical cues from either species should indicate risk to the other. It can be said, therefore, that minnows and darters are in the same prey guild.

We placed minnow traps baited with small blocks of cellulose sponge along the shore of Feedlot Pond (Chivers et al. 1995a). Sponges in half of the traps were soaked in skin extract of Iowa darters. Sponges in the remaining traps were soaked in skin extract of swordtails *Xiphophorus helleri* (Pisces: Poeciliidae). We used swordtail skin instead of water to determine if minnows avoid general cues associated with any injured fish or if minnows respond to specific cues associated with injured darters. Swordtails were chosen as the control because they are endemic to Mexico; thus, fathead minnows in our test population could not have had any previous exposure to them, and swordtails are not phylogenetically related to either minnows or darters (see below).

After 2 h, traps baited with cues from injured darter skin caught significantly fewer fathead minnows than traps baited with cues from injured swordtail skin, indicating avoidance specific to the darter cue (Fig. 9).

We repeated this experiment in nearby Lakeview Pond, where Iowa darters do not occur. In this population, fathead minnows showed no aversion to the darter cue (Fig. 9). This result indicated that the avoidance response by minnows in Feedlot Pond was not attributable to general noxiousness of darter skin. Could the population difference in the recogni-

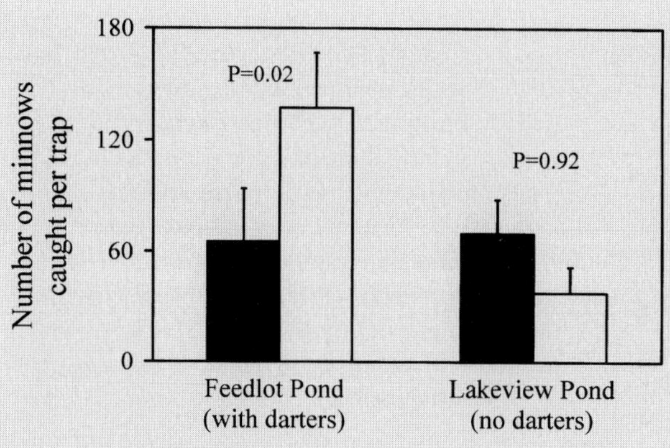

Figure 9. *Mean±SE number of minnows caught per trap in Feedlot Pond and Lakeview Pond. Solid bars: traps baited with injury-released chemical cues from darters; open bars: traps baited with swordtail cue. (Modified from Chivers et al. 1995a, Copyright by Brill, reprinted by permission).*

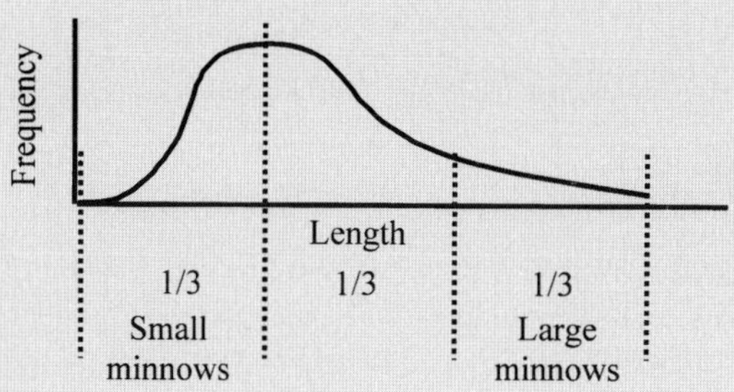

Figure 10. *Method used in testing for learned recognition of cross-species reactions. The catches for darter and swordtail traps were combined. This length distribution was divided into thirds and used as the criterion for small and large minnows. Minnow catch for each pond was calculated separately.*

tion of injured darter skin as an indictor of danger be attributed to learning? A closer examination of the catch data suggested that this may be the case, at least in part. For each pond, we plotted the length frequency distribution for minnow catches of traps baited with darter and swordtail cues combined, and partitioned it into thirds (Fig. 10). Fish in the first third were deemed to be small minnows, fish in final third were deemed to be large minnows. If avoidance of injured darter skin is acquired over the lifetime of individual minnows, minnows caught in darter traps should be represented by a relatively large proportion of small, inexperienced minnows, and a relatively small proportion of large, experienced minnows compared to those caught in swordtail traps. This was indeed the case for the experiment in Feedlot Pond where darters are present (Fig. 11A), but did not hold for the experiment in Lakeview Pond where darters are absent (Fig. 11B). This evidence supports the hypothesis that learning is involved in the recognition of cross-species alarm cues.

Learned recognition of predation odours

Predator pressure varies temporally: seasonally, annually, and over evolutionary time (Huntingford & Wright 1989, 1992). It also varies spatially across ontogenetic niche shifts in habitat and ultimately over the area in which an individual and its progeny potentially disperse. Small aquatic animals are potential prey for myriad predators. One

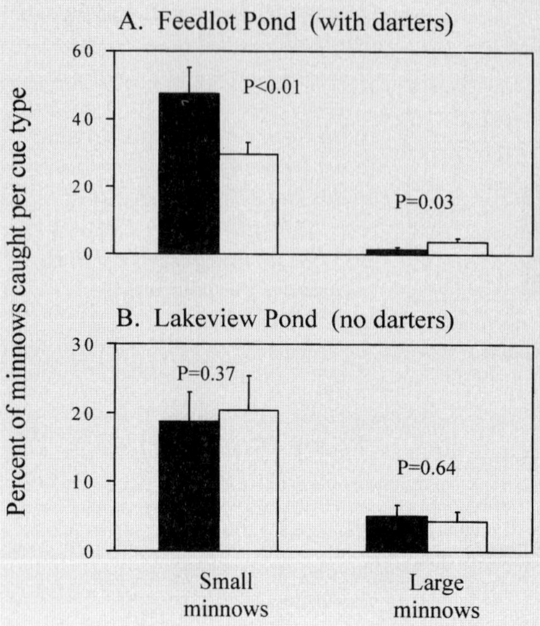

Figure 11. *The percentage of small and large minnows caught per trap in Feedlot Pond (**A**) and Lakeview Pond (**B**). Solid bars: traps baited with injury-released cues from darters; open bars: traps marked with the swordtail cue. (From Chivers et al. 1995a, Copyright by Brill, reprinted by permission).*

evolutionary solution to this problem is a genetically-based template of each predator that an individual and all its descendants might encounter. A more robust solution is to evolve the ability to form permanent associations between a novel cue and risk, after a single simultaneous presentation of a novel cue and a cue known to indicate risk. This learning paradigm affords to an individual tremendous flexibility in adapting to variation in the predator landscape.

One population, from Pike Lake near Saskatoon, co-occur with predatory northern pike. The second population, from nearby Feedlot Pond, does not encounter pike. We tested individual larvae from each population for antipredator responses to injury-released chemical cues from conspecifics, pike odour or water (Wisenden et al. 1997). Damselflies were given some live brine shrimp (*Artemia salina*) immediately before testing. We recorded feeding bites, head bends and walking movements associated with foraging for 5 min before and 5 min after the introduction of test stimuli. Damselflies from Pike Lake responded to injury-released cues from conspecifics or pike odour by significantly reducing foraging activity (Fig. 12A). However, damselflies from Feedlot Pond, where pike are absent, responded with antipredator behaviour only to cues from injured conspecifics, but not pike odour or water (Fig. 12B). Could this population difference in recognition of pike odour be a result of learning? To test this we designed an experiment to train pike-naïve damselflies from Feedlot Pond to recognise pike odour as dangerous (Wisenden et al. 1997). We presented pike-naïve damselflies with a combination of pike odour plus water (control), or a combination of pike odour plus chemical cues from injured conspecific damselflies. As in the first experiment, live brine shrimp were introduced into the test chambers just before testing. Damselflies given pike odour plus water did not change their

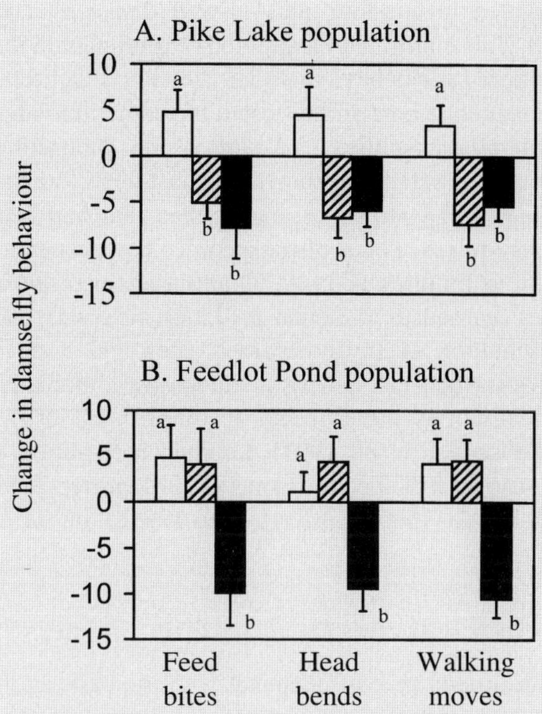

Figure 12. *Mean±SE change in feeding behaviours per 5 min in two populations of damselflies after introduction of water (open bars), pike odour (hatched bars) or chemical cues from injured conspecific damselflies (solid bars). Different letters indicate P < 0.05. (Modified from Wisenden et al. 1997).*

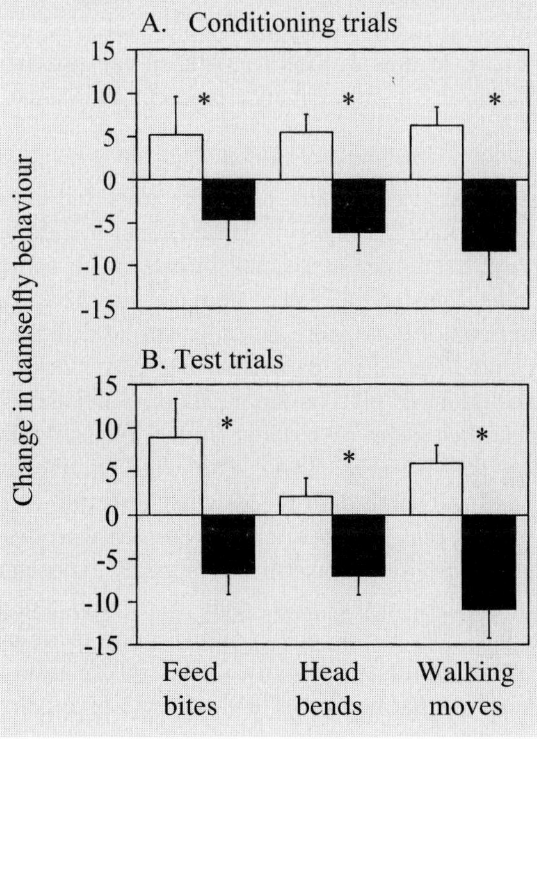

Figure 13. *Change in foraging activity per 5 min by pike-naïve damselflies from Feedlot Pond after (**A**) the addition of pike odour plus water (open bars) or pike odour plus injured damselfly cues (solid bars), and (**B**) subsequent re-testing with pike odour alone. Bar fill in **(B)** indicates the stimulus received in **(A)**. (Modified from Wisenden et al. 1997).*

rate of foraging behaviour, but those given pike odour plus injured damselfly cues significantly decreased activity (Fig. 13A). This confirmed the previous experiment in that pike odour on its own did not induce antipredator behaviour in Feedlot Pond damselfly larvae. We then changed the water in the test chambers and re-tested the same individual damselflies two days later, this time giving all damselflies pike odour alone. Damselflies that had been 'conditioned' with pike odour plus water still showed no change in foraging activity after exposure to pike odour (Fig. 13B). However, damselflies that had been previously exposed to pike odour plus injury-released cues of conspecifics now responded to pike odour on its own with a reduction in foraging activity; an antipredator response (Fig. 13 B). Thus, pike-naïve damselflies learned to associate predation risk with pike odour after a single simultaneous presentation of injury-released cues and a novel stimulus. Using this learning paradigm, damselflies can quickly adapt to any combination of predators (and spatial and temporal variations therein) and accomplish all of this simply on the basis of chemical cues (Wisenden et al. 1997). Learned recognition of chemical cues has been demonstrated in several fish species (Göz 1941; Magurran 1989; Mathis & Smith 1993c; Chivers & Smith 1994b, 1995a; Chivers et al. 1995b), but never before in an invertebrate.

Detection of post-ingestion chemical cues by damselfly larvae

I stressed in the first part of this paper the importance of chemical warning early in the predation sequence, when risk is still low and the chances of executing effective antipredator behaviour is high. Disturbance pheromones and predator odours accomplish this, while injury-released cues from fellow prey arrive only at advanced stages of the predation sequence. However, aquatic prey that detect conspecific injury-released cues emanating from the gut of their predators (e.g. Mathis & Smith 1993b, c) can use these cues as a form of early warning of risk and thereby greatly enhance predator avoidance.

We performed an experiment to test for post-ingestion chemical cues of *Enallagma* damselflies from the gut of pike (Chivers et al. 1996a). Juvenile pike were fed either damselfly larvae or mealworms (Insecta: Coleoptera, *Tenebrio*), rinsed, then transferred to separate holding tanks to avoid injury-released cues associated directly with eating. Water samples from holding tanks where pike digested their meal were used as the source of pike odour. Pike-naïve damselflies from Feedlot Pond were given live brine shrimp, as described for the previous experiments, and then either damselfly-diet pike odour or mealworm-diet pike odour. Those given the odour of pike on the mealworm diet showed no change in foraging behaviour, but damselflies given pike odour on the damselfly diet showed significant reductions in foraging activity (Fig. 14A). The water in the test chambers was replaced with fresh water and the same damselflies were re-tested using pike odour on the mealworm diet. Damselflies that previously received mealworm diet pike odour still showed no change in foraging activity. Damselflies previously exposed to damselfly-diet pike odour now recognised mealworm-diet pike odour as an indicator of risk (Fig. 14B). These data indicate that damselfly larvae detect and respond to injury-released chemical cues from the diet of pike and that they can use this information to learn to recognise a novel predator. Damselflies that had never encountered pike before

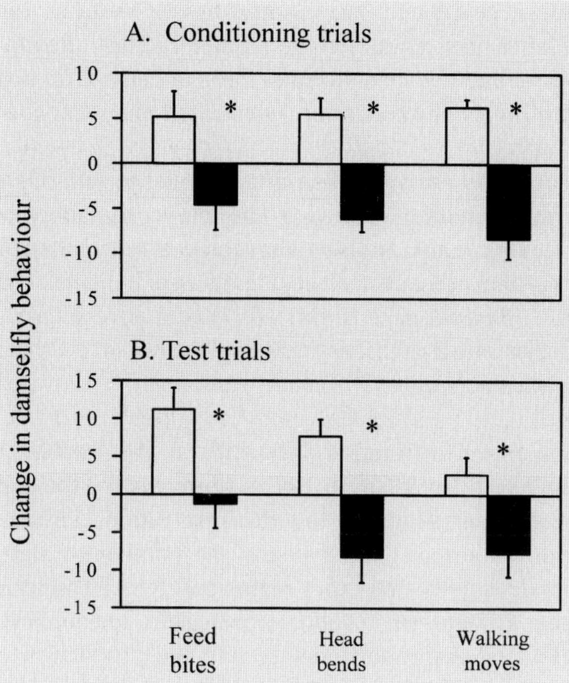

*Figure 14. Change in foraging activity per 5 min by pike-naïve damselflies from Feedlot Pond after (**A**) the addition of mealworm-diet pike odour (open bars) or damselfly-diet pike odour (solid bars), and (**B**) subsequent re-testing with odour of pike on a mealworm diet only. Bar fill in (**B**) indicates the stimulus received in (**A**). (Modified from Chivers et al. 1996a).*

recognised pike as dangerous, and remembered it for at least two days, after a single presentation of injury-released cues emanating from the predator's gut (Chivers et al. 1996a).

Alarm substance cells: olfactory amplifiers of injury-released cues

The preceding examples demonstrate how chemical cues are used at each step of the predation sequence to ameliorate risk. However, publicly available chemical cues do not in themselves qualify as signals because they are released passively, or involuntarily, and do not necessarily benefit the sender. A notable exception occurs in the fishes of the superorder Ostariophysi (minnows, catfish, characins, etc.). This group is speciose representing 64% of all freshwater fish species (Nelson 1994). They possess specialised epidermal cells containing a chemical, alarm substance (*Schreckstoff*), that when detected induces a fright reaction (*Schreckreaktion*), or antipredator behaviour (von Frisch 1938, 1941; Smith 1992). These alarm substance cells are thin-walled and positioned on the surface of the epidermis. They lack a connection to the external environment; thus, the alarm substance can only be released when the skin is damaged. This typically occurs during a predator attack; thus, the presence of alarm substance serves as a reliable indicator of predation risk.

Selection pressures that maintain the alarm substance – alarm reaction system of ostariophysan fishes has long puzzled evolutionary ecologists (Smith 1982; Williams 1992). How does the signaller receive fitness benefits from its investment into its own alarm substance cells if it typically dies when sending the signal? A number of hypotheses have been proposed (Smith 1992) and recent evidence has demonstrated at least one mechanism by which direct fitness benefits may accrue to the signaller. The following experiments build the case for the argument that the specialised epidermal cells of ostariophysan fishes is a signal evolved as an amplifier of general injury-released chemical cues.

Alarm substance cells may benefit the sender by attracting secondary predators (Mathis et al. 1995) that interfere with the primary predator allowing the prey (signaller) an opportunity to escape (Chivers et al. 1996b). Thus, individuals accrue a selfish benefit from their own investment into alarm substance cells. Recent evidence indicates that these cells are metabolically expensive (Wisenden & Smith 1997) and investment is facultative, modified by social context that affects the cost/benefit tradeoff of chemical signalling to secondary predators (Wisenden & Smith 1998).

Fathead minnows fed low or high food ration differed significantly in the number of alarm substance cells in their epidermis (Wisenden & Smith 1997). Minnows fed the high ration were in better physical condition than minnows fed the low ration. Physical condition was positively correlated with epidermal thickness and the number of alarm substance cells, but not the number of mucus cells (the other major cell type in minnow epidermis). This indicated that as the amount of resources available to minnows increased, resource investment was allocated disproportionately into the production of alarm substance cells (Fig. 15).

A second factor in the experiment was shoalmate familiarity. Familiar shoalmates are defined as those that are collected shoaling together in nature. Non-familiar shoalmates are artificial shoals created in the laboratory by placing together individual minnows collected from disparate shoals.

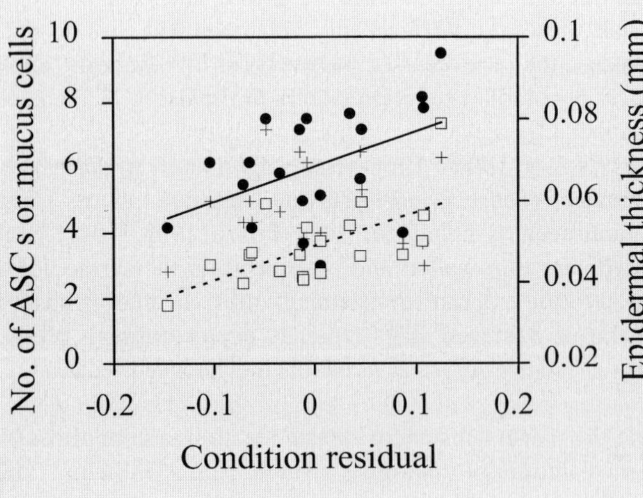

Figure 15. The number of alarm substance cells (ASCs) (solid line, circles), mucus cells (plus signs) and epidermal thickness (dashed lines, squares) relative to physical condition for minnows fed a high ration. Physical condition was determined by residuals from regression of Ln(weight) versus Ln(length). (Modified from Wisenden & Smith 1997).

Antipredator behaviour by minnows when they detect an alarm pheromone may be either a group-based response or an individual-based response. Both tactics include a reduction in movement and a movement toward the bottom. Group responses involve increased shoal cohesion while individual responses involve dashing behaviour (erratic swimming) and seeking shelter. When in the company of familiar shoalmates, minnows engage in more group-level antipredator responses but when in the company of non-familiar shoalmates they engage in more individual-level antipredator responses (Chivers et al. 1994). In choice tests, fathead minnows prefer to associate with familiar shoalmates over non-familiar shoalmates, and recognise them solely on the basis of chemical cues (Brown & Smith 1994). One possible reason is that groups of familiar minnows exhibit a higher degree of coordination and cooperation in response to predator risk than non-familiar shoalmates (Chivers et al. 1994).

Wisenden and Smith (1998) showed that high-ration minnows held with non-familiar shoalmates produced significantly more alarm substance cells than minnows held with familiar shoalmates (Fig. 16). Minnows held with no shoalmates had intermediate numbers of alarm substance cells. The number of mucus cells was not affected by epidermal thickness. Minnows fed the low ration showed no effect of the shoalmate treatment. These data suggest that minnows make fewer alarm substance cells when with familiar shoalmates because they can rely on an effective group-level antipredator response. When in the company of non-familiar shoalmates, minnows make more alarm substance cells because they cannot rely on an effective group response and instead must rely more on their ability to attract secondary predators with their alarm substance cells. Fathead minnows appear to adjust investment into alarm substance cells facultatively based upon the level of perceived risk, but do this only when resources are abundant (Wisenden & Smith 1998). As such, alarm substance cells of the Ostariophysan fishes may be considered olfactory ornaments that have been selected as signals to predators (receivers) for the benefit of individual prey (senders).

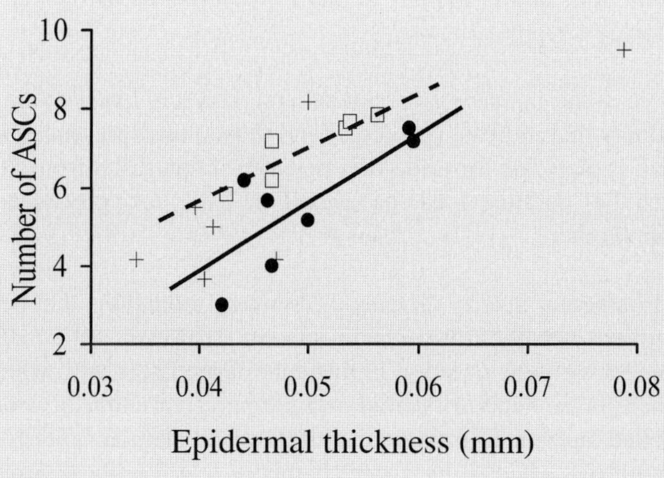

Figure 16. *The number of alarm substance cells in minnow epidermis for fish on the high ration diet. Open squares and dashed line: fish kept with non-familiar shoalmates; closed circles and solid line: fish kept with familiar shoalmates; plus signs: fish kept with no shoalmates. (Modified from Wisenden & Smith 1998).*

Synthesis and future directions for research

The preceding examples illustrate the pervasive use of chemical information for the assessment of predation risk. Aquatic environments are well suited for the elaboration of chemical cues as indicators of predation risk and the evolution of physiological receptors to perceive these cues. For the most part, the cues discussed here are not released voluntarily but are passive consequences of an encounter with a predator. These chemical cues present 'public information' about predation risk and are not necessarily signals. The specialised epidermal cells of the Ostariophysi, however, do satisfy the criteria for signals: signallers (minnows) benefit from signalling to receivers (predators). This is further supported by evidence that investment into the olfactory signal is sensitive to social context.

Predation is a powerful selection agent, especially among small, fecund species at the base of the trophic hierarchy. Animals that respond adaptively to chemical information in their environment accrue greater fitness than individuals that do not. This is particularly evident in single-trial learning of novel predator odours. A great fitness cost would be paid by individuals that required repeated brushes with death to learn a predator's identity.

Future research efforts should continue to explore the range of ways in which prey use chemical information for reducing predation risk, and ways in which predators may manipulate information available to prey (e.g. Brown et al. 1995). Attention needs to be directed toward the nature, specificity and reliability of chemical signalling. Little is known about the specific active compounds in the chemical cues used by aquatic animals. An important component of disturbance pheromones may be ammonia (Crustacea, Hazlett 1990; Amphibia, Kiesecker et al. 1999; fish, K. L. Young & B. D. Wisenden, unpublished). Some workers have reported that the specialised alarm pheromone of ostariophysan fishes is hypoxanthine – 3(N) – oxide, but further investigation indicates that this compound is not unique in its ability to elicit an alarm response (G. Brown & J. Adrian, personal communication).

Cross-species reactions occur in some instances but not others. There is likely to be a range of trade-offs between odours that indicate risk (cost) and those that represent food, or mates (benefits). This system is also ripe for trade-offs between information from different sensory modalities. When they conflict, does one sensory modality take precedence or is deemed more reliable than another?

How robust is olfaction? Environmental quality dictates the sensory modalities that best assess the environment and govern behavioural decision-making. However, all aquatic environments are not the same. As with the variation in the suite of predators with which an individual may contend, there is tremendous spatial and temporal variation in water chemistry (e.g. pH, dissolved and suspended organic materials, trace elements, etc.). Is chemical assessment of predation risk affected by water chemistry (e.g. Lemly & Smith 1987)? Do anthropogenic changes in water chemistry impact chemical assessment in aquatic environments?

Finally, experimenters whose primary interests are not in chemical ecology should nevertheless be mindful of chemical cues that may be inadvertently released during an experiment. In particular, whether or not study animals have been exposed to predators in their lifetime, and the diet of predators restrained behind a mesh partition could effect behaviours such as foraging, mate choice and other social interactions.

Acknowledgements

I thank Yngve Espmark, Gunilla Rosenqvist, Trond Amundsen, Turid Fredagsvik and the Royal Norwegian Society of Sciences and Letters, The Foundation, for organising and funding a wonderful meeting that stimulated my thinking and led to this manuscript. I also thank my co-authors of the works described here, namely Douglas Chivers, Jan Smith, Grant Brown, Karen Wudkevich, Alecia Cline and Tim Sparkes. Funding for these works, including this manuscript, were provided by the Natural Sciences and Engineering Research Council of Canada, the University of Saskatchewan, the University of Kentucky, Moorhead State University and the Royal Norwegian Society of Sciences and Letters, The Foundation.

References

Bengtsson, G. 1982. Energetic costs of amino acids exudation in the interaction between the predator *Gammarus pulex* L. and the prey *Asellus aquaticus* L. *Journal of Chemical Ecology*, **8**, 1271-1281.

Brown, G. E. & Smith, R. J. F. 1994. Fathead minnows use chemical cues to discriminate natural shoalmates from unfamiliar conspecifics. *Journal of Chemical Ecology*, **20,** 3051-3061.

Brown, G. E., Chivers, D. P. & Smith, R. J. F. 1995. Localized defecation by pike: a response to labelling by cyprinid alarm pheromone. *Behavioral Ecology and Sociobiology*, **36**, 105-110.

Chivers D. P. & Smith R. J. F. 1994a. Fathead minnows (*Pimephales promelas*) acquire predator recognition when alarm substance is associated with the sight of unfamiliar fish. *Animal Behaviour*, **48**, 597-605.

Chivers D. P. & Smith R. J. F. 1994b. The role of experience and chemical alarm signalling in predator recognition by fathead minnows, *Pimephales promelas*. *Journal of Fish Biology*, **44**, 273-285.

Chivers D. P. & Smith R. J. F. 1995a. Fathead minnows (*Pimephales promelas*) learn to recognize chemical stimuli from high-risk habitats by the presence of alarm substance. *Behavioral Ecology*, **6**, 155-158.

Chivers D. P. & Smith R. J. F. 1995b. Free-living fathead minnows rapidly learn to recognize pike as predators. *Journal of Fish Biology*, **46**, 949-954.

Chivers, D. P. & Smith, R. J. F. 1998. Chemical alarm signalling in aquatic predator/prey systems: a review and prospectus. *Écoscience*, **5**, 338-352.

Chivers, D. P., Brown, G. E. & Smith, R. J. F. 1994. Familiarity and shoal cohesion in fathead minnows (*Pimephales promelas*): implications for antipredator behaviour. *Canadian Journal of Zoology*, **73,** 955-960.

Chivers, D. P., Wisenden, B. D. & Smith, R. J. F. 1995a. The role of experience in the response of fathead minnows (*Pimephales promelas*) to skin extract of Iowa darters (*Etheostoma exile*). *Behaviour*, **132**, 665-674.

Chivers, D. P., Brown, G. E. & Smith, R. J. F. 1995b. Acquired recognition of chemical stimuli from pike, *Esox lucius*, by brook sticklebacks, *Culaea inconstans* (Osteichthyes, Gasterosteidae). *Ethology*, **99**, 234-242.

Chivers, D. P., Wisenden, B. D. & Smith, R. J. F. 1996a. Damselfly larvae learn to recognize predators from chemical cues in the predator's diet. *Animal Behaviour*, **52**, 315-320.

Chivers, D. P., Brown, G. E. & Smith, R. J. F. 1996b. The evolution of chemical alarm signals: Attracting predators benefits alarm signal senders. *American Naturalist*, **148**, 649-659.

Dodson, S. I., Crowl, T. A., Peckarsky, B. L., Covich, A. P. & Culp, J. M. 1994. Non-visual communication in freshwater benthos: an overview. *Journal of the North American Benthological Society*, **13**, 268-282.

Frisch, K. von. 1938. Zur Psychologie des Fische-Schwarmes. *Naturwissenschaften*, **26**, 601-606.

Frisch, K. von. 1941. Über einen Schreckstoff der Fischhaut und seine Biologische Bedeutung. *Zeitschrift für Vergleichende Physiologie*, **29**, 46-145.

Göz, H. 1941. Über den Art-und individualgeruch bei Fischen. *Zeitschrift für Vergleichende Physiologie*, **29**, 1-45.

Hazlett, B. A. 1990. Source and nature of disturbance-chemical system in crayfish. *Journal of Chemical Ecology*, **6**, 381-385.

Hews, D. K. 1988. Alarm response in larval western toads, *Bufo boreas*: release of larval chemicals by a natural predator and its effect on predator capture efficiency. *Animal Behaviour*, **36**, 125-133.

Huntingford, F. A. & Wright, P. J. 1989. How sticklebacks learn to avoid dangerous feeding patches. *Behavioural Processes*, **19**, 181-189.

Huntingford, F. A. & Wright, P. J. 1992. Inherited population differences in avoidance conditioning in three-spined sticklebacks, *Gasterosteus aculeatus*. *Behaviour* **122**, 264-273.

Kats, L. & Dill, L. M. 1998. The scent of death: chemosensory assessment of predation risk by prey animals. *Écoscience*, **5**, 361-394.

Kiesecker, J. M., Chivers, D. P., Marco, A., Quilchano, C., Anderson, M. T. & Blaustein, A. R. 1999. Identification of a disturbance signal in larval red-legged frogs (*Rana aurora*). *Animal Behaviour*, **57**, 1295-1300.

Lemly, A. D. & Smith, R. J. F. 1985. Effect of acute exposure to acidified water on the behavioral response of fathead minnows, *Pimephales promelas*, to chemical feeding stimuli. *Aquatic Toxicology*, **6,** 25-36.

Lemly, A. D. & Smith, R. J. F. 1987. Effects of chronic exposure to acidified water on chemoreception and ecological implications. *Environmental Toxicology and Chemistry*, **6**, 225-238.

Lima, S. L. & Dill, L. M. 1990. Behavioural decisions made under the risk of predation: a review and prospectus. *Canadian Journal of Zoology*, **68**, 610-640.

Magurran, A. E. 1989. Acquired recognition of predator odour in the European minnow (*Phoxinus phoxinus*). *Ethology*, **82**, 216-223.

Mathis, A. & Smith, R. J. F. 1993a. Intraspecific and cross-superorder responses to chemical alarm signals by brook stickleback. *Ecology*, **74**, 2395-2404.

Mathis, A. & Smith, R. J. F. 1993b. Chemical labeling of northern pike (*Esox lucius*) by the alarm pheromone of fathead minnows (*Pimephales promelas*). *Journal of Chemical Ecology*, **19**, 1967-1979.

Mathis, A. & Smith, R. J. F. 1993c. Fathead minnows, *Pimephales promelas*, learn to recognize northern pike, *Esox lucius*, as predators on the basis of chemical stimuli from minnows in the pike's diet. *Animal Behaviour*, **46**, 645-656.

Mathis, A. & Smith, R. J. F. 1993d. Chemical alarm signals increase the survival time of fathead minnows (*Pimephales promelas*) during encounters with northern pike (*Esox lucius*). *Behavioral Ecology*, **4**, 260-265.

Mathis, A., Chivers, D. P. & Smith, R. J. F. 1995. Chemical alarm signals: predator deterrents or attractants? *American Naturalist*, **146**, 994-1005.

Nelson, J. S. 1994. *Fishes of the World.* 4th edn. New York: Wiley.

Pfeiffer, W., Riegelbauer, G., Meier, G. & Scheibler, B. 1985. Effect of hypoxanthine-3(N)-oxide and hypoxanthine-1(N)-oxide on central nervous excitation of the black tetra *Gymnocorymbus ternetzi* (Characidae, Ostariophysi, Pisces) indicated by dorsal light response. *Journal of Chemical Ecology*, **11**, 507-523.

Smith, R. J. F. 1982. The adaptive significance of the alarm substance - fright reaction system. In: *Chemoreception in Fishes* (Ed. by T. J. Hara), pp. 327-342. Amsterdam: Elsevier.

Smith, R. J. F. 1992. Alarm signals in fishes. *Reviews in Fish Biology and Fisheries*, **2**, 33-63.

Williams, G. C. 1992. *Natural Selection: Domains, Levels, and Challenges.* Oxford: Oxford University Press.

Wisenden, B. D., Chivers, D. P., Brown, G. E & Smith, R. J. F. 1995a. The role of experience in risk assessment: avoidance of areas chemically labelled with fathead minnow alarm pheromone by conspecifics and heterospecifics. *Écoscience*, **2**, 116-122.

Wisenden, B. D., Chivers, D. P. & Smith, R. J. F. 1995b. Early warning in the predation sequence: a disturbance pheromone in Iowa darters (*Etheostoma exile*). *Journal of Chemical Ecology*, **21**, 1469-1480.

Wisenden, B. D., Chivers, D. P. & Smith, R. J. F. 1997. Learned recognition of predation risk by *Enallagma* damselfly larvae (Odonata, Zygoptera) on the basis of chemical cues. *Journal of Chemical Ecology*, **23**, 137-151.

Wisenden, B. D., Cline, A. & Sparkes, T. C. 1999. Survival benefit to antipredator behavior in the amphipod *Gammarus minus* (Crustacea: Amphipoda) in response to injury-released chemical cues from conspecifics and heterospecifics. *Ethology*, **105**, 407-414.

Wisenden, B. D. & Smith, R. J. F. 1997. The effect of physical condition and shoalmate familiarity on proliferation of alarm substance cells in the epidermis of fathead minnows. *Journal of Fish Biology*, **50,** 799-808.

Wisenden, B. D. & Smith, R. J. F. 1998. A re-evaluation of the effect of shoalmate familiarity on the proliferation of alarm substance cells in ostariophysan fishes. *Journal of Fish Biology*, **53**, 841-846.

Wudkevich, K., Wisenden, B. D., Chivers, D. P. & Smith, R. J. F. 1997. Reactions of *Gammarus lacustris* to chemical stimuli from natural predators and injured conspecifics. *Journal of Chemical Ecology*, **23**, 1163-1173.

IN: ESPMARK,Y., AMUNDSEN,T. & ROSENQVIST,G. (EDS.) 2000
ANIMAL SIGNALS:
SIGNALLING AND SIGNAL DESIGN IN ANIMAL COMMUNICATION, PP. 387-405
TAPIR ACADEMIC PRESS, TRONDHEIM, NORWAY

20

Signals and Multi-modal Signaling in Spider Communication

GEORGE W. UETZ

Department of Biological Sciences, University of Cincinnati,
Cincinnati, OH 45221-0006, USA

E-mail: uetzgw@email.uc.edu

Abstract

Spiders are predatory and potentially cannibalistic, and selection pressures for effective signaler-receiver communication and species recognition are strong, resulting in a diversity of signaling behaviors used in courtship, agonistic interactions, parent-offspring contact and social behavior. Spiders utilize various modes of communication, including acoustic/vibratory mechanisms, visual displays, and pheromones, either singly or in multi-modal communication. In this brief review of the breadth of research on spider signaling, I present examples from the literature and provide an overview of recent work done in my laboratory on wolf spiders (Lycosidae: Schizocosa*), which use multi-modal communication (visual and vibratory cues) in courtship. Studies have examined female responses to components of male courtship, in species-typical and in conspecific/heterospecific combinations, for two closely-related* Schizocosa *species (*S. ocreata *and* S. rovneri*). Results of cue isolation experiments suggest that the relative roles of courtship modes have diverged in these species. Cue-conflict experiments (mixed conspecific/heterospecific components) generally show that females respond to mixed information at the same rate as they do to heterospecific males. Video playback experiments manipulating decorative tufts in male* S. ocreata*, while controlling for behavior and size, show that female receptivity* *varies with presence/absence and size of tufts. As tufts vary with condition in live males, this visual signal may also serve as an indicator trait in female mate choice. The role of various signaling modes, and the evolution of multiple modes is discussed.*

As in other animals, the exchange of signals between spiders serve many purposes, and these frequently overlooked animals provide a wealth of opportunites for research into various modes of communication, including acoustic/vibratory mechanisms, visual displays, and pheromones (Witt & Rovner 1982; Uetz & Stratton 1983). Moreover, since spiders are predators and potentially cannibalistic, selection pressures for effective signaler-receiver communication and species recognition are strong, resulting in a diversity of signaling behaviors used in courtship, agonistic interactions, parent-offspring contact and social behavior. In this presentation, I will briefly review the breadth of research on spider signaling, present a few examples for illustration and provide an overview of recent work done in my laboratory.

Spider signaling modes

In many spiders, the sensory modes used in prey detection also serve in detecting and receiving signals from conspecifics (Krafft 1982; Barth 1985; Rovner 1996). Spiders perceive prey and signals of other spiders through three sensory modes: (1) mechanical senses that detect acoustic/vibratory cues, (2) chemical/olfactory senses that detect pheromones and prey odors, and (3) visual senses that detect motion and/or images.

Vibratory cues

The primary mode of prey detection for most spiders would appear to be by substratum-borne mechanical (vibratory) cues, either through environmental surfaces such as leaves, or through silk threads on webs (Barth 1982). Spiders possess an array of sensory organs for vibration detection (Barth 1985 and references therein), including slit sensillae and lyriform organs (which respond to substratum-borne vibration) and specialized hairs such as trichobothria (which respond to airborne vibration). Hence, communication via web-plucking, percussive sound production, substratum-coupled stridulation or body vibration is common in spiders (Rovner 1967, 1975, 1980; Rovner & Barth 1981; Uetz & Stratton 1982; Barth 1993; Kronestedt 1996).

Chemical cues

Spiders also use chemical communication, although available information for spiders is not nearly as extensive as for insects and vertebrate animals (Tietjen & Rovner 1982; Jackson 1987; Pollard et al. 1987). Spiders may use chemical cues in detection of prey and/or recognition of preferred or unpalatable prey types (Persons & Uetz 1996; Punzo & Kukoyi 1997). Many spider species are known to produce pheromones, which play a variety of roles in social recognition (Krafft 1982; Pollard et al. 1987) as well as species recognition and reproductive isolation (Jackson 1987; Barth 1993; Costa et al. 1997). In most cases, pheromones are deposited in silk, e.g. female webs or draglines, which allows mate location and releases male courtship behavior (Tietjen 1977, 1979; Tietjen & Rovner 1980, 1982; Suter & Renkes 1982; Roland 1984; Barth 1993; Costa et al. 1997). Male spiders may also use the presence and/or composition of female pheromones in silk to determine her reproductive status (Jackson 1981a, 1987) and alter courtship behavior accordingly. Male spiders may also produce pheromones, which serve to mediate agonistic interactions (Ayyagari & Tietjen 1987).

Visual cues

While most spiders lack keen vision and rely primarily on vibration and chemical senses, communication in spider taxa possessing well-developed visual senses (e.g. Salticidae, Lycosidae) involves greater reliance on visual displays (Crane 1949; Rovner 1968; Forster 1982; Jackson 1982). The structure and visual capabilities of spiders' eyes vary between families, and the nature of motion detection, image resolution and color perception may facilitate or constrain evolution of both predatory and signalling behavior. For example, the jumping spiders (Salticidae) are able to perceive color, and detect motion while moving, whereas wolf spiders (Lycosidae) do not perceive color and must remain stationary to detect motion (Forster 1982; Land 1985; Rovner 1989, 1996). As a consequence, jumping spiders hunt prey by stalking and utilize complex dances and bright color patterns in signaling (Crane 1949; Jackson 1982; Richman & Jackson 1982). Wolf spiders, in contrast, use a sit-and-wait foraging strategy, primarily rely upon leg raises and waves as signals, and have contrasting dark pigmentation or hairs as decorations (den Hollander 1970; Miller et al. 1998; Hebets & Uetz 1999a).

Detection of motion and discrimination from prey are critical elements in mate recognition and sexual selection among the Salticidae and Lycosidae, as selection pressures on the signaler are strong, and male displays must be instantly recognizable at a distance. Several studies have shown that attracting the attention of females at a distance results in a higher probability of mating success, as females are more likely to mate with the male they see move first (Jackson 1982; Clark & Uetz 1990, 1992; Scheffer et al. 1996). A review of numerous species within the family Lycosidae, and the genus *Schizocosa* in particular, shows a strong correlation between the presence of leg decoration - either pigmentation alone, or pigmentation and tufts of bristles - and the use of legs in visual displays (Hebets & Uetz 1999a). These studies provide support for the ideas of Hasson (1989, 1991) that male secondary characters serve as 'amplifiers' that increase the detectability of displays (Hebets & Uetz 1999a).

Context and content of spider signals

Courtship and mating

One of the most important contexts of spider communication is in pre-copulatory behavior and courtship, where species recognition is critical, and selection pressure from potentially cannibalistic females has resulted in a great diversity of male displays. Signaling may serve multiple functions in this context, including mate location, recognition, suppression of predatory behavior and sexual arousal (Rovner 1967, 1968; Robinson 1980, 1982; Jackson 1982; Barth 1993). Signaling has been demonstrated as a reproductive isolating mechanism in sympatric spider species (den Hollander et al. 1973; Stratton & Uetz 1981, 1983, 1986; Uetz & Stratton 1982; Barth & Schmitt 1991), as well as a criterion used in female mate choice (Jackson 1977, 1982; Watson 1991, 1993; Clark & Uetz 1992; Schmitt et al. 1994; McClintock & Uetz 1996; Uetz et al. 1996).

The selection pressures shaping spider communication behaviors during courtship are unique: a trade-off between mating success and risk of predation, with selection from each imposed by females of the same species (Elgar & Nash 1988, Elgar 1991; Prenter et al. 1994; M. H. Persons & G. W. Uetz, unpublished). For both male and female spiders, the potential fitness costs of errors in species recognition are higher for acceptance errors (mistaken identity) than for rejection errors (lost mating opportunity). Moreover, there are differences in the nature of these fitness costs between the sexes. For females, the costs of acceptance errors include wastage of gametes and brooding time (reduced reproductive success), while for males, the costs of acceptance errors include the risk of predation (reduced survival probability). Selection on females may have favored a strategy of choosiness; females respond to males as prey unless convinced otherwise by specific cues from the male. Consequently, selection on males has resulted in a strategy of conspicuous, species-specific display to enable females to distinguish them from prey. Since female spiders generally adopt a sit-and-wait foraging mode, and males an active mate searching mode, it follows that the criteria and cues used in species recognition should operate at a distance and allow discrimination of males from prey and/or heterospecifics (e.g. vibratory signals, morphological characteristics, visual displays).

The extensive work of Friederich Barth and colleagues on tropical wandering spiders of the genus *Cupiennius* serves as a good illustration of the mate recognition process (see review in Barth 1993). Males search for females at night on the broad leaves of the monocotyledonous plants on which they forage, and courtship behavior is initiated upon contact with pheromone-laden female dragline silk. Males produce species-specific courtship signals by body vibration, which are transmitted through leaves to females nearby (Rovner & Barth 1981). Female recognition and response is based upon reception of vibration cues within a narrow range of spectral and temporal properties, specific for each of five species (Schuch & Barth 1985, 1990; Barth & Schmitt 1991). Females signal receptivity with a brief pulse of vibration, which males use for spatial orientation and location of the female (Rovner & Barth 1991).

In addition to recognition, signaling behavior in spiders serves as a criterion used in female mate choice. While spider signaling behavior was among the first examples offered in support of Darwin's theory of sexual selection (Peckham & Peckham 1889), the study of female mate choice in spiders has received attention only recently. While most have asserted that elaborate courtship displays of male jumping spiders (Salticidae) function in species recognition/isolation (Crane 1949; Richman 1982), Jackson (1981b, 1982) suggested that intersexual selection might serve as an explanation. He found intraspecific variation in male behavioral complexity and evidence of female mate preference for more complex dances in several salticid species, and suggested that females might use male dancing behaviors as indicators of male quality (Jackson 1978, 1981b, 1982). As has recently been shown in studies with the European wolf spider *Hygrolycosa rubrofasciata*, females choose males on the basis of signaling (drumming) rates, which are good predictors of male viability (Kotiaho et al. 1996, 1998; Parri et al. 1997). Female *Schizocosa* wolf spiders vary in receptivity responses based on the degree of asymmetry in a male visual signaling character (tufts of bristles on forelegs), which may arise from leg loss and regeneration or developmental instability, and serve as an indicator of condition or reduced fighting success (Uetz et al. 1997; Uetz & Smith 1999).

Aggressive interactions

Communication is also an important part of agonistic interactions between spiders and ritualized signaling behavior has been suggested as an alternative to more injurious combat behaviors (Riechert 1982). Signaling is frequently the primary activity during territorial encounters over webs (Riechert 1978, 1986; Hodge & Uetz 1996) or mates (Christenson & Goist 1979; Austad 1983; Watson 1990, 1998). Hunting spiders without territories, such as wolf spiders, also show a diversity of signals during agonistic interactions (Rovner 1967; Aspey 1977a; Nossek & Rovner 1984; Dodson & Beck 1993; Kronestedt 1996), which may also serve to maintain dominance status (Aspey 1977b).

Vibratory signals produced by stridulation in the wolf spider *Rabidosa rabida* (Walckenaer) vary between aggressive and courtship contexts (Rovner 1967). However, for other lycosid species, e.g. *Schizocosa*, signals used in male-male encounters are identical to those seen in courtship (Aspey 1977a). Recent work (K. Delaney & G. W. Uetz, unpublished) has shown that even though signals are the same, signaling rates and sequences of behavior in male *Schizocosa* vary significantly depending on aggressive or courtship contexts, and may be used in assessment of opponents or potential mates.

Predation

The 'darker side' of spider communication involves the use of aggressive mimicry. For instance, the bolas spider emits a chemical mimic of moth sex pheromones to lure prey (Stowe et al. 1987; Yeargan 1994). Other spiders, like the araneophagic jumping spider *Portia*, use vibrations that mimic those of prey insects, or even male courtship displays, to lure spiders out onto webs and eat them (Jackson & Wilcox 1990, 1993; Wilcox et al. 1996).

Multi-modal signaling - the *Schizocosa* model system

Wolf spiders use multi-modal communication (visual and vibratory cues) in courtship to varying degrees; some species rely on a single mode, while others use both simultaneously. Within the genus *Schizocosa*, the *ocreata* clade contains six species that apparently have arisen from rapid speciation via behavioral isolation driven by sexual selection. Members of this clade are similar in size and coloration, have nearly identical genitalia, and females are largely indistinguishable. Males, however, vary in the degree of decoration of their forelegs (ranging from partial pigmentation to dark pigment and tufts of bristles) and courtship behavior, which serve as pre-mating isolation barriers.

Experimental studies on a number of species in this clade and an outgroup (*S. duplex*) have isolated channels of communication between male and female (Fig. 1), and demonstrate that these species differ in their reliance on visual and vibratory communication during courtship (Stratton & Uetz 1983; Scheffer et al. 1996; Hebets & Uetz 1999b). While, in some species (*duplex*, *rovneri, uetzi*), females respond predominately to vibration cues, in others (*stridulans, crassipes, ocreata*) females exhibit equal or greater

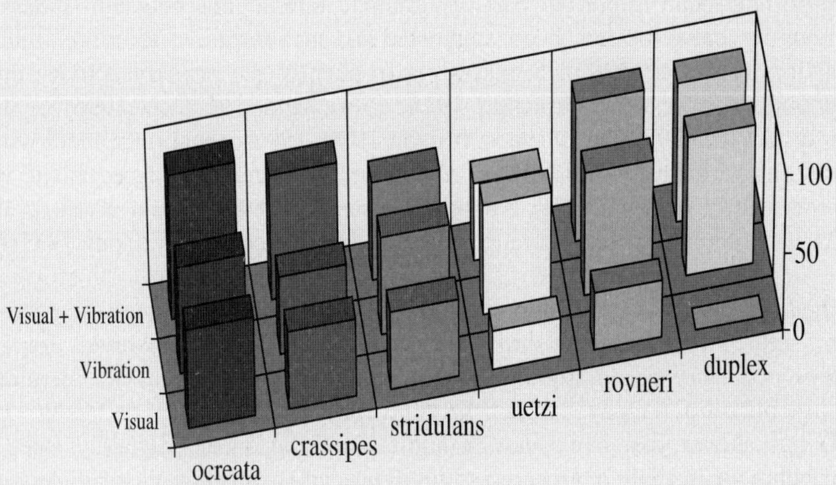

Figure 1. *Female response (%) to signals from male courtship of several* Schizocosa *species in cue isolation experiments (see text for details). Modified from Stratton & Uetz (1983), Scheffer et al. (1996), and Hebets and Uetz (1998b).*

receptivity to visual cues (Fig. 1). In the latter three species, males have pigmentation (*stridulans*) or pigmentation and decorative bristles (*crassipes, ocreata*) on the forelegs, and engage in visual displays. The other species do not use visual displays.

The species currently studied in my lab is the brush-legged wolf spider, *Schizocosa ocreata*, a well-studied model species in which visual and vibratory communication serves as a pre-mating reproductive isolating mechanism separating *S. ocreata* from a sympatric sibling species, *S. rovneri*. These species differ in male morphology: male *S. ocreata* possess conspicuous tufts of bristles on the forelegs (used in visual courtship displays). Tufts and visual displays are lacking in *S. rovneri*. Females of these species are indistinguishable, and males court conspecific and heterospecific females with equal frequency (Uetz & Denterlein 1979; Stratton & Uetz 1981). Despite the highly effective behavioral barrier, heterospecific matings with anesthetized females revealed these species to be interfertile and capable of producing hybrids (Stratton & Uetz 1981, 1986; M. Orr & G. W. Uetz, unpublished).

In current research, the role(s) of visual vs. vibratory cues is being studied by isolating sensory channels of communication (with live spiders), and by cue-conflict experiments (with both live spiders and videotape/audiotape visual/vibratory playback). Melissa Orr and I have examined the responses of female *S. ocreata* and females of the closely related *S. rovneri,* to male courtship, by separating components of courtship communication from male conspecific and heterospecific spiders (in species-typical and conspecific/heterospecific combinations). In cue-isolation experiments, the experimental arena allowed females to perceive cues from courting males behind a clear partition on separate substrata (visual only), an opaque partition on a common substratum (vibration only) or a common substratum with a clear partition (both). In conspecific pairings, either visual or vibratory cues were sufficient to elicit receptivity in female *Schizocosa ocreata* (Fig. 2), and the proportion of females responding with receptivity did not differ significantly between cues (Scheffer et al. 1996). Vibratory cues alone were more important for *S. rovneri* (Fig. 3), as differences in proportion of females exhibiting receptivity responses were significant between cues (Stratton & Uetz 1983). In heterospecific pairings, the response of *S. ocreata* females was not independent of stimulus ($G = 20.28$; $P < 0.01$), as there were some responses to *S. rovneri* males when only vibratory cues were present, but few responses with visual cues alone or visual/vibratory cues together (Fig. 2). The response of female *S. rovneri* to heterospecific cues was also not independent of stimulus ($G = 21.05$; $P < 0.01$), as there were some responses to *S. ocreata* males with visual cues alone, but no responses with vibration alone or visual/vibratory cues together (Fig. 3). These results suggest that the relative roles of courtship modes might have diverged in these species, as the vibratory component appears primary for *S. rovneri,* and the visual component appears primary (or equal to vibration) for *S. ocreata* (M. Orr & G. W. Uetz, unpublished).

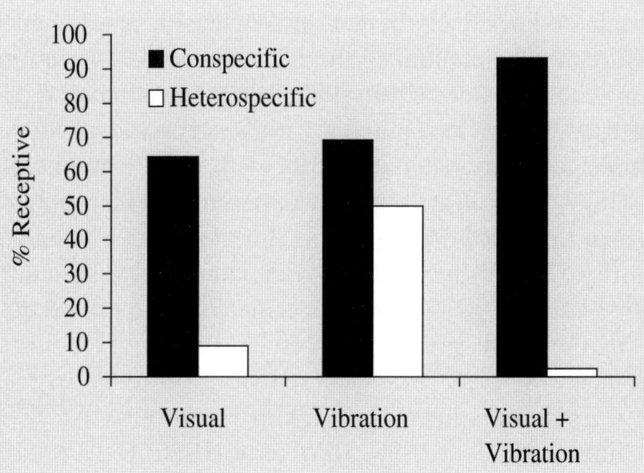

***Figure 2.** Response of female* Schizocosa ocreata *(%) to isolated male courtship signals in cue-isolation experiments (see text for details). Modified from Scheffer et al. (1996) and M. Orr & G. W. Uetz (unpublished).*

The role of multiple cues in female mate choice was considered further by Scheffer et al. (1996), in which the presence/absence of male decorative tufts was manipulated by shaving live males. Results demonstrated that females discriminated between shaved and intact males only when visual cues were presented in isolation; when vibratory cues were presented simultaneously, there was no difference. One possible explanation for this result is that, in the absence of vibration cues, females failed to recognize shaved males as conspecifics. This is doubtful, as females approached shaved males with equal frequency, but without displaying receptivity. It is also important to note that in pairings with visual and vibratory cues present, only the visual cue of the tufts was manipulated – male body size, courtship vigor and vibratory cues were not. As a consequence, females could still evaluate multiple cues, and the other unmanipulated signals (and the information contained therein) did not vary between shaved and unshaved males. On the other hand, it is possible that variation in a single trait - in this case the visual cue provided by tufts - may have greater importance for female choice when presented alone. However, experiments like this also raise questions about the impact of manipulation, as shaved males may exhibit subtle behavioral differences from intact males that are more obvious to female spiders (but not necessarily human observers) when visual cues alone are used in mate discrimination.

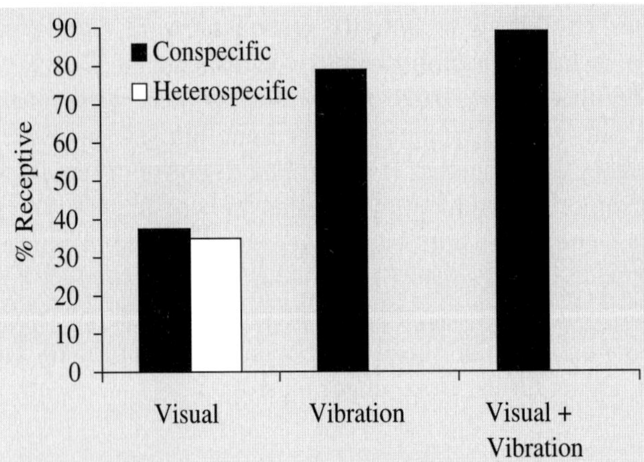

Figure 3. *Response of female* Schizocosa rovneri *(%) to isolated male courtship signals in cue-isolation experiments (see text for details). Modified from Stratton & Uetz (1983) and M. Orr & G. W. Uetz (unpublished).*

Image digitization and video playback studies have proven useful in teasing apart elements of visual display in spiders, and eliminate the potential confound of behavioral variation as a consequence of surgical alteration (Clark & Uetz 1991, 1992, 1993; McClintock & Uetz 1996; Uetz et al. 1996; Uetz & Smith 1999). In video playback studies, continuous loops of digitized videotape sequences of male courtship behavior (some with modifications) were presented to females in an arena with a Sony Watchman® television. In video playback studies with female *S. ocreata*, removal of tufts from conspecific male video stimuli reduced receptivity, but addition of tufts to heterospecific (*S. rovneri*) male video stimuli had no effect (McClintock & Uetz 1996). This result suggests that tufts alone are not necessarily a species recognition signal. Previous work with live males, as well as video stimuli, has shown that the size and symmetry of tufts influence female receptivity to visual courtship signals from males; i.e., receptivity is lower with males whose tufts are absent, small in size or asymmetric (McClintock & Uetz 1996; Uetz et al. 1996; Uetz & Smith 1999). These results suggest the possibility that this male visual signaling character is important in mate choice as well as species recognition. I used animated video images of the same courting male to manipulate and control male tuft size as a means of testing whether variation in the visual signaling character alone - without concomitant body size, behavioral or vibration cues - influenced female receptivity. Female receptivity was not independent of video stimulus ($G = 15.80$; $P < 0.001$); females exhibited lower receptivity to video males with tufts removed or reduced in size, but receptivity to a video male with enlarged tufts was equal to controls (Fig. 4). Responses of female spiders to these video images were similar to those with live males (not significantly different, $G = 2.169$, $P > 0.50$). Studies with live males demonstrated female preference for the top third ('large') and middle third ('intermediate') of the tuft size distribution ($G = 19.92$; $P < 0.001$).

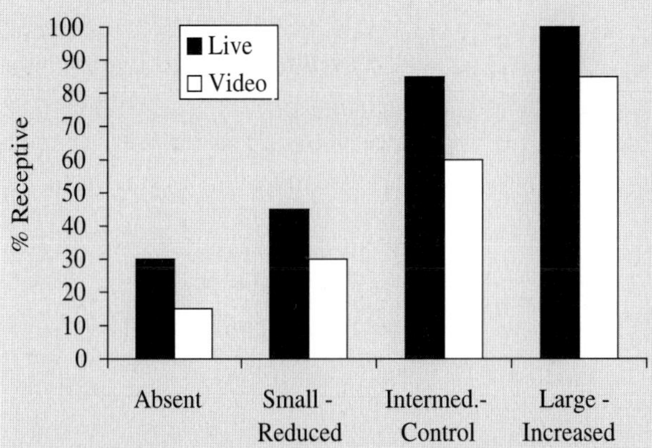

Figure 4. *Response (%) of female* Schizocosa ocreata *to variation in male tuft size in live and video males. In live males, tuft size categories 'small', 'intermediate' and 'large' represent thirds of the distribution of naturally occuring tuft size, and 'removed' indicates shaved males (modified from Scheffer et al. 1996). Live males were presented in a 'visual only' experimental arena, and video males in an arena with a Sony Watchman® TV (see text for details).*

Studies with field-collected males have demonstrated that the size of these secondary sexual characters varies with body condition. A commonly-used ratio index of body condition for spiders (Abdomen volume/cephalothorax width) was calculated for a sample of male spiders collected from the field site at the peak of the breeding season in June. Male tuft size was positively correlated with body condition (Fig. 5), suggesting that tufts may serve as quality indicator traits or condition-dependent handicaps in mate choice.

Another means of addressing questions about the relative importance of multiple signals is through cue-conflict experiments, in which conspecific and heterospecific cues in different modes are presented simultaneously. The apparatus allowed females to receive signals from two males – one which could be seen through a clear partition but on a separate substratum (visual only), and one behind an opaque barrier on the same substratum (vibration only). In this study (G. W. Uetz & M. Orr, unpublished), females of both species responded positively to the conspecific control, but all other combinations of visual and vibratory cues had the same level of response as the heterospecific control (Fig. 6). Differences between treatments were significant for both species (*S. ocreata*: $G = 16.47$, $P < 0.01$; *S. rovneri*: $G = 15.23$, $P < 0.01$), but differences between species were not significant. In this experiment, however, male communication was not always synchronized, and in many cases one male was courting while the other was not.

This might have resulted in the female perceiving signals from only one of the males at certain times, and responding appropriately.

To account for playback synchrony, J. Andrew Roberts and I (unpublished) combined video and audio playback (by modifying the video presentation arena with a miniature speaker mounted below the substratum) and presented continuous loops of visual and

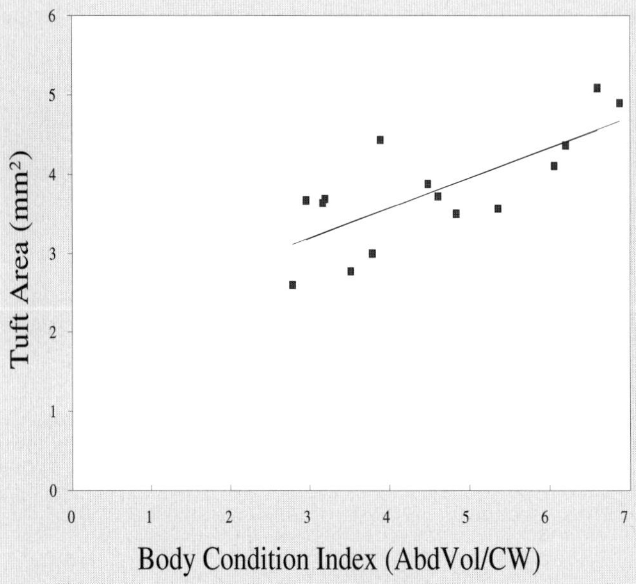

Figure 5. *Relationship between foreleg tuft area and an index of body condition (see text) in field-collected male* Schizocosa ocreata.

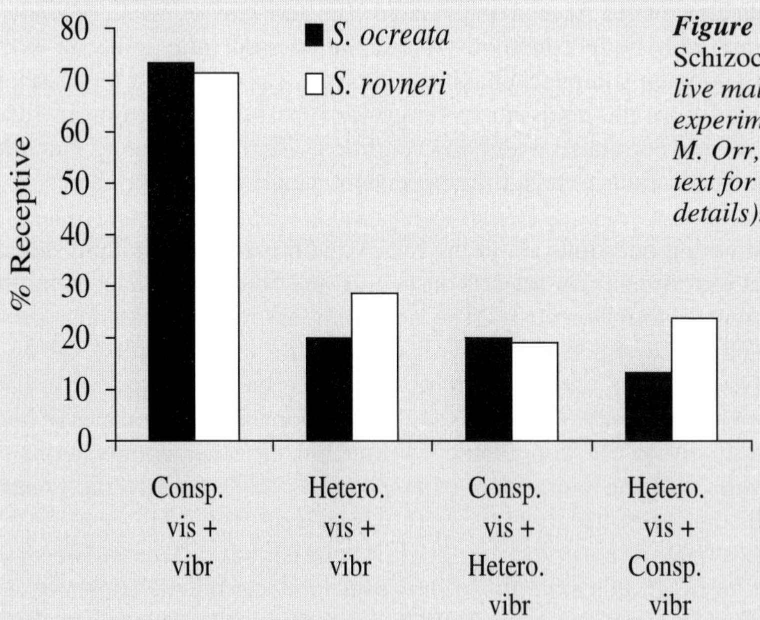

Figure 6. *Response of female* Schizocosa *to signals from live males in cue-conflict experiments (G. W. Uetz & M. Orr, unpublished) (see text for explanation and details).*

vibratory signals together. The results of these experiments (Fig. 7) were similar to those with live males: females responded positively to conspecific cues when presented together, and far less to other combinations; differences between treatments were significant for both species *S. ocreata*: $G = 17.36$, $P < 0.01$; *S. rovneri*: $G = 19.06$, $P < 0.01$). However, results of this experiment were somewhat different in that one combination (female *S. rovneri* presented with conspecific vibrations and *S. ocreata* visual cues) produced a higher level of receptivity response (Fig. 7). The response of female *S. rovneri* to this combination of stimuli seems too high to represent mistaken identity. One possibility is that female *S. rovneri* responded to conspecific vibratory cues alone and that the visual cues were ignored. However, these data also support an alternative possibility suggested by McClintock and Uetz (1996). In video playback studies (without vibration), female *S. rovneri* exhibited higher receptivity to male *S. ocreata* video stimuli than to video images of conspecific males. However, when tufts were added to male *S. rovneri* video stimuli, the response of female *S. rovneri* increased. As *S. rovneri* lack tufts and visual signals, which would appear to be closer to the ancestral state (McClintock & Uetz 1996; Hebets & Uetz 1999a), an argument could be made that *S. rovneri* may have a pre-existing bias for visual cues.

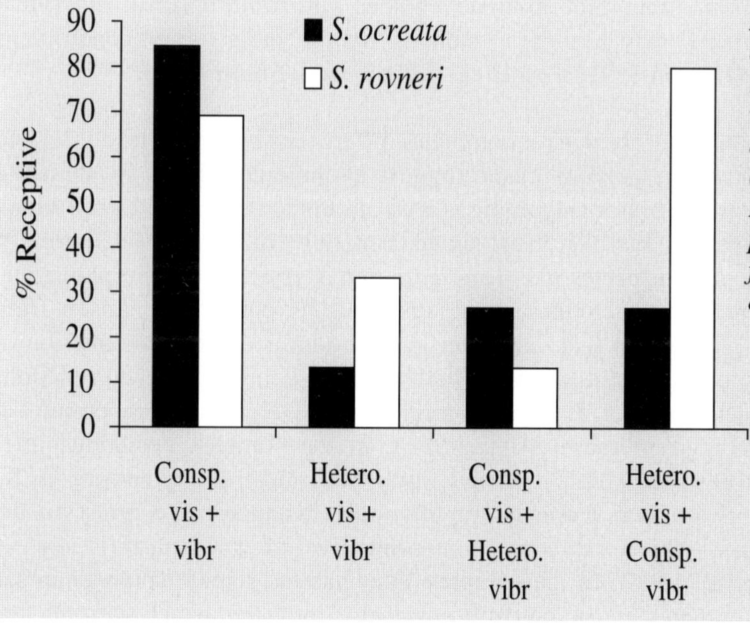

Figure 7. *Response of female* Schizocosa *to male signals in video/audio cue-conflict experiments (G. W. Uetz & J. A. Roberts, unpublished) (see text for explanation and details).*

Conclusions

Spiders exhibit considerable diversity in signaling behavior and although reliance on sensory channels varies among taxa, many species use multiple modes of communication. One major question remains: if single modes of communication are sufficient, why do spiders use multiple modes? The answer is far from clear. One explanation is that different modes are useful at different stages as pre-copulatory behavior proceeds, as suggested by Barth (1993). The distances at which females detect and recognize males may vary considerably, and the role of signals in sequential stages can therefore be critical. Redundancy in signaling also reduces the possibility of error in courtship communication, preventing wastage of gametes (Hebets & Uetz 1999b) and/or reduced fitness as a consequence of mate choice (Møller & Pomiankowski 1993; Johnstone 1995). As mentioned before, and as suggested by numerous others (Guilford & Dawkins 1991; Endler 1992a, b, 1993; Endler & Basolo 1998), the signaling environment may affect transmission, and use of multi-modal signaling may compensate for variability in environmental constraints. Finally, it is also possible that, by the signaler exploiting multiple sensory channels of the receiver with different signal modes simultaneously, a stronger signal may be produced (Møller & Pomiankowski 1993; Johnstone 1995; Hebets & Uetz 1999b). Multi-modal signaling in spiders may therefore be a response to multiple selection pressures.

Another question arising from current research is whether different signaling modes serve different purposes. Even if the redundancy of multiple modes allows species or mate recognition under the contingency of varying environmental constraints, single modes (e.g. visual) may be more important in mate choice. For some species, individual channels (vibration or visual) may be more reliable as 'honest' indicators of condition as well as species identity (Hasson 1991; Uetz et al. 1996; Uetz & Smith 1999).

Although phylogenetic analyses are not yet complete, differences between closely-related species within the *Schizocosa ocreata* clade suggest evolutionary shifts in signaling modes. While it is clear that vibratory signaling is plesiomorphic (Hebets & Uetz 1999a), the possibility of a pre-existing bias for visual signals cannot be ruled out (McClintock & Uetz 1996). In the case of ethospecies like *S. ocreata* and *S. rovneri*, the morphological and behavioral differences between these sympatric species could have arisen from divergent microhabitat preferences (e.g. leaf litter structure influence on transmission of vibratory cues used in species identification). Visual signaling behaviors and conspicuous secondary sexual characteristics of *S. ocreata* might have evolved to compensate for environmental attenuation of vibration (Uetz 1991). In the complex deciduous litter habitat of the upland forests in which it occurs throughout most of the eastern U. S., vibrations of courting *S. ocreata* attenuate rapidly over distance, in contrast to the floodplain litter habitats, where vibratory communication of *S. rovneri* is not so constrained (Scheffer et al. 1996). In the complex litter habitat, female spiders can see males at distances greater than they can perceive courtship vibrations, which suggests that selection should favor a shift in tactical design of signals from vibratory to visual cues. The potential for cannibalism has been shown to be an important factor in the population ecology of this species (Wise & Wagner 1992; M. H. Persons & G. W. Uetz, unpublished), and this would clearly increase selection for signal efficacy.

I hope that, from this review, the reader will come to appreciate the diversity of signaling behaviors exhibited by spiders and the enormous potential for research in this often-neglected taxon. While we are still far from answers to a number of key questions, the development of the experimental techniques demonstrated here has provided some insights to spider communication. For these reasons, spiders in general, and the *Schizocosa* model system in particular, provide an opportunity to examine the interaction of sensory cues, behavior, and habitat in the evolution of signaling behavior.

Acknowledgements

This research was supported by grant IBN 9414239 from the National Science Foundation. I am grateful to the Cincinnati Nature Center, Milford, Ohio, for allowing us to collect spiders at their Rowe Woods facility. I would also like to acknowledge the assistance of my students - J. Calos, A. DeLay, K. Delaney, E. Hebets, B. Kilinc, D. Kroeger, H. Kuyper, R. Mendoza, M. Orr, M. Persons, A. Pruden, J. A. Roberts, H. Roberts, E. Sanborn, D. Safriet and E. Smith - in collecting and maintaining spiders for this research project and in running some critical experiments. I sincerely appreciate comments on the manuscript from J.A. Roberts, M. Orr and two anonymous reviewers.

References

Aspey, W. P. 1977a. Wolf spider sociobiology: I. Agonistic display and dominance-subordinance relations in adult male *Schizocosa crassipes*. *Behaviour*, **62**, 103-141.

Aspey, W. P. 1977b. Wolf spider sociobiology: II. Density parameters unfluencing agonistic behavior in *Schizocosa crassipes*. *Behaviour*, **62**, 142-162.

Austad, S. N. 1983. A game theoretical interpretation of male combat in the bowl and doily spider (*Frontinella pyramitela*). *Animal Behaviour*, **31**, 59-73.

Ayyagari, L. R. & Tietjen, W. J. 1987. Preliminary isolation of male-inhibitory pheromone of the spider *Schizocosa ocreata* (Araneae, Lycosidae). *Journal of Chemical Ecology*, **13**, 237-244.

Barth, F. G. 1982. Spiders and vibratory signals: sensory reception and behavioral significance. In: *Spider Communication: Mechanisms and Ecological Significance* (Ed. by P. W. Witt & J. S. Rovner), pp. 67-120. Princeton, NJ: Princeton University Press.

Barth, F. G. 1993. Sensory guidance in spider pre-copulatory behaviour. *Comparative Biochemistry and Physiology*, Series A, **104**, 717-733.

Barth, F. G. 1985. *Neurobiology of Arachnids*. Berlin: Springer-Verlag.

Barth, F. G. & Schmitt, A. 1991. Species recognition and species isolation in wandering spiders (*Cupiennius* spp., Ctenidae). *Behavioral Ecology and Sociobiology*, **29**, 333-339.

Christenson, T. E. & Goist, K. C. Jr. 1979. Costs and benefits of male-male competition in the orb weaving spider, *Nephila clavipes*. *Behavioral Ecology and Sociobiology*, **5**, 87-92.

Clark, D. L. & Uetz, G. W. 1991. Video image recognition by jumping spiders (Araneae: Salticidae). *Animal Behaviour*, **40**, 884-890.

Clark, D. L. & Uetz, G. W. 1992. Morph-independent mate selection in a dimorphic jumping spider: demonstration of movement bias in female choice using video-controlled courtship behavior. *Animal Behaviour*, **43**, 247-254.

Clark, D. L. & Uetz, G. W. 1993. Signal efficacy and the evolution of male dimorphism in the jumping spider, *Maevia inclemens* (Araneae: Salticidae). *Proceedings of the National Academy of Sciences, USA*, **90**, 11954-11957.

Costa, F. G., Viera, C. & Francescoli, G. 1997. Male sexual behavior elicited by a hybrid pheromone: a comparative study on *Lycosa thorelli*, *L. carbonelli*, and their hybrid progeny. *Canadian Journal of Zoology*, **75**, 1845-1856.

Crane, J. 1949. Comparative biology of salticid spiders at Rancho Grande, Venezuela, Part IV: an analysis of display. *Zoologica*, **34**, 159-215.

Dodson, G. N. & Beck, M. W. 1993. Pre-copulatory mate-guarding of penultimate females by male crab spiders, *Misumenoides formosipes*. *Animal Behaviour*, **46**, 951-959.

Elgar, M. 1991. Sexual cannibalism, size dimorphism, and courtship behavior in orb-weaving spiders (Araneidae). *Evolution*, **45**, 444-448.

Elgar, M. & Nash, D. R. 1988. Sexual cannibalism in the garden spider *Araneus diadematus*. *Animal Behaviour*, **36**, 1511-1517.

Endler, J. A. 1992a. Sensory drive: does sensory biology bias or constrain the direction of evolution? *American Naturalist*, **139** (supplement), s1-s3.

Endler, J. A. 1992b. Signals, signal conditions and the direction of evolution. *American Naturalist*, **139** (supplement), s125-s153.

Endler, J. A. 1993. Some general comments on the evolution and design of animal communication systems. *Philosophical Transactions of the Royal Society of London, Series B*, **340**, 215-225.

Endler, J. A. & Basolo, A. L. 1998. Sensory ecology, receiver biases and sexual selection. *Trends in Ecology and Evolution*, **13**, 415-420.

Forster, L. 1982. Visual communication in jumping spiders (Salticidae). In: *Spider Communication: Mechanisms and Ecological Significance* (Ed. by P. W. Witt & J. S. Rovner), pp. 161-212. Princeton, NJ: Princeton University Press.

Guilford, T. & Dawkins, M. S. 1991. Receiver psychology and the evolution of animal signals. *Animal Behaviour*, **42**, 1-14.

Hasson, O. 1989. Amplifiers and the handicap principle in sexual selection: a different emphasis. *Proceedings of the Royal Society of London, Series B*, **235**, 383-406.

Hasson, O. 1991. Sexual displays as amplifiers: practical examples with an emphasis on feather decorations. *Behavioral Ecology*, **2**, 189-197.

Hebets, E. A. & Uetz, G. W. 1999a. Leg ornamentation and the efficacy of courtship display in four species of wolf spider (Araneae: Lycosidae). *Behavioral Ecology and Sociobiology.* In press.

Hebets, E. A. & Uetz, G. W. 1999b. Female responses to isolated signals from multi-modal male courtship displays in the wolf spider genus *Schizocosa* (Araneae: Lycosidae). *Animal Behaviour*, **57**, 865-872.

Hodge, M. A. & Uetz, G. W. 1995. A comparison of agonistic behaviour of colonial web-building spiders from desert and tropical habitats. *Animal Behaviour*, **50**, 963-972.

den Hollander, J. 1970. Morphological variation in *Pardosa prativaga* L. Koch, 1870, *P. prativaga* var. *fulvipes* Collet, 1875, and *P. pullata* Clerck, 1757 (Araneae, Lycosidae). *Tijdschrift voor Entomologie*, **113**, 283-290.

den Hollander, J., Dijkstra, H., Alleman, H. & Vlijm, L. 1973. Courtship behavior as a species barrier in the *Pardosa pullata* group. *Tijdschrift voor Entomologie*, **116**, 1-22.

Jackson, R. R. 1977. Courtship versatility in the jumping spider, *Phidippus johnsoni* (Araneae: Salticidae). *Animal Behaviour*, **25**, 952-957.

Jackson, R. R. 1978. An analysis of alternative mating tactics of the jumping spider *Phidippus johnsoni* (Araneae: Salticidae). *Journal of Arachnology*, **5**, 185-230.

Jackson, R. R. 1981a. Nest-mediated sexual discrimination by a jumping spider (*Phidippus johnsoni*). *Journal of Arachnology*, **9**, 87-92

Jackson, R. R. 1981b. Relationship between reproductive security and intersexual selection in a jumping spider *Phidippus johnsoni* (Araneae: Salticidae). *Evolution*, **35**, 601-604.

Jackson, R. R. 1982. The behaviour of communicating in jumping spiders (Salticidae). In: *Spider Communication: Mechanisms and Ecological Significance* (Ed. by P. W. Witt & J. S. Rovner), pp. 213-245. Princeton, NJ: Princeton University Press.

Jackson, R. R. 1986. Use of pheromones by males of *Phidippus johnsoni* (Araneae, Salticidae) to detect subadult females that are about to molt. *Journal of Arachnology*, **14**, 137-139.

Jackson, R. R. 1987. Comparative study of releaser pheromones associated with the silk of jumping spiders (Araneae: Salticidae). *New Zealand Journal of Zoology*, **14**, 1-10.

Jackson, R. R. & Wilcox, R. S. 1990. Aggressive mimicry, prey-specific predatory behavior and predator-recognition in predator-prey interactions of *Portia fimbriata* and *Euryattus* sp. jumping spiders from Queensland. *Behavioral Ecology and Sociobiology*, **26**, 111-119.

Jackson R. R. & Wilcox, R. S. 1993. Spider flexibly chooses aggressive mimicry signals for different prey by trial and error. *Behaviour*, **127**, 21-36.

Johnstone, R. A. 1995. Honest advertisement of multiple qualities using multiple signals. *Journal of Theoretical Biology*, **177**, 87-94.

Kotiaho, J. S., Alatalo, R. V., Mappes, J., Nielsen, M. G. & Parri, S. 1996. Sexual selection in a wolf spider: Male drumming activity, body size and viability. *Evolution*, **50**, 1977-1981.

Kotiaho, J. S., Alatalo, R. V., Mappes, J., Nielsen, M. G., Parri, S. & Rivero, A., 1998. Energic costs of size and sexual signalling in a wolf spider. *Proceedings of the Royal Society of London, Series B*, **265**, 2203-2209.

Krafft, B. 1982. The significance and complexity of communication in spiders. In: *Spider Communication: Mechanisms and Ecological Significance* (Ed. by P. W. Witt & J. S. Rovner), pp. 17-66. Princeton, NJ: Princeton University Press.

Kronestedt, T. 1996. Vibratory communication in the wolf spider *Hygrolycosa rubrofasciata* (Araneae: Lycosidae). *Revue Suisse de Zoologie, vol. hors serie*, 341-354.

Land, M. F. 1985. The morphology and optics of spider eyes. In: *Neurobiology of Arachnids* (Ed. by F.G. Barth), pp. 53-78. Berlin: Springer-Verlag.

McClintock, W.J. & Uetz, G. W. 1996. Female choice and preexisiting bias: visual cues during courtship in two *Schizocosa* wolf spiders (Araneae: Lycosidae). *Animal Behaviour*, **52**, 167-181.

Miller, G. L., Stratton, G. E., Miller, P. E. & Hebets, E. A. 1998. Geographic variation in male courtship behavior and sexual isolation in wolf spiders of the genus *Schizocosa. Animal Behaviour*, **56**, 937-951.

Møller, A. P. & Pomiankowski, A. 1993. Why have birds got multiple sexual ornaments? *Behavioral Ecology and Sociobiology*, **32**, 167-176.

Nossek, M. E. & Rovner, J. S. 1984. Agonistic behavior in female wolf spiders (Araneae: Lycosidae). *Journal of Arachnology*, **11**, 407-422.

Parri, S., Alatalo, R. V., Kotiaho, J. & Mappes, J. 1997. Female choice for male drumming in the wolf spider *Hygrolycosa rubrofasciata. Animal Behaviour*, **53**, 305-312.

Peckham, G. W. & Peckham, E. G. 1889. Observations on sexual selection in spiders of the family Attidae. *Occasional Papers of the Wisconsin Natural History Society*, **1**, 3-60.

Persons, M. H. & Uetz, G. W. 1996. Wolf spiders vary patch residence time in the presence of chemical cues from prey (Araneae, Lycosidae). *Journal of Arachnology*, **24**, 76-79.

Pollard, S. D., Macnab, A. M. & Jackson, R. R. 1987. Communication with chemicals: pheromones and spiders. In: *Ecophysiology of Spiders* (Ed. by W. Nentwig), pp. 133-141. Berlin: Springer-Verlag.

Prenter, J., Elwood, R. W. & Montgomery, I. 1994. Male exploitation of female predatory behaviour reduces sexual cannibalism in male autumn spiders, *Metellina segmentata. Animal Behaviour*, **47**, 235-236.

Punzo, F. & Kukoyi, O. 1997. The effects of prey chemical cues on patch residence time in the wolf spider *Trochosa parthenus* (Chamberlin) (Lycosidae) and the lynx spider *Oxyopes salticus* Hentz (Oxyopidae). *Bulletin of the British Arachnological Society*, **10**, 323-326.

Richman, D. 1982. Epigamic display in jumping spiders (Araneae: Salticidae) and its use in systmatics. *Journal of Arachnology*, **10**, 47-67.

Richman, D. & Jackson, R. R. 1992. A review of the ethology of jumping spiders. *Bulletin of the British Arachnological Society*, **9**, 33-37.

Riechert, S. E. 1978. Games spiders play: behavioral variability in territorial disputes. *Behavioral Ecology and Sociobiology*, **3**:135-162.

Riechert, S. E. 1982. Spider interaction strategies: communication vs. coercion. In: *Spider Communication: Mechanisms and Ecological Significance* (Ed. by P. W. Witt & and J. S. Rovner). pp. 281-315. Princeton, NJ: Princeton University Press.

Riechert, S. E. 1986. Spider fights as a test of evolutionary game theory. *American Scientist*, **74**, 604-610.

Robinson, M. H. 1980. Comparative studies of the courtship and mating behavior of tropical araneid spiders. *Pacific Insects Monograph*, **36**, 1-218.

Robinson, M. H. 1982. Courtship and mating behaviour in spiders. *Annual Review of Entomology*, **27**, 1-20.

Roland, C. 1984. Chemical signals bound to the silk in spider communication (Arachnida, Araneae). *Journal of Arachnology*, **11**, 309-314.

Rovner, J. S. 1967. Acoustic communication in a lycosid spider (*Lycosa rabida* Walckenaer). *Animal Behaviour*, **15**, 273-281.

Rovner, J. S. 1968. An analysis of display in the lycosid spider *Lycosa rabida* Walkenaer. *Animal Behaviour*, **20**, 133-128.

Rovner, J. S. 1975. Sound production by Nearctic wolf spiders: a substratum-coupled stridulatory mechanism. *Science*, **190**, 1309-1310.

Rovner, J. S. 1980. Vibration in *Heteropoda venatoria* (Sparassidae): a third method of sound production in spiders. *Journal of Arachnology*, **8**, 193-200.

Rovner, J. S. 1989. Wolf spiders lack mirror-image responsiveness seen in jumping spiders. *Animal Behaviour*, **38**, 526-533.

Rovner, J. S. 1996. Conspecific interactions in the lycosid spider *Rabidosa rabida*: the roles of different senses. *Journal of Arachnology*, **24**, 16-23.

Rovner, J. S. & Barth, F. G. 1981. Vibratory communication through living plants by a tropical wandering spider. *Science*, **214**, 464-466.

Scheffer, S. J., Uetz, G. W. & Stratton, G. E. 1996. Sexual selection, male morphology, and the efficacy of courtship signalling in two wolf spiders. (Araneae: Lycosidae). *Behavioral Ecology and Sociobiology*, **38**, 17-24.

Schmitt, A., Schuster, M. & Barth, F. G. 1994. Vibratory communication in a wandering spider, *Cupiennius getazi*: female and male preferences for features of the conspecific male's releaser. *Animal Behaviour*, **48**, 1155-1171.

Schuch, W. & Barth, F. G. 1985. Temporal patterns in the vibratory courtship signals of the wandering spider *Cupiennius salei* (Araneae: Ctenidae). *Behavioral Ecology and Sociobiology*, **16**, 263-271.

Schuch, W. & Barth, F. G. 1990. Vibratory communication in a spider: female responses to synthetic male vibrations. *Journal of Comparative Physiology*, **166**, 817-826.

Stowe, M. K., Tumlinson, J. H. & Heath, R. R. 1987. Chemical mimicry: bolas spiders emit components of moth prey species sex pheromones. *Science*, **236**, 964-967.

Stratton, G.E. & Uetz, G. W. 1981. Acoustic communication and reproductive isolation in two species of wolf spiders (Araneae: Lycosidae). *Science*, **214**, 575-577.

Stratton, G. E. & Uetz, G. W. 1983. Communication via substratum-coupled stridulation and reproductive isolation in wolf spiders (Araneae: Lycosidae). *Animal Behaviour*, **31**, 164-172.

Stratton, G. E. & Uetz, G. W. 1986. The inheritance of courtship behavior and its role as a reproductive isolating mechanism in two species of *Schizocosa* wolf spiders (Araneae: Lycosidae). *Evolution*, **40**, 129-141.

Suter, R. B. & Renkes, G. 1982. Linyphiid spider courtship: releaser and attractant functions of a contact sex pheromone. *Animal Behaviour*, **30**, 714-718.

Tietjen, W. J. 1977. Dragline-following by male lycosid spiders. *Psyche*, **84**, 165-178.

Tietjen, W. J. 1979. Is the sex pheromone of *Lycosa rabida* (Araneae: Lycosidae) deposited on a substratum? *Journal of Arachnology*, **6**, 207-212.

Tietjen, W. J. & Rovner, J. S. 1980. Trail-following behavior in two species of wolf spiders: sensory and etho-ecological concomitants. *Animal Behavior*, **28**, 735-741.

Tietjen, W. J. & Rovner, J. S. 1982. Chemical communication in lycosids and other spiders. In: *Spider Communication: Mechanisms and Ecological Significance* (Ed. by P. N. Witt & J. S. Rovner), pp. 249-279. Princeton, NJ: Princeton University Press.

Uetz, G. W. 1991. Habitat structure and spider foraging. In: *Habitat Structure: The Physical Arrangement of Objects in Space* (Ed. by S. S. Bell, E. D. McCoy & H. R. Mushinsky), pp. 325-348. London: Chapman and Hall.

Uetz, G. W. & Denterlein, G. J. 1979. Courtship behavior, habitat, and reproductive isolation in *Schizocosa rovneri* (Uetz and Dondale) (Araneae: Lycosidae). *Journal of Arachnology*, **7**, 86-88.

Uetz, G. W., McClintock, W. J., Miller, D., Smith, E. I. & Cook, K. K. 1996. Limb regeneration and subsequent asymmetry in a male secondary sexual character influences sexual selection in wolf spiders. *Behavioral Ecology and Sociobiology*, **38**, 253-257.

Uetz, G. W. & Smith, E. I. 1999. Asymmetry in a visual signaling character and sexual selection in a wolf spider. *Behavioral Ecology and Sociobiology*, **45**, 87-93.

Uetz, G. W. & Stratton, G. E. 1982. Acoustic communication in spiders. In: *Spider Communication: Mechanisms and Ecological Significance* (Ed. by P.N. Witt & J.S. Rovner), pp. 123-159. Princeton, NJ: Princeton University Press.

Uetz, G. W. & Stratton, G. E. 1983. Communication in spiders. *Endeavour*, **7**, 13-18.

Watson, P. 1990. Female-enhanced male competition determines the first mate and principal sire in the spider *Linyphia litigiosa*. *Behavioral Ecology and Sociobiology*, **26**, 77-90.

Watson, P. 1998. Multiple mating and female choice increase offspring growth in the spider *Nereine litigiosa* (Linyphiidae). *Animal Behaviour*, **55**, 387-403.

Wilcox, R. S., Jackson, R. R. & Gentile, K. 1996. Spiderweb smokescreens: spider trickster uses background noise to mask stalking movements. *Animal Behaviour*, **51**, 313-326.

Wise, D. S. & Wagner, J. 1992. Evidence of exploitative competition among young stages of the wolf spider *Schizocosa ocreata*. *Oecologia*, **91**, 7-13.

Witt, P. W. & Rovner, J. S. (Eds) 1982. *Spider Communication: Mechanisms and Ecological Significance.* Princeton, NJ: Princeton University Press.

Yeargan, K. V. 1994. Biology of bolas spiders. *Annual Review of Entomology*, **39**, 81-99.

IN: ESPMARK,Y., AMUNDSEN,T. & ROSENQVIST,G. (EDS.) 2000
ANIMAL SIGNALS:
SIGNALLING AND SIGNAL DESIGN IN ANIMAL COMMUNICATION, PP. 407-419
TAPIR ACADEMIC PRESS, TRONDHEIM, NORWAY

21

Signals, Parasites and the Immune System

MARLENE ZUK* & KURT A. McKEAN

Department of Biology, University of California, Riverside, CA 92521, USA

**E-mail: mzuk@citrus.ucr.edu*

Abstract

Signaling is ubiquitous in biological systems, whether the signals are transmitted between cells within an organism or among individual organisms of the same or different species. Although the evolution of animal signals such as sexually-selected ornaments is usually viewed as quite separate from the origin of cellular signals, many parallels exist between the two types, particularly when the role of parasites in mediating such signals is considered. A crucial feature is the honesty of the signal, which depends on the costs and benefits to signaler and receiver. Both signal exploitation, in which a signal benefits a recipient but is costly to the sender, and deceitful signaling, in which a signal benefits the sender but is costly to the recipient, are common among animals as well as in intercellular communication involving pathogens and the immune system. Deceit, particularly via molecular mimicry, appears to be more common than signal exploitation in intercellular signaling. Consequences of this dishonest signaling include signal redundancy, so that multiple signals convey the same information, and a potential for rapid signal evolution, leading in the case of animal signals to an increased likelihood of signal divergence and speciation.

Signaling is ubiquitous in biological systems, whether the signals are transmitted between cells within an organism or among individual organisms of the same or different species. We usually think of signals between organisms as very different from those occurring within a body, and the latter, with a few exceptions, are not considered the purview of evolutionary biologists. We suggest, however, that many parallels exist between animal signaling and the kinds of signals that are exchanged between cells in an organism during the immune response. Here we want to provide a link between these types of signals via the role of parasites, pointing out some of the similarities and analogies that exist and then suggesting how these resemblances can be useful in understanding the evolution of signaling itself.

For a long time, researchers have focused on conflicts of interest between sender and receiver (Krebs & Dawkins 1978; Wiley 1994), which makes good evolutionary sense. At first glance, cells of a body should not be expected to show conflict, and hence not be subject to the same types of analysis; cells within a multicellular organism are all (barring mutations) related by a factor of 1, and the separation of the germ line from the soma ensures that cells within that multicellular organism must cooperate to ensure the success of the indiviual (McKean & Zuk 1995). Nevertheless, we believe that parasites can play a role in the evolution of intercellular signaling if they take advantage of host intercellular communication to aid in their own survival and reproduction. The potential conflicts of interest between parasite and host are obvious, and through those conflicts of interest the signals and the system of communication become subject to evolution.

The signaling process

The signaling process can be seen as involving three units: a sender, a receiver, and an information processing unit (IPC) (Fig. 1). The signals of the sender act as selective factors that influence the evolution of the receptor and of the processing center. A change in a signal making the information more reliable will favor receptors that accurately receive the change, as well as processors that can make appropriate decisions based on the new information. Receptors also act as selective agents, influencing the evolution of both the signal and the IPC. For example, receptors may have biases making certain types of signals easier to receive. Finally, the IPC also influences the other two components, because selection will favor rapid collection and transfer of accurate information from receptors and the ability to discern differences between reliable and unreliable signals. All of these units are influenced by the environment, whether that is the internal environment of a body or the ecological conditions in which an organism occurs, and all are constrained by various elements as well. Figure 1 emphasizes the role of deceitful signalers (which provide dishonest signals in an effort to manipulate receivers) and exploitative receivers of signals (which use information provided by the signaler to benefit themselves at a cost to the signaler).

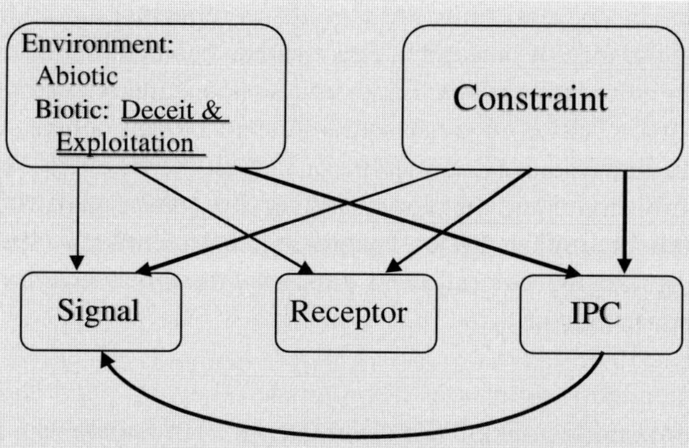

Figure 1. *Selective factors and constraints in communication. Communication requires that a signaler provides information to a receiver that has a means of receiving that signal (the receptor), in addition to decision rules which govern the response to the signal (the information processing center – IPC). Arrows represent potential avenues for selection on components of communication. The response of the receiver promotes the evolution of reliability in signaling. Errors in communication can be caused by constraint in the design of the signals, receptors or IPC's or by the environment. (Modified from McKean and Zuk).*

Table 1. *Honest and dishonest signals.*

Benefit to Sender	Benefit to Receiver: Positive	Benefit to Receiver: Zero (or Negative)
Positive	**Honesty**	**Deceit**
Zero (or Negative)	**Exploitation**	**Spite**

If we consider the costs and benefits to either sender or receiver, we can see that the type of signal depends on this relationship (Table 1). An honest signal occurs when both parties benefit, a deceitful signal when the sender benefits but the receiver does not, perhaps because the sender mimics a poisonous prey item when it is in fact palatable. Signal exploitation happens when a signal is 'overheard' by an unintended recipient, as might be the case when a bat uses the mating call of a frog to find its prey. Finally, a signal is spiteful if it is costly to both the signaler and receiver, a situation that should not ordinarily evolve. We will consider both exploitation and deceit separately in signals exchanged between individuals as well as between cells.

Signal exploitation

First, let us consider signal exploitation by parasites and predators in animals. Most researchers agree that because many types of signals, especially sexual signals, are by definition conspicuous to their intended recipient, they may also be detected by enemies, and animals producing mate attraction signals are therefore faced with a conflict between mating success and being eaten (Burk 1982; Sakaluk 1990; Verrell 1991; Endler 1992;

Zuk & Kolluru 1998). Many secondary sexual traits are thought to represent a compromise between attractiveness and avoidance of detection. This risk has been examined in a wide range of taxa using several signaling modalities. For example, acoustically-orienting parasitoid flies use the song of a male cricket or other orthopterans to locate a host; the female fly then deposits larvae on the cricket (Cade 1975; Zuk et al. 1993, 1995). The larvae burrow into the cricket's body cavity and develop inside for 7-10 days, after which they emerge and pupate in the soil. In another mode of signaling, John Endler's (1992) classic work has shown that bright orange coloration in guppies *Poecilia reticulata* is associated with the absence of visual predators.

Otte (1974) referred to such unintended recipients as 'illegitimate receivers', and Dicke and Sabelis (1992) discuss further subdivisions of signal interception, including 'spies', 'stowaways' and 'boasters'; we do not go into the semantic distinctions here. The metaphors virtually always involve dishonesty and cheating, and in many species of animals, the continued evolution of elaborate ornamental traits like the classic peacock's tail are assumed to be constrained or checked by the effects of predators or parasites that take advantage of the very extravagance favored by sexual selection.

What affects the likelihood that particular types of signals will or will not be exploited? In animal signaling, researchers have long noted the paucity of exploitation of olfactory signals such as pheromones (Zuk & Kolluru 1998). Interestingly, although chemical signals are the primary means of sexual attraction in most Lepidoptera and other insects, relatively few examples exist of natural enemies using those signals to find prey or hosts. In addition, and equally interestingly, the type of signal used - acoustic, visual, or olfactory - is also associated with a sex difference in who produces the signal. Acoustic and visual signals are most often produced by males, with females traveling to stationary groups or territorial individuals, as evidenced in taxa as diverse as crickets and katydids (Gwynne & Morris 1983; Thornhill & Alcock 1983), fruit flies *Drosophila* spp. (Spieth 1974), anuran amphibians (Howard 1988; Sullivan 1989) and lekking birds and mammals (Höglund & Alatalo 1995). Even when males do not signal from a fixed position, they are still the sex that pays the price of conspicuous ornaments. For example, singing male crickets use several times more energy than animals at rest (Prestwich & Walker 1981), and the metabolic power output of several species of hylid frogs is much greater than resting metabolic rate (Prestwich et al. 1989). Energetic costs of visual signals are more difficult to determine, but studies of barn swallows *Hirundo rustica* also suggest that the long tail of males, a sexually-selected trait, is costly (Møller 1994).

Most authors who addressed the role reversal in olfactory signaling concluded that producing pheromones is not particularly costly, whereas responding to the odor and traveling to its source involves relatively more risks (Cardé & Baker 1984; Dicke & Sabelis 1992, Svensson 1996; Phelan 1997). The risky behavior is therefore taken on, as usual, by males, while females are not in danger of detection or exploitation by predators and parasites because they emit only minute amounts of highly specific chemicals. Indeed, the intensity of long range pheromones produced by females is dramatically less than that of male-produced pheromones. Greenfield (1981) suggested that female moths are not competing among themselves by signaling in the way that male peacocks or crickets may be; by emitting such low intensity signals, females might even be presenting a passive filter to test male response, such that only those males able to detect minute concentrations of odors can find a mate. This test would only work, of course, if detection ability is linked to the viability of the male, perhaps because of the ability to locate a host plant via chemical cues.

It seems plausible that selection has acted on females to produce appropriately alluring signals, although the lack of exaggeration of those signals remains intriguing. What would constitute an elaborated scent? Our own relative insensitivity to olfactory cues may hinder our ability to imagine the odor equivalent of a bird of paradise's (Paradisaeidae) plumage (Zuk & Kolluru 1998). Certainly little is known about the ancestral state of pheromones among those insects employing them, which makes comparative studies even more problematic.

If the pheromone is not costly to produce, and if males compete with one another by searching and incur costs as they travel to the female, then Greenfield's (1981) idea about using pheromones as a filter to test males is appealing and sex roles are no different from those in most animals. The problem is that virtually no data on the energetic costs of pheromone production are available. Interestingly, female moths sometimes produce greater concentrations of pheromone as they age (Greenfield 1981), consistent with the idea that odor production is expensive, because females should be more willing to pay costs as their reproductive value decreases and less of their reproductive lifespan is left (Williams 1975). If increased signal intensity is not costly, but attracts males more effectively, why has selection not increased signal intensity at all ages?

The other cost besides an energetic one is, of course, exploitation by predators or parasites. If odors are not likely to attract natural enemies, then females take no risks by producing them. Greenfield (1981) suggested that the apparent rarity of parasitoids locating female-emitted pheromones may reflect the rarity with which parasitoids attack adult Lepidoptera; most such parasitoids attack eggs or larvae. It may not pay a parasitoid to locate a pheromone-emitting female and wait until she mates and lays eggs. Thus, except for a few special cases such as the attraction of an egg parasitoid to a moth at oviposition, most sex pheromones which attract natural enemies are male-produced pheromones in aggregating species such as bark beetles (Zuk & Kolluru 1998). This argument does not, however, explain why predators on adult insects are thought to be unlikely to use odor cues.

Why are there so few examples of odor detection by predators compared to detection of other sensory cues? It has been suggested that predators might find it more difficult to exploit odors because of the precise composition of most pheromones. We are skeptical of this explanation given the remarkable adaptations seen in other natural enemies with special sensitivities to acoustic frequencies or other properties of signals (Zuk & Kolluru 1998). The difference may lie in a distinction drawn by Maynard Smith (1958, 1991) between 'notices' and 'advertisements'. If the interests of signaler and receiver do not coincide, the evolution of a costly advertisement is likely. Most interactions between males and females fall into this category because of the disparity of parental investment between the sexes and subsequent male competition for females. However, some signals, such as the railway timetable or bee waggle dance, are not selected to be costly because both sender and receiver benefit by accurate transfer of information. If female moths do not compete amongst themselves for males, and if odors are not energetically costly to produce, sex pheromones may qualify as notices, and hence not be conspicuous to natural enemies in the manner of, for example, cricket song. The disparity in parental investment between the sexes suggests that males may end up shouldering more of the burden, whether they play the role of signaler or receiver.

Cellular signals are also exploited by parasites, although the examples of deceit in intercellular communication seem to be more common than exploitation. One classic case is the interaction between the European rabbit *Oryctolagus cuniculus* and its flea *Spilopsyllus cuniculi*. The fleas use hormonal signals produced by the hosts in their own reproduction to obtain information about when the female rabbits will be producing young. The fleas become reproductive in synchrony with the pregnancy, and flea eggs are ripe at the time when young rabbits are available as new hosts. As the mother tends her young, the adult fleas move onto the newborn rabbits and feed voraciously, mate, and lay eggs. After about 12 days, the fleas leave the young and return to the mother rabbit. This ensures a good food supply for the young fleas and allows the colonization of new hosts (Schmidt & Roberts 1989).

Deceitful signaling

Next let us consider deceitful signals, signals produced by the parasite or predator that are dishonest, as opposed to the parasite merely capitalizing on a signal intended for a different recipient. We will consider two types of deceit in intercellular signaling: molecular mimicry and what we will call signal attenuation. A well-known example of the former is the trypanosome, a unicellular parasite that evades detection by the host by imitating the host's antigen coat and appearing as 'self' to the host immune system (Schmidt & Roberts 1989). Here we will concentrate on more subtle examples. A number of DNA viruses (especially poxviruses and herpes viruses) and RNA retroviruses possess viral homologs of host signals, and they use these signals to manipulate the host immune response via mimicry and thereby prevent their own recognition and destruction.

For example, Epstein-Barr virus is a member of the herpes virus family, and infects the B-cells of the immune system, the cells used to produce antibodies specific to particular pathogens. Diseases associated with Epstein-Barr include infectious mononucleosis, Burkitt's lymphoma, and nasopharyngeal carcinoma. A protein in the virus, the BCRF1 protein, shows homology to a human interleukin, Il-10 (Crowe 1994). Interleukins are a type of cytokine, chemicals produced by the immune system to aid in fighting disease. By mimicking Il-10, the virus manipulates the host's intercellular communication system, and elicits an inappropriate immune response. Ordinarily, the host's cellular immune response would be crucial in fighting Epstein-Barr. But the 'mock' Il-10 seems to circumvent the cellular immune response by signaling to T cells to cease cytokine production, promote proliferation of B cells, and suppress cellular immunity in favor of humoral immunity (Crowe 1994). Therefore, the deceitful signal can bind Il-10 receptors on the surface of T helper cells, resulting in an immune response that is less effective at clearing the virus and also results in the proliferation of the B cells that serve as new host cells for the virus to infect.

Another example of deceitful signaling by a virus comes from Kaposi's Sarcoma-associated herpes virus, a virus closely related to Epstein-Barr and one which causes the previously rare cancerous lesions made familiar by individuals with HIV. This virus possesses a number of proteins which manipulate intercellular signaling pathways (Moore et al. 1996). For example, KSHV vIl-6, a homologue of human interleukin 6, is a cytokine involved in many aspects of the immune response. In particular, Il-6 can inhibit apoptosis, the breaking up of DNA and subsequent destruction of infected cells. It is likely that the mimic similarly plays a role in preventing protective host cell mediated apoptosis following infection.

Viruses are not the only type of parasite capable of producing deceitful signals through molecular mimicry. *Mycobacterium* is a genus of pathogenic bacteria that resides in the macrophages of the host. The two most notable mycobacterial infections in humans are *M. tuberculosis* and *M. leprae*, which cause tuberculosis and leprosy, respectively. Defense against mycobacterial infections is cell-mediated, and involves the induction of infected macrophages to release lysosomal enzymes which can kill the bacteria inside the cell. It appears that *Mycobacterium* can cause the sustained release of Il-6 by infected macrophages (Russell et al. 1997). In addition to inhibiting apoptosis of virally-infected cells, Il-6 is also implicated in B-cell proliferation, differentiation, and immunoglobulin production. This enhancement of the humoral component of the immune response appears to come at a cost to the cell-mediated arm of immunity, preventing the release of cytokines by T cells which activate macrophages. In this case, it is possible that the *Mycobacterium* actually induces the host cell to produce a deceitful signal, rather than producing a signal of its own.

Another way in which deception occurs at a cellular level is via the active attenuation of host signals. Hasson (this volume) discusses signal amplification; on the other hand, signal attenuation or dampening can result in decreased effectiveness of the signal. We view this as deception because the parasite is altering the signal emitted by the host for a particular purpose. Some viruses may diminish the host immune signal, thereby preventing the induction of an anti-viral process. For example, the T2 protein in poxvirus mimic the receptor for tumor necrosis factor, another cytokine with numerous functions in the immune system. In an infected cell, new virus is made along with the T2 protein. Then, during the lytic phase, this new virus is released along with the T2, which binds the tumor necrosis factor and prevents the killing of virus-infected cells (Barinaga 1992).

Interestingly, a number of poxviruses, which cause acute cytolytic infections, have molecular mimics of receptors for the cytokines Il-1 and TNF, which the virus secretes in an apparent effort to manipulate host immune regulation. Vaccinia virus produce a protein, B15R, which is an Il-1 receptor type II homolog (Spriggs et al. 1992). But the protein is homologous only to the binding portion of the Il-1 receptor, and lacks the membrane anchor and cytoplasmic signaling domain. Following recognition of a virus, the host phagocytes secrete Il-1 and tumor necrosis factor to initiate the immune response. The virus appears to respond by providing its own unbound receptor, which binds the signal before it can reach the appropriate receptor and initiate an immune response.

Animal signaling also contains many examples of deception either to escape detection by predators or find prey or hosts. Perhaps the most common is classic Batesian mimicry, in which a palatable species resembles a sympatric toxic one (Alcock 1998). The frequency with which receivers (the predators) can respond to deceptive signals and still be at an advantage over their unresponsive competitors depends on the costs of responding to a deceptive signaler.

Another example involves North American fireflies in the genera *Photinus* and *Photuris* (Lloyd 1997). *Photinus* males and females flash species-specific light patterns to each other when they are ready to mate. The females of the predatory *Photuris* take advantage of this signal by mimicking the female response pattern of the other species, luring males in and then eating them (Lloyd 1997).

Consequences of signal dishonesty

Signal redundancy

Deceit and signal exploitation appear to be rampant in both intercellular signaling and between individuals, with parasites as the driving force. What are the consequences of this dishonesty for signal evolution? There are many common principles by which both immune systems and whole organisms avoid such exploitation. First, signal redundancy and the use of multiple signals are widespread among animals. Multiple ornaments occur in a variety of species, and several explanations have been offered for their occurrence, including the idea that different ornaments mean different things, that the various signals reinforce each other, or that they are used under different circumstances (Møller & Pomiankowski 1993). Multiple signals may also be a way in which animals avoid exploitation by predators or parasites; for example, in a risky environment for visual predators, animals could rely more on acoustic or chemical cues.

What is perhaps not so commonly known is that such use of multiple or redundant signals is also common within the immune system. Experiments with knock-out mice, strains of mice with particular genes disabled, have found that mice lacking genes to produce various cytokines can still exhibit an apparently normal immune response (Abbas et al. 1994). Perhaps the best example is that the cytokines Il-1 and TNF show a great deal of functional redundancy (Abbas et al. 1994). Both are involved in regulating local inflammation and in the induction of systemic responses such as fever and the production of acute phase proteins. The functions of several of the white blood cells are also overlapping, and anyone beginning a study of the immune system is impressed by how incredibly complex and multifaceted it can be.

This redundancy has some interesting implications. From a practical and clinical point of view, it means that one can still defend against pathogens even if part of the immune system is damaged, so that the various components may provide back-up for each other. From the point of view of evolutionary biologists trying to see how to use immune function or immune competence as a gauge of condition or as a focus of female choice in sexual selection, the redundancy provides some caveats. First, it makes the idea of finding a single immune system parameter that by itself signals 'overall health' or 'disease resistance' unlikely – the system simply is not built that way. Second, because the immune system is not itself a signal, we cannot realistically expect to measure one of its components as if it were, and gain the same understanding that we do when we measure, say, tail length in a bird. The tail is a signal; the T cells are not. They can be one element of a conglomerate trait, which we can use in a given context to evaluate the ability to resist disease. But they do not provide a way for us to assess health any more than trying to measure the distance between the barbules on the tail feathers would substitute for measuring the tail itself.

Rapid evolution

The second consequence of dishonest signaling driven by parasites is that it can cause both animal signals and cellular signals to undergo rapid evolutionary change. In the case of animal signals, this change as the host signal alters to avoid detection by the parasite can lead to reproductive isolation and hence speciation. The role of signal exploitation in speciation was discussed by Verrell (1991), who pointed out that arms races between signalers and natural enemies can lead to rapid divergence of populations of both. If populations of signalers are subject to different exploiters, evasion of the predator or parasitoid might be expected to generate isolation from other populations of signalers as the signal changes. Although several authors have suggested that sexual selection can drive rapid speciation in theory, or in certain groups, such as the Hawaiian *Drosophila* (Lande 1981; West-Eberhard 1983), less attention has been paid to signal exploitation as a part of the sexual selection process.

An interesting and little-considered aspect of escape from predation or parasitism by signaling animals is the possibility that such escape may include the production of novel signals, which could instigate or facilitate speciation via sexual selection on the new signal (Lande 1981; West-Eberhard 1983; Verrell 1991). If female preference for a signal is correlated with the genes for the production of the signal, as many models of sexual selection suggest, females may 'follow' males as they evade detection by natural enemies. Alternatively, if a risky signal contains elements favored by females, either because they exploit her sensory systems or because the risk constitutes a test of male fitness, such rapid isolation of populations is less likely.

Similarly, rapid evolutionary change is likely to occur at the level of the immune system, potentially reinforcing or counteracting the changes mentioned above. A recent paper by Murphy (1993) analyzed the percentage sequence divergence of a large data base of human and rodent (mouse and rat) orthologous amino acid sequences. By far the most divergent genetic material is that associated with the host defense ligands and receptors. These do not include the immunoglobulins, T cell receptors, and MHC proteins, but they do include other proteins such as the cytokines involved in intercellular communication referred to earlier. These proteins are three times more divergent than the average protein in the data set. In addition, the plasma and exocrine proteins, which appear just above the host defense sequences, contain a number of proteins known or suggested to be involved in the inflammatory response, such as complement proteins. Murphy (1993) suggests that the greater degree of sequence divergence is due, in part, to molecular mimicry by viruses, driving the divergence of host defense ligands and their receptors. Will this divergence lead to genetic differentiation among populations exposed to different pathogens?

Another form of deceptive signaling that can lead to rapid evolution is male manipulation of female reproductive effort through mechanisms such as the production of substances that alter female physiology included in the ejaculate. For example, in the grasshopper *Locusta migratoria*, males transfer a substance similar to a neurotransmitter which then promotes ovulation and/or oviposition (Lafon-Cazal et al. 1987). These substances may exact a cost from the female, and are a way in which the male manipulates her reproductive system through deception. Genes directly involved in promoting reproductive isolation might thus be subject to selection and rapid evolution as male and female interests conflict.

Conclusions

Finally, both redundancy and the idea of rapid evolutionary change suggest that equilibrium states are unlikely in signal evolution. Because of continued arms races between hosts and pathogens, signals are unlikely to settle at a single point. We do not suggest that selection results in the evolution of very costly signals that cannot be copied by the parasite, a view that Zahavi (1975) might champion. Instead, we suggest a never-ending arms race where the signals and their cellular receptors are always changing in an effort to escape the pathogens.

Acknowledgements

M.Z. is supported by grants from the National Science Foundation (US).

References

Abbas, A. K., Lichtman, A. H. & Pober, J. S. 1994. *Cellular and Molecular Immunology*. Philadelphia: Saunders

Ahuja, A. K., Gao, J. & Murphy, P. M. 1994. Chemokine receptors and molecular mimicry. *Immunology Today*, **15**, 281-287.

Alcock, J. 1998. *Animal Behavior: an Evolutionary Approach*. Sunderland, MA: Sinauer.

Barinaga, M. 1992. Viruses launch their own 'Star Wars'. *Science*, **258**, 1730-1731.

Burk, T. 1982. Evolutionary significance of predation on sexually signalling males. *Florida Entomologist*, **65**, 90-104.

Cade, W. 1975. Acoustically orienting parasitoids: fly phonotaxis to cricket song. *Science*, **190**, 1312-1313.

Cardé, R. T. & Baker, T. C. 1984. Sexual communication with pheromones. In: *Chemical Ecology of Insects* (Ed. by W. J. Bell & R. T. Cardé), pp. 355-383. New York: Chapman and Hall.

Crowe, S. & Mills, J. 1994. Virus infections of the immune system. In: *Basic and Clinical Immunology* (Ed. by D. P. Stites, A. I. Terr & T. G. Parslow), pp. 689-705. Norwalk, CT: Appleton and Lange.

Dicke, M. & Sabelis, M. W. 1992. Costs and benefits of chemical information conveyance: proximate and ultimate factors. In: *Insect Chemical Ecology* (Ed. by B. D. Roitberg & M. B. Isman), pp 122-156. New York: Chapman and Hall.

Endler, J. A. 1992. Signals, signal conditions, and the direction of evolution. *American Naturalist*, **139** (supplement), s125-s153.

Greenfield, M. D. 1981. Moth sex pheromones: an evolutionary perspective. *Florida Entomologist*, **64**, 4-17.

Gwynne, D. T. & Morris, G. K. (Eds) 1983. *Orthopteran Mating Systems: Sexual Competition in a Diverse Group of Insects*. Boulder, CO: Westview.

Höglund, J. & Alatalo, R. V. 1995. *Leks*. Princeton: Princeton University Press.

Howard, R. D. 1988. Reproductive success in two species of anurans. In: *Reproductive Success* (Ed. by T. H. Clutton-Brock), pp. 99-117. Chicago: University of Chicago Press.

Krebs, J. R. & Dawkins, R. 1978. Animal signals: information or manipulation? In: *Behavioural Ecology: an Evolutionary Approach* (Ed. by J. R. Krebs & N. B. Davies), pp. 282-309. Oxford: Blackwell.

Lande, R. 1981. Models of speciation by sexual selection on polygenic traits. *Proceedings of the National Academy of Sciences, USA,* **78**, 3721-3725.

Lloyd, J. E. 1997. Firefly mating ecology, selection and evolution. In: *The Evolution of Mating Systems in Insects and Arachnids* (Ed. by J. C. Choe & B. J. Crespi), pp. 184-192. Cambridge: Cambridge University Press.

Lafon-Cazal, M., Gallois, D., Lehouelleur, J. & Bochaert, J. 1987. Stimulatory effects of male accessory-gland extracts on the myogenicity and the adenylate cyclase activity of the oviduct of *Locusta migratoria. Journal of Insect Physiology,* **33**, 909-915.

Maynard Smith, J. 1958. Sexual selection. In: *A Century of Darwin* (Ed. by S. A. Barnett), pp 231-244. London: Heinemann.

Maynard Smith, J. 1991. Theories of sexual selection. *Trends in Ecology and Evolution,* **6**, 146-151.

McKean, K. A. & Zuk, M. 1995. An evolutionary perspective on signaling in behavior and immunology. *Naturwissenschaften,* **82**, 509-516.

Møller, A. P. 1994. *Sexual Selection and the Barn Swallow*. Oxford: Oxford University Press

Møller, A. P. & Pomiankowski, A. 1993. Why have birds got multiple sexual ornaments? *Behavioral Ecology and Sociobiology*, **3**, 167-176.

Moore, P. S., Bochoff, C., Weiss, R. A. & Chang, Y. 1996. Molecular mimicry of human cytokine and cytokine response pathway genes by KSHV. *Science*, **274**, 1739-1744.

Murphy, P. M. 1993. Molecular mimicry and the generation of host defense protein diversity. *Cell*, **72**, 823-826.

Otte, D. 1974. Effects and functions in the evolution of signaling systems. *Annual Review of Ecology and Systematics,* **5**, 385-417.

Phelan, P. L. 1997. Evolution of mate-signaling in moths: phylogenetic considerations and predictions from the asymmetric tracking hypothesis. In: *The Evolution of Mating Systems in Insects and Arachnids* (Ed. by J. C. Choe & B. J. Crespi), pp. 240-256. Cambridge: Cambridge University Press.

Prestwich, K. N., Brugger, K. E. & Topping, M. 1989. Energy and communication in three species of Hylid frogs: power input, power output and efficiency. *Journal of Experimental Biology,* **144**, 53-80.

Prestwich, K. N. & Walker, T. J. 1981. Energetics of singing crickets: effects of temperature in three trilling species (Orthoptera: Gryllidae). *Journal of Comparative Physiology, Series A,* **143**, 199-212.

Russell, D. G., Sturgill-Koszycki, S., Vanheyningen, T., Collins, H. & Schaible, U. E. 1997. Why intracellular parasitism need not be a degrading experience for *Mycobacterium. Philosophical Transactions of the Royal Society of London, Series B,* **352**, 1303-1310.

Sakaluk, S. K. 1990. Sexual selection and predation: balancing reproductive and survival needs. In: *Insect Defenses* (Ed. by D. L. Evans & J. O. Schmidt), pp. 63-90. Albany: State University of New York.

Schmidt, G. R. & Roberts, L. S. 1989. *Foundations of Parasitology*. 4th edn. St. Louis: Times Mirror/Mosby.

Spieth, H. T. 1974. Mating behavior and evolution of the Hawaiian *Drosophila*. In: *Genetic Mechanisms of Speciation in Insects* (Ed. by M. J. D. White), pp. 94-101. Dordrecht, Holland: Reidel.

Spriggs, M. K., Hruby, D. E., Maliszewski, C. R., Pickup, D. J., Buller, J. E. & van Slyke, J. 1992. Vaccinia and cowpox viruses encode a novel secreted interleukin-1-binding protein. *Cell*, **71**, 145-152.

Sullivan, B. K. 1989. Passive and active female choice: a comment. *Animal Behaviour*, **37**, 692-694.

Svensson, M. 1996. Sexual selection in moths: the role of chemical communication. *Biological Reviews,* **71**, 113-135.

Thornhill, R. & Alcock, J. 1983. *The Evolution of Insect Mating Systems*. Cambridge, MA: Harvard University Press.

Verrell, P. A. 1991. Illegitimate exploitation of sexual signalling systems and the origin of species. *Ethology, Ecology and Evolution*, **3**, 273-283.

West-Eberhard, M. J. 1983. Sexual selection, social competition, and speciation. *Quarterly Review of Biology*, **58**, 155-183.

Wiley, R. H. 1994. Errors, exaggeration, and deception in animal communication. In: *Behavioral Mechanisms in Evolutionary Ecology* (Ed. by L. A. Real), pp. 157-189. Chicago: University of Chicago Press.

Williams, G. C. 1975. *Sex and Evolution*. Princeton, NJ: Princeton University Press.

Zahavi, A. 1975. Mate selection - a selection for a handicap. *Journal of Theoretical Biology,* **53,** 205-214.

Zuk, M. & Kolluru, G. R. 1998. Exploitation of sexual signals by predators and parasitoids. *Quarterly Review of Biology*, **73**, 415-438.

Zuk, M., Simmons, L. W., & Cupp, L. 1993. Song characteristics of parasitized and unparasitized populations of the field cricket *Teleogryllus oceanicus*. *Behavioral Ecology and Sociobiology*, **33**, 339-343.

Zuk, M., Simmons, L. W., & Rotenberry, J. T. 1995. Acoustically-orienting parasitoids in calling and silent males of the field cricket *Teleogryllus oceanicus*. *Ecological Entomology*, **20**, 380-383.

IN: ESPMARK,Y., AMUNDSEN,T. & ROSENQVIST,G. (EDS.) 2000
ANIMAL SIGNALS:
SIGNALLING AND SIGNAL DESIGN IN ANIMAL COMMUNICATION, PP. 421-436
TAPIR ACADEMIC PRESS, TRONDHEIM, NORWAY

22

MHC Genotype and Ornamentation

MATS GRAHN

Department of Biosciences, Södertörns University College, Karolinska Institute,
S-141 57 Huddinge, Sweden

E-mail: mats.grahn@cnt.ki.se

Abstract

Differences among males in immunocompetence can affect resistance to parasites and diseases, and sexual ornaments have been suggested to function as indicators of additive genetic variation in disease resistance, health and condition in the good gene models of sexual selection. In this paper the role of genetic variation in the immune system is discussed as an alternative route to explore whether or not immunological variation influences condition and sexual ornamentation. In vertebrates, the effect of the genes located in the major histocompatibility complex (MHC) in shaping immunodominance, lasting immunity, and the risk of developing pathological immune responses are suggested as the link between genetic variation at the MHC and male ornaments. It is also suggested that general immune responsiveness may be optimized at an individual basis depending on the MHC associated risk of autoimmune disorders.

The expression of ornaments is highly variable among males within a species and can also differ both between and within seasons for individual males. The variation in ornaments is often correlated with condition and survival (canine teeth in primates: Manning & Chamberlain 1993; carotenoid pigmentation in fish and birds: Hill 1990, 1991; Milinski & Bakker 1990; Niccoletto 1993; combs in red jungle fowl *Gallus gallus*: Zuk et al. 1990; see also a general review by Andersson 1994). Theoretical models have suggested that the intensity of male ornaments and displays are used by the females in many animal species through behaviours and other adaptations aimed at picking out high quality males to sire their young (Zahavi 1975, 1977; Andersson 1982, 1986, 1994; Hamilton & Zuk 1982; Kodric-Brown & Brown 1984). By basing mate choice on the information displayed by the males, a female can increase the quality, viability, and future reproductive success of her offspring even in the absence of non-genetic direct benefits (Pomiankowski & Møller 1995; see also Kirkpatrick & Ryan 1991). Studies of animals, from vertebrates to insects, in the lab and the wild, support the concept of honest indicators of good genes by showing that females indeed can increase offspring fitness by mating with more ornamented males (Norris 1993; Møller 1994; von Schantz et al. 1994; Hasselquist et al. 1996; Hoikkala et al. 1998; Welch et al. 1998).

In search of the link between ornamentation and condition of males, controlled infection experiments have shown that ornaments are sensitive to diseases (Houde & Torio 1992), even more so than other morphological traits (Hillgarth 1990; Zuk et al. 1990). Several different methods have been used to evaluate an individual's capacity to control infections. Antibody response to novel or familiar antigens, levels of serum globulin, size of immune organs and lymphocyte counts have all been used to estimate variation in immunocompetence. From an immunological perspective it is not always evident that these assays accurately reflect individual variation in immunocompetence and how these estimates relates to variation in fitness (Sheldon & Verhulst 1996; Penn & Potts 1998). The emerging picture is less than clear: more ornamented males have been interpreted to show weaker responses to pathogens, as measured by counts of white blood cells (Zuk et al. 1995; Skarstein & Folstad 1996) but stronger immune reaction to novel antigens when assayed with delayed hypersensitivity tests (Zuk & Johnsen 1998) or total immunoglobulin response (Saino & Møller 1996).

A major problem when exploring the connection between variation in disease resistance and ornamentation comes from the observation that immune function and disease resistance is itself dependent on condition. Individuals with a poor immune system will be in poor condition but it is equally true that individuals in poor condition will have a weakened immune system (Lord et al. 1998). Hence, the causality works in both directions and firm evidence that link disease resistance and ornamentation, independently of an underlying correlation to general condition, can be difficult to establish by exploring the relationship between quantitative variables of the immune system and ornamentation (Fig. 1).

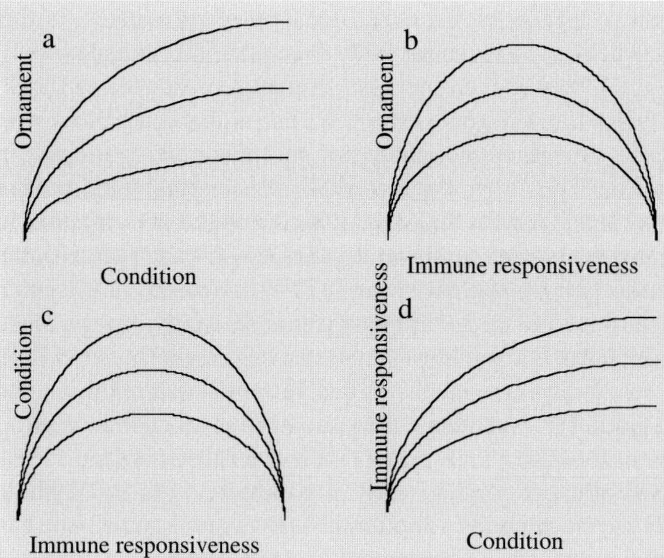

***Figure 1.** Hypothetical relationships underlying an observed correlation between male ornamentation and condition (three lines shown for each relationship indicating individual variation in reaction norms). (**a**) Ornaments are supposed to increase with condition, but only up to a given level above which increased condition will not lead to a further increase. (**b**) Increased immune responsiveness gives bigger ornaments, but an overly aggressive immune system can lead to immunopathological diseases. (**c**) Accordingly, condition will show a similar relationship with immune responsiveness. (**d**) A correlation between immune responsiveness and ornamentation as in (b) may however be confounded by an underlying relationship between condition and immune function.*

Genetic variation in the immune system

An alternative approach to explore how differences between males in resistance to diseases may affect ornamentation, that circumvents much of the complexity of the immune system, is to directly study if genetic variation in disease resistance and susceptibility correlate with ornamentation. This approach depends on the straightforward assumption that differences in condition may depend on the genes of the immune system but not the other way around. Both acquired (adaptive) and preimmune (innate) host resistance shows extensive genetic variation in mice (Malo & Skamene 1994) and humans (Hill 1998). Disease resistance is in most cases a multigenic quantitative trait and most species are challenged with a multitude of different pathogens. Variation among males in genetic resistance to parasites and diseases will influence the likelihood of developing a disease when exposed to pathogens and thus also affect the expression of condition-dependent ornaments.

At the level of populations, much of the genetic variation in the immune system of vertebrates is located in the major histocompatibility complex (MHC). The MHC region contains several linked genes of central importance to antigen processing and immune function (Malo & Skamene 1994; Hill 1998). The genomic organization and function of MHC has been revealed in both humans and other vertebrates (Trowsdale 1993, 1995). These genes are often inherited together on one chromosome as one haplotype and are critical for the function and control of the innate and acquired immune functions of vertebrates. In this genetic region there are two kinds of slightly different MHC receptors, class I and class II, which present peptides from normal host metabolism and invading pathogens to the T-cells. In most organisms MHC class I and II genes exists in several duplicated loci with different but overlapping degrees of function and allelic polymorphism. Also situated in the MHC class II region are two genes for transporters associated with antigen processing, TAP1 and TAP2, that interact with the MHC class I antigen presentation. In the MHC class III region several complement component genes C2, C4 and factor B, the heat shock protein HSP70 and the tumour necrosis factor (TNF-A and -B), are located and code for proteins with critical functions in the immune response. Many, maybe most, of the genes in the MHC region seem not to show an extensive polymorphism within populations and probably do not contribute much to the variation in condition and ornamentation in nature. The remarkable exceptions are some of the loci encoding MHC class I and class II and possibly the TAP and TNF genes (Trowsdale 1993; Poland 1998), but extensive population data for other genes than MHC I and MHC II is still lacking even in humans and mice. However, by studying variation at MHC class I and II, one can get a handle also on the linked immunogenetic variation in the MHC region.

Cellular function of the MHC class I and II genes

MHC class I receptors are expressed on all nucleated cells and present intracellular peptides to cytotoxic T-cells (CTL). Hence, MHC class I is mainly involved in protection against viruses and intracellular parasites, and mark infected cells for destruction by the immune system. Class II MHC is expressed on specialized phagocytic white blood cells (e.g. macrophages), specialized antigen presenting cells (e.g. dendritic cells), but also on B-cells, and presents peptides of extracellular origin to regulatory T-helper (Th) cells. MHC class II is thus involved in the activation of the major regulatory cells in the immune system. Activated Th-cells (T-cells whose receptors have attached to an antigen-presenting MHC molecule) release immune regulating hormones (cytokines) that attract other immune cells to start an inflammatory response (Roitt et al. 1996). Th-cells also interact with B-cells (Clark & Ledbetter 1994) to induce the clonal selection and proliferation of those particular B-cells that produce antibodies with high affinity for the presented antigen (Rajewsky 1996). Unlike most of the variability of the T- and B-cell receptors, which is generated anew at the individual level by genetic rearrangements and other mutations during somatic development, the variation of MHC is inherited from the parents' gametes and is dependent on the variation in the local population (Roitt et al. 1996). The MHC class II receptor is a dimeric protein formed by two subunits, a and b. The antigen binding part of the b chain is coded by the second exon of the MHC class II b gene (Roitt et al. 1996). MHC class II b genes have been found to be highly polymorphic in the majority of species studied, with most of the variation located in the second exon. In mice, up to 20 different alleles have been found in local populations in the wild (Potts & Wakeland 1990).

The MHC phenotype

A useful description of the phenotypic effect of MHC could be that the variable MHC class I and II genes to a large extent shapes the adaptive immune system and sets the boundary conditions for its function. While the target antigens of innate immunity do not change qualitatively during the lifetime of the individual they can do so for adaptive immunity. Hence, the adaptive immune system can respond to, and somewhat keep track of, the rapid evolutionary change of many pathogens, both within a host and between hosts in a population. However, with this flexibility comes the associated risk of developing immunity to self, autoimmunity. Most host cells express several thousands of different genes. Peptide fragments from this pool creates blind spots in the immune defense and an opportunity for molecular mimicry and detection evasion strategies of pathogens. The main problem to be solved by the adaptive immune system is how to detect a set of peptides from invading pathogens against a much larger background set of self generated peptides (Fig. 2). In normal cases the adaptive immune response is blocked by self peptides presented by MHC by the selective destruction of self responsive T-cells in the thymus (Roitt et al. 1996). Hence, peptides from pathogens that readily bind to a MHC molecule can not elicit an immune response since there are no responding T-cells (Jones et al. 1993; Gelder et al. 1998). Each variant of the MHC receptor is able to bind only a limited

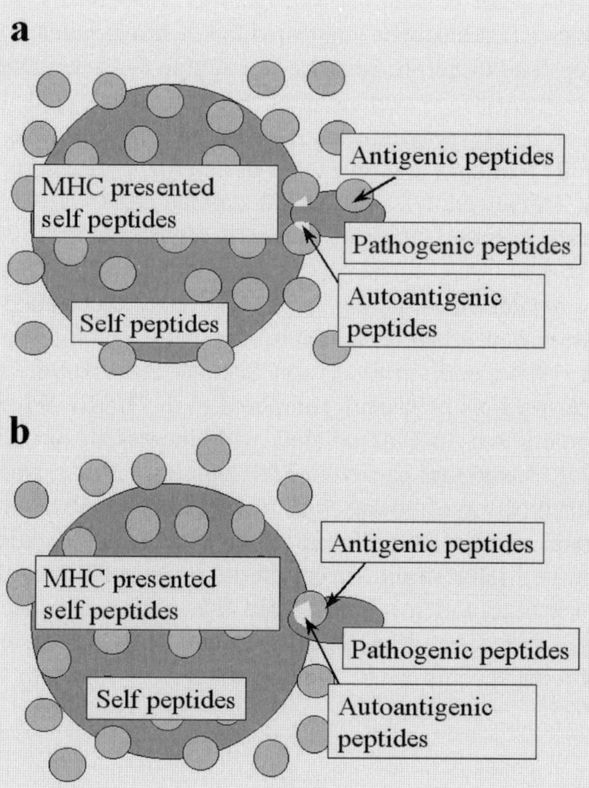

Figure 2. *The Swiss cheese model of the MHC effect on the adaptive immune system. Different MHC genotypes (**a**) and (**b**) will differ in their presentation of self peptides (small circles overlapping with the big circle), creating a specific pattern of 'blind spots' in the adaptive immune system. The MHC genotypes will also differ in their presentation of peptides derived from an invading pathogen (small circles overlapping the oval). Some of the peptides from the pathogen will be similar to self derived peptides and may be presented by the MHC (white sections). The area of this overlap will differ between MHC genotypes and this cross reactivity may elicit an autoimmune response with different probability depending on pathogen, host genetic background, and host MHC*

number of different peptides, and different MHC alleles thus allow the presentation of a different set of pathogenic antigens and self peptides, creating a different pattern of presentation and blind spots (Fig. 2). From this simple model follows some important observations. Both MHC dependent resistance to infection and susceptibility to autoimmunity are codominant traits. Any particular MHC allele can provide protection or disease depending on the genetic background variation in the host or in the pathogen.

MHC and fitness

Several studies have found an association between specific MHC haplotypes and resistance to infectious diseases and autoimmune disorders in mammals and birds (Briles et al. 1983; Tiwari & Teraski 1985; Hill et al. 1991; Nepom & Erlich 1991; Han et al. 1992; McGuire et al. 1994; Apanius et al. 1997; Paterson et al. 1998). Therefore, it has been suggested that the high degree of within-population variability of MHC genes is the result of selection from pathogens. From the pathogens' point of view, different hosts are different 'food patches', and even if a pathogen evolves the ability to avoid adverse immune responses in hosts of one specific MHC genotype it may still be less effective to infect other hosts in a population with high MHC variation (Zinkernagel 1996; see also Hamilton et al. 1990). In this system, uncommon MHC genotypes will be at an advantage compared to common ones, leading to frequency-dependent selection favoring rare and disfavoring common MHC alleles (Clarke & Kirby 1966; Hedrick 1994; Potts & Slev 1995; Apanius et al. 1997). Furthermore, it has been suggested that heterozygotes may be able to detect a larger set of antigens and therefore be more resistant to diseases (Doherty & Zinkernagel 1975; Black & Salzano 1981; Wakeland et al. 1989; Kroemer et al. 1990; Nevo & Beiles 1992). Studies of MHC polymorphism in some mammal and bird populations have found a high allelic variation and an excess of heterozygotes, compared to Hardy-Weinberg expectations (reviewed by Potts & Wakeland 1990, 1993; von Schantz et al. 1996; but see Edwards & Potts 1996; Ellegren et al. 1996).

In wild male pheasants *Phasianus colchicus*, MHC genotype was found to correlate both with spur length, which is a condition-dependent ornament in this species (von Schantz et al. 1989; Grahn and von Schantz 1994), and survival (von Schantz et al. 1996, 1997). There was an excess of MHC heterozygous pheasants compared to the Hardy-Wienberg expectation but MHC heterozygosity was not associated with increased survival or ornamentation. Paterson et al. (1998) found that particular MHC alleles, rather than MHC heterozygosity, correlated with survivorship of young Soay sheep (*Ovis aries*). Both von Schantz et al. (1997) and Paterson et al. (1998) found some evidence indicative of frequency dependent selection on the MHC. In the sheep, the two most common alleles were associated with decreased survival and in the pheasant the relationship between MHC genotype and spur length differed between years. Survival to very old age in humans are positively associated with specific MHC class II alleles but not with MHC heterozygosity (Ivanova et al. 1998).

MHC and immunocompetence

A basic definition of immunocompetence often given in medical literature is the strength of the response to a random antigen. Although this definition may be useful when studying immunology as a separate subject, immunocompetence needs a more detailed definition to be helpful in ecological and evolutionary models. The definition could be recast in more detailed immunological functions:

- Rapid detection of pathogens.
- Activation of relevant effector cells.
- Avoidance of immunopathology.
- Response to evolutionary constrained epitopes.

Infection with complex pathogens elicits T-cell responses to only a small subset of all possible peptide antigens from any given pathogen (the so called epitopes), an effect called immunodominance. The major effect of MHC on immunocompetence is probably a function of how MHCs affect immunodominance and the establishment of lasting immunity. In acute stages of viral infections more than 25% and up to 50% of all activated CTLs are specific to as few as three major epitopes (McMichael & O'Callaghan 1998). Immunodominance is a puzzling phenomena. Large numbers of different peptides ($>10^3$) can be presented by class I molecules on cells, largely derived from self-proteins. However, dominant epitopes do not always bind to MHC molecules with high affinity or come from the most abundant pathogenic proteins. Yet immunodominance is not a chance event, because most individuals with the same MHC molecules respond to the same epitope peptides, regardless of genetic background (McMichael & Phillips 1997; McMichael & O'Callaghan 1998). From studies of the specificity of the response after vaccination with whole pathogens or complex antigens it seems clear that different MHC class I (Goddeeris et al. 1990; Taracha et al. 1995) and MHC class II (Hayney et al. 1996; Gelder et al. 1998; Poland 1998) genotypes differ in their pattern of immunodominance and that this affect both the repertoire of CTLs, antibody response and establishment of immunological memory. Furthermore, T-helper cells can be categorized into two functional subsets, Th1 and Th2 cells, which produce different regulatory signals, lymphokines. In general, Th1 cells mediate cellular immune responses and Th2 cells mediate humoral immunity. MHC seems to influence the activation of different effector cells, via its effect on the balance between Th1/Th2 types of immune response (Milich et al. 1995; Schountz et al. 1996; Murray 1998). In humans, the pattern of Th1 or Th2 activation may affect the risk for immunopathology after infection of pathogens (De Carli et al. 1994). These and other findings have led to an increasing recognition that genetic factors, such as the genes in the MHC, and their interaction with host and parasite genetic variation, and environmental factors have a significant role in disease susceptibility, resistance and progression (Poland 1998).

The presentation by MHC of evolutionary conserved epitopes to CTLs and Th-cells may be an adaptation in hosts to constrain the adaptation of the pathogens by lowering the fitness of escape mutants (McMichael & Phillips 1997; Gelder et al. 1998; da Silva & Hughes 1998). Similarly, the preferential presentation by MHC of evolutionary constrained self epitopes (Hughes & Hughes 1995) can lower the risk for autoimmunity while maintaining the tolerance for evolutionary change in host background genes. Studies of heat shock proteins (HSP) provide a clear example of the interaction between host background genes, pathogens and the MHC (Lamb et al. 1989; Jones et al. 1993). Immune responses to microbial HSP could lead to autoimmune responses because of cross-reactivity with host HSP. The risk of cross-reactivity may depend on an individual's MHC class II genotype (Lamb et al. 1989; Jones et al. 1993). As infection or stress is liable to increase the risk of autoimmune responses, individuals with low-risk MHC genotypes could maintain a higher immune responsiveness at a given point than those carrying a high-risk genotype (see also Råberg et al. 1998; Westneat & Birkhead 1998). Thus, there is a trade-off between the benefit of resistance to parasites and the cost of immunopathology that will differ between MHC genotypes. Immune responsiveness may be optimized at an individual basis with respect to this trade-off. A simple graphical cost-benefit model (Behnke et al. 1992; Råberg et al. 1998) can illustrate this trade off (Fig. 3).

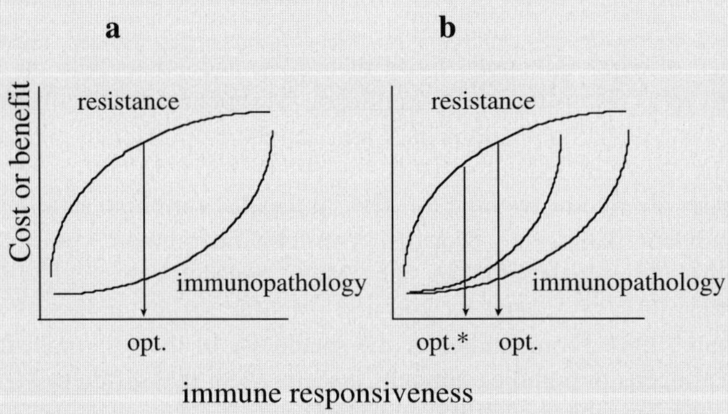

Figure 3. *(**a**) Hypothetical cost (immunopathology) and benefit (resistance to parasites) curves as functions of immune responsiveness, and the optimal solution to this trade-off (opt.). (**b**) The cost function (immunopathology) varies depending on the MHC genotype of the organism. During infection there is a change in antigen repertoire that stimulates the immune system and could cause immunopathological responses. Consequently, the cost curve will be steeper and optimal immune responsiveness (opt.*) will be reduced. High risk genotype and low risk genotype refer to difference in factors predisposing individuals to developing autoimmunity (redrawn from Råberg et al. 1998).*

MHC variation, disease resistance and ornaments

It is likely that differences in nutritional status, exposure to pathogens and other environmental factors directly affect males' investments in ornaments and immune system in a condition dependent manner. Low nutritional status caused by, for instance, inferior competitive ability, may leave some males less to allocate to either immune function or reproductive effort, or both (e.g. Sheldon & Verhulst 1996; Westneat & Brikhead 1998). Males may even have the ability to grow larger ornaments by transferring resources, such as calories, proteins or carotenoids, from investment in the immune system to investment in ornaments (Folstad & Karter 1992; Lozano 1994). The influence of MHC genes on male ornamentation, suggested here, is not an alternative explanation to the condition dependence of male ornaments. It is an attempt to explain how genetic variation in the immune system can translate into condition dependent information revealed in male ornaments.

Genetic variation in the immune system may affect male sexual signals if development of ornaments compete with pathogens or the immune system for shared resources, for instance energy or nutrients, and these resources are in short supply. Specifically, MHC genes may affect the strength and effectiveness of the immune response and establishment of lasting immunity that may influence how often a male contract a disease and how well he will recover.

The supply of energy and nutrients need not be the only connection between MHC and ornaments. The effect of oxidative stress induced by free radicals on ornamentation has recently been suggested to provide a reliable measure of the genotype-environment interactions of the immune system (Grahn et al. 1998; von Schantz et al. 1999). The cytotoxic effects of free radicals are used by phagocytes and other cells in innate immunity when killing pathogens in the inflammatory response (Klebanoff & Clark 1978). The MHC-mediated recognition is essential for the adaptive immune system to mount a specific and more effective response (shorter time lag, higher antibody titre, and higher antibody affinity) on subsequent infections. Some of the activated B-cells develop into long-lived memory B-cells which, upon a secondary infection of the same pathogen, present antigen very efficiently to T-cells (Rajewsky 1996). Hence, individuals whose battery of MHC molecules fail to bind efficiently to the peptide fragments from the pathogen will produce a less efficient, low-affinity, antibody-mediated response (Roitt et al. 1996). A less specific defense, or a misdirected immune response leading to an autoimmune disease, may be costly in that it generates prolonged periods of sickness and extensive oxidative stress.

Genetic resistance may not always result in 'parasite-free' individuals, but can be manifested as a greater tolerance to common parasites (Zinkernagel 1996), and mate choice based on condition-dependent ornaments may promote matings with parasite tolerant partners (Skarstein & Folstad 1996). The double edged nature of the immune system have prompted a broader analysis considering both benefits and costs associated with the immune system and the relation to fitness traits, discussing not only the energetic and nutritional constraints on the immune system, but also the balance between immune responsiveness and immunopathology (Råberg et al. 1998; Westneat & Birkhead 1998). A strong immune response does not always imply higher fitness. For instance, several diseases are immunopathological where a misdirected immune response causes more harm than good to the organism (Penn & Potts 1998; Råberg et al. 1998; Westneat & Birkhead 1998). It is conceivable that phenotypic condition-dependent adjustments in male ornamentation are made at a lower cost to the male if his immune system is operating smoothly and at a lower risk for misfired immune responses (Fig. 3). This provides a direct link between mate choice based on condition-dependent ornaments and the population's ability to track co-evolving diseases (Grahn et al. 1998).

In conclusion, it is likely that MHC genotype can affect male ornaments through its effects on the immune system. Beside phenotypic condition-dependent adjustments, part of the variation among males in immunity may depend on the effects of the MHC on immunodominance, the balance between Th1/Th2 type response, and on the risk of developing autoimmune disorders or pathological responses to otherwise harmless pathogens. The MHC genes may constitute an important justification for the good genes implied in the Hamilton and Zuk (1982) model for the evolution of female choice and male condition-dependent ornamentation. The operation of female choice can, via mating preferences for males possessing large ornaments, select for MHC alleles that are both less sensitive to autoimmune disorders and more protective against common pathogens.

Acknowledegments

I thank D. Hasselquist, L. Råberg and two anonymous referees for valuable discussions and comments on earlier drafts. This work was supported by the Swedish Research Council for Forestry and Agriculture (SJFR).

References

Andersson, M. 1982. Sexual selection, natural selection, and quality advertisement. *Biological Journal of the Linnean Society*, **17**, 375-393.

Andersson, M. 1986. Evolution of condition-dependent sex ornaments and mating preferences: sexual selection based on viability differences. *Evolution*, **40**, 804-816.

Andersson, M. 1994. *Sexual Selection.* Princeton, NJ: Princeton University Press.

Apanius, V., Penn, D., Slev, P. R., Ruff, L. R. & Potts, W. K. 1997. The nature of selection on the major histocompatibility complex. *Critical Reviews in Immunology*, **17**, 179-224.

Behnke, J. M., Barnard, C. J. & Wakelin, D. 1992. Understanding chronic nematode infections: evolutionary considerations, current hypotheses and the way forward. *International Journal for Parasitology*, **22**, 861-907.

Black, F. L. & Salzano, F. M. 1981. Evidence for heterosis in the HLA system. *American Journal of Human Genetics*, **33**, 894-899.

Briles, W. E., Briles, R. W., Taffs, R. E. & Stone, H. A. 1983. Resistance to a malignant lymphoma in chickens is mapped to subregion of major histocompatibility (B) complex. *Science*, **219**, 977-979.

Clarke, B. & Kirby, D. R. S. 1966. Maintenance of histocompatibility polymorphisms. *Nature*, **211**, 999-1000.

Clark, E. A. & Ledbetter, J. A. 1994. How B and T cells talk to each other. *Nature*, **367**, 425-428.

De Carli, M., D'Elios, M. M., Zancuoghi, G., Romagnani, S. & Del Prete, G. 1994. Human Th1 and Th2 cells: functional properties, regulation of development and role in autoimmunity. *Autoimmunity*, **18**, 301-308.

Doherty, P. C. & Zinkernagel, R. M. 1975. Enhanced immunological surveillance in mice heterozygous at the H-2 gene complex. *Nature*, **256**, 50-52.

Edwards, S. V. & Potts, W. K. 1996. Polymorphism of genes in the major histocompatibility complex: implications for conservation genetics of vertebrates. In: *Molecular Genetic Approaches in Conservation*. (Ed. by T.B. Smith & R. K. Wayne), pp. 214-237. Oxford: Oxford University Press.

Ellegren, H., Mikko, S., Wallin, K. & Andersson, L. 1996. Limited polymorphism at major histocompatibility complex (MHC) loci in the Swedish moose *A. alces*. *Molecular Ecology*, **5**, 3-9.

Folstad, I. & Karter, A. J. 1992. Parasites, bright males and the immunocompetence handicap. *American Naturalist*, **139**, 603-622.

Gelder, C., Davenport, M., Barnardo, M., Bourne, T., Lamb, J., Askonas, B., Hill, A. & Welsh, K. 1998. Six unrelated HLA-DR-matched adults recognize identical CD4+ T cell epitopes from influenza A haemagglutanin that are not simply peptides with high HLA-DR binding affinities. *International Immunology*, **10**, 211-222.

Goddeeris, B. M., Morrison, W. I., Toye, P. G. & Bishop, R. 1990. Strain specificity of bovine *Theileria parva*-specific cytotoxic T cells is determined by the phenotype of the restricting class I MHC. *Immunology*, **69**, 39-44.

Grahn, M., Langefors, Å. & von Schantz, T. 1998. The importance of mate choice in improving viability in captive populations. In: *Behavioral Ecology and Conservation Biology* (Ed. by T. Caro), pp. 341-363. Oxford: Oxford University Press.

Grahn, M. & von Schantz, T. 1994. Fashion and age in pheasants: age differences in mate choice. *Proceedings of the Royal Society of London, Series B*, **255**, 237-241.

Hamilton, W. D., Axelrod, R. & Tanese, R. 1990. Sexual reproduction as an adaptation to resist parasites: a review. *Proceedings of the National Academy of Sciences, USA*, **87**, 3566-3573.

Hamilton, W. D. & Zuk, M. 1982. Heritable true fitness and bright birds: a role for parasites? *Science*, **218**, 384-387.

Hayney, M. S., Poland, G. A., Jacobson, R. M., Schaid, D. J. & Lipsky, J. J. 1996. The influence of the HLA-DRB1*13 allele on measles vaccine response. *Journal of Investigative Medicine*, **44**, 261-263.

Han, R., Breitburd, F., Marche, P. N. & Orth, G. 1992. Linkage regression and malignant conversion of rabbit viral papillomas to MHC class II genes. *Nature*, **356**, 66-68.

Hasselquist, D., Bensch, S. & von Schantz, T. 1996. Correlation between male song repertoire, extra-pair paternity and offspring survival in the great reed warbler. *Nature*, **381**, 229-232.

Hedrick, P. W. 1994. Evolutionary genetics of the major histocompatibility complex. *American Naturalist*, **143**, 945-964.

Hill, A. V. S. 1998. The immunogenetics of human infectious disease. *Annual Review of Immunology*, **16**, 593-617.

Hill, A. V. S., Allsopp, C. E. M., Kwiatkowski, D., Anstey, N. M., Twumasi, P., Rowe, P. A., Bennett, S., Brewster, D., McMichael, A. J. & Greenwood, B. M. 1991. Common West African HLA antigens are associated with protection from severe malaria. *Nature*, **352**, 595-600.

Hill, G. E. 1990. Female house finches prefer colorful males: sexual selection for a condition-dependent trait. *Animal Behaviour*, **40**, 563-572.

Hill, G. E. 1991. Plumage coloration is a sexually selected indicator of male quality. *Nature*, **350**, 337-339.

Hillgarth, N. 1990. Parasites and female choice in the ring-necked pheasant. *American Zoologist*, **30**, 227-233.

Hoikkala, A., Aspi, J. & Suvanto, L. 1998. Male courtship song frequency as an indicator of male genetic quality in an insect species, *Drosophila montana. Proceedings of the Royal Society of London, Series B*, **265**, 503-508.

Houde, A. E. & Torio, A. J. 1992. Effect of parasitic infection on male color pattern and female choice in guppies. *Behavioral Ecology*, **3**, 346-351.

Hughes, A. L. & Hughes, M. K. 1995. Self peptides bound by HLA class I molecules are derived from highly conserved regions of a set of evolutionary conserved proteins. *Immunogenetics*, **41**, 257-262.

Ivanova, R., Henon, N., Lepage, V., Charron, D., Vicaut, E. & Schachter, F. 1998. HLA-DR alleles display sex-dependent effects on survival and discriminate between individual and familial longevity. *Human Molecular Genetics*, **7**, 187-194.

Jones, D. B., Coulson, A. F. & Duff, G. W. 1993. Sequence homologies between hsp60 and autoantigens. *Immunology Today*, **14**, 115-118.

Kirkpatrick, M. & Ryan, M. J. 1991. The evolution of mating preferences and the paradox of the lek. *Nature*, **350**, 33-38.

Klebanoff, S .J. & Clark, R.A. 1978 *The Neutrophil: Function and Clinical Disorders*. Amsterdam: North-Holland.

Kodric-Brown, A. & Brown, J. H. 1984. Truth in advertising: the kinds of traits favored by sexual selection. *American Naturalist*, **124**, 309-323.

Kroemer, G., Bernot, A., Béhar, G., Chaussé, A-M., Gastinel, L-N., Guillemot, F., Park, I., Thoraval, P., Zoorob, R. & Auffray, C. 1990. Molecular genetics of the chicken MHC: current status and evolutionary aspects. *Immunological Reviews*, **113**, 119-145.

Lamb, J.R., Bal, V., Rothbard, J.B., Mehlert, A., Mendez-Samperio, P. & Young, D. B. 1989. The mycobacterial stress protein: a common target of T-cell recognition in infection and autoimmunity. *Journal of Autoimmunology*, **2** (supplement), 93-100.

Lord, G. M., Matarese, G., Howard, J. K., Baker, R. J., Bloom, S. R. & Lechler, R. I. 1998. Leptin modulates the T-cell immune response and reverses starvation-induced immunosuppression. *Nature*, **394**, 897-901.

Lozano, G. A. 1994. Carotenoids, parasites, and sexual selection. *Oikos*, **70**, 309-311.

Malo, D. & Skamene, E. 1994. Genetic control of host resistance to infection. *Trends in Genetics*, **10**, 365-371.

Manning, J. T. & Chamberlain, A. T. 1993. Fluctuating asymmetry, sexual selection and canine teeth in primates. *Proceedings of the Royal Society of London, Series B*, **251**, 83-87.

McGuire, W., Hill, A. V. S., Allsopp, C. E. M., Greenwood, B. M. & Kwiatkowski, D. 1994. Variation in the TNF - a promoter region associated with susceptibility to cerebral malaria. *Nature*, **371**, 508-511.

McMichael, A. J. & O'Callaghan, C. A. 1998. A new look a T cells. *Journal of Experimental Medicine*, **187**, 1367-1381.

McMichael, A. J. & Phillips, R. E. 1997. Escape of human immunodeficiency virus from immune control. *Annual Review of Immunology*, **15**, 271-296

Milich, D. R., Peterson, D. L., Schodel, F., Jones, J. E. & Hughes, J. L. 1995. Preferential recognition of hepatitis B nucleocapsid antigens by Th1 or Th2 cells is epitope and major histocompatibility complex dependent. *Journal of Virology*, **69**, 2776-2785.

Milinski, M. & Bakker, T. C. M. 1990. Female sticklebacks use male coloration in mate choice and hence avoid parasitized males. *Nature*, **344**, 330-333.

Møller, A. P. 1994. *Sexual Selection and the Barn Swallow*. New York: Oxford University Press.

Murray, J. S. 1998. How the MHC selects Th1/Th2 immunity. *Immunology Today*, **19**, 157-163.

Nepom, G. T. & Erlich, H. 1991. MHC class-II molecules and autoimmunity. *Annual Review of Immunology*, **9**, 493-525.

Nevo, E. & Beiles, A. 1992. Selection for class II MHC heterozygosity by parasites in subterranean mole rats. *Experimentia*, **48**, 512-515.

Norris, K. 1993. Heritable variation in a plumage indicator of viability in male great tits *Parus major*. *Nature*, **362**, 537-539.

Nicoletto, P. F. 1993. Female sexual response to condition-dependent ornaments in the guppy, *Poecilia reticulata*. *Animal Behaviour*, **46**, 441-450.

Paterson, S., Wilson K. & Pemberton, J. M. 1998. Major histocompatibility complex variation associated with juvenile survival and parasite resistance in a large unmanaged ungulate population *(Ovis aries* L.*)*. *Proceedings of the National Academy of Sciences, USA*, **95**, 3714-3719.

Penn, D. & Potts, W. K. 1998. Chemical signals and parasite-mediated sexual selection. *Trends in Ecology and Evolution*, **13**, 391-396.

Poland, G. A. 1998. Variability in immune response to pathogens: using measles vaccine to probe immunogenetic determinants of response. *American Journal of Human Genetics*, **62**, 215-20.

Pomiankowski, A. & Møller, A. P. 1995. A resolution of the lek paradox. *Proceedings of the Royal Society of London, Series B*, **260**, 21-29.

Potts, W. K. & Slev, P. R. 1995. Pathogen-based models favoring MHC genetic diversity. *Immunological Reviews*, **143**, 181-197.

Potts, W. K. & Wakeland, E. K. 1990. Evolution of diversity at the major histocompatibility complex. *Trends in Ecology and Evolution*, **5**, 181-187.

Potts, W. K. & Wakeland, E. K. 1993. Evolution of MHC genetic diversity: a tale of incest, pestilence, and sexual preference. *Trends in Genetics*, **9**, 408-412.

Råberg, L., Grahn, M., Hasselquist, D. & Svensson, S. 1998. On the adaptive significance of stress-induced immuno-suppression. *Proceedings of the Royal Society of London, Series B*, **265**, 1637-1641.

Rajewsky, K. 1996. Clonal selection and learning in the antibody system. *Nature*, **381**, 751-758.

Roitt, I., Brostoff, J. & Male, D. 1996. *Immunology*. 4th edn. London: Mosby.

Saino, N. & Møller, A. P. 1996. Sexual ornamentation and immunocompetence in the barn swallow. *Behavioral Ecology*, **7**, 227-232.

von Schantz, T., Bensch, S., Grahn, M., Hasselquist, D. & Wittzell, H. 1999. Good genes, oxidative stress and condition-dependent sexual signals. *Proceedings of the Royal Society of London, Series B*, **266**, 1-12.

von Schantz, T., Göransson, G., Andersson, G., Fröberg, I., Grahn, M., Helgée, A. & Wittzell, H. 1989. Female choice selects for a viability-based male trait in pheasants. *Nature*, **337**, 166-169.

then used inbred strains of mice with different MHC-haplotypes to examine whether gamete fusion deviated from random with respect to the MHC, and we tested whether such a deviation was influenced by the presence or absence of an infection by mouse hepatitis virus (MHV).

Studies on humans (the 'T-shirt studies')

In a first study (Wedekind et al. 1995), 49 female and 44 male students were typed for their HLA-A, -B and -DR (i.e., loci within the MHC). The men wore a T-shirt during two nights. On the following day, the women were asked to judge the odours of six T-shirts each. The body odours were judged as more pleasant when they originated from men whose MHC genotype was different from that of the judging woman. In contrast, the odours were judged less pleasant when the MHC genotype of the odour-producing males and that of the judging woman were similar. This difference in odour assessment was reversed when the women were taking oral contraceptives. Furthermore, the odours of MHC-dissimilar men were more frequently reminding the women of their own present or former partners than did the odours of MHC-similar men. This suggests that the MHC or linked genes influence human mate choice today. Both MHC-class I and class II genes seemed to be similarily involved in the correlation to body odour and odour preferences (C. Wedekind et al., unpublished). However, all these findings did not yet allow for a discrimination between the parasite hypotheses and the inbreeding avoidance hypotheses.

Certain MHC combinations (normally heterozygous ones) may be superior to others in conferring better resistance against pathogens. This could influence mate choice either in a way to maximise MHC-heterozygosity of the offspring or, alternatively, to create specific allele-combinations that are beneficial under given environmental conditions through their complementary or epistatic effects. The latter (but not the former) possibility would specifically support the parasite hypotheses, since a preference for specific allele-combinations would not be expected by the inbreeding avoidance hypothesis.

We performed a second study to test this hypothesis, but also to test for gender effects, and to get a first estimate of the amount of variance in pleasantness scorings that is correlated to the MHC (Wedekind & Füri 1997). We asked 58 women and 63 men, all typed for their HLA-A, -B and -DR, to score the odours of six T-shirts, worn always by the same two women and four men. The pleasantness scorings correlated again negatively with the degree of MHC similarity between smeller and T-shirt wearer in men and in women who were not using the contraceptive pill (but not in pill users). Depending on the T-shirt wearer, the amount of variance in the scorings of odour pleasantness that was explained by the degree of MHC similarity (r^2) varied between nearly 0 and 23 %. However, the median of all r^2-values was only 1.1%. The six T-shirt-wearers differed significantly from each other in the degree to which pleasantness scorings correlated to the MHC. There was no significant effect of gender on the correlation between pleasantness and MHC similarity: the highest r^2 was actually reached with one of the male odours sniffed by male smellers. Men and women who were reminded of their own mate/ex-mate when sniffing a T-shirt had significantly fewer MHC-alleles in common with this T-shirt wearer than expected by chance. This suggests again that the MHC or linked genes influence human mate choice. We found no significant effect when we tested for an influence of the MHC on odour preferences after the degree of similarity between T-shirt wearer and smeller was statistically controlled for. This suggests that body odour preferences are mainly influenced by the degree of similarity or dissimilarity at the MHC in our study population. The observed preferences would increase heterozygosity in the progeny, without producing specific combinations at the MHC. As mentioned above, a more specific choice of particular alleles would have provided strong evidence in favour of the first parasite hypothesis, as opposed to the inbreeding avoidance hypothesis. However, this second study did not provide such evidence.

In both studies, the analysis of memory associations suggests that the odour preference reflects the actual choice of mates. However, MHC-correlated mate preferences may vary between populations, and do not seem to be a general phenomenon in mammals. In a field study with American Hutterites, Ober et al. (1997) confirmed that there are MHC-correlated mate preferences, but Hedrick & Black (1997) did not find such an effect in South Amerindians. Although the latter study had a lower statistical power (the sample size in their field study was much smaller than in the study by Ober et al., while the heterogeneity on the MHC seemed to be larger, and fewer alleles had been analysed), the possibility remains that there could be population differences with respect to MHC-correlated mating. A recent study by Paterson & Pemberton (1998) tested for MHC-correlated mate preferences in another mammal, the Soay sheep *Ovis aries*. They could not find any indication of such mate preferences in an unmanaged population on a Scottish island. Their analysis was conservative with respect to their conclusion, and their sample size was comparatively large, which strongly suggests that MHC-correlated mate choice is not universal among mammals.

Cryptic female choice: studies on mice

Mate choice is only one level of sexual selection. Further possible levels include selection of sperm type by the female reproductive tract, selective fertilization, or selective support of the embryo or the born offspring (Wedekind 1994). Many studies in humans from many different regions of the world indicate a connection between the risk to experience a spontaneous abortion and the degree of MHC similarity between the pregnant woman and the father of her embryo (see Gill 1994 for a review citing 26 different studies on MHC-sharing in recurrent spontaneous abortions). The more similar the couple is at the MHC, the higher the risk that an apparently healthy embryo is going to be aborted. There is even evidence that the degree of MHC-similarity between couples has an effect on earlier stages of pregnancy, before an abortion would be detected. Weckstein et al. (1991) found that the success rate of *in vitro* fertilization and tubal embryo transfer correlates negatively with the degree of MHC similarity, and Ober et al. (1988) found that longer intervals between successive births are again associated with increased similarities at the MHC in American Hutterites who proscribe contraception.

All the evidence on cryptic female selection in humans is of course correlational, i.e. causes and effects remain unclear since experiments on that topic are not possible in our species for obvious reasons. Other possible levels of MHC-correlated sexual selection have not been studied much. Therefore, we designed an experiment to test whether the MHC also influences fertilization itself.

In a first series of *in vitro* fertilization experiments (Wedekind et al. 1996) we used two inbred mouse strains that were congenic with respect to their MHC, i.e. they differed on their MHC but were bred to be identical on the rest of the genome. We crossed these two strains to produce F1 males and females that were heterozygous on their MHC but virtually identical to either parental strain on most other loci. These heterozygous F1 mice and the homozygous mice of the parental strains were used in two different experimental set-ups. First, we used sperm of heterozygous males to fertilise eggs of homozygous females. This way we tested for potential choice by the egg. Then, we used sperm of homozygous males to fertilise eggs of heterozygous females. This way we tested whether the second meiotic division in the egg (which takes place only after the sperm has entered the egg) could be influenced by the MHC of the fertilising egg. We actually found non-random combination of MHC-haplotypes as the outcome of both selection mechanisms: (i) the eggs selected the fertilizing sperm, and (ii) the second meiotic division in the eggs was influenced by the type of sperm that had entered the egg.

A surprising further finding was that the proportion of MHC-heterozygous embryos varied over time. Since this could not be explained by differential mortality, it suggested that the formation of specific combinations of maternal and paternal genomes was depending on an external factor that had changed over time. A *post-hoc* analysis of the experimental conditions revealed that the higher frequency of heterozygous individuals correlated with an uncontrolled epidemic by mouse hepatitis virus in the animal facility. To test this we performed a second series of experiments (Rülicke et al. 1998), in which we carefully controlled for mouse hepatitis virus infection. These experiments confirmed that virus-infected mice produced more MHC-heterozygous embryos than sham-infected ones. This finding can now explain the variability in the results of a number of previous studies where infection condition was not controlled (Gorer & Mikulska 1959; Palm 1969, 1970; Hings & Billingham 1981, 1983, 1985; Potts et al. 1991).

The finding that parents are able to promote specific combinations of MHC-haplotypes during fertilization according to the presence or absence of a viral infection would not have been expected from the inbreeding avoidance hypothesis. It therefore provides evidence for the parasite hypotheses, although a number of questions still remain to be answered. The most important questions may be: (i) what is the fitness benefit of heterozygosity at the MHC when an epidemic is present, and (ii) what is the fitness benefit of homozygosity at the MHC under epidemic-free conditions.

Acknowledgements

We thank the anonymous reviewers for their comments. Our projects were supported by the Roche Research Foundation and the Swiss National Science Foundation.

References

Apanius, V., Penn, D., Slev, P. R., Ruff, L. R. & Potts, W. K. 1997. The nature of selection on the major histocompatibility complex. *Critical Reviews in Immunology*, **17**, 179-224.

Brown, J. L. & Eklund, A. 1994. Kin recognition and the major histocompatibility complex: an integrative review. *American Naturalist*, **143**, 435-461.

Egid, K. & Brown, J. L. 1989. The major histocompatibility complex and female mating preferences in mice. *Animal Behaviour*, **38**, 548-550.

Gorer, P. A. & Mikulska, Z. B. 1959. Some further data on the H-2 system of antigens. *Proceedings of the Royal Society of London, Series B*, **151**, 57-69.

Gill, T. J. III. 1994. Reproductive immunology and immunogenetics. In: *The Physiology of Reproduction*. 2nd edn (Ed. by E. Knobil & J. D. Neill), pp. 783-812. New York: Raven Press.

Hedrick, P. W. 1994. Evolutionary genetics of the major histocompatibility complex. *American Naturalist*, **143**, 945-964.

Hedrick, P. W. & Black, F. L. 1997. HLA and mate selection: no evidence in South Amerindians. *American Journal of Human Genetics*, **61**, 505-511.

Hings, I. M. & Billingham, R. E. 1981. Splenectomy and sensitization of Fischer female rats favors histoincompatibility of R2 back-cross progeny. *Transplantation Proceedings*, **13**, 1253-1255.

Hings, I. M. & Billingham, R. E. 1983. Parity-induced changes in the frequency of RT1 heterozygotes in an R2 backcross. *Transplantation Proceedings*, **15**, 900-902.

Hings, I. M. & Billingham, R. E. 1985. Maternal-fetal immune interactions and the mainenace of major histocompatibility complex polymorphism in the rat. *Journal of Reproductive Immunology*, **7**, 337-350.

Jordan, W. C. & Bruford, M. C. 1998. New perspectives on mate choice and the MHC. *Heredity*, **81**, 127-133.

May, R. M. 1979. When to be incestuous. *Nature*, **279**, 192-194.

Ober, C., Elias, S., O'Brien, E., Kostyu, D. D., Hauck, W. W. & Bombard, A. 1988. HLA sharing and fertility in Hutterite couples: evidence for prenatal selection against compatible fetuses. *American Journal of Reproductive Immunology and Microbiology*, **18**, 111-115.

Ober, C., Weitkamp, L. R., Cox, N., Dytch, H., Kostyu, D. & Elias, S. 1997. HLA and mate choice in humans. *American Journal of Human Genetics*, **61**, 497-505.

Palm, J. 1969. Association of maternal genotype and excess heterozygosity for Ag-B histocompatibility antigens among male rats. *Transplantation Proceedings*, **1**, 82-84.

Palm, J. 1970. Maternal-fetal interaction and histocompatibility antigen polymorphisms. *Transplantation Proceedings*, **2**, 162-173.

Penn, D. & Potts, W. K. 1999. The evolution of MHC-dependent mate preferences. *American Naturalist*, **153**, 145-164.

Potts, W. K., Manning, C. J. & Wakeland, E. K. 1991. Mating patterns in seminatural populations of mice influenced by MHC genotype. *Nature*, **352**, 619-621.

Potts, W. K., Manning, C. J. & Wakeland, E. K. 1994. The role of infectious disease, inbreeding and mating preferences in maintaining MHC genetic diversity: an experimental test. *Philosophical Transactions of the Royal Society of London, Series B*, **346**, 369-378.

Potts, W. K. & Wakeland, E. K. 1993. Evolution of MHC genetic diversity: a tale of incest, pestilence and sexual preference. *Trends in Genetics*, **9**, 408-412.

Rülicke, T., Chapuisat, M., Homberger, F. R., Macas, E. & Wedekind, C. 1998. MHC-genotype of progeny influenced by parental infection. *Proceedings of the Royal Society of London, Series B*, **265**, 711-716.

Spiro, M. E. 1958. *Children of the Kibbutz.* Cambridge, MA: Harvard University Press.

Weckstein, L. N., Patrizio, P., Balmaceda, J. P., Asch, R. H. & Branch, D. W. 1991. Human leukocyte antigen compatibility and failure to achieve a viable pregnancy with assisted reproductive technology. *Acta Europaea Fertilitatis*, **22**, 103-107.

Wedekind, C. 1994. Mate choice and maternal selection for specific parasite resistances before, during and after fertilization. *Philosophical Transactions of the Royal Society of London, Series B*, **346**, 303-311.

Wedekind, C., Chapuisat, M., Macas, E. & Rülicke, T. 1996. Non-random fertilization in mice correlates with the MHC and something else. *Heredity*, **77**, 400-409.

Wedekind, C. & Füri, E. 1997. Body odour preference in men and women: do they aim for specific MHC combinations or simply heterozygosity? *Proceedings of the Royal Society of London, Series B*, **264**, 1471-1479.

Wedekind, C., Seebeck, T., Bettens, F. & Paepke, A. J. 1995. MHC-dependent mate preferences in humans. *Proceedings of the Royal Society of London, Series B*, **260**, 245-249.

Wolf, A. P. 1966. Childhood association, sexual attraction and the incest taboo: a Chinese case. *American Anthropologist*, **68**, 883-898.

Wolf, A. P. 1970. Childhood association and sexual attraction: a further test of the Westermark hypothesis. *American Anthropologist*, **72**, 503-515.

Yamazaki, K., Beauchamp, G. K., Shen, F.-W., Bard, J. & Boyse, E. A. 1994. Discrimination of odortypes determined by the major histocompatibility complex among outbred mice. *Proceedings of the National Academy of Sciences, USA*, **91**, 3735-3738.

Yamazaki, K., Beauchamp, G. K., Wysocki, C. J., Bard, J., Thomas, L. & Boyse, E. A. 1983. Recognition of H-2 types in relation to the blocking of pregnancy in mice. *Science*, **221**, 186-188.

Yamazaki, K., Boyse, E. A., Miké, V., Thaler, H. T., Mathieson, B. J., Abbott, J., Boyse, J., Zayas, Z. A. & Thomas, L. 1976. Control of mating preference in mice by genes in the major histocompatibility complex. *Journal of Experimental Medicine*, **144**, 1324-1335.

Yamazaki, K., Yamaguchi, M., Baranoski, L., Bard, J., Boyse, E. A. & Thomas, L. 1979. Recognition among mice. Evidence from the use of a Y-maze differentially scented by congenic mice of different major histocompatibility types. *Journal of Experimental Medicine*, **150**, 755-760.

In: Espmark, Y., Amundsen, T. & Rosenqvist, G. (Eds.) 2000
Animal Signals:
Signalling and Signal Design in Animal Communication, Pp. 445-463
Tapir Academic Press, Trondheim, Norway

24

Knowledge, Information, Biases and Signal Assemblages

OREN HASSON

P.O. Box 11598, Ramat-Gan 52015, Israel

E-mail: ohasson@hotmail.com

Abstract

Information has been usually treated by Information Theory that provides tools to measure the capacity of signals to store data. However, it often makes more biological and empirical sense to measure information stored in a signal by the changes it induces on the recipient's knowledge. Knowledge is defined here as the individual's estimate of the mean values and variances of the variables that affect its decision about its future actions, relatively to their true values when it acts. Given this definition, knowledge, hence also information, are best estimated by using tools that are derived from Signal Detection Theory, rather than those provided by Information Theory, because the former measures the precision of decision making. In order to improve their knowledge, organisms have developed tools that improve their perception of the relevant parameter values, and corresponding biases that improve performance. Because knowledge is required on many different variables, it only makes sense that so are biases for certain estimated values of these variables. Each bias can lead to the evolution of at least one signal. It can lead to an exaggeration of a trait, which becomes either Fisherian or a handicap, and/or to signals that function as amplifiers and improve its perception. Hence, if organisms evolve multiple biases in response to the assessment of different variables, we should expect that these biases will lead to assemblages of multiple co-existing signals. I present here an analysis of an observation on pursuit-deterrent signals of the spur-winged plover, Hoplopterus spinosus. *I use it to explore some of the possible parameters, detectable in prey, which might be important to predators. Predators that evaluate these parameters may evolve biases, to which prey are likely to respond by evolving assemblages of signals. Indeed, we often find that a prey sends to its predator several types of signals that operate in concert, and include behavioral displays (such as chronologically oriented head-bobs, described here), color patterns that improve spatial orientation, alarm calls, or even duetting.*

The Trollveggen fallacy

Not too far from Kongsvoll, there is a cliff called Trollveggen. It is the tallest vertical wall in Europe. It is over one kilometer high, with a hangover of about 50 m (see: http://www.stud.ntnu.no/~olegram/Romsdalen.html, for interesting photographs). If you think of jumping over the rim of Trollveggen, your intuition will tell you that you are bound to fall down and get killed. You can, if you like, be more specific and build a detailed verbal description of everything that might happen to you along the long way down. You can describe how the wind would blow through your hair, the feel of your shirt flapping in the wind on your skin and stroking it, while its sound is dimmed by the sound of the wind in your ears. You can go on and on describing all the possible little details of your fall. Some of these details, especially the qualitative ones, might be accurate. Some other details, especially the quantitative ones, might be false. If you remember a few elementary equations in physics, you can probably build a simple mathematical model that will accurately predict the moment you hit the ground, and the speed of your fall at that moment. If you are very good in mathematical modeling, you will be able to also construct the precise recursive equations that describe, for example, the movements of your shirt in the wind and your exact position at each moment. This model will also inform you how sensitive are your predictions to the exact angle and momentum of each of your limbs, as you start falling down. One thing is certain, however. In order to build a full mathematical model of your fall, you need a detailed verbal description. The verbal model will list all the possible options for you, and will guide you in constructing the model's equations. When you are done, you will be able to know exactly how it would be falling down from Trollveggen. The conclusion is an empirical one: you don't want to do it.

Despite the (seemingly) undoubtable consequences of such a fall, a number of people, each year, used to leap over the rim of Trollveggen. In fact, there is even a ledge, near the top of Trollveggen, which is marked with the word 'Exit', pointing at a convenient leaping point. However, this sign was not intended for people who wanted to commit suicide. Instead, it marks the exit point for skydivers. People who leaped there were parachuting their way down (a practice now forbidden by the Norwegian authorities). The interesting lesson from this little story is that the thinking method that we used had no effect on our conclusion. We used intuition, we used a verbal description and we used a mathematical model, and all gave us the wrong conclusion. This is because we were imprisoned by a certain conception, and our initial assumptions were wrong or incomplete. So, let us give homage to Trollveggen, and call the fallacy of using a wrong or incomplete set of assumptions, the 'Trollveggen Fallacy'.

A few years ago I received a manuscript to review. It was based on a mathematical model and it was written by an excellent modeler. The math was a bit over my head, but I thought the model was pretty good and presented an interesting point, so I sent it back to the editor, recommending its publication. Something bothered me, though. A couple of days later I realized what it was. It was the Trollveggen fallacy, or so I thought. In mathematical models assumptions are more explicit than in verbal ones because one must give them a certain value, either a variable or a constant. Often, at least some assumptions in mathematical models are clearly unrealistic. We nevertheless use them because we need solutions. Unrealistic assumptions simplify models enough to give us solutions. However, it is important for a modeler to make sure that unrealistic assumptions will not result in solutions that are the opposite of what one expects from realistic ones. And this is what I felt happened in that particular model. Hence, I rushed an e-mail letter to the author. I told this modeler that it seemed to me that by making one of the assumptions more realistic, the model's conclusion would be reversed. In reply, the modeler told me that my intuition was correct.

The manuscript was, eventually, turned down by the editor on the ground of being too mathematical for that particular journal. However, the important point about this case is that it shows us that the Trollveggen fallacy can be fairly common. Take the saga of the handicap idea, for example. Amotz Zahavi first published it in 1975 (Zahavi 1975), and since then it was crucified by modelers for almost 15 years. Zahavi presented it as a verbal model. In response to the strong criticism, he later expanded and explained it further in a number of additional papers, using other verbal descriptions of the handicap mechanism. The first mathematical models did not agree with Zahavi's conclusions (Davis & O'Donald 1976; Maynard Smith 1976), and only at about the turn of the 90's theorists have finally and fully accepted it (Maynard Smith 1991). What went wrong? There are two possibilities: either Zahavi's verbal model was erroneous, and the mathematical models explored options beyond Zahavi's description until they found those assumptions that made it work, or, alternatively, Zahavi made the correct assumptions in his verbal description, though these were not included in the early mathematical models. The truth leans toward the second possibility. After receiving the first wave of criticism, in 1976, Zahavi modified and clarified his assumptions (Zahavi 1977). The main core of the handicap mechanism, i.e., the assumptions that made it tick in the genetic models 10 years later (Pomiankowski 1987), were already found in his verbal description in 1977.

The complexity of the mathematical models of the handicap mechanisms increased over time. Assumptions of the first models were simple, and their solutions were mathematically neat. Indeed, the assumption that a handicap is condition-dependent complicates the model. It was an assumption that was ignored in the first models. When it was used, in later models, it was first assumed that condition or the handicap or both vary, but also that they were not heritable (Kirkpatrick 1986). Pomiankowski (1987) was the first modeler who put together, in genetic models, all the assumptions that made the handicap mechanism work. Simple optimization models of Andersson (1982), and of Nur and Hasson (1984), or game theory models starting with Grafen (1990), made the same assumptions, and arrived at the same conclusions.

What we should learn from the saga of the handicap mechanism is that we cannot rely exclusively on mathematical models. They alone will not prevent us from sinking again into the abyss of the Trollveggen fallacy. It is in the hands of naturalists, empiricists and modelers to put forward verbal explanations that cover all the reasonable options. The role of mathematical modelers is to use their models to explore the logic of these explanations. It is their duty to expose hidden assumptions of the verbal models, and to make sure their own assumptions do not remain hidden. It is their duty to explain, as much as they can, what unrealistic assumptions do to the model's conclusions. And it is in the hands of us all to use our intuition, each of us based on his or her own knowledge and skills, and ensure the conclusions make sense. Given the appropriate assumptions, they must, at least as an aftermath, make intuitive sense, or we do not fully understand the theory.

Signal diversity

Signals refer to traits (such as behavioral, acoustical, chemical or structural patterns) whose specific evolutionary benefit stems from their potential to change the behavior of others (Hasson 1994, 1997; Maynard Smith & Harper 1995). However, to make decisions, organisms use, in addition to signals, also characteristics, termed 'cues', which are found in living organisms or in non-living objects, and which have evolved (in living organisms) for other reasons. In fact, they may have evolved despite their effect on recipients (in such cases, organisms might evolve attenuators that hide these cues; see Hasson et al. 1992). In signaling theory, because we are interested in understanding signals rather than cues, we often tend to ignore the perception and assessment of cues. Most of this paper is dedicated to the argument that we should not.

The potential of signals to coexist is a relatively new topic which, in my opinion, has already slipped a few times into the Trollveggen fallacy. Evelyn Hutchinson gave a famous plenary lecture in the annual meeting of the American Society of Naturalists, 1959, entitled 'Homage to Santa Rosalia, or why are there so many kinds of animal?' He talked mainly about competition and limiting similarities, and ended his lecture with the sentence: 'And this is where solutions may be found to questions ecologists should ask, such as: How similar can species be, yet coexist? and Why are there so many kinds of animals?' This is the kind of questions we should ask today about assemblages of signals found in gene pools (Hasson 1997).

The study of signal diversity may benefit from the parallelism between communication and community ecology. This analogy hints upon competition that occurs between signals over a certain resource or resources. Like species in ecological communities, signals may be arranged in 'guilds' according to their contexts and recipients. Furthermore, some signals are generalists and exploit perceptions and biases of several recipients, or in several contexts, whereas others are specialists and exploit perceptions and biases of only a single recipient at a very specific context. By way of analogy, perceptions and biases are the resources upon which signals compete, and they exist mainly because potential recipients of signals need information.

Information and knowledge

What is information? The answer is not as trivial as it may sound. Information is described, by Information Theory (Shannon & Weaver 1949), as the capacity of a signal to store data, which is a positive function of the number of bits that one can put in it. Given that signals store and transmit information, recipients make judgements of how many perceived bits in a signal are turned on or off. Note, however, that according to Information Theory, information is not an asset of recipients. According to Information Theory information is not even an asset of signalers but, instead, it is an asset of signals. Other than in the context of Information Theory, however, information has been used for both signals and individuals. In contrast, 'knowledge' is a term assigned for individuals only. To improve clarity, I use the term 'information' for signals, and 'knowledge' for individuals.

Hence, according to Information Theory, a signaler fills signal's bits with zeros and ones. This information, however, is only as efficient as its perception by recipients. It makes a practical sense, therefore, to handle this by using perceived bits rather than submitted bits (e.g. Bradbury & Vehrencamp 1998). Another difficulty with Information Theory is that the mathematical treatment it offers cannot handle negative values of information stored in signals. This is important, because signals can change recipients' knowledge either positively (improving their performance) or negatively (diminishing their performance). Although it is theoretically possible to assume, a priori, that a signal is a cheating signal, and then measure its potential to deceive by measuring its capacity to store 'cheating units' in cheating-bits, filled with zeros and ones, this would be rather cumbersome. Information Theory does not handle well cheating signals. Hence, it makes more intuitive sense to measure information by its effect on the performance of recipients (Hasson 1994, 1997; Maynard Smith & Harper 1995). By using the above distinction between 'information' (of signals) and 'knowledge' (of individuals), we can define reliable signals as signals that induce a positive change in the recipients' knowledge. As a result, recipients of reliable signals make better judgements about their future actions, and their fitness increases. Similarly, cheating signals are defined as signals that decrease recipients' knowledge, thereby decreasing their fitness (Hasson 1994). We can use this distinction to also improve the definition of signals (Hasson 1994, 1997; Maynard Smith & Harper 1995) as characters that evolve because they change the recipient's knowledge, hence, their behavior.

Although both Maynard Smith and Harper (1995) and I (Hasson 1994, 1997) thought that the information content of a signal should be evaluated by the signal's effect on its recipients' knowledge, this approach was made on the intuitive level only. Neither Maynard Smith and Harper, nor I, defined 'information' and 'knowledge'. We know that the more knowledge individuals have, the more 'precise' their behavior would be. In other words, additions to existing knowledge increase fitness. Hence, knowledge refers to the values an individual assigns to all variables that affect its decisions about its future actions. The better the correspondence between the estimate of these values with their real values, the more knowledge an organism has about its environment, and the more precise its behavior would be. Consequently, cheating signals decrease correspondence of the relevant set of variables with their true values, and reliable signals improve this correspondence.

Most of the time, knowledge is not complete. For some variables, for example, when features of the environment or the individuals' qualities that are sought by recipients fluctuate, knowledge can never be complete. Hence, knowledge is better defined as the individual's estimate of the mean values and variances of variables that affect its decision about its future action, relatively to their true values when it acts. Knowledge does not necessarily imply consciousness, but it determines performance. Hence, by observing an individual's performance, we can make inferences about its knowledge. I hereon restrict the term 'information' to the potential of cues and signals to change their recipients' knowledge.

Although this definition is new, it can be treated within a framework of an already existing theory. However, this is not the framework of Information Theory, but that of Signal Detection Theory. In a review on pursuit deterrent signals (Hasson 1991a), I introduced an approach, which I thought was new, to explain communication. It was, in my mind, the simplest way to explain why a predator should respond to a pursuit-deterrent signal. It recognized that when a predator makes a decision of whether or not to pursue an individual prey, it can make two types of errors. One is the error of being too anxious, pursuing prey that it should not pursue, the other is the error of being too cautious, namely, not pursuing a prey that it should pursue. When a prey estimates that the predator is committing or is about to commit the error of being too anxious, it should benefit by increasing the predator's knowledge. Consequently, it sends reliable signals that improve the predator's estimate of the prey's alertness, health and vigor. Hence, in order to deter predators, prey benefits by informing the predator of its alertness and strength, minimizing the predator's error of being too anxious. Later, I realized that the same logic applies to other communication systems, though, if they are aimed to attract rather than to deter (e.g. courtship signals), they are targeted to reduce the recipient error of being too cautious (Hasson 1994). In terms of Signal Detection Theory, to which I was introduced only a short time before the latter paper was published, overanxiousness is a 'false alarm' and overcautiousness is a 'miss".

Signal Detection Theory is founded on these two types of errors. It also recognizes two types of correct choices: 'a correct rejection' and a 'hit'. Unlike Information Theory, which is concerned with the capacity of signals to store data, Signal Detection Theory is interested in choices individuals make based on their current knowledge. Signals change knowledge by either increasing it (if reliable) or decreasing it (if cheating). Both of these effects can be easily included, as different parameter values, in Signal Detection Theory (see Getty et al. 1987, for the use of Signal Detection Theory, with examples of cheating signals).

The most appropriate way to measure knowledge is probably indirectly, by measuring the precision of an individual's choice of actions. This can be done by calculating the relationships between the individual's choice of hits and false alarms, relative to the expected or to the empirical choice of actions the individual makes when it is fully informed. Signal Detection Theory provides mathematical tools for describing the relationships between hits and false alarms. The actual information content of a signal, as it is perceived by recipients, is therefore defined as the change in knowledge it induces in the recipient (which may be either positive or negative). It can be measured by monitoring the average change in the relationships between hits and false alarms induced by the signal.

The role of multiple biases

By definition (Hasson 1997), signals survive or perish because recipients' perceptions and biases change fitness of signalers, and give direct benefits or losses to alleles that code for signals. These perceptions and biases are the evolutionary resources of signals upon which they compete. The more efficient are signals in exploiting biases of recipients, the more likely they are to eliminate competing signals. Hence, in order to understand signaling, we must study the evolution of perceptions and biases, where perceptions are the tools that organisms use to gather information (from cues and signals), and biases are tendencies of organisms to make certain choices based on different values of parameters in their knowledge space. It seems to me that the role of the animal's quest for information, especially via modifications of biases and perceptions, was heavily undermined in the theory of communication. The source of this negligence is probably the fact that the theory of communication developed mainly through studies of sexual selection. Sexual selection theory, including handicap models, is mainly based on the (Fisherian) genetic linkage between choice and male trait, which results in a simultaneous evolution of both signals and biases. With, perhaps, one exception (Hasson 1994), such a linkage does not exist in other communication systems. The study of the evolution of biases independently of signals has been, nevertheless, ignored too frequently.

This negligence is probably limited to studies of communication. Take, for example, studies of foraging behavior. It is a trivial assumption in optimal foraging models that animals are capable of gathering information based on many different cues. It is an assumption based on a plethora of empirical studies. Similarly, animals were found to be capable of using many environmental cues in order to navigate in it. Yet, if we look into theoretical studies of communication, either verbal or mathematical ones, we find that most of them assume that recipients are incapable of receiving information in the absence of signals. Relative to studies of animal behavior in other fields, this is a naïve assumption. Worse, it hinders our understanding of signaling.

I should give credit here to Amotz Zahavi. It was from him that I absorbed, as a young student at Tel Aviv University, the concept that communication is based on information recipients need. Although, as far as I can appreciate it, this was his working assumption, he has never explicitly published it. Since then, this concept has guided my understanding of communication (admittedly, at first unconsciously). The possibility that signals function as amplifiers (Hasson 1989, 1991b), for example, is based on the assumptions that individuals need information, are capable of perceiving it, though not perfectly, and make choices regardless of the existence of signals. I show, below, why this assumption matters, and how multiple biases can lead to multiple signals.

Alertness and chronological orientation

Predators need information about several prey characteristics, such as alertness, escape potential, nutritional value and defense mechanisms. Let us focus on alertness alone. One way of perceiving it is by looking at the prey facial symmetry. If a prey is vigilant and knows where the predator is, its face would appear symmetric to the predator, at least significantly more so than expected by a random distribution of facial orientations. This can lead to the evolution of symmetrical facial color patterns that function as amplifiers of the previously perceived symmetry (Hasson 1991a). Sometimes, however, prey does not use binocular vision. This is often the case in birds and in some mammals such as hares, rabbits and other Lagomorpha. In such cases, other cues may be used, such as, for mammals, the orientation of their ears. Cues refer to characteristics upon which biases are based, but which are not signals (Hasson 1994). Organisms use cues, such as daylength, light or prey movements because, on the average, they add information. In contrast with signals, however, if they are found in living organisms, they have not evolved because they change information but, sometimes, despite the fact that they add information. Hares and other mammalian prey must be attentive to their predators, hence, just like facial orientation of antelopes and deer, their ear orientation is a cue predators can use for detecting alertness in such prey. Ear markings, such as colored ear tips or margins that are found on many mammal's ears, are amplifiers of the ears' orientation. Because they better reveal the ears' orientation, the predator's knowledge of their prey's attentiveness, increases: the predator knows better when its prey is alert, but also when it is not.

Both facial symmetry and ear orientation can be categorized as 'spatial orientation'. Predators detect alertness by perceiving orientation of the prey attentiveness in space. However, there is more in vigilance than spatial orientation alone. Predators can use cues such as prey movements and flight distance. If prey respond to the predator's attempts to stalk them by moving away to maintain safe distance between them, predators should be able to learn that they have been spotted. Hence, they should pay attention to their prey's movements. In other words, they should respond to their prey's 'chronological orientation' in addition to their spatial orientation. For species, such as many small birds, whose face may be too small for its orientation to be perceived by a mammalian predator, and their head lacks conspicuous appendages, signals that use chronological orientation make more sense. For example, alarm calls and head-bobbing, which are common in birds, such as waders, may be chronologically oriented if given at a higher frequency immediately following their predator's movements.

Head-bobbing as chronological orientation

I tested this possibility by approaching the spur-winged plover *Hoplopterus spinosus*, in Israel, in September 1989, following the bird's breeding season. When approached, it uses both sharp alarm calls and head-bobs. Upon further approach, the plover walks away from the observer while continuing head-bobbing and calling and, eventually, it escapes by flying away. This sequence of behavioral displays is a typical response of spur-winged plovers toward human and natural predators alike (Simmons 1955; Cramp 1983; Walters 1990). To test the chronological orientation of the spur-winged plover's head-bobs I recorded the birds' reactions to my movements, which included only alternate walking toward the birds, and standing. Chronological orientation of head-bobs was measured by their proximity in time to the initiation of my walking.

To collect data I used a Casio PB-1000 hand-held computer (435 g), which I programmed to perform two operations: (i) to instruct me, by beeps given at random intervals of 3 to 15 seconds, to start walking toward the plover, or to stop, and (ii) record the timing of the plovers' head-bobs and, based on the numbers I pressed on the key-pad, also the estimated distance between me and the bird in units of ten meters.

Randomizing my walking movements ensured that I was not responding to the birds' behavior. By dividing each recording session into five equal time intervals, I found that the total number of bobs significantly increased as the recording session proceeded (x_1=69, x_2=113, x_3=124, x_4=129, x_5=172; Spearman rank-order correlation coefficient, r_S=1, one-tailed test, P<0.01, Siegel & Castellan 1989). This indicates that the spur-winged plovers were bobbing in response to my approach (see Appendix for other technical details).

During each uninterrupted walk, i.e. after beginning walking and before stopping, the spur-winged plovers bobbed more than once. After they bobbed, they waited a little, and bobbed again. Such a response that was not related to my change of behavior peaked at the third second (Fig. 1). This confirms that head-bobbing is not exclusively used for chronological orientation, and shows that head-bobbing is given periodically. My test of chronological orientation of head-bobs therefore considered only the first head-bobs made after each initiation of walking.

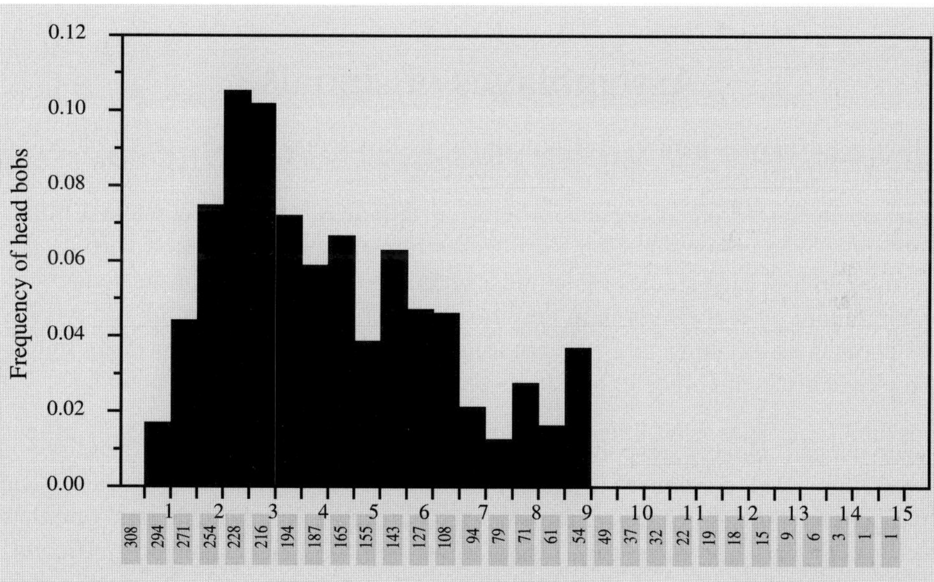

Figure 1. *Distribution of head bobbing intervals between two successive head bobs while I was walking uninterruptedly toward the plovers. Because long intervals were less available than short intervals (e.g. 5 sec or longer intervals were not available if only 4 sec remained between a head bob and the next beep) frequencies were calculated as the number of times each interval between two successive head bobs occurred, divided by the number of times this time interval was available (given by vertical numbers at the bottom).*

To make sure the analyses created no bias (see Appendix), Figure 2 presents two identical analyses. The first is the frequency of first head-bobs that were given forward in time, starting at the moment I heard the beep (0 to 15 sec) and while I was approaching the plovers (Fig. 2A), and the second is the frequency of first head-bobs that were given backward in time (0 to minus 15 sec), prior to the beep, while I was standing (Fig. 2B). Figure 2A shows a significant change over time, not seen in 2B, which indicates a response to my initiation of walking. Furthermore, the frequency of head-bobs during the first or second seconds, respectively, were significantly higher than the frequency of bobs during most or all other seconds (Fig. 2A). The second second was significantly different from any later second at a level of at least $P<0.05$ (Pearson chi-square), and the first second from all but the seconds marked with crossed squares. The first and second seconds were not significantly different from each other. Considering that I was slow to respond to the beep, and had a small delay of about three tenth of a second in pressing the key in response to the plover's head-bobbing, I was responsible for a delay of about a second between time zero (beep time) and the recorded first bob (Appendix). As a result, if the plovers use chronological orientation, their peak response is expected, indeed, to be found at around the beginning of the second second. In addition, the null hypothesis (no chronological orientation), given periodical head-bobs with a mode during the third second, and beeps that were given at intervals that were never shorter than three seconds, predicts no difference among the first three seconds. However, bobbing at the third second was significantly lower than that at the second second, indicating chronological orientation of head-bobbing.

Assemblages of signals

This little field study is hardly a breakthrough. It shows that signals can be chronologically oriented at predators, which is an interesting point, but one that hardly justifies its presentation in this forum. I present it here, nevertheless, in order to make a point. In order to understand communication, such as the plover's communication with its predator, we must follow such studies by taking a broader view and look for the signal's place in the community of signals given in the same context. Hence, we should ask: what is so special about head-bobbing and its chronological orientation, and why does the spur-winged plover also send other signals to its predator in addition to head-bobbing?

The spur-winged plovers responded to my initiation of walking by head-bobbing. However, they also continued to head-bob, periodically, in a non chronologically-oriented pattern. This must have diminished my perception of chronological orientation. Why do the plovers bob more than seems to be both necessary and most efficient for advertising their alertness toward a particular predator? Moreover, why do the plovers bob at all if, at the same time, they emit sharp, loud alarm calls? Alarm calls seem to be perfect for the job because, by being more conspicuous they are more informative about alertness. Also, as vocal rather than visual signals, they are less informative about the plover's exact location, hence they are also less risky.

The answers to both questions probably lie in the type of information that is transferred by these signals. True, repeating head-bobs in the absence of further changes in the

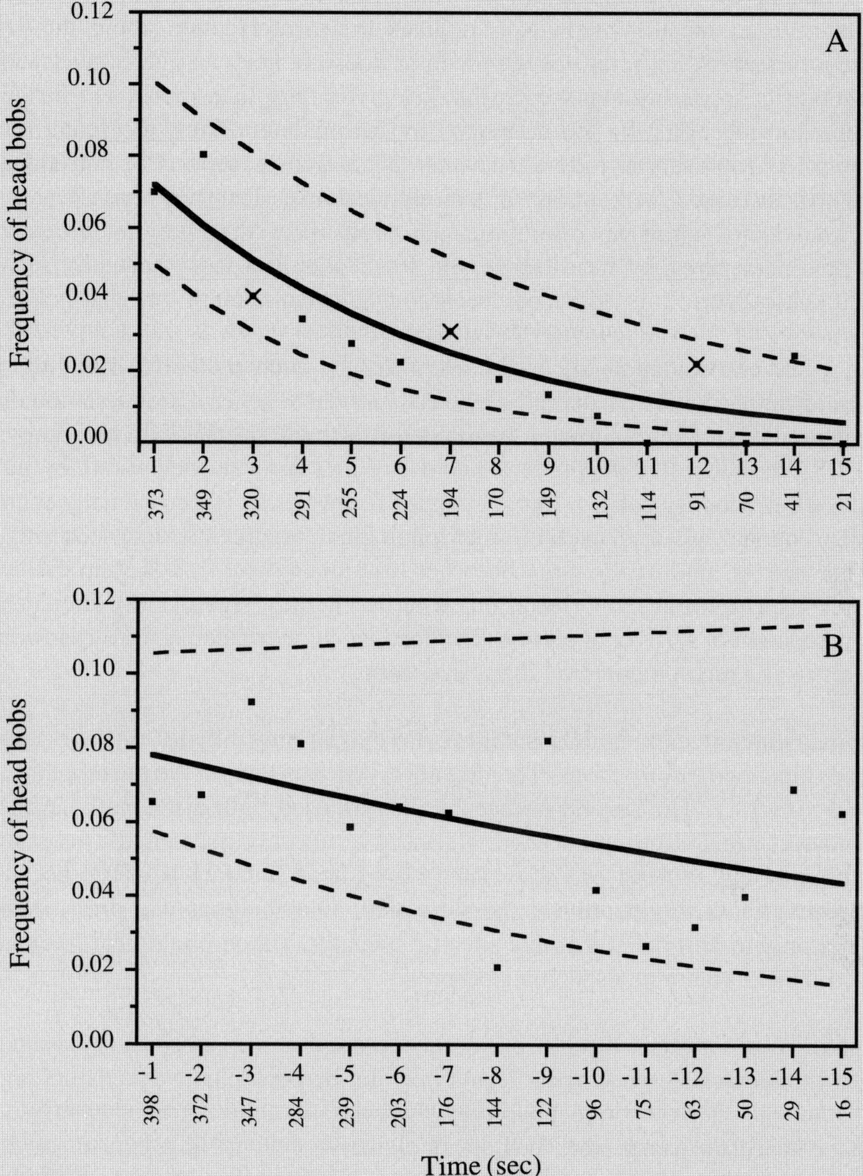

Figure 2. *Frequency of first head bobs as a function of time. Time zero is beep time, which was followed by my initiation of walking toward the plovers. (**A**) represents a forward time sequence and (**B**) represents a backward time sequence, starting at the beep. Lines represent the fitted curve (thick) and 95% confidence limits, calculated by the logistic model (Wilkinson 1988) and presented in proportions. Vertical numbers at the bottom of each second represent its availability for a first head bob. Due to the small number of expected bobs, data of the last three seconds (13-15) was pooled for the chi-square test.*

predator's behavior decreases perception of the signals' chronological orientation. As a result, prey impairs the information it adds to the predator about its knowledge of the predator identity and location. However, it gives more information about the prey's overall state of alertness, especially if the predator does not keep an eye contact with its prey throughout its approach. Head-bobs that are given only in response to changes in predator behavior may often be left unnoticed by the predator. Thus, the actual rate of head-bobbing may represent a trade-off between decreased perception of chronological orientation and increased perception of overall alertness. The two variables in the predator's knowledge space are therefore: the probability that prey is aware of a predator's presence (given by head-bobs), and the probability that it actually sees the predator and knows where it is (given by chronological orientation of the head-bobs).

In contrast, vocal alarm signals are less constrained by such tradeoffs. They are very effective at indicating the prey's overall alertness, but where several prey individuals are present, they are ineffective in allowing the predator to single out the individual prey that is alert (i.e., that has identified the predator and uses chronological orientation), because it is more difficult to identify callers. Hence, at least for prey that is occasionally found in groups, vocal signals are less useful for chronological orientation. Consequently, by increasing the rate of alarm calls, prey does not lose much in decreasing the predator's perception of its chronological orientation, whereas it may significantly increase the predator's perception of its overall alertness. Indeed, vocal alarm calls of the spur-winged plover are given at a much higher rate than head-bobs.

A predator may not quit stalking unless it is certain that all prey individuals have detected it. This can explain two observations. The first observation is that all the plovers that were present were signaling. The second is that the plovers used both alarm calls and head-bobs. If predators need information about alertness of all prey individuals, vocal alarm signals alone may fail to deter predators. The visual head-bobbing is useful because it adds information about the identity of the alert prey. Visual signals are probably used, therefore, not only to aim the signal at a specific predator, via chronological orientation, but also to aim the predator at a specific alert prey.

Finally, unlike vocal signals, visual displays may be conspiratorial, especially when prey is cryptic to its predator. A weak vocal signal may not be perceived by predators regardless of whether they have spotted their prey. In contrast, an weak and inconspicuous visual signal can be used as a private message of alertness received only by a predator that has already established an eye contact with its prey. Such a signal decreases the prey's risk of exposure to predators that are unaware of its presence or location.

Wilson (1975) wrote that if there is one thing that is certain about signals, it is their redundancy. If we recognize the limitations and advantages of different types of signals, we may find they are not redundant. We should also recognize that repetition of signals is itself a signal if it evolves, like head-bobbing and calling, to change the predator's behavior. Hence, instead of seeing redundancy, we may see that each signal, including signals' chronological orientation or repetitiveness, specializes on providing a different piece of information that is important to the predator. Each signal changes a variable, or sometimes several variables in the recipient's knowledge.

Duetting

It is mentioned above that predators may not quit unless they are certain that all prey individuals are fully alert. Some prey species are found, almost always, in pairs. Predators may learn that if they see and hear only one individual, the other individual may be in a vulnerable state. This is the case, for example, for klipspringers *Oreotragus oreotragus*, who inhabit little rocky islands in the African savanna. Sometimes they feed in the open, but when a predator approaches, they flee to the rocky hills where they are safe. It is there, on the rocky hill, that they make their alarm calls. The interesting twist here is that they duet. The male gives a call, and the female immediately follows. Why duetting? What additional information is given by duetting? The answer to this puzzle first occurred to me when I looked at spectograms of the alarm duetting of the klipspringer (Tilson & Norton 1981). These spectograms show that the calls of the male and the female are about one and a half tenths of a second apart. I suddenly realized that, considering the speed of sound, there is no way they could duet like this if they were more than 50 m apart (assuming the predator is at equal distances from both). This is because, in order to duet, the female has to first hear the male call before she responds with hers (note that if the female responds to a certain visual signal given by the male then, depending on the actual distances from each other and from the predator, her call might overlap or even precede that of the male; since the female call consistently follows the male call, then they must call only when close to each other, or use a very sophisticated calculation method). The time interval between her call and that of the male gives the predator an estimate of the maximum distance they are from each other. It is an efficient way of informing the predator that both of them are together on the hill, attentive and safe. Hence, there is no reason for the predator to stay.

By considering the selective forces that operate on duetting, i.e., the potential responses by recipients, we can identify its signal type. If predators, such as hyenas, have a bias to quit if the klipspringers' alarm calls are synchronized, then klipspringers always benefit by minimizing the interval between their alarm calls. The result, duetting, should therefore be classified as an index (Maynard Smith & Harper 1995; Hasson 1997) of the actual distance between pairs of klipsringers.

Conclusion

Perceptions and biases play a crucial role in the evolution of signals. Perceptions evolve for the purpose of gathering information about specific variables, and biases as appropriate, adaptive responses to certain values of these variables. However, biases may also exist as a result of side-effects and constraints. We know something about the latter (Ryan & Keddy-Hector 1992), although there are reasons to think that biases of this type play only a minor role in the evolution of signaling, at least in a stable state (Hasson 1997).

However, in studies of signaling we have not even scratched the surface of the understanding of what is the kind of information that interacting individuals (hence, potential signal recipients) need and, especially, how they gather it: what cues are they using, what is the role of signals in adding or subtracting knowledge, and how do signals do it? Instead, we have focused on signals and signalers. This might change if we acknowledge that knowledge is important to interacting individuals or to individuals that consider interacting, and that the major selective force that drives the evolution of signals is their power to change it.

The second point is that we must give more credit to verbal models. We need them to explore possible solutions. My purpose here is not to undermine the role of mathematical models, but to give more room for people who know their system and use their intuition to understand it and bring about reasonable options. Then, such options or assumptions should be rigorously tested using both mathematical models and experiments. While exploring all these options, we should remember that there is more in communication than handicaps and Fisherian traits alone. Take for example a female preference for a long tail and the elongated male tail that corresponds with this bias. The tail could be either a handicap or a Fisherian 'aesthetic' trait. One way or another, at least two possible amplifiers may evolve and coexist as a result of the bias for a long tail: one is conspicuous tail colors, and the other is behavioral displays, such as tail fanning, that expose the tail to females. Both are signals, because their sole benefit stems from the fact that they change the female's behavior, and they do it by improving the female's perception of differences in tail length. On average, males with short tails lose by adding information to females, males with long tails benefit by doing so. By definition, such signals are amplifiers. Hence, a single bias can lead to the evolution of not only one signal, but to an assemblage of at least three signals.

Pursuit-deterrent signals are examples of a relatively simple communication system. It is simple, because there are relatively few variables for which predators need to make an estimate, perhaps because the interaction between predators and prey is usually very short. As a result, predators evolve relatively few biases, to which prey respond by evolving only small assemblages of reliable signals that correspond with them. Add to these a few possible cheating signals such as camouflage (a response to the perception of prey's presence), attenuators (of cues that reveal information useful to predators but harmful to prey), or mimicry (a response to biases regarding prey recognition), and we can see that a prey-predator communication system, despite its relative simplicity, can maintain a number of coexisting signals aimed by prey at their predators.

In other systems, such as sexual selection, recipients often need to fit values to a larger array of variables. Hence a female should find information about male features such as the accumulation of his deleterious mutations, his resources, his resource holding power, his feeding efficiency, his health etc., using many different cues. Some of her biases may operate simultaneously, others sequentially. She might find a provider, and then look for a sire, she might put more emphasis on some cues under some conditions, and on other cues at different conditions. We should be very naïve to think that females cannot assess males in the absence of signals, or that all these biases will not create many different signals.

Finally, if signals evolve in response to biases and not biases in response to signals, then another term might be obsolete. It is the alleged recipient's 'cost of communication'. This concept was probably inspired by two ideas. One is the underlined traditional assumption that perceptions and biases follow the evolution of signals. The other, and maybe the dominant one, is Information Theory that assumes information is accumulated in signals in distinct bits, and should, therefore, be extracted the same way. Hence, it follows that 'The amount of information also affects the costs of communicating. Usually the greater the amount of information, the more it costs senders to encode it into signals and receivers to extract it from signals.' (Bradbury & Vehrencamp 1998, p. 388). The second argument is often false. There is no a-priori reason to believe that it is more difficult to make a distinction between two tails, one is 70 cm long and the other is 72 cm long, than it is for a similar pair of tails of 70 and 72 mm long. However, this is only a minor criticism on the concept of the recipient's 'cost of communication', based on empirical grounds. The other comes from a much more fundamental difference in the understanding of communication. It challenges the assumption that perceptions and biases only follow the evolution of signals. This assumption is often subtle, but its effect on the understanding of signals is dramatic. By being more conscious to it, we may avoid yet another Trollveggen fallacy.

Indeed, recipients make a costly investment. However, they would make such an investment regardless of the existence of signals. Their cost is not the cost of signaling, but the cost of information. It costs to gather information, and they would pay this cost if benefits, by the new choices they make as a result of the additional knowledge, exceed costs of gathering the information that changed it. If this is how it works, then reliable signals which, by definition, add information, can only decrease this cost. If it costs more to add a given piece of information via signals than it does via other cues, then biases for such signals will be eliminated by natural selection. After all, if biases evolve largely independently of signals, then the use of signals, from the recipient point of view, is only an option. Hence, the role that the search for information plays in determining recipients' perceptions and biases should not be ignored. Without perceptions and biases as selective agents, signals cannot endure.

Acknowledgements

I thank Rufus A. Johnstone, Phil Taylor and an anonymous referee for their useful comments.

References

Andersson, M. 1982. Sexual selection, natural selection and quality advertisement. *Biological Journal of the Linnean Society*, **17**, 375-393.

Bradbury, J. W. & Vehrencamp, S. L. 1998. *Principles of Animal Communication*. Sunderland, MA: Sinauer Associatiates.

Cramp, S. (ed.)1983. *The Birds of the Western Palearctic. Vol III: Waders to Gulls.* Oxford: Oxford University Press.

Davis, G. W. C. & O'Donald, P. 1976. Sexual selection for a handicap: a critical analysis of Zahavi's model. *Journal of Theoretical Biology*, **57**, 345-354.

Grafen, A. 1990. Biological signals as handicaps. *Journal of Theoretical Biology*, **144**, 517-546.

Getty, T., Kamil, A. C. & Real, P. G. 1987. Signal detection theory and foraging for cryptic or mimetic prey. In: *Foraging Behavior*, Vol. 2 (Ed. by A. C. Kamil, J. R. Krebs & H.R. Pulliam), pp. 525-548. New York: Plenum Press.

Hasson, O. 1989. Amplifiers and the handicap principle in sexual selection: a different emphasis. *Proceedings of the Royal Society of London, Series B*, **235**, 383-406.

Hasson, O. 1991a. Pursuit-deterrent signals: the communication between prey and predator. *Trends in Ecology and Evolution*, **6**, 325-329.

Hasson, O. 1991b. Sexual displays as amplifiers: practical examples with an emphasis on feather decorations. *Behavioral Ecology*, **2**, 189-197.

Hasson, O. 1994. Cheating signals. *Journal of Theoretical Biology*, **167**, 223-238.

Hasson, O. 1997. Toward a general theory of biological signaling. *Journal of Theoretical Biology*, **185**, 139-156.

Hasson, O., Cohen, D. & Shmida, A. 1992. Providing or hiding information: on the evolution of amplifiers and attenuators of perceived quality differences. *Acta Biotheoretica*, **40**, 269-283.

Hutchinson, E. 1959. Homage to Santa Rosalia, or why are there so many kinds of animal? *American Naturalist*, **93**, 145-159.

Kirkpatrick, M. 1986. The handicap mechanism of sexual selection does not work. *American Naturalist*, **127**, 222-240.

Leger, D. W. & Didrichsons, I. A. 1994. An assessment of data pooling and some alternatives. *Animal Behaviour*, **48**, 823-832.

Maynard Smith, J. 1976. Sexual selection and the handicap principle. *Journal of Theoretical Biology*, **57**, 239-242.

Maynard Smith, J. 1991. Theories of sexual selection. *Trends in Ecology and Evolution*, **6**, 146-151.

Maynard Smith, J. & Harper, D. G. C. 1995. Animals signals: models and terminology. *Journal of Theoretical Biology*, **177**, 305-311.

Nur, N. & Hasson, O. 1984. Phenotypic plasticity and the handicap principle. *Journal of Theoretical Biology*, **110**, 275-297.

Ryan, M. J. & Keddy-Hector, A. 1992. Directional patterns of female mate choice and the role of sensory biases. *American Naturalist* (supplement), **139**, s4-s35.

Shannon, C. E. & Weaver, W. 1949. *The Mathematical Theory of Communication.* Urbana, Illinois: University of Illinois Press.

Siegel, S. & Castellan, N. J. Jr. 1989. *Nonparametric Statistics for the Behavioral Sciences.* 2nd edn. New York: McGraw-Hill.

Simmons, K. E. L. 1955. The nature of the predator-reaction of waders towards humans; with special reference to the role of the aggressive-, escape- and brooding-drives. *Behaviour*, **8**, 130-173.

Tilson, R. L. & Norton, P. M. 1981. Alarm duetting and pursuit deterrent in an African antelope. *American Naturalist*, **118**, 455-462.

Walters, J. R. 1990. Anti-predatory behavior of lapwings: field evidence of discriminative abilities. *Wilson Bulletin*, **102**, 49-70.

Wilkinson, L. 1988. *SYSTAT: The System for Statistics.* Evanston, Illinois: SYSTAT Inc.

Wilson, E. O. 1975. *Sociobiology: the New Synthesis.* Cambridge, Mass.: Belknap Press.

Zahavi, A. 1975. Mate selection – a selection for a handicap. *Journal of Theoretical Biology*, **53**, 205-214.

Zahavi, A. 1977. Reliability in communication systems and the evolution of altruism. In: *Evolutionary Ecology* (Ed. by B. Stonehouse & C. M. Perrins), pp. 253-259. London: MacMillan .

Appendix: Techniques and additional data

This appendix adds some information about the field study of head-bobbing of the spur-winged plover.

Techniques

Observations were made in September 1989, following the end of the plovers' breeding season, near kibbutzes Ma'agan Michael and Ma'ayan Zvi's artificial fish ponds, along dirt roads, most of which were surrounded by short shrubs and annuals. I focused on birds that were alone or in small groups (of five or less; mean 2.13, SD=1.04, n=48) and that appeared to be isolated from other groups (apparently, still in territories).

For each observation I focused on an individual bird, the one closest to me. I slowly approached it, alternating walking and stopping until the bird flew away. As a result of the apparent territorial behavior of the birds that I chose to watch, and of my chosen path in the large study area during the entire field study, it is highly unlikely that a plover was recorded more than once.

The numbers pressed on the key-pad indicated the estimated distance between myself and the bird, truncated to the nearest 10 m. The times (in tenths of seconds) of beep instructions and of keys pressed, as well as the estimated distances, were automatically filed. These files were later transferred to an MS-DOS computer for data analysis. However, although the use of a computer in the field improved data handling, it also introduced a measure of imprecision due to my inability to immediately respond to stimuli. To take this into account, I estimated my average delay in initiating walking in response to the beep, and in pressing keys in response to a visual stimulus. Based on daily measurements in the field, using a similar program modified to simulate and measure my response, I found the average delay in my initiation of walking to be 0.66 sec (SD=0.17, n=82). I also measured a delay of 0.30 sec (SD=0.03, n=40) in my key-pressing response to a visual signal that appeared on the computer screen at random intervals of 2 to 6 sec, approximating the plover's maximal bobbing rate (see below). The recording procedure therefore resulted in a total estimated delay of almost a full second, independently of the plover's behavior.

When head-bobs are given periodically, the statistical conclusions of the experiment may be seriously biased. For example, if, regardless of my behavior, head-bobbing is given every n seconds, any movement of mine will be followed by a head-bob within the next n seconds (with an equal probability for each second smaller or equal to n). Thus, if n is small, periodical head-bobbing will indicate an apparent response even if none exists.

To overcome this difficulty I used two statistical procedures: one compared the timing of head-bobs given immediately following my initiation of walking with that given immediately before my initiation of walking (expected to have the same bias). This analysis is shown in Figure 2. The other tested for a difference in response among seconds fewer than n (the cycle of periodical head-bobbing). For the first procedure, the null hypothesis (no chronological orientation) predicts no significant differences between the frequency of head-bobbing expressed backward or forward in time, starting with my initiation of walking. For the second procedure, it predicts no significant difference in the frequency of head-bobs among seconds that are equal to or smaller than n, following my initiation of walking.

One final complexity that I needed to take into account is the availability of time intervals: long intervals are not available for head-bobbing as often as short intervals. Furthermore, once a first head-bob is given following my initiation of walking, the rest of that interval is no longer available for a first head-bob. Thus, the frequency of head-bobs at each time interval (either a half or full second) was calculated as the number of times head-bobs were given at that interval divided by the number of times that interval was available for first head-bobs (which are given as vertical numbers at the bottom of the figures). Available intervals in Figure 2B were similarly calculated, using intervals of standing prior to initiations of walking.

Figure 2B is useful to ascertain there were no other hidden statistical or procedural biases in Figure 2A. For example, to construct Figure 2A, I have pooled data of 48 different individuals, each had 1 to 11 observations. This might create a bias, if between-subject variance exceeds intra-subject variance (Leger and Didrichsons 1994). However, we are interested here in distributions of timing of head-bobs rather than in their averages, and there is no sufficient data for each individual to compare between-subject variations in skewness. Luckily, all potential biases, including this one, if it exists, should be included in Figure 2B. Hence, any pattern that is found in both Figures 2A and 2B may be attributed to such biases. In contrast, patterns that exit in Figure 2A but are not found in Figure 2B could only be attributed to my initiation of walking toward the plovers.

Additional data

The average estimated distance at which I detected the plovers was 45.3 m (range 25-65, SD=9.49, n=48), and the flight distance, 34.5 m (range 18-55, SD=9.34). The difference between detection and flight distances was small because the plovers usually walked away from me during the observations (many observations ended when the plovers seemed to have reached the border of their territory and flew above or around me, to land behind me). The average duration of recording sessions was 2 min 26 sec (SD=1 min 42 sec), and the number of head-bobs per session, 12.9 (SD=10.8).

IN: ESPMARK,Y., AMUNDSEN,T. & ROSENQVIST,G. (EDS.) 2000
ANIMAL SIGNALS:
SIGNALLING AND SIGNAL DESIGN IN ANIMAL COMMUNICATION, PP. 465-485
TAPIR ACADEMIC PRESS, TRONDHEIM, NORWAY

25

Conflicts of Interest in Signal Evolution

RUFUS A. JOHNSTONE

Department of Zoology, University of Cambridge, Downing Street,
Cambridge CB2 3EJ, UK

E-mail: raj1003@hermes.cam.ac.uk

Abstract

Recent models of signalling have assumed that the expenditure required to ensure detection of a display is negligible, and have concentrated instead on the costs that may be necessary to maintain honesty. Such models predict that individuals who share the same interests are likely to communicate using 'conspiratorial whispers', signals that are cheap and inconspicuous. Here, I present a game-theoretical model of signal detection (in a noisy environment, in the presence of potential eavesdroppers) which demonstrates that the idea of conspiratorial whispers is far too simplistic. It is true that in 'cooperative' signalling systems (where signallers attempt to elicit responses that are beneficial for receivers) signal cost is not required to maintain honesty. However, some level of expenditure is still needed to ensure that a signal is reliably detected. Moreover, there exists a conflict of interest between signallers and receivers over the division of this expenditure. To predict the stable level of display in such cases one needs to know how this conflict of interest will be resolved. The model reveals that the outcome may range from a whisper to a conspicuous and costly (though still conspiratorial) display. The more closely related the receiver is to the signaller, the greater the level of signal exaggeration that is expected - the opposite prediction to that of honest signalling models.

Biological signals are the products of coevolution between individuals whose interests may conflict. This simple observation has profound implications. In particular, it has led many authors to ask how selection might maintain a reliable link between signalling behaviour and the underlying state of the signaller. If signallers stand to gain by misleading receivers as to their condition, how can honesty prove stable?

Most readers will be familiar with at least one answer to this question: honesty can be maintained by signal cost. This idea, better known as the 'handicap principle', was proposed by Amotz Zahavi (1975, 1987). If signals are costly to produce, then they may prove worthwhile only for signallers who are of high quality, or who stand to gain a great deal by eliciting a favourable response. Strategic models of signal evolution have confirmed this possibility (e.g. Grafen 1990; Godfray 1991; Maynard Smith 1991a; Johnstone & Grafen 1992; Vega-Redondo & Hasson 1993), and a great deal of theoretical research has been devoted to exploring further implications of the handicap principle for signal evolution (reviewed in Johnstone 1998a).

Unfortunately, strategic signalling models have focused on the problem of honesty perhaps too closely. They assume, in most cases, that signal cost is only ever needed to maintain reliability. Consequently, they predict that when there is no incentive for deceit, animals will employ 'conspiratorial whispers' (i.e., signals that are cheap and inconspicuous - Dawkins & Krebs 1984; Godfray 1991; Maynard Smith 1991a; Johnstone & Grafen 1992; Reeve 1997). This is to overlook the problems of signal detection. Animal communication often takes place over long ranges, in a noisy environment. Considerable expenditure may be required simply to produce a detectable display. Signal cost may therefore serve to enhance the efficacy, rather than the reliability of communication (Wiley 1983, 1994; Harper 1991; Dusenbery 1992; Johnstone 1997, 1998b; Bradbury & Vehrencamp 1998).

There does exist a large body of biological literature dealing with the problems of signal detection (see Bradbury & Vehrencamp 1998 for a review). A number of published analyses, for instance, focus on the challenge faced by receivers of distinguishing a signal from random noise, or ask how the level of noise should influence the effort that signallers invest in display (e.g. Getty 1985; Reeve 1989). But just as strategic models ignore the problems of signal detection, models of signal detection in turn ignore the possible conflicts of interest with which the strategic analyses are concerned. Instead, they treat communication as an essentially cooperative endeavour, in which both signallers and receivers share the same goal of effective signal transmission, or else focus on the behaviour of only one party (either signaller or receiver) in isolation (though see Wiley 1994 for an exception).

This is a mistake. Even in an apparently cooperative signalling system, where signallers stand to gain by eliciting a response that is advantageous for receivers, there still exists a conflict of interest between the two. Signallers must balance the benefits of effective communication against the costs of attracting eavesdroppers, receivers against the costs of overeagerness leading to false alarms and inappropriate responses (Hasson 1991;

Wiley 1994; Bradbury & Vehrencamp 1998). Each does best when the other bears the greater part of the total cost of communication. To predict the outcome of signal evolution, one must determine how this conflict is resolved.

Below, I present a simple model of cooperative communication, in which I attempt to do just that. I assume that signallers stand to gain by eliciting a response that benefits receivers. Consequently, the problem of honesty does not arise. However, I also assume that communication takes place in a noisy environment in the presence of potential eavesdroppers (McGregor 1993) or illegitimate receivers (Otte 1974), so that some level of cost must be incurred simply to permit detection. Signallers and receivers thus have conflicting interests, since each does best if the other bears the greater part of this cost. As in strategic models of signalling, I adopt an explicitly game-theoretical approach in order to determine the resolution of the conflict.

A game-theoretical model of signal detection

The model features three types of player: signallers, receivers and eavesdroppers. Signallers must choose what magnitude of signal to adopt when they display. Their goal is to maximise the probability of eliciting a response from a receiver, while minimising the probability of eliciting a response from an eavesdropper. Receivers and eavesdroppers, in turn, must choose what response thresholds to adopt (i.e. what is the minimum level of perceived signal magnitude to which they will respond). Their goal is to maximise the probability of responding if a signal is present, while minimising the probability of responding if a signal is absent. Both types of recipient are constrained by environmental noise (or, equivalently, by perceptual error), which leads to overlap between these two stimulus categories.

Formally, I assume that signallers adopt some non-negative signal magnitude s when displaying. Receivers and eavesdroppers each adopt some threshold level of perceived signal magnitude above which they will respond, denoted t_r for the former and t_e for the latter. Perceived magnitude is equal to true magnitude (s if a signal is present, 0 if it is absent) plus a random noise or error term that is normally distributed with mean 0 and variance σ^2.

This simple model can serve to represent many different situations. Signallers could represent offspring trying to draw the attention of their parents, or parents attempting to alert their offspring to the presence of a predator. Alternatively, they might be males attempting to attract conspecific females, or individuals of a social species trying to establish contact with relatives or fellow group members. The eavesdroppers could represent predators (e.g. Endler 1980; Tuttle and Ryan 1981; Sakaluk & Belwood 1984; Haskell 1994), parasites (e.g. Cade 1975, 1979; Zuk et al. 1995), potential competitors (e.g. McGregor 1993; Oliveira et al. 1998) or even prey (e.g. Grether & Grey 1996). The signal itself could be visual, auditory or olfactory, morphological or behavioural.

The three key features that characterise the model are as follows: First, the signalling system is 'cooperative', in the sense that signallers are attempting to elicit a response that is beneficial for receivers (although, as emphasised in the introduction, their evolutionary interests nevertheless differ). Second, I am concerned only with signal detection, not with discrimination between signallers of different qualities or types (the focus of most recent game-theoretical models of signalling). Thirdly, I do not explicitly address the issue of competition between different signallers (or, for that matter, between different receivers or different eavesdroppers). My focus is on the conflict of interest between signallers and receivers, and how it is resolved.

The optimal signaller strategy (given particular receiver and eavesdropper strategies) depends upon the probabilities that an eavesdropper and that a receiver are present when a signal is given, and on the relative costs and benefits of eliciting a response from each. The optimal receiver or eavesdropper strategy (given a particular signaller strategy) depends upon the probability that a signal is present when a decision is made about whether or not to respond, and on the relative costs (to recipients of that type) of responding when a signal is absent and of failing to respond when a signal is present.

Below, I first determine the optimal response thresholds for receivers and eavesdroppers given a particular signaller strategy, and then the optimal signal magnitude for signallers given particular receiver and eavesdropper strategies. Finally, I use these two results to calculate the model's evolutionarily stable solution, i.e., the (unique) set of strategies for which each component strategy is optimal given the other two.

The recipients' perspective

From the point of view of a recipient (either receiver or eavesdropper), the decision to respond or to refrain from doing so yields one of four possible outcomes, summarised in Table 1: correct detection, missed detection, false alarm and correct rejection. I shall write the inclusive fitness payoffs to the recipient contingent on these outcomes as w_{CD}, w_{MD}, w_{FA} and w_{CR}, respectively, where $w_{CD} > w_{MD}$, i.e., a correct detection is better for the recipient than a missed detection, and $w_{FA} < w_{CR}$, i.e., a false alarm is worse for the recipient than a correct rejection. I assume that all costs and benefits are additive, so that a recipient does best to maximise the expected payoff following its decision.

Table 1. *Possible outcomes of a signal recipient's decision whether or not to respond.*

	Signal Present	**Signal Absent**
Positive Response	Correct Detection	False Alarm
Negative Response	Missed Detection	Correct Rejection

Assuming that $s > 0$, i.e. that signallers employ a signal of non-zero magnitude, the optimal strategy for the recipient is to respond whenever perceived magnitude exceeds some threshold value. Let p denote the probability that a signal is present. Then the expected inclusive fitness payoff to a recipient who responds whenever intensity exceeds a threshold of t, denoted $w(t)$, is given by

$$w(t)=\frac{p}{\sqrt{2\pi\sigma^2}}\left[w_{MD}\int_{x<t}\exp\left(\frac{-(x-s)^2}{2\sigma^2}\right)dx+w_{CD}\int_{x>t}\exp\left(\frac{-(x-s)^2}{2\sigma^2}\right)dx\right]$$

$$+\frac{(1-p)}{\sqrt{2\pi\sigma^2}}\left[w_{CR}\int_{x<t}\exp\left(\frac{-x^2}{2\sigma^2}\right)dx+w_{FA}\int_{x>t}\exp\left(\frac{-x^2}{2\sigma^2}\right)dx\right] \quad (1)$$

(where x, the variable of integration in the above equation, denotes perceived signal intensity). Differentiating, we obtain

$$w'(t)=\frac{(1-p)(w_{CR}-w_{FA})}{\sqrt{2\pi\sigma^2}}\exp\left[\frac{-t^2}{2\sigma^2}\right]-\frac{p(w_{CD}-w_{MD})}{\sqrt{2\pi\sigma^2}}\exp\left[\frac{-(t-s)^2}{2\sigma^2}\right] \quad (2)$$

Solving for $w'(t)=0$ yields a unique stationary value, $\hat{t}(s)$, given by

$$\hat{t}(s)=\frac{s}{2}-\frac{\sigma^2\ln K}{s} \quad (3a)$$

where K, which I shall refer to as the recipient's 'incentive' to respond, is defined as

$$K=\frac{p(w_{CD}-w_{MD})}{(1-p)(w_{CR}-w_{FA})} \quad (3b)$$

It is then easy to show (given $s > 0$) that $w''(\hat{t}(s))<0$, and hence that $\hat{t}_s$ represents the optimal, fitness-maximising choice of t for the recipient given s.

Since the fitness consequences of a missed detection and a false alarm are likely to differ for the intended receiver of a signal and for an eavesdropper, I will write K_r and $\hat{t}_r(s)$ for receivers' incentive and optimal threshold given s, and K_e and $t_e(s)$ for eavesdroppers'. I assume that $K_e < K_r < 1$, i.e., that the intended receiver of a signal has a greater incentive to respond than an eavesdropper, but that neither recipient's incentive is sufficiently great to favour a response in the absence of communication.

The signaller's perspective

From a signaller's point of view, production of a signal brings potential benefits, in that it may serve to elicit a response from a receiver. At the same time, it brings potential costs, in that it may attract the attention of an eavesdropper (I assume that the physical cost of producing the signal is negligible by comparison with this risk). I shall assume that the inclusive fitness benefits and costs of signalling are additive (so that a signaller does best to maximise the expected payoff from signalling), and are proportional on the one hand to the probability of eliciting a response from a receiver, and on the other to the probability of eliciting a response from an eavesdropper. The probability that a receiver is present when the signal is given will be denoted q, while the probability that an eavesdropper is present will be denoted q'. The benefit to be gained by eliciting a response from the former will be denoted b, and the cost of eliciting a response from the latter c.

If receivers and eavesdroppers respond whenever perceived signal magnitude exceeds a threshold value of t_r and t_e, respectively, then the expected fitness payoff to a signaller who produces a signal of magnitude s, denoted v(s), is given by

$$v(s)=\frac{qb}{\sqrt{2\pi\sigma^2}}\int_{x>t_r} exp\left(\frac{-(x-s)^2}{2\sigma^2}\right)dx-\frac{q'c}{\sqrt{2\pi\sigma^2}}\int_{x>t_e} exp\left(\frac{-(x-s)^2}{2\sigma^2}\right)dx \quad (4)$$

(where x once again denotes perceived signal intensity). Equation (4), by a change of variable, may be written as

$$v(s)=\frac{qb}{\sqrt{2\pi\sigma^2}}\int_{y>t_r-s} exp\left(\frac{-y^2}{2\sigma^2}\right)dy-\frac{q'c}{\sqrt{2\pi\sigma^2}}\int_{y>t_e-s} exp\left(\frac{-y^2}{2s^2}\right)dy \quad (5)$$

Differentiating, we obtain

$$v'(s)=\frac{qb}{\sqrt{2\pi\sigma^2}}exp\left[\frac{-(t_r-s)^2}{2\sigma^2}\right]-\frac{q'c}{\sqrt{2\pi\sigma^2}}exp\left[\frac{-(t_e-s)^2}{2\sigma^2}\right] \quad (6)$$

Solving for $v''(s)=0$ yields a unique stationary value, $\hat{s}(t_r,t_e)$, given by

$$\hat{s}(t_r,t_e)=\frac{t_r+t_e}{2}+\frac{\sigma^2 \ln K_s}{t_e-t_r} \quad (7a)$$

where K_s, which I shall refer to as the signaller's 'incentive' to display, is defined as

$$K_s=\frac{qb}{q'c} \quad (7b)$$

It is then easy to show (given $t_r < t_e$) that $v''(\hat{s}(t_r,t_e))<0$, and hence that $\hat{s}(t_r,t_e)$ represents the optimal, fitness-maximising choice of s for the signaller given t_r and t_e.

I will assume that $K_s < 1$, i.e., that benefits of attracting a response from a receiver are outweighed by the costs of attracting a response from an eavesdropper, so that signallers stand to gain from display only if they can achieve a higher probability of attracting the attention of the former than of the latter.

Solving the model

A solution to the model is a set of strategies (s^*, t^*_r, t^*_e) that satisfies

$$s^* = \hat{s}(t^*_r, t^*_e)$$

$$t^*_r = \hat{t}_r(s^*) \qquad (8)$$

$$t^*_e = \hat{t}_e(s^*)$$

in other words, for which each strategy represents a unique best response to the other two. Substitution of (3) and (7) into (8) gives

$$s^* = \frac{t^*_r + t^*_e}{2} + \frac{\sigma^2 \ln K_s}{t^*_e - t^*_r} \qquad (9a)$$

$$t^*_r = \frac{s^*}{2} - \frac{\sigma^2 \ln K_r}{s^*} \qquad (9b)$$

$$t^*_e = \frac{s^*}{2} - \frac{\sigma^2 \ln K_e}{s^*} \qquad (9c)$$

Substituting (9b) and (9c) into (9a), and solving for s^*, we obtain (subject to the conditions on K_s, K_r and K_e mentioned above) the following unique, non-negative solution

$$s^* = \sigma \sqrt{\frac{\ln(K_r K_e)\ln(K_r / K_e)}{\ln(K_s^2 K_e / K_r)}} \qquad (10a)$$

This represents the evolutionarily stable level of display adopted by signallers. The corresponding response thresholds adopted by receivers and by eavesdroppers (obtained by substituting (10a) into (9b) and (9c), respectively) are given by

$$t^*_r = \sigma \left[\frac{\ln(K_r K_e)\ln(K_r / K_e) - 2\ln K_r \ln(K_s^2 K_e / K_r)}{2\sqrt{\ln(K_r K_e)\ln(K_r / K_e)\ln(K_s^2 K_e / K_r)}} \right] \qquad (10b)$$

and

$$t^*_e = \sigma \left[\frac{\ln(K_r K_e)\ln(K_r / K_e) - 2\ln K_e \ln(K_s^2 K_e / K_r)}{2\sqrt{\ln(K_r K_e)\ln(K_r / K_e)\ln(K_s^2 K_e / K_r)}} \right] \qquad (10c)$$

Communication between relatives

The payoffs to receivers and to signallers following their various possible courses of action are measured in terms of inclusive fitness. Consequently, the precise values of these payoffs, and hence of K_r (receivers' incentive to respond) and K_S (signallers incentive to display), depend on the coefficient of relatedness between a signaller and the receiver to which its signal is directed, which I will denote r.

All other things being equal, b, the benefit that a signaller stands to gain by eliciting a response from the receiver, is likely to increase with r. The reason for this is that in addition to any direct fitness payoff from signalling, there is also an indirect fitness benefit to be gained by eliciting a response that is advantageous for the receiver. The magnitude of this indirect benefit is proportional to r (since the more closely related the receiver is to the signaller, the more the latter gains by provoking a response that benefits the former). For the same reasons, (w_{CD} - w_{MD}), the advantage that a receiver stands to gain by responding when a signal is present, is also likely to increase with r. In addition to the direct fitness payoff from a correct response, the receiver also stands to reap an indirect fitness gain by boosting the fitness of the signaller (who will benefit from a response). Once again, the magnitude of this indirect benefit is proportional to r.

As a consequence of the above, K_r (receivers' incentive to respond) and K_S (signallers' incentive to display) both increase with r. Formally, writing b_0 for the benefit to a signaller of attracting a receiver response when $r = 0$ (i.e., for the direct fitness gain from attracting such a response), and (w_{CD0} - w_{MD0}) for the benefit to a receiver of responding to a signal when $r = 0$ (i.e., for the direct fitness gain from responding), we have

$$K_s = \frac{q(b_0 + r(w_{CD0} - w_{MD0}))}{q'c} \tag{11a}$$

and

$$K_r = \frac{p((w_{CD} - w_{MD}) + rb_0)}{(1 - p)(w_{CR} - w_{FA})} \tag{11b}$$

These equations, in conjunction with (10), can be used to determine the effects of changes in r on the evolutionarily stable outcome of the model (all other parameters being held constant).

Results of the model

In this section I will illustrate graphically the solution specified by (10), focusing on the level of signallers' display, and on the probability that receivers and eavesdroppers respond to a signal when one is present.

From (10a), it is clear that the level of display adopted at equilibrium is proportional to the level of noise or perceptual error, σ. In other words, the greater the level of noise (or the less accurate perception), the more exaggerated the displays that signallers are predicted to employ (a pattern for which there is empirical support - e.g. Marchetti 1993). Below, when discussing the effects of the model's other parameters on the evolutionarily stable signalling strategy, I shall focus on signal magnitude measured in units of size s (which may be thought of as 'perceptible' magnitude).

Signal magnitude

Figure 1 shows s^* (ESS signal magnitude, measured in units of size ∂) as a function of K_r (receivers' incentive to respond) for several different values of K_e (eavesdropper's incentive to respond), assuming that K_s (signallers' incentive to display) is fixed at 0.5. Successively lower curves correspond to successively higher values of K_e, indicating that signal magnitude is predicted to decrease with eavesdroppers' incentive to respond (dropping to 0 for $K_e = K_r$). This trend reflects the fact that as eavesdroppers' incentive to respond approaches that of receivers, it becomes increasingly difficult to elicit a response from the latter without attracting the attention of the former. Consequently, the net benefit to be gained by signalling diminishes, and selection favours lower levels of display. For the same reason, the predicted ESS signal magnitude tends to increase with K_r, the receivers' incentive to respond. At values of K_r much greater than K_e, however, this increase begins to level off, and s^* may even start to decrease. At this point, the effects of

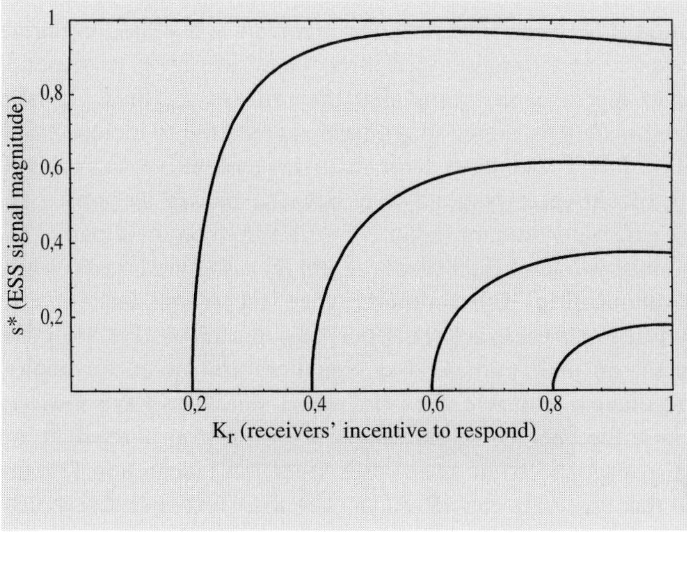

Figure 1. *s^* (ESS signal magnitude, measured in units of size σ) as a function of K_r (receivers' incentive to respond), for several different values of K_e (eavesdroppers' incentive to respond). K_s (signallers' incentive to display) is fixed at 0.5, and successively lower curves correspond to successively higher values of K_e : 0.2, 0.4, 0.6 and 0.8. (From Johnstone 1998; Evolution 52, 1554-1563).*

the widening gap between K_r and K_e (which makes it easier to attract receivers' attention without drawing that of eavesdroppers) is outweighed by the increasing 'adaptive gullibility' of receivers, which allows signallers to elicit a response with a display of reduced magnitude and hence lower cost (see below for further discussion).

Figure 2 shows s^* as a function of K_s (signallers' incentive to display) for several different values of K_e (eavesdroppers' incentive to respond), assuming that K_r (receivers' incentive to respond) is fixed at 0.8. Successive curves correspond to successively higher values of K_e, demonstrating once again that (as discussed above) an increase in eavesdroppers' incentive to respond leads to a decrease in predicted signal magnitude. By contrast, the graph shows that s^* is predicted to increase with K_s. The greater the signallers' incentive to display, the greater the magnitude of the signals that they employ.

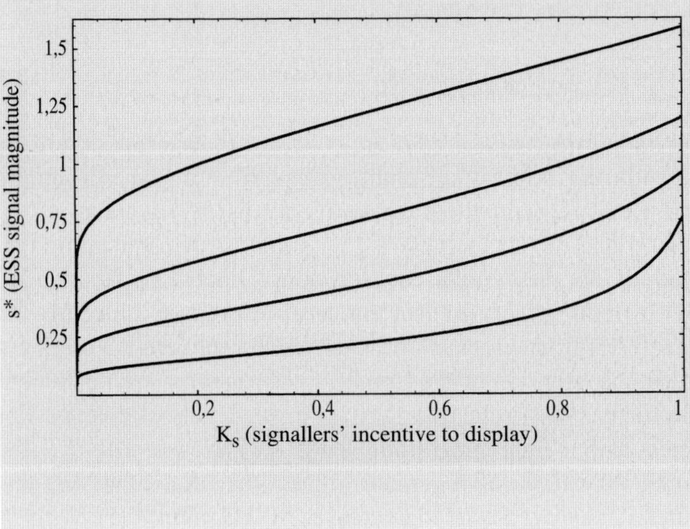

***Figure 2**. s* (ESS signal magnitude, measured in units of size σ) as a function of K_s (signallers' incentive to display), for several different values of K_e (eavesdroppers' incentive to respond). K_r (receivers' incentive to respond) is fixed at 0.8, and successively lower curves correspond to successively higher values of K_e : 0.1, 0.3, 0.5 and 0.7. (From Johnstone 1998; Evolution 52, 1554-1563).*

The fundamental conflict of interest between signallers and receivers is revealed in Figure 3, which shows a contour plot of s^* as a function of K_r (receivers' incentive to respond) and K_s (signallers' incentive to display), assuming that the ratio of K_e to K_r is held constant at 0.5. Under these circumstances, signal magnitude is predicted to increase with signallers' incentive to display, but to decrease with receivers' incentive to respond. When receivers have a strong incentive to respond (e.g. because missed responses are costly, or false alarms cheap), selection favours 'adaptive gullibility' on their part. This allows signallers to reduce the magnitude of their display (and its associated costs) while still achieving effective communication. By contrast, when receivers' incentive to respond is low (e.g. because missed responses are cheap, or false alarms costly), selection favours 'adaptive fastidiousness' on their part, so that signallers are forced to employ highly exaggerated displays to elicit a response. Equally, when signallers have a strong incentive to display (e.g. because the benefits of eliciting a response from a receiver are high, or the costs of eliciting a response from an eavesdropper low), selection favours more exaggerated displays, so that receivers can afford to raise their response thresholds.

When, on the other hand, the incentive to display is weak, selection favours low level displays, so that receivers are forced to lower their response threshold.

Probability of receiver response

Figure 4 shows the probability that a receiver responds to the signal when present, as a function of K_r (receivers' incentive to respond) for several different values of K_e (eavesdroppers' incentive to respond). Successively lower curves correspond to successively higher values of K_e, indicating that receiver response is less probable when eavesdroppers have a greater incentive to respond. This reflects the fact that, as discussed above, signallers tend to employ displays of lower magnitude (which are harder to detect) under these circumstances. By contrast, the probability of response increases with K_r

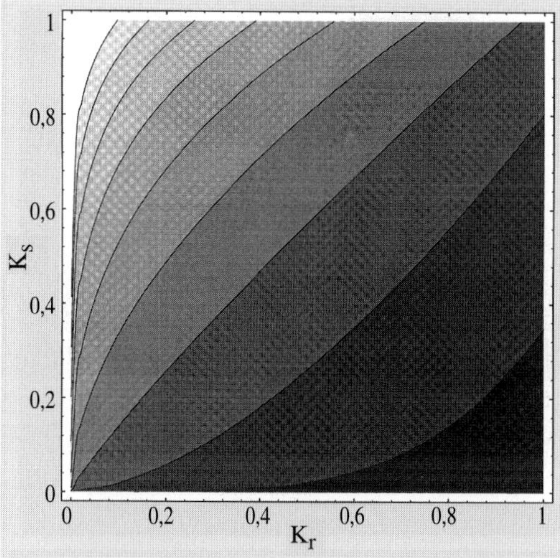

Figure 3. *Contour plot of s* (ESS signal magnitude, measured in units of size σ) as a function of K_s (signallers' incentive to display), and K_r (receivers' incentive to respond), assuming that the ratio of K_e (eavesdroppers' incentive to respond) to K_r is fixed at 0.5. Darker regions of the plot represent lower values of s*, and lighter regions higher values. (From Johnstone 1998; Evolution 52, 1554-1563).*

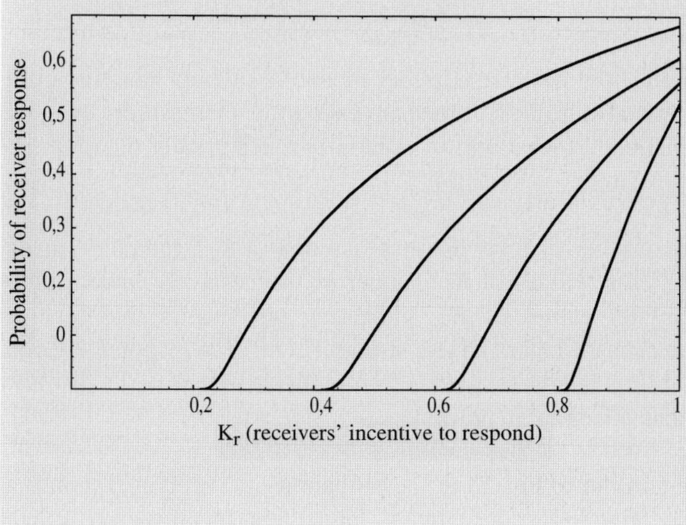

Figure 4. *Probability that a receiver responds to the signal when present, as a function of K_r (receivers' incentive to respond), for several different values of K_e (eavesdroppers' incentive to respond). K_s (signallers' incentive to display) is fixed at 0.5, and successively lower curves correspond to successively higher values of K_e : 0.2, 0.4, 0.6 and 0.8. (From Johnstone 1998; Evolution 52, 1554-1563).*

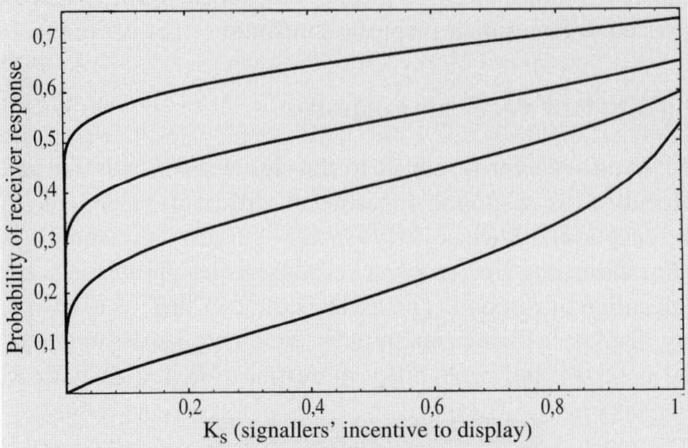

Figure 5. *Probability that a receiver responds to the signal when present, as a function of K_s (signallers' incentive to display), for several different values of K_e (eavesdroppers' incentive to respond). K_r (receivers' incentive to respond) is fixed at 0.8, and successively lower curves correspond to successively higher values of K_e: 0.1, 0.3, 0.5 and 0.7.(From Johnstone 1998; Evolution 52, 1554-1563).*

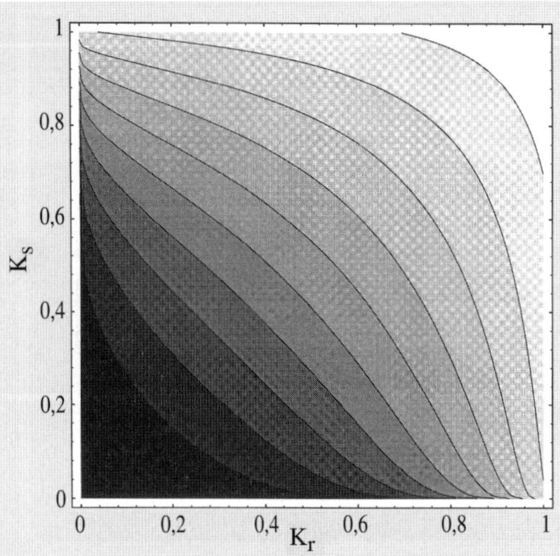

Figure 6. *Contour plot showing the probability that a receiver responds to the signal when present, as a function of K_s (signallers' incentive to display), and K_r (receivers' incentive to respond), assuming that the ratio of K_e (eavesdroppers' incentive to respond) to K_r is fixed at 0.5. Darker regions of the plot represent lower probabilities of response, and lighter regions higher probabilities. (From Johnstone 1998; Evolution 52, 1554-1563).*

(receivers' incentive to respond), both because a higher incentive favours adaptive gullibility on the part of receivers, and also because it tends to encourage increasingly exaggerated display on the part of signallers (at least until K_r becomes much greater than K_e - see above).

Figure 5 shows the probability of receiver response as a function of K_s (signallers' incentive to display), for several different values of K_e. As before, successively lower curves correspond to successively higher values of K_e, indicating once again that receiver response is less probable when eavesdroppers have a strong incentive to respond. By contrast, receiver response probability increases with K_s (signallers' incentive to display), due to the positive effect this has on signal magnitude.

Figure 6 shows a contour plot (similar to Fig. 3) of the probability of receiver response as a function of K_r (receivers' incentive to respond) and K_s (signallers' incentive to display), assuming that the ratio of K_e to K_r is held constant at 0.5. Under these circumstances, response probability increases with both K_s and K_r. The effect of the former parameter is unsurprising: as mentioned above, signallers employ more exaggerated displays when their incentive is higher, making signal detection easier. By contrast, signal magnitude is actually predicted to decrease as K_r increases (assuming that the ratio of K_e to K_r is held constant). Response is nevertheless more probable when receivers' incentive is greater, because selection favours greater 'gullibility' on their part (and this increase outweighs the effects of the reduction in signal magnitude).

Probability of eavesdropper response

Figure 7 shows the probability that an eavesdropper responds to the signal when present, as a function of K_r (receivers' incentive to respond), for several different values of K_e (eavesdroppers' incentive to respond). The graph shows that the probability of eavesdropper response initially increases rapidly with receivers' incentive to respond, but then starts to level off and, for values of K_r much greater than K_e, starts to decrease once again (correspondingly, therefore, the probability of response increases with K_e for low values of K_r, but decreases for higher values). This reflects the corresponding changes in ESS signal magnitude: signallers' displays increase with K_r up to a point (as it becomes easier to attract the attention of receivers without alerting eavesdroppers), but then start to decrease as K_r becomes much larger than K_e, due to increasing gullibility on the part of receivers.

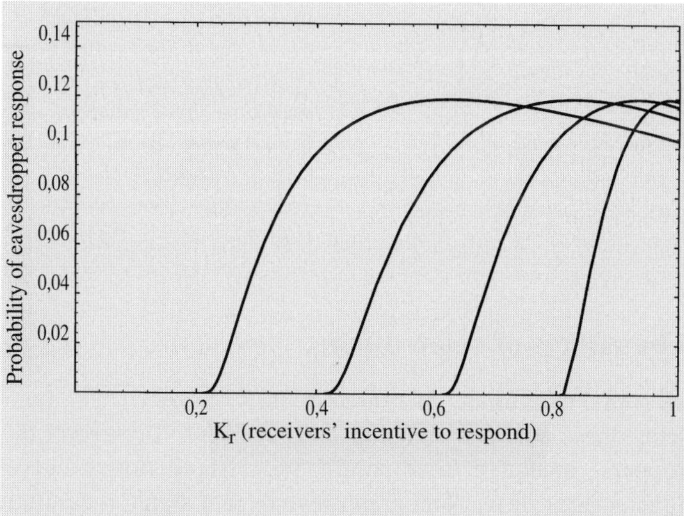

Figure 7. *Probability that an eavesdropper responds to the signal when present, as a function of K_r (receivers' incentive to respond), for several different values of K_e (eavesdroppers' incentive to respond). K_s (signallers' incentive to display) is fixed at 0.5, and successively lower (i.e., rightward shifted) curves correspond to successively higher values of K_e : 0.2, 0.4, 0.6 and 0.8. (From Johnstone 1998; Evolution 52, 1554-1563).*

A corresponding pattern can be seen in Figure 8, which shows the probability of eavesdropper response as a function of K_e for several different values of K_s (signallers' incentive to display), assuming that K_r is held constant at 0.8. The graph shows that eavesdropper response peaks for intermediate values of K_e. When eavesdroppers' incentive is small, their probability of response is low simply because selection favours adaptive fastidiousness on their part. As K_e increases, the probability of response also grows, up to a point. But as K_e approaches the value of K_r, it becomes increasingly difficult for signallers to attract the attention of receivers without drawing a response from eavesdroppers, so that they lower their level of display. The result is that the probability of eavesdropper response starts to decline once more, eventually reaching zero for $K_e = K_r$ (at which point signallers cease to display at all).

Figure 8 also shows that the probability of eavesdropper response increases with K_s, since greater signaller incentives lead to more exaggerated, and hence more readily detectable signals. Moreover, the greater the value of K_s, the higher the value of K_e at which response probability peaks. This reflects the fact that when signallers' incentive is large, signal magnitude declines more slowly as K_e approaches K_r.

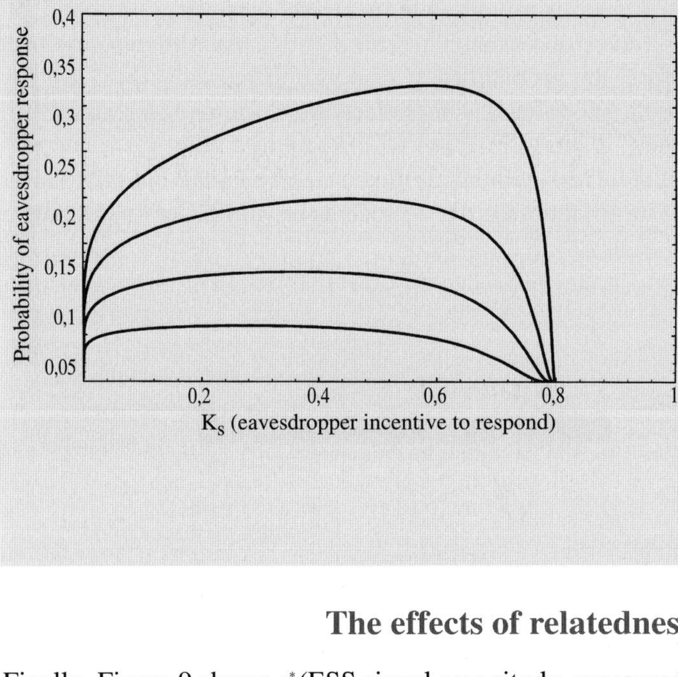

Figure 8. *Probability that an eavesdropper responds to the signal when present, as a function of K_e (eavesdroppers' incentive to respond), for several different values of K_s (signallers' incentive to display). K_r (receivers' incentive to respond) is fixed at 0.8, and successively lower curves correspond to successively lower values of K_s: 0.3, 0.5, 0.7 and 0.9. (From Johnstone 1998; Evolution 52, 1554-1563).*

The effects of relatedness

Finally, Figure 9 shows s^* (ESS signal magnitude, measured in units of size σ), as a function of r (the coefficient of relatedness between signaller and receiver), for several different values of K_e (eavesdroppers' incentive to respond). Successively lower curves correspond to successively higher values of K_e. The graph reveals that signal magnitude is predicted to increase with the degree of relatedness between signaller and receiver, this effect being more pronounced the greater the eavesdroppers' incentive to respond. As r increases, so too do K_s (signallers' incentive to display) and K_r (receivers' incentive to respond), as signallers and receivers stand to gain greater indirect fitness benefits from signalling or responding. Both effects favour increased signal magnitude, at least up to a point. The reason why the effects of relatedness are less marked for lower values of K_e, however, is that (as previously mentioned) when the difference between K_e and K_r becomes very large, signal magnitude starts to decrease with increasing K_r. Under these circumstances, the effects of an increase in r on K_s (greater incentives for signallers favouring displays of greater magnitude) are partially cancelled out by the effects on K_r (greater incentives for receivers leading to adaptive gullibility, which favours signals of lower magnitude).

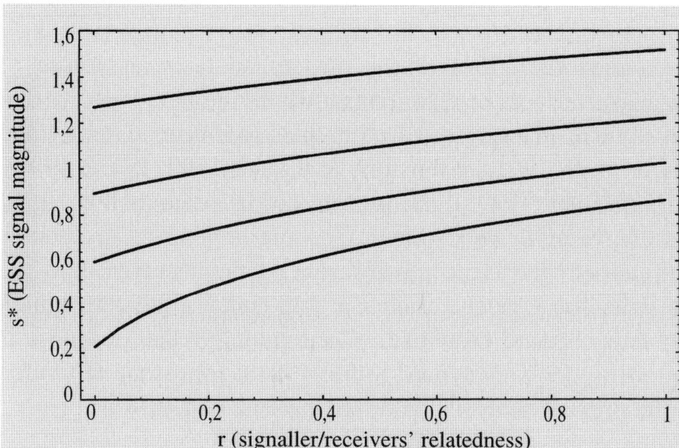

Figure 9. *s* (ESS signal magnitude) as a function of r (the coefficient of relatedness between signaller and receiver), for several different values of K_e (eavesdroppers' incentive to respond). Signallers' incentive to display and receivers' incentive to respond are both equal to 0.5 when r = 0. Successively lower curves correspond to successively higher values of K_e: 0.1, 0.225, 0.35 and 0.475. (From Johnstone 1998; Evolution 52, 1554-1563).*

Discussion

The model that I have presented demonstrates that the idea of 'conspiratorial whispers' is far too simplistic. It is true that in 'cooperative' signalling systems (where signallers attempt to elicit responses that are beneficial for receivers) signal cost is not required to maintain honesty. However, some level of expenditure is still needed to ensure that a signal is reliably detected. Moreover, there exists a conflict of interest between signallers and receivers over the division of this expenditure. To predict the stable level of display in such cases one needs to know how this conflict of interest will be resolved. As the model reveals, the result may range from a whisper to a conspicuous and costly (though still conspiratorial) display.

The main parameters that influence the outcome are: (i) the relative costs to receivers and to eavesdroppers of false alarms and missed detections; (ii) the relative benefits and costs to signallers of attracting receivers and eavesdroppers; (iii) the degree of relatedness between signallers and receivers. A high cost to signallers of attracting eavesdropper attention increases the risks of exaggerated display, as does a relatively low cost to eavesdroppers of false alarms (which favours a low response threshold on their part). Both of these factors thus tend to select for conspiratorial whispers. A high benefit to signallers of attracting receiver attention, by contrast, favours conspicuous display, as (up to a point) does a low relative cost to receivers of false alarms. The latter gives receivers a greater incentive to respond, which makes it easier for a signaller to attract their attention without drawing that of eavesdroppers. If the receiver's incentive is much greater than eavesdroppers' incentive to respond, however, further increases may instead favour reduced expenditure on display, because the receiver's greater adaptive gullibility makes it easy for the signaller to communicate using a cheaper signal.

One striking difference between the predictions of the present analysis and those of honest signalling models concerns the effects of relatedness. While the latter predict that signal cost should tend to decrease with the degree of relatedness between signaller and receiver (Godfray 1991; Maynard Smith 1991a; Johnstone & Grafen 1992; Reeve 1997), or between competing signallers (Godfray 1995), the present model predicts that signal cost should increase with signaller/receiver relatedness. The predictions of the honest signalling models are easily explained: the more closely related the receiver is to the signaller, the less incentive the latter has for deceit (since it will suffer a greater indirect fitness penalty if it elicits an inappropriate response). Consequently, the less cost is required to maintain honesty (Godfray 1991; Maynard Smith 1991a; Johnstone & Grafen 1992; Reeve 1997). Why does the present analysis yield the opposite conclusion?

As signaller/receiver relatedness increases, so does the signaller's incentive to display (K_s in the model) and the receiver's incentive to respond (K_r). The signaller stands to gain an indirect fitness payoff by provoking a response that is beneficial for the receiver, while the latter stands to gain an indirect fitness payoff by adopting a response that benefits the signaller. In both cases, the magnitude of the indirect fitness gain is proportional to the degree of relatedness between the two. An increase in the signaller's incentive obviously favours higher expenditure on display; so too (up to a point) does an increase in the receiver's incentive to respond. If the receiver's incentive is much greater than eavesdroppers' incentive to respond, however, increases in K_r may (as discussed above) favour reduced expenditure on display. Consequently, the increase in signal cost with relatedness will be less marked when eavesdroppers are less likely to respond to a signal (i.e., when signalling is less risky).

A good example of a situation in which the model's predictions should be met is that of alarm calling. Here, the signal serves to elicit a response that is beneficial for the receiver (at least when employed to warn of an actual predator; the deceptive use of alarm calls - e.g. Munn 1986; Møller 1988, 1990 - is a different matter), and its production is often thought to entail some risk (e.g. Alatalo & Helle 1990). Running through the predictions given above, we would therefore expect signal intensity to be lower the greater the risk of calling (i.e., the closer the predator is to the signaller, or the more likely it is to be attracted by a call) and the greater the cost of a false alarm to receivers. Both the intensity of the signal and its cost (in terms of risk) are predicted to be greater the more closely related potential receivers are to the signaller (this effect being more marked when signalling is more dangerous). In addition, receivers are predicted to exhibit a lower response threshold under these circumstances. Existing data suggest that in a number of cases, individuals are in fact more likely to give alarm calls in the presence of relatives (e.g.

Hoogland 1983; Davis 1984), and are less likely to do so when a predator is close to them (e.g. Alatalo & Helle 1990), but there is little evidence regarding the nature of the signals given in different circumstances (e.g. their length or volume, or the position from which they are given) and even less regarding the responsiveness of potential receivers. There is thus still considerable scope for tests of the model (especially since similar predictions can also be made in other instances of 'cooperative' communication, e.g. recruitment of conspecifics to foraging sites or for group defence, or mutual location of parents and offspring).

The conflict of interest that is the focus of the model becomes relevant whenever signal detection poses a significant challenge, entailing costs for signallers or receivers (or both). Such costs could take many different forms. I have focused on the risks to signallers of attracting eavesdropper attention, and to receivers of false alarms and inappropriate responses. This is a convenient way in which to demonstrate the conflict of interest between the two parties, and to model its resolution, since it yields a simple, analytical solution. Moreover, there is evidence from a wide range of different signalling systems that the risk of attracting unwanted attention can often represent a significant cost to the signaller (e.g. Cade 1979; Endler 1980; Tuttle & Ryan 1981; Sakaluk & Belwood 1984; Haskell 1994; Grether & Grey 1996). The production of a detectable signal may also, however, involve considerable energetic expenditure, while effective detection may, on the part of receivers, require investment in receptor structures, or expenditure of time. It would be of great interest to incorporate such 'intrinsic' costs of signal production or detection into the model.

Even where communication is not 'cooperative' and signallers have an incentive to deceive, expenditure on signal detection cannot be overlooked. Sexual signalling to attract the attention of mates, mentioned earlier, is a case in point. This is the context in which ideas and models of honest signalling were first developed, and the emphasis in recent studies has been on the extent to which females (usually the choosier sex - see Trivers 1972; Clutton-Brock & Parker 1992; Johnstone et al. 1996) can discriminate between desirable and undesirable males on the basis of their displays, and on the factors that can maintain the reliability of sexual signals (Andersson 1994; Johnstone 1995). Yet discrimination is not the only challenge that females face; their most basic task is simply to detect a mate. From this perspective, male displays can be viewed as 'cooperative' signals that serve to facilitate detection, an outcome that is beneficial to both male and female. The costly nature of many sexual signals may thus reflect expenditure on efficacy rather than a means to ensure reliability (Wiley 1994).

The relative importance of detection and of discrimination are likely to vary from one signalling system to another. As a result, so too are the relative importance of the predictions made by the present model (which emphasises detection) and by strategic analyses of honest signalling (which emphasise discrimination). When detection represents the chief problem faced by receivers, for instance, we might expect signal intensity to increase with relatedness, as the present model predicts. Conversely, when discrimination poses a greater challenge than detection, we might expect signal intensity to decrease with relatedness, as honest signalling models predict. A study by Briskie et al. (1994) of chick begging behaviour in 11 species of passerine birds is often cited in support of the latter prediction - it revealed that nestlings of species with higher levels of extra-pair paternity, and hence lower relatedness among brood members, called more loudly (when controlling for the effects of brood size and body mass). Intriguingly, however, this represents a signalling system in which the chief difficulty faced by the receiver (parent) is that of discrimination between chicks, rather than that of detection (as the signal is given at very close range).

Ultimately, a comprehensive theory of animal signalling (if one is possible - see Dawkins 1993) will require the development of models that simultaneously address both signal detection (and associated conflicts over expenditure), and also discrimination (and the maintenance of honesty). Maynard Smith's (1991b) division of signals into 'notices', where cost serves to facilitate detection, and 'advertisements', where cost serves to provide evidence of quality or need, is a useful aid to clear thinking. It seems likely, however, that most biological displays combine elements of both. Strategic models of honest advertisement thus cannot, alone, provide a full account of signal evolution.

References

Alatalo, R. V. & Helle, P. 1990. Alarm calling by individual willow tits, *Parus montanus*. *Animal Behaviour*, **40**, 437-442.

Andersson, M. B. 1994. *Sexual Selection*. Princeton, NJ: Princeton University Press.

Bradbury, J. W. & Vehrencamp, S. L. 1998. *Principles of Animal Communication*. Sunderland, MA: Sinauers.

Briskie, J. V., Naugler, C. T. & Leech, S. M. 1994. Begging intensity of nestling birds varies with sibling relatedness. *Proceedings of the Royal Society of London, Series B*, **258**, 73-78.

Cade, W. 1975. Acoustically orienting parasitoids: fly phonotaxis to cricket song. *Science*, **190**, 1312-1313.

Cade, W. 1979. The evolution of alternative male reproductive strategies in field crickets. In: *Sexual Selection and Reproductive Competition in Insects* (Ed. by M. Blum & N. Blum), pp. 343-380. New York: Academic Press.

Clutton-Brock, T. H. & Parker, G. A. 1992. Potential reproductive rates and the operation of sexual selection. *Quarterly Review of Biology*, **67**, 437-455.

Davis, L. S. 1984. Alarm calling in Richardson's ground-squirrels *(Spermophilus richardsonii)*. *Zeitschrift für Tierpsychologie*, **66**,152-164.

Dawkins, M. S. 1993. Are there general principles of signal design? *Philosophical Transactions of the Royal Society of London, Series B*, **340**, 251-255.

Dusenbery, D. B. 1992. *Sensory Ecology*. New York: Freeman.

Endler, J. A. 1980. Natural selection on colour patterns in *Poecilia reticulata*. *Evolution*, **34**, 76-91.

Getty, T. 1995. Search, discrimination and selection: mate choice by pied fly-catchers. *American Naturalist*, **145**, 146-154.

Getty, T. 1996. Mate selection by repeated inspection: more on pied flycatchers. *Animal Behaviour*, **51**, 739-745.

Godfray, H. C. J. 1991. Signalling of need by offspring to their parents. *Nature*, **376**, 133-138.

Godfray, H. C. J.. 1995. Signaling of need between parents and young: parent-offspring conflict and sibling rivalry. *American Naturalist*, **146**, 1-24.

Grafen, A. 1990. Biological signals as handicaps. *Journal of Theoretical Biology*, **144**, 517-546.

Grether, G. F. & Grey, R. M. 1996. Novel cost of a sexually selected trait in the rubyspot damselfly *Hetaerina americana*: conspicuousness to prey. *Behavioral Ecology*, **7**, 465-473.

Harper, D. G. C. 1991. Communication. In*: Behavioural Ecology: an Evolutionary Approach*, 3rd edn (Ed. by J. R. Krebs & N. B. Davies), pp. 374-398. Oxford: Blackwell.

Haskell, D. 1994. Experimental evidence that nestling begging incurs a cost due to nest predation. *Proceedings of the Royal Society of London, Series B*, **257**, 161-164.

Hasson, O. 1991. Pursuit - deterrent signals - communication between prey and predator. *Trends in Ecology and Evolution*, **6**, 325-329.

Hoogland, J. L. 1983. Nepotism and alarm calling in the black-tailed prairie dog (*Cynomys ludovicianus*). *Animal Behaviour*, **31**,472-479.

Johnstone, R. A. 1995. Sexual selection, honest advertisement and the handicap principle: reviewing the evidence. *Biological Reviews*, **70**, 1-65.

Johnstone, R. A. 1997. The evolution of animal signals. In: *Behavioural Ecology: an Evolutionary Approach,* 4th edn (Ed. by J. R. Krebs & N. B. Davies), pp. 155-178. Oxford: Blackwell.

Johnstone, R. A. 1998a. Game theory and communication. In*: Game Theory and Animal Behaviour* (Ed. by L. A. Dugatkin & H. K. Reeve), pp. 94-117. Oxford: Oxford University Press.

Johnstone, R. A. 1998b. Efficacy and honesty in communication between relatives. *American Naturalist*, **152**, 45-58.

Johnstone, R. A. & Grafen, A. 1992. The continuous Sir Philip Sidney Game: a simple model of biological signalling. *Journal of Theoretical Biology*, **156**, 215-234.

Johnstone, R. A., Reynolds, J. D. & Deutsch, J. C. 1996. Mutual mate choice and sex differences in choosiness. *Evolution*, **50**, 1382-1391.

Krebs, J. R. & Dawkins, R. 1984. Animal signals: mind-reading and manipulation. In: *Behavioural Ecology: an Evolutionary Approach,* 2nd edn (Ed. by J. R. Krebs & N. B. Davies), pp. 380-402. Oxford: Blackwell.

Marchetti, K. 1993. Dark habitats and bright birds illustrate the role of the environment in species divergence. *Nature*, **362**, 149-152.

Maynard Smith, J. 1991a. Honest signalling - the Philip Sidney game. *Animal Behaviour*, **42**, 1034-1035.

Maynard Smith, J. 1991b. Theories of sexual selection. *Trends in Ecology and Evolution*, **6**, 146-151.

McGregor, P. K. 1993 Signalling in territorial systems: a context for individual identification, ranging and eavesdropping. *Philosophical Transactions of the Royal Society of London, Series B*, **340**, 237-244.

Møller, A. P. 1988. False alarm calls as a means of resource usurpation in the great tit *Parus major*. *Ethology*, **79**, 25-30.

Møller, A. P. 1990. Deceptive use of alarm calls by male swallows, *Hirundo rustica*: a new paternity guard. *Behavioural Ecology*, **1**, 1-6.

Munn, C. A. 1986. Birds that cry ‘wolf’. *Nature*, **319**, 1433-1435.

Oliveira, R. F., McGregor, P. K. & Latruffe, C. 1998. Know thine enemy: fighting fish gather information from observing conspecific interactions. *Proceedings of the Royal Society of London, Series B*, **265**, 1045-1049.

Otte, D. 1974. Effects and functions in the evolution of signalling systems. *Annual Review of Ecology and Systematics*, **5**, 385-417.

Reeve, H. K. 1989. The evolution of conspecific acceptance thresholds. *American Naturalist*, **133**, 407-435.

Reeve, H. K. 1997. Evolutionarily stable communication between kin: a general model. *Proceedings of the Royal Society of London, Series B*, **264**, 1037-1040.

Sakaluk, S. K. & Belwood, J. J. 1984. Gecko phonotaxis to cricket calling song - a case of satellite predation. *Animal Behaviour*, **32**, 659-662.

Trivers, R. L. 1972. Parental investment and sexual selection. In: *Sexual Selection and the Descent of Man* (Ed. by B. Campbell), pp. 136-179. Chicago: Aldine.

Tuttle, M. D. & Ryan, M. J. 1981. Bat predation and the evolution of frog vocalisations in the Neotropics. *Science*, **214**, 677-678.

Vega-Redondo, F. & Hasson, O. 1993. A game-theoretic model of predator-prey signalling. *Journal of Theoretical Biology*, **162**, 309-319.

Wiley, R. H. 1983. The evolution of communication: information and manipulation. In: *Communication* (Ed. by T. R. Halliday & P. J. B. Slater), pp. 156-189. Oxford: Blackwell.

Wiley, R. H. 1994. Errors, exaggeration and deception in animal communication. In: *Behavioral Mechanisms in Evolutionary Ecology* (Ed. by L. A. Real), pp. 157-189. Chicago: University of Chicago Press.

Zahavi, A. 1975. Mate selection - a selection for a handicap. *Journal of Theoretical Biology*, **53**, 205-214.

Zahavi, A. 1987. The theory of signal selection and some of its implications. In: *International Symposium of Biological Evolution* (Ed. by V. P. Delfino), pp. 305-327. Bari: Adriatica Editrice.

Zuk, M., Simmons, L. W. & Rotenberry, J. T. 1995. Acoustically-orienting parasitoids in calling and silent males of the field cricket *Teleogryllus oceanicus*. *Ecological Entomology*, **20**, 380-383.

Subject Index

Taxonomic index

For reef fishes, see also pp. 106–111.

DerFUCHS

Leitgedanken für Leitwölfe
Die fundamental anders denken wollen
und es auch können